Table of Contents

D0123525

Chapter P
Exercises P.1

2. (a) $T = 70 - 0.003h = 70 - 0.003(1500) = 65.5°\text{F}$

    (b) $64 = 70 - 0.003h \quad \Leftrightarrow \quad 0.003h = 6 \quad \Leftrightarrow \quad h = 2000 \text{ ft.}$

4. (a) $V = 9.5S = 9.5(4.5 \text{ km}^3) = 42.75 \text{ km}^3$

    (b) $8 \text{ km}^3 = 9.5S \quad \Leftrightarrow \quad S = 0.84 \text{ km}^3$

6. $C = 500 + 0.35(1000) = \$850.$

8. $V = \pi r^2 h = \pi(3^2)(5) = 45\pi \approx 141.4 \text{ in}^3$

10. The number N of cents in q quarters is $N = 25q$.

12. The average A of three numbers, a, b, and c, is $A = \dfrac{a + b + c}{3}$.

14. The sum of three consecutive integers is $S = (n - 1) + n + (n + 1) = 3n$, where n is the middle integer.

16. The sum S of the squares of two numbers, n and m, is $S = n^2 + m^2$

18. The product P of an integer n and twice the integer is $P = n(2n) = 2n^2$.

20. The distance d in a miles that a car travels in t hour at a speed of r miles per hour is $d = rt$.

22. The speed r of a boat that travels d miles in t hours is $r = \dfrac{d}{t}$.

24. The length is $4 + x$, so the area A is $A = (4 + x)x = 4x + x^2$.

26. The perimeter P of an equilateral triangle of side a is $P = a + a + a = 3a$.

28. The volume V of a box with a square base of side x and height $2x$ is $V = x \cdot x \cdot 2x = 2x^3$.

30. A box of length l, width w, and height h has two sides of area $l \cdot w$, two sides of area $l \cdot h$, and two ends with area $w \cdot h$. So the surface area A is $A = 2lw + 2lh + 2wh$.

32. The race track consists of a rectangle of length x and width $2r$ and two semicircles of radius r. Each semi circle has length πr and two straight runs of length x. So the length L is $L = 2x + 2\pi r$.

34. The area A of a triangle is $A = \frac{1}{2}\text{base} \cdot \text{height}$. Since the base is twice the height h we get
$A = (\frac{1}{2})(2h)(h) = h^2$.

36. The volume V of a cylinder is $V = \pi r^2 h$. Since the radius is x and the height is $2x$ we get
$V = \pi(x^2)(2x) = 2\pi x^3$.

38. (a) $3(38) + 280(0.15) = 114 + 42 = \$156.$

     (b) $Cost = \left(\dfrac{daily}{rental}\right) \times \left(\dfrac{days}{\text{rented}}\right) + \left(\dfrac{cost}{per\,mile}\right) \times \left(\dfrac{miles}{driven}\right)$ so $C = 38n + 0.15m.$

     (c) So we have $C = 149.25$ and $n = 3$. Substituting we get $149.25 = 38(3) + 0.15m$ $\Leftrightarrow$
          $149.25 = 114 + 0.15m$ $\Leftrightarrow$ $35.25 = m$ $\Leftrightarrow$ $m = 235$. So 235 miles was drive.

40. (a) The area that she can mow is $150\,\dfrac{\text{ft}^2}{\min} \times 30\,\min = 4{,}500\text{ ft}^2.$

     (b) $Area = 150\,\dfrac{\text{ft}^2}{\min} \times T\min = 150T\text{ ft}^2.$

     (c) The area of the lawn $= 80\text{ ft} \times 120\text{ ft} = 9{,}600\text{ ft}^2$. If T is the time required, then from part (b),
          $150T = 9600$, so $T = \dfrac{9600}{150} = 64\text{ min}.$

     (d) If R is the rate of mowing, then $60R = 9600$, so $R = \dfrac{9600}{60} = 160\,\dfrac{\text{ft}^2}{\min}.$

42. (a) The 10 minute call cost $0.25 + 0.08(10) = 0.25 + 0.80 = \$1.05.$

     (b) $Cost = \left(\dfrac{connect}{fee}\right) + \left(\dfrac{cost\,per}{minute}\right) \times (minutes)$ so $C = 0.25 + 0.08t.$

     (c) When $C = 73$ we have $73 = 25 + 8t$ $\Leftrightarrow$ $48 = 8t$ $\Leftrightarrow$ $t = 6$. The call lasted 6
          minutes.

     (b) Since $Cost = \left(\dfrac{connect}{fee}\right) + \left(\dfrac{cost\,per}{minute}\right) \times (minutes)$, when connection fee is F cents and the
          rate is r cents then the cost (in cents) is given by $C = F + rt.$

Exercises P.2

2. (a) natural number $11, \sqrt{16}$

    (b) integers $-11, 11, \sqrt{16}$

    (c) rational numbers $1.001, 0.333\ldots, -11, 11, \frac{13}{15}, \sqrt{16}, 3.14, \frac{15}{3}$

    (d) irrational numbers $-\pi$

4. Commutative Property for multiplication.   6. Distributive Property.

8. Distributive Property.   10. Distributive Property.

12. $7(3x) = (7 \cdot 3)x$ 14. $5x + 5y = 5(x + y)$

16. $(a - b)8 = 8a - 8b$ 18. $\frac{4}{3}(-6y) = \left[\frac{4}{3}(-6)\right]y = -8y$

20. $(3a)(b + c - 2d) = 3ab + 3ac - 6ad$

22. (a) $\dfrac{2}{3} - \dfrac{3}{5} = \dfrac{10}{15} - \dfrac{9}{15} = \dfrac{1}{15}$ (b) $1 + \dfrac{5}{8} - \dfrac{1}{6} = \dfrac{24}{24} + \dfrac{15}{24} - \dfrac{4}{24} = \dfrac{35}{24}$

24. (a) $(3 + \frac{1}{4})(1 - \frac{4}{5}) = (\frac{12}{4} + \frac{1}{4})(\frac{5}{5} - \frac{4}{5}) = \frac{13}{4} \cdot \frac{1}{5} = \frac{13}{20}$

    (b) $(\frac{1}{2} - \frac{1}{3})(\frac{1}{2} + \frac{1}{3}) = (\frac{3}{6} - \frac{2}{6})(\frac{3}{6} + \frac{2}{6}) = \frac{1}{6} \cdot \frac{5}{6} = \frac{5}{36}$

26. (a) $\dfrac{2 - \frac{3}{4}}{\frac{1}{2} - \frac{1}{3}} = \dfrac{2 - \frac{3}{4}}{\frac{1}{2} - \frac{1}{3}} \cdot \dfrac{12}{12} = \dfrac{24 - 9}{6 - 4} = \dfrac{15}{2}$

    (b) $\dfrac{\frac{2}{5} + \frac{1}{2}}{\frac{1}{10} + \frac{3}{15}} = \dfrac{\frac{2}{5} + \frac{1}{2}}{\frac{1}{10} + \frac{1}{5}} = \dfrac{\frac{2}{5} + \frac{1}{2}}{\frac{1}{10} + \frac{1}{5}} \cdot \dfrac{10}{10} = \dfrac{4 + 5}{1 + 2} = \dfrac{9}{3} = 3$

28. (a) Since $3 \cdot \frac{2}{3} = 2$ and $3 \cdot 0.67 = 2.01$ so $\frac{2}{3} < 0.67$

    (b) $\frac{2}{3} > -0.67$

    (c) $|0.67| = |-0.67|$

30. (a) True. $\dfrac{10}{11} = \dfrac{10}{11} \cdot \dfrac{13}{13} = \dfrac{130}{143}$ and $\dfrac{12}{13} = \dfrac{12}{13} \cdot \dfrac{11}{11} = \dfrac{132}{143}$. Therefore $\dfrac{10}{11} < \dfrac{12}{13}$, because $\dfrac{130}{143} < \dfrac{132}{143}$.

    (b) False.

32. (a) False   (b) True.

34. (a) $y < 0$ (b) $z > 1$

    (c) $b \le 8$ (d) $0 < w \le 17$

    (e) $|y - \pi| \ge 2$

36. (a) $B \cup C = \{2, 4, 6, 7, 8, 9, 10\}$ (b) $B \cap C = \{8\}$

38. (a) $A \cup B \cup C = \{1, 2, 3, 4, 5, 6, 7, 8, 9, 10\}$
     (b) $A \cap B \cap C = \emptyset$

40. (a) $A \cap C = \{x \mid -1 < x \le 5\}$ (b) $A \cap B = \{x \mid -2 \le x < 4\}$

42. $(2, 8] = \{x \mid 2 < x \le 8\}$

44. $\left[-6, -\tfrac{1}{2}\right] = \left\{x \mid -6 \le x \le -\tfrac{1}{2}\right\}$

46. $(-\infty, 1) = \{x \mid x < 1\}$

48. $1 \le x \le 2 \quad \Leftrightarrow \quad x \in [1, 2]$

50. $x \ge -5 \quad \Leftrightarrow \quad x \in [-5, \infty)$

52. $-5 < x < 2 \quad \Leftrightarrow \quad x \in (-5, 2)$

54. (a) $[0, 2)$ (b) $(-2, 0]$

56. $(-2, 0) \cap (-1, 1) = (-1, 0)$

58. $[-4, 6] \cup [0, 8) = [-4, 8)$

60. $(-\infty, 6] \cap (2, 10) = (2, 6]$

62. (a) $\left|\sqrt{5} - 5\right| = -\left(\sqrt{5} - 5\right) = 5 - \sqrt{5}$, since $5 > \sqrt{5}$.
     (b) $|10 - \pi| = 10 - \pi$, since $10 > \pi$.

64. (a) $\left|2 - |-12|\right| = |2 - 12| = |-10| = 10$.
     (b) $-1 - \left|1 - |-1|\right| = -1 - |1 - 1| = -1 - |0| = -1$.

66. (a) $\left|\tfrac{-6}{24}\right| = \left|\tfrac{-1}{4}\right| = \tfrac{1}{4}$.   (b) $\left|\tfrac{7-12}{12-7}\right| = \left|\tfrac{-5}{5}\right| = |-1| = 1$.

68. $|-2.5 - 1.5| = |-4| = 4$

70. (a) $\left|\dfrac{7}{15} - \left(-\dfrac{1}{21}\right)\right| = \left|\dfrac{49}{105} + \dfrac{5}{105}\right| = \left|\dfrac{54}{105}\right| = \left|\dfrac{18}{35}\right| = \dfrac{18}{35}$.
     (b) $|-38 - (-57)| = |-38 + 57| = |19| = 19$.
     (c) $|-2.6 - (-1.8)| = |-2.6 + 1.8| = |-0.8| = 0.8$.

72. (a) Let $x = 5.2323\ldots$. So,
     $$100x = 523.2323\ldots$$
     $$\underline{1x = 5.2323\ldots}$$
     $$99x = 518$$ Thus $x = \dfrac{518}{99}$

(b) Let $x = 1.3777\ldots$. So,

$$100x = 137.7777\ldots$$
$$10x = 13.7777\ldots$$
$$\overline{90x = 124}$$

Thus $x = \frac{124}{90} = \frac{62}{45}$

(c) Let $x = 2.13535\ldots$. So,

$$1000x = 2135.3535\ldots$$
$$10x = 21.3535\ldots$$
$$\overline{990x = 2114}$$

Thus $x = \frac{2114}{990} = \frac{1057}{495}$

74.

Say	T_O	T_G	$T_O - T_G$	$|T_O - T_G|$
Sunday	68	77	−9	9
Monday	72	75	−3	3
Tuesday	74	74	0	0
Wednesday	80	75	5	5
Thursday	77	69	8	8
Friday	71	70	1	1
Saturday	70	71	−1	1

$T_O - T_G$ give more information because it tells us which city had the higher (or lower) temperature.

76. (a) When $L = 60$, $x = 8$, and $y = 6$, we have $L + 2(x + y) = 60 + 2(8 + 6) = 60 + 28 = 88$, since $88 \leq 108$ the post office will accept this package.

NO, when $L = 48$, $x = 24$, and $y = 24$, we have
$L + 2(x + y) = 48 + 2(24 + 24) = 48 + 96 = 144$, since $144 \nleq 108$, the post office will NOT accept this package.

(b) If $x = y = 9$, then $L + 2(9 + 9) \leq 108 \quad \Leftrightarrow \quad L + 36 \leq 108 \quad \Leftrightarrow \quad L \leq 72$. So the length can be as long as 72 in. = 6 ft.

78. Let $x = \dfrac{m_1}{n_1}$ and $y = \dfrac{m_2}{n_2}$ be rational numbers. Then

(i) $x + y = \dfrac{m_1}{n_1} + \dfrac{m_2}{n_2} = \dfrac{m_1 n_2 + m_2 n_1}{n_1 n_2}$

(ii) $x - y = \dfrac{m_1}{n_1} - \dfrac{m_2}{n_2} = \dfrac{m_1 n_2 - m_2 n_1}{n_1 n_2}$

(iii) $x \cdot y = \dfrac{m_1}{n_1} \cdot \dfrac{m_2}{n_2} = \dfrac{m_1 m_2}{n_1 n_2}$

This shows that the sum, difference, and product of two rational numbers are again rational numbers. However the product of two irrational numbers is not necessarily irrational; for example, $\sqrt{2} \cdot \sqrt{2} = 2$, which is rational. Also, the sum of two irrational numbers is not necessarily irrational; for example, $\sqrt{2} + (-\sqrt{2}) = 0$ which is rational.

80.

x	1	2	10	100	1000
$\dfrac{1}{x}$	1	$\dfrac{1}{2}$	$\dfrac{1}{10}$	$\dfrac{1}{100}$	$\dfrac{1}{1000}$

As x gets large, the fraction $\dfrac{1}{x}$ gets small. Mathematically, we say that $\dfrac{1}{x}$ goes to zero.

x	1	0.5	0.1	0.01	0.001
$\dfrac{1}{x}$	1	$\dfrac{1}{0.5} = 2$	$\dfrac{1}{0.1} = 10$	$\dfrac{1}{0.01} = 100$	$\dfrac{1}{0.001} = 1000$

As x gets small, the fraction $\dfrac{1}{x}$ gets large. Mathematically, we say that $\dfrac{1}{x}$ goes to infinity.

82.   (a)   Subtraction is not commutative since $5 - 1 \neq 1 - 5$.

       (b)   Division is not commutative since $5 \div 1 \neq 1 \div 5$.

       (c)   Only if you like to wear your socks over your shoes.

       (d)   Yes, since you get the same result when you put on your coat and then you hat as when you put on your hat and then your coat.

       (e)   No, you must wash the laundry before dry it.

       (f)   Answers vary.

       (g)   Answers vary.

Exercises P.3

2. $2^3 \cdot 2^2 = 2^{3+2} = 2^5 = 32$

4. $(2^3)^0 = 1$

6. $-6^0 = -(6^0) = -(1) = -1$

8. $(-3)^2 = 9$

10. $5^2 \cdot (\frac{1}{5})^3 = 5^2 \cdot 5^{-3} = 5^{2-3} = 5^{-1} = \frac{1}{5}$

12. $\dfrac{3}{3^{-2}} = 3^{1-(-2)} = 3^3 = 27$

14. $\dfrac{3^{-2}}{9} = \dfrac{3^{-2}}{3^2} = \dfrac{1}{3^{2-(-2)}} = \dfrac{1}{3^4} = \dfrac{1}{81}$

16. $(\frac{2}{3})^{-3} = (\frac{3}{2})^3 = \frac{3^3}{2^3} = \frac{27}{8}$

18. $(\frac{1}{2})^4 \cdot (\frac{5}{2})^{-2} = (\frac{1}{2})^4 \cdot (\frac{2}{5})^2 = \frac{1}{2^4} \cdot \frac{2^2}{5^2} = \frac{1}{2^{4-2}} \cdot \frac{1}{5^2} = \frac{1}{2^2 \cdot 5^2} = \frac{1}{100}$

20. $\dfrac{3^2 \cdot 4^{-2} \cdot 5}{2^{-4} \cdot 3^3 \cdot 25} = \dfrac{3^2 \cdot (2^2)^{-2} \cdot 5}{2^{-4} \cdot 3^3 \cdot 5^2} = \dfrac{3^2 \cdot 2^{-4} \cdot 5}{2^{-4} \cdot 3^3 \cdot 5^2} = 2^{-4-(-4)} \cdot 3^{2-3} \cdot 5^{1-2} = 2^0 \cdot 3^{-1} \cdot 5^{-1}$
$= \dfrac{1}{3 \cdot 5} = \dfrac{1}{15}$

22. $3^{-1} - 3^{-3} = \frac{1}{3} - \frac{1}{3^3} = \frac{1}{3} - \frac{1}{27} = \frac{9}{27} - \frac{1}{27} = \frac{8}{27}$

24. $\frac{1}{6}x(3x^2)^3 = \frac{1}{6}x(3^3 x^{2 \cdot 3}) = \frac{3^3}{2 \cdot 3} \cdot x \cdot x^6 = \frac{3^2}{2} \cdot x^{1+6} = \frac{9}{2}x^7$

26. $(5x^2)^3(\frac{1}{25}x^4)^2 = (5^3 x^6)\left[(\frac{1}{5^2})^2 x^8\right] = (5^3 x^6)(\frac{1}{5^4}x^8) = 5^{3-4} \cdot x^{6+8} = 5^{-1}x^{14} = \frac{1}{5}x^{14}$

28. $(2z^2)^{-5}z^{10} = 2^{-5}z^{-10} \, z^{10} = 2^{-5}z^{-10+10} = \frac{1}{2^5}z^0 = \frac{1}{32}$

30. $\left(\dfrac{3}{x}\right)^4 \left(\dfrac{4}{x}\right)^{-2} = \left(\dfrac{3}{x}\right)^4 \left(\dfrac{x}{4}\right)^2 = \dfrac{3^4}{x^4} \cdot \dfrac{x^2}{4^2} = \dfrac{3^4}{4^2} \cdot x^{2-4} = \dfrac{81}{16}x^{-2} = \dfrac{81}{16x^2}$

32. $\dfrac{[2(r-s)]^2}{(r-s)^3} = \dfrac{2^2(r-s)^2}{(r-s)^3} = \dfrac{2^2}{(r-s)^{3-2}} = \dfrac{4}{r-s}$

34. $(3y^2)(4y^5) = 3 \cdot 4y^{2+5} = 12y^7$

36. $(6y)^3 = 6^3 \, y^3 = 216y^3$

38. $\dfrac{a^{-3}b^4}{a^{-5}b^5} = a^{-3-(-5)}b^{4-5} = a^2 b^{-1} = \dfrac{a^2}{b}$

40. $\left(2s^3 t^{-1}\right)\left(\dfrac{1}{4}s^6\right)(16t^4) = \dfrac{2 \cdot 16}{4} \, s^{3+6} \, t^{-1+4} = 8s^9 t^3$

42. $\left(2u^2 v^3\right)^3 \left(3u^3 v\right)^{-2} = 2^3 u^6 v^9 \cdot 3^{-2}u^{-6}v^{-2} = \dfrac{2^3}{3^2}u^{6-6}v^{9-2} = \dfrac{8}{9}v^7$

44. $\dfrac{(2x^3)^2(3x^4)}{(x^3)^4} = \dfrac{2^2 x^6 \cdot 3x^4}{x^{12}} = 12x^{6+4-12} = 12x^{-2} = \dfrac{12}{x^2}$

46. $\left(\dfrac{c^4 d^3}{cd^2}\right)\left(\dfrac{d^2}{c^3}\right)^3 = \dfrac{c^4 d^3}{cd^2} \cdot \dfrac{d^6}{c^9} = c^{4-1-9} d^{3+6-2} = c^{-6} d^7 = \dfrac{d^7}{c^6}$

48. $\left(\dfrac{xy^{-2} z^{-3}}{x^2 y^3 z^{-4}}\right)^{-3} = \dfrac{x^{-3} y^6 z^9}{x^{-6} y^{-9} z^{12}} = x^{-3-(-6)} y^{6-(-9)} z^{9-12} = x^3 y^{15} z^{-3} = \dfrac{x^3 y^{15}}{z^3}$

50. $(3ab^2 c)\left(\dfrac{2a^2 b}{c^3}\right)^{-2} = 3ab^2 c \cdot \dfrac{2^{-2} a^{-4} b^{-2}}{c^{-6}} = \dfrac{3}{4} a^{1-4} b^{2-2} c^{1-(-6)} = \dfrac{3}{4} a^{-3} b^0 c^7 = \dfrac{3c^7}{4a^3}$

52. $7,200,000,000,000 = 7.2 \times 10^{12}$ **54.** $0.0001213 = 1.213 \times 10^{-4}$

56. $7,259,000,000 = 7.259 \times 10^9$ **58.** $0.0007029 = 7.029 \times 10^{-4}$

60. $2.721 \times 10^8 = 272,100,000$ **62.** $9.999 \times 10^{-9} = 0.000000009999$

64. $6 \times 10^{12} = 6,000,000,000,000$ **66.** $6.257 \times 10^{-10} = 0.0000000006257$

68. (a) $93,000,000 \text{ mi} = 9.3 \times 10^7 \text{ mi}$

    (b) $0.000000000000000000000053 \text{ g} = 5.3 \times 10^{-23} \text{ g}$

    (c) $5,970,000,000,000,000,000,000,000 \text{ kg} = 5.97 \times 10^{24} \text{ kg}$

70. $(1.062 \times 10^{24})(8.61 \times 10^{19}) = 1.062 \times 8.61 \times 10^{24} \times 10^{19} \approx 9.14 \times 10^{43}$

72. $\dfrac{(73.1)(1.6341 \times 10^{28})}{0.0000000019} = \dfrac{(7.31 \times 10)(1.6341 \times 10^{28})}{1.9 \times 10^{-9}}$

$= \dfrac{7.31 \times 1.6341}{1.9} \times 10^{1+28-(-9)} \approx 6.3 \times 10^{38}$

74. $\dfrac{(3.542 \times 10^{-6})^9}{(5.05 \times 10^4)^{12}} = \dfrac{(3.542)^9 \times 10^{-54}}{(5.05)^{12} \times 10^{48}} = \dfrac{87747.96}{275103767.10} \times 10^{-54-48} \approx 3.19 \times 10^{-4} \times 10^{-102}$

$\approx 3.19 \times 10^{-106}$

76. (a) $\dfrac{a^m}{a^n} = \dfrac{a \cdot a \cdot a \cdot \;\cdots\; \cdot a \;(m \text{ factors})}{a \cdot a \cdot a \cdot \;\cdots\; \cdot a \;(n \text{ factors})} = a \cdot a \cdot a \cdot \;\cdots\; \cdot a \;(m-n \text{ factors}) = a^{m-n} \;(a \neq 0)$.

    (b) $\left(\dfrac{a}{b}\right)^n = \dfrac{a}{b} \cdot \dfrac{a}{b} \cdot \;\cdots\; \cdot \dfrac{a}{b} \;(n \text{ factors}) = \dfrac{a \cdot a \cdot a \cdot \;\cdots\; \cdot a \;(n \text{ factors})}{b \cdot b \cdot b \cdot \;\cdots\; \cdot b \;(n \text{ factors})} = \dfrac{a^n}{b^n} \;(b \neq 0)$.

    (c) $\left(\dfrac{a}{b}\right)^{-n} = \left[\left(\dfrac{a}{b}\right)^{-1}\right]^n = \left[\dfrac{1}{\left(\dfrac{a}{b}\right)}\right]^n = \left(\dfrac{b}{a}\right)^n = \dfrac{b^n}{a^n} \;(a \neq 0, b \neq 0)$.

    The first equality follows from Law 3. The second equality is given by the Negative Exponents Law. The third equality is simplification, and the fourth equality follows from Law 5.

78.　$9.3 \times 10^7 \text{mi} = 186,000 \dfrac{\text{mi}}{\text{sec}} \times t \text{ sec} \quad \Leftrightarrow \quad t = \dfrac{9.3 \times 10^7}{186,000} \text{ sec} = 500 \text{ sec} = 8\dfrac{1}{3} \text{ min.}$

80.　$\dfrac{each\ person's}{share} = \dfrac{national\ debt}{population} = \dfrac{5.736 \times 10^{12}}{2.83 \times 10^8} \approx 2.03 \times 10^4 = \$20,300.$

82.　(a)

Person	Weight	Height	$\text{BMI} = 703\dfrac{W}{H^2}$	
Brain	295 lb	5 ft 10 in. = 70 in	42.32	obese
Linda	105 lb	5 ft 6 in. = 66 in.	16.95	underweight
Larry	220 lb	6 ft 4 in. = 76 in.	26.78	overweight
Helen	110 lb	5 ft 2 in. = 62 in.	20.12	normal

　　(b)　Answers vary.

84.　Since $10^6 = 10^3 \cdot 10^3$ it would take 1000 days $= 2.74$ years to spend the million dollars.
Since $10^9 = 10^3 \cdot 10^6$ it would take $10^6 = 1,000,000$ days $= 2739.72$ years to spend the billion dollars.

86.　$|10^{50} - 10^{10}| < 10^{50}$, whereas $|10^{101} - 10^{100}| = 10^{100}|10 - 1| = 9 \times 10^{100} > 10^{50}$. So 10^{10} is closer to 10^{50} than 10^{100} is to 10^{101}.

Exercises P.4

2. $\sqrt[3]{7^2} = 7^{2/3}$

4. $11^{-3/2} = (11^{3/2})^{-1} = (\sqrt{11^3})^{-1} = \dfrac{1}{\sqrt{11^3}}$

6. $2^{-1.5} = 2^{-3/2} = \dfrac{1}{\sqrt{2^3}} = \dfrac{1}{\sqrt{8}}$

8. $\dfrac{1}{\sqrt{x^5}} = \dfrac{1}{x^{5/2}} = x^{-5/2}$

10. (a) $\sqrt{64} = \sqrt{8^2} = 8$

    (b) $\sqrt[3]{-64} = \sqrt[3]{(-4)^3} = -4$

    (c) $\sqrt[5]{-32} = \sqrt[5]{(-2)^5} = -2$

12. (a) $\sqrt{7}\sqrt{28} = \sqrt{7}\sqrt{4\cdot7} = \sqrt{7}\cdot2\cdot\sqrt{7} = 2\cdot7 = 14$

    (b) $\dfrac{\sqrt{48}}{\sqrt{3}} = \sqrt{\dfrac{48}{3}} = \sqrt{16} = 4$

    (c) $\sqrt[4]{24}\sqrt[4]{54} = \sqrt[4]{8\cdot3}\sqrt[4]{2\cdot27} = \sqrt[4]{2^3\cdot3\cdot2\cdot3^3} = \sqrt[4]{2^4\cdot3^4} = 2\cdot3 = 6$

14. (a) $1024^{-0.1} = (2^{10})^{-0.1} = 2^{-1} = \frac{1}{2}$

    (b) $\left(-\dfrac{27}{8}\right)^{2/3} = \left(\sqrt[3]{-\dfrac{27}{8}}\right)^2 = \left(-\dfrac{3}{2}\right)^2 = \dfrac{9}{4}$

    (c) $\left(\dfrac{25}{64}\right)^{3/2} = \left(\sqrt{\dfrac{25}{64}}\right)^3 = \left(\dfrac{5}{8}\right)^3 = \dfrac{125}{512}$

16. (a) $(-1000)^{-2/3} = [(-10)^3]^{-2/3} = (-10)^{-2} = \dfrac{1}{100}$

    (b) $(10{,}000)^{-3/2} = (10^4)^{-3/2} = 10^{4\cdot(-3/2)} = 10^{-6} = \dfrac{1}{1{,}000{,}000}$

    (c) $(-8000)^{4/3} = [(-20)^3]^{4/3} = (-20)^{3\cdot(4/3)} = (-20)^4 = 160{,}000$

18. (a) $\left(\dfrac{1}{16}\right)^{-0.75} = \left(\dfrac{1}{2^4}\right)^{-3/4} = (2^{-4})^{-3/4} = 2^{(-4)\cdot(-3/4)} = 2^3 = 8$

    (b) $0.25^{-0.5} = \left(\dfrac{1}{4}\right)^{-1/2} = \left(\dfrac{1}{2^2}\right)^{-1/2} = (2^{-2})^{-1/2} = 2^{(-2)\cdot(-1/2)} = 2^1 = 2$

    (c) $9^{1/3}\cdot15^{1/3}\cdot25^{1/3} = (3^2)^{1/3}\cdot(3\cdot5)^{1/3}\cdot(5^2)^{1/3} = 3^{2/3}\cdot(3^{1/3}\cdot5^{1/3})\cdot5^{2/3}$
        $= 3^{2/3+1/3}\cdot5^{2/3+1/3} = 3^1\cdot5^1 = 15$

20. When $x = 3$, $y = 4$, $z = -1$ we have
    $\sqrt[4]{x^3 + 14y + 2z} = \sqrt[4]{3^3 + 14(4) + 2(-1)} = \sqrt[4]{27 + 56 - 2} = \sqrt[4]{81} = \sqrt[4]{3^4} = 3.$

22. When $x = 3$, $y = 4$, $z = -1$ we have $(xy)^{2z} = (3\cdot4)^{2\cdot(-1)} = 12^{-2} = \frac{1}{144}$.

24. $\sqrt{75} + \sqrt{48} = \sqrt{25 \cdot 3} + \sqrt{16 \cdot 3} = \sqrt{5^2 \cdot 3} + \sqrt{4^2 \cdot 3} = 5\sqrt{3} + 4\sqrt{3} = 9\sqrt{3}$

26. $\sqrt[3]{54} - \sqrt[3]{16} = \sqrt[3]{2 \cdot 3^3} - \sqrt[3]{2^3 \cdot 2} = 3\sqrt[3]{2} - 2\sqrt[3]{2} = \sqrt[3]{2}$

28. $\sqrt{8} + \sqrt{50} = \sqrt{2^2 \cdot 2} + \sqrt{5^2 \cdot 2} = 2\sqrt{2} + 5\sqrt{2} = 7\sqrt{2}$

30. $\sqrt[3]{24} - \sqrt[3]{81} = \sqrt[3]{2^3 \cdot 3} - \sqrt[3]{3^4} = 2\sqrt[3]{3} - 3\sqrt[3]{3} = -\sqrt[3]{3}$

32. $\sqrt[4]{48} - \sqrt[4]{3} = \sqrt[4]{16 \cdot 3} - \sqrt[4]{2^3 \cdot 2} = \sqrt[4]{2^4 \cdot 3} - \sqrt[4]{3} = 2\sqrt[4]{3} - \sqrt[4]{3} = \sqrt[4]{3}$

34. $\sqrt[5]{x^{10}} = (x^{10})^{1/5} = x^2$

36. $\sqrt[3]{x^3 y^6} = (x^3 y^6)^{1/3} = xy^2$ 38. $\sqrt{x^4 y^4} = (x^4 y^4)^{1/2} = x^2 y^2$

40. $\sqrt[3]{a^2 b} \, \sqrt[3]{a^4 b} = \sqrt[3]{a^6 b^2} = (a^6 b^2)^{1/3} = a^2 b^{2/3} = a^2 \sqrt[3]{b^2}$

42. $\sqrt[4]{x^4 y^2 z^2} = \sqrt[4]{x^4} \, \sqrt[4]{y^2 z^2} = |x| \sqrt[4]{y^2 z^2}$

44. $(2x^{3/2})(4x)^{-1/2} = (2x^{3/2})(2^2 x)^{-1/2} = (2x^{3/2})(2^{-1} x^{-1/2}) = 2^{1-1} \cdot x^{3/2-1/2} = 2^0 \cdot x = x$

46. $\left(-2a^{3/4}\right)\left(5a^{3/2}\right) = -10a^{3/4} a^{3/2} = -10a^{(3/4+6/4)} = -10a^{9/4}$

48. $\left(8x^6\right)^{-2/3} = \left(2^3\right)^{-2/3} \left(x^6\right)^{-2/3} = 2^{-6/3} x^{-12/3} = 2^{-2} x^{-4} = \dfrac{1}{4x^4}$

50. $(4x^6 y^8)^{3/2} = (2^2)^{3/2} (x^6)^{3/2} (y^8)^{3/2} = 2^{6/2} x^{18/2} y^{24/2} = 2^3 x^9 y^{12} = 8x^9 y^{12}$

52. $\left(a^{2/5}\right)^{-3/4} = a^{(2/5) \cdot (-3/4)} = a^{-6/20} = a^{-3/10} = \dfrac{1}{a^{3/10}}$

54. $\left(x^{-5} y^3 z^{10}\right)^{-3/5} = x^{15/5} y^{-9/5} z^{-30/5} = x^3 y^{-9/5} z^{-6} = \dfrac{x^3}{y^{9/5} z^6}$

56. $\left(\dfrac{-2x^{1/3}}{y^{1/2} z^{1/6}}\right)^4 = \dfrac{(-2)^4 x^{4/3}}{y^{4/2} z^{4/6}} = \dfrac{16 x^{4/3}}{y^2 z^{2/3}}$

58. $\dfrac{(y^{10} z^{-5})^{1/5}}{(y^{-2} z^3)^{1/3}} = \dfrac{y^2 z^{-1}}{y^{-2/3} z} = y^{2-(-2/3)} z^{-1-1} = y^{8/3} z^{-2} = \dfrac{y^{8/3}}{z^2}$

60. $\left(\dfrac{a^2 b^{-3}}{x^{-1} y^2}\right)^3 \left(\dfrac{x^{-2} b^{-1}}{a^{3/2} y^{1/3}}\right) = \dfrac{a^6 b^{-9}}{x^{-3} y^6} \cdot \dfrac{x^{-2} b^{-1}}{a^{3/2} y^{1/3}} = a^{6-(3/2)} b^{-9-1} x^{-2-(-3)} y^{-6-1/3}$

$= a^{9/2} b^{-10} x^1 y^{-19/3} = \dfrac{a^{9/2} x}{b^{10} y^{19/3}}$

62. (a) $\dfrac{12}{\sqrt{3}} = \dfrac{12}{\sqrt{3}} \cdot \dfrac{\sqrt{3}}{\sqrt{3}} = \dfrac{12\sqrt{3}}{3} = 4\sqrt{3}$ (b) $\dfrac{5}{\sqrt{2}} = \dfrac{5}{\sqrt{2}} \cdot \dfrac{\sqrt{2}}{\sqrt{2}} = \dfrac{5\sqrt{2}}{2}$

(c) $\dfrac{2}{\sqrt{6}} = \dfrac{2}{\sqrt{6}} \cdot \dfrac{\sqrt{6}}{\sqrt{6}} = \dfrac{2\sqrt{6}}{6} = \dfrac{\sqrt{6}}{3}$

64. (a) $\dfrac{1}{\sqrt[5]{2^3}} = \dfrac{1}{\sqrt[5]{2^3}} \cdot \dfrac{\sqrt[5]{2^2}}{\sqrt[5]{2^2}} = \dfrac{\sqrt[5]{2^2}}{2} = \dfrac{\sqrt[5]{4}}{2}$

(b) $\dfrac{2}{\sqrt[4]{3}} = \dfrac{2}{\sqrt[4]{3}} \cdot \dfrac{\sqrt[4]{3^3}}{\sqrt[4]{3^3}} = \dfrac{2\sqrt[4]{3^3}}{3} = \dfrac{2\sqrt[4]{27}}{3}$

(c) $\dfrac{3}{\sqrt[4]{2^3}} = \dfrac{3}{\sqrt[4]{2^3}} \cdot \dfrac{\sqrt[4]{2}}{\sqrt[4]{2}} = \dfrac{3\sqrt[4]{2}}{2}$

66. (a) $\dfrac{1}{\sqrt[3]{x^2}} = \dfrac{1}{\sqrt[3]{x^2}} \cdot \dfrac{\sqrt[3]{x}}{\sqrt[3]{x}} = \dfrac{\sqrt[3]{x}}{x}$ (b) $\dfrac{1}{\sqrt[4]{x^3}} = \dfrac{1}{\sqrt[4]{x^3}} \cdot \dfrac{\sqrt[4]{x}}{\sqrt[4]{x}} = \dfrac{\sqrt[4]{x}}{x}$

(c) $\dfrac{1}{\sqrt[3]{x^4}} = \dfrac{1}{\sqrt[3]{x^3 \cdot x}} = \dfrac{1}{x\sqrt[3]{x}} = \dfrac{1}{x\sqrt[3]{x}} \cdot \dfrac{\sqrt[3]{x^2}}{\sqrt[3]{x^2}} = \dfrac{\sqrt[3]{x^2}}{x\sqrt[3]{x^3}} = \dfrac{\sqrt[3]{x^2}}{x^2}$

68. (a) Using $f = 0.4$ and substituting $d = 65$, we obtain $s = \sqrt{30fd} = \sqrt{30 \times 0.4 \times 65} \approx 28$ mi/h.

(b) Using $f = 0.5$ and substituting $s = 50$, we find d. Thus gives $s = \sqrt{30fd}$ $\Leftrightarrow$
$50 = \sqrt{30 \cdot (0.5)d}$ $\Leftrightarrow$ $50 = \sqrt{15d}$ $\Leftrightarrow$ $2500 = 15d$ $\Leftrightarrow$ $d = \dfrac{500}{3} \approx 167$ feet.

70.

Planet	d	$\sqrt{d}$	T	$\sqrt[3]{T}$
Mercury	0.387	0.622	0.241	0.622
Venus	0.723	0.850	0.615	0.850
Earth	1.000	1.000	1.000	1.000
Mars	1.523	1.234	1.881	1.234
Jupiter	5.203	2.281	11.861	2.281
Saturn	9.541	3.089	29.457	3.088

$\sqrt{d} \approx \sqrt[3]{T}$ or $d^3 \approx T^2$

72. (a) Substituting the given values we get $V = 1.486 \dfrac{75^{2/3} \cdot 0.050^{1/2}}{24.1^{2/3} \cdot 0.040} \approx 17.707$ ft/s.

(b) Since the volume of the flow is $V \cdot A$ we have the canal discharge $= 17.707 \cdot 75 = 1328.0$ ft^3/s.

74. (a) Since $\frac{1}{2} > \frac{1}{3}$, $2^{\frac{1}{2}} > 2^{\frac{1}{3}}$.

(b) $\left(\dfrac{1}{2}\right)^{\frac{1}{2}} = 2^{-\frac{1}{2}}$ and $\left(\dfrac{1}{2}\right)^{\frac{1}{3}} = 2^{-\frac{1}{3}}$. Since $-\frac{1}{2} < -\frac{1}{3}$, we have $\left(\dfrac{1}{2}\right)^{\frac{1}{2}} < \left(\dfrac{1}{2}\right)^{\frac{1}{3}}$.

(c) We find a common root: $7^{\frac{1}{4}} = 7^{\frac{3}{12}} = (7^3)^{\frac{1}{12}} = 343^{\frac{1}{12}}$; $4^{\frac{1}{3}} = 4^{\frac{4}{12}} = (4^4)^{\frac{1}{12}} = 256^{\frac{1}{12}}$. So $7^{\frac{1}{4}} > 4^{\frac{1}{3}}$.

(d) We find a common root: $\sqrt[3]{5} = 5^{\frac{1}{3}} = 5^{\frac{2}{6}} = (5^2)^{\frac{1}{6}} = 25^{\frac{1}{6}}$; $\sqrt{3} = 3^{\frac{1}{2}} = 3^{\frac{3}{6}} = (3^3)^{\frac{1}{6}} = 27^{\frac{1}{6}}$. So $\sqrt[3]{5} < \sqrt{3}$.

Exercises P.5

2.   Type: binomial; Terms: $2x^5$ and $4x^2$; Degree 5

4.   Type: monomial; Terms: $\frac{1}{2}x^7$; Degree 7

6.   Type: binomial; Terms: $\sqrt{2}x$ and $-\sqrt{3}$; Degree 1

8.   Not a polynomial.

10.   Polynomial, degree 5.

12.   Polynomial, degree 1.

14. $(5 - 3x) + (2x - 8) = -x - 3$

16. $(3x^2 + x + 1) - (2x^2 - 3x - 5) = 3x^2 + x + 1 - 2x^2 + 3x + 5 = x^2 + 4x + 6$

18. $3(x - 1) + 4(x + 2) = 3x - 3 + 4x + 8 = 7x + 5$

20. $4(x^2 - 3x + 5) - 3(x^2 - 2x + 1) = 4x^2 - 12x + 20 - 3x^2 + 6x - 3 = x^2 - 6x + 17$

22. $5(3t - 4) - (t^2 + 2) - 2t(t - 3) = 15t - 20 - t^2 - 2 - 2t^2 + 6t = -3t^2 + 21t - 22$

24. $3x^3(x^4 - 4x^2 + 5) = 3x^3x^4 - 3x^3(4x^2) + 3x^3(5) = 3x^7 - 12x^5 + 15x^3$

26. $x^{3/2}(\sqrt{x} - 1/\sqrt{x}) = x^{3/2}(x^{1/2} - x^{-1/2}) = x^{3/2}x^{1/2} - x^{3/2}x^{-1/2} = x^2 - x$

28. $y^{1/4}(y^{1/2} + 2y^{3/4}) = (y^{1/4})(y^{2/4}) + y^{1/4}(2y^{3/4}) = y^{3/4} + 2y$

30. $(4x - 1)(3x + 7) = 12x^2 + 28x - 3x - 7 = 12x^2 + 25x - 7$

32. $(4x - 3y)(2x + 5y) = 8x^2 + 20xy - 6xy - 15y^2 = 8x^2 + 14xy - 15y^2$

34. $(3x + 4)^2 = (3x)^2 + 2(3x)(4) + (4)^2 = 9x^2 + 24x + 16$

36. $\left(c + \dfrac{1}{c}\right)^2 = (c)^2 + 2(c)\left(\dfrac{1}{c}\right) + \left(\dfrac{1}{c}\right)^2 = c^2 + 2 + \dfrac{1}{c^2}$

38. $(1 + 2x)(x^2 - 3x + 1) = x^2 - 3x + 1 + 2x^3 - 6x^2 + 2x = 2x^3 - 5x^2 - x + 1$

40. $\left(x^{1/2} + y^{1/2}\right)\left(x^{1/2} - y^{1/2}\right) = \left(x^{1/2}\right)^2 - \left(y^{1/2}\right)^2 = x - y$

42. $\left(\sqrt{h^2 + 1} + 1\right)\left(\sqrt{h^2 + 1} - 1\right) = \left(\sqrt{h^2 + 1}\right)^2 - (1)^2 = h^2 + 1 - 1 = h^2$

44. $(1 - 2y)^3 = (1)^3 - 3(1)^2(2y) + 3(1)(2y)^2 - (2y)^3 = 1 - 6y + 12y^2 - 8y^3$

46. $(3x^3 + x^2 - 2)(x^2 + 2x - 1) = 3x^3(x^2 + 2x - 1) + x^2(x^2 + 2x - 1) - 2(x^2 + 2x - 1)$
      $= 3x^5 + 6x^4 - 3x^3 + x^4 + 2x^3 - x^2 - 2x^2 - 4x + 2 = 3x^5 + 7x^4 - x^3 - 3x^2 - 4x + 2$

48. $(1 + x + x^2)(1 - x + x^2) = [(x^2 + 1) + x][(x^2 + 1) - x] = (x^2 + 1)^2 - (x)^2$
      $= x^4 + 2x^2 + 1 - x^2 = x^4 + x^2 + 1$

50. $(1 - b)^2(1 + b)^2 = [(1 - b)(1 + b)]^2 = (1 - b^2)^2 = 1 - 2b^2 + b^4$

52. $(x^4 y - y^5)(x^2 + xy + y^2) = x^6 y + x^5 y^2 + x^4 y^3 - x^2 y^5 - xy^6 - y^7$

54. $(x^2 + y - 2)(x^2 + y + 2) = [(x^2 + y) - 2][(x^2 + y) + 2] = (x^2 + y)^2 - 2^2 = x^4 + 2x^2 y + y^2 - 4$

56. $(x^2 - y + z)(x^2 + y - z) = [x^2 - (y - z)][x^2 + (y - z)] = (x^2)^2 - (y - z)^2$
      $= x^4 - (y^2 - 2yz + z^2) = x^4 - y^2 + 2yz - z^2$

58.   (a)   The *width* is the width of the lot minus the "set backs" of 10 feet each. Thus the
          width $= x - 20$. Likewise the *length* $= y - 20$. Since *Area* $=$ *width* $\times$ *length* we get
          $A = (x - 20)(y - 20)$.

      (b) $A = (x - 20)(y - 20) = xy - 20x - 20y + 400$

      (c)   For the 100×400 lot the building envelope $A = (100 - 20)(400 - 20) = 80(380) = 30,400$.
          For the 200×200 lot the building envelope
          $A = (200 - 20)(200 - 20) = 180(180) = 32,400$, it has the larger building envelope.

60.   (a) $P = R - C = (50x - 0.05x^2) - (50 + 30x - 0.1x^2) = 50x - 0.05x^2 - 50 - 30x + 0.1x^2$
          $= 0.05x^2 + 20x - 50$.

      (b)   The profit on 10 calculators is $P = 0.05(10^2) + 20(10) - 50 = \$155$. The profit on 20
          calculators is $P = 0.05(20^2) + 20(20) - 50 = \$370$.

62.   (a)   The degree of the product is the sum of the degrees.

      (b)   The degree of a sum is *at most* the largest of the degrees, it could be lower than either.

      (c)   Product:
          $(2x^3 + x - 3)(-2x^3 - x + 7) = -4x^6 - 2x^4 + 14x^3 - 2x^4 - x^2 + 7x + 6x^3 + 3x - 21$
          $= -4x^6 - 4x^4 + 20x^3 - x^2 + 10x - 21$.
          Sum: $(2x^3 + x - 3) + (-2x^3 - x + 7) = 4$.

Exercises P.6

2. $-3b + 12 = -3(b - 4)$

4. $2x^4 + 4x^3 - 14x^2 = 2x^2(x^2 + 2x - 7)$

6. $(z + 2)^2 - 5(z + 2) = (z + 2)[(z + 2) - 5] = (z + 2)(z - 3)$

8. $-7x^4y^2 + 14xy^3 + 21xy^4 = 7xy^2(-x^3 + 2y + 3y^2)$

10. $x^2 - 6x + 5 = (x - 5)(x - 1)$

12. $24 - 5t - t^2 = (8 + t)(3 - t)$

14. $6y^2 + 11y - 21 = (y + 3)(6y - 7)$

16. $2(a + b)^2 + 5(a + b) - 3 = [(a + b) + 3][2(a + b) - 1] = (a + b + 3)(2a + 2b - 1)$

18. $(x + 3)^2 - 4 = (x + 3)^2 - 2^2 = [(x + 3) - 2][(x + 3) + 2] = (x + 1)(x + 5)$

20. $a^3 - b^6 = a^3 - (b^2)^3 = (a - b^2)[a^2 + ab^2 + (b^2)^2] = (a - b^2)(a^2 + ab^2 + b^4)$

22. $1 + 1000y^3 = 1 + (10y)^3 = (1 + 10y)[1 - 10y + (10y)^2] = (1 + 10y)(1 - 10y + 100y^2)$

24. $16z^2 - 24z + 9 = (4z)^2 - 2(4z)(3) + 3^2 = (4z - 3)^2$

26. $3x^3 - x^2 + 6x - 2 = x^2(3x - 1) + 2(3x - 1) = (3x - 1)(x^2 + 2)$

28. $-9x^3 - 3x^2 + 3x + 1 = -3x^2(3x + 1) + 1(3x + 1) = (3x + 1)(-3x^2 + 1) = (3x + 1)(1 - 3x^2)$.
    If irrational coefficients are permitted, then this can be further factored as
    $(3x + 1)(1 - \sqrt{3}x)(1 + \sqrt{3}x)$

30. $x^5 + x^4 + x + 1 = x^4(x + 1) + 1(x + 1) = (x + 1)(x^4 + 1)$

32. $30x^3 + 15x^4 = 15x^3(2 + x)$

34. $5ab - 8abc = ab(5 - 8c)$

36. $x^2 - 14x + 48 = (x - 8)(x - 6)$

38. $z^2 + 6z - 16 = (z - 2)(z + 8)$

40. $2x^2 + 7x - 4 = (2x - 1)(x + 4)$

42. $8x^2 + 10x + 3 = (4x + 3)(2x + 1)$

44. $6 + 5t - 6t^2 = (3 - 2t)(2 + 3t)$

46. $4x^2 + 4xy + y^2 = (2x + y)^2$

48. $25s^2 - 10st + t^2 = (5s - t)^2$

50. $4x^2 - 25 = (2x - 5)(2x + 5)$

52. $4t^2 - 9s^2 = (2t - 3s)(2t + 3s)$

54. $\left(1 + \dfrac{1}{x}\right)^2 - \left(1 - \dfrac{1}{x}\right)^2 = \left[\left(1 + \dfrac{1}{x}\right) - \left(1 - \dfrac{1}{x}\right)\right]\left[\left(1 + \dfrac{1}{x}\right) + \left(1 - \dfrac{1}{x}\right)\right]$

$$= \left(1 + \dfrac{1}{x} - 1 + \dfrac{1}{x}\right)\left(1 + \dfrac{1}{x} + 1 - \dfrac{1}{x}\right) = \left(\dfrac{2}{x}\right)(2) = \dfrac{4}{x}$$

56. $(a^2 - 1)b^2 - 4(a^2 - 1) = (a^2 - 1)(b^2 - 4) = (a - 1)(a + 1)(b - 2)(b + 2)$

58. $x^3 - 27 = x^3 - 3^3 = (x - 3)(x^2 + 3x + 9)$

60. $x^6 + 64 = x^6 + 2^6 = (x^2)^3 + (4)^3 = (x^2 + 4)\left[(x^2)^2 - 4(x^2) + (4)^2\right]$

$$= (x^2 + 4)(x^4 - 4x^2 + 16)$$

62. $27a^3 + b^6 = (3a)^3 + (b^2)^3 = (3a + b^2)\left[(3a)^2 - (3a)(b^2) + (b^2)^2\right]$

$$= (3a + b^2)(9a^2 - 3ab^2 + b^4)$$

64. $3x^3 - 27x = 3x(x^2 - 9) = 3x(x - 3)(x + 3)$

66. $x^3 + 3x^2 - x - 3 = (x^3 + 3x^2) + (-x - 3) = x^2(x + 3) - (x + 3) = (x + 3)(x^2 - 1)$
$= (x + 3)(x - 1)(x + 1)$

68. $y^3 - y^2 + y - 1 = y^2(y - 1) + 1(y - 1) = (y^2 + 1)(y - 1)$

70. $3x^3 + 5x^2 - 6x - 10 = x^2(3x + 5) - 2(3x + 5) = (x^2 - 2)(3x + 5)$

72. $(x + 1)^3 x - 2(x + 1)^2 x^2 + x^3(x + 1) = x(x + 1)\left[(x + 1)^2 - 2(x + 1)x + x^2\right]$

$$= x(x + 1)\left[(x + 1) - x\right]^2 = x(x + 1)(1)^2 = x(x + 1)$$

74. $n(x - y) + (n - 1)(y - x) = n(x - y) - (n - 1)(x - y) = (x - y)\left[n - (n - 1)\right] = x - y$

76. $(a^2 + 2a)^2 - 2(a^2 + 2a) - 3 = [(a^2 + 2a) - 3][(a^2 + 2a) + 1]$
$= (a^2 + 2a - 3)(a^2 + 2a + 1) = (a - 1)(a + 3)(a + 1)^2$

78. $3x^{-1/2} + 4x^{1/2} + x^{3/2} = x^{-1/2}(3 + 4x + x^2) = \left(\dfrac{1}{\sqrt{x}}\right)(3 + x)(1 + x)$

80. $(x - 1)^{7/2} - (x - 1)^{3/2} = (x - 1)^{3/2}\left[(x - 1)^2 - 1\right] = (x - 1)^{3/2}[(x - 1) - 1][(x - 1) + 1]$
$= (x - 1)^{3/2}(x - 2)(x)$

82. $x^{-1/2}(x+1)^{1/2} + x^{1/2}(x+1)^{-1/2} = x^{-1/2}(x+1)^{-1/2}[(x+1)+x] = \dfrac{2x+1}{\sqrt{x}\sqrt{x+1}}$

84. $3x^{-1/2}(x^2+1)^{5/4} - x^{3/2}(x^2+1)^{1/4} = x^{-1/2}(x^2+1)^{1/4}[3(x^2+1) - x^2(1)]$

$= x^{-1/2}(x^2+1)^{1/4}(3x^2+3-x^2) = x^{-1/2}(x^2+1)^{1/4}(2x^2+3) = \dfrac{\sqrt[4]{x^2+1}(2x^2+3)}{\sqrt{x}}$

86. $5(x^2+4)^4(2x)(x-2)^4 + (x^2+4)^5(4)(x-2)^3$

$= 2(x^2+4)^4(x-2)^3[(5)(x)(x-2) + (x^2+4)(2)] = 2(x^2+4)^4(x-2)^3(5x^2-10x+2x^2+8)$

$= 2(x^2+4)^4(x-2)^3(7x^2-10x+8)$

88. $\frac{1}{3}(x+6)^{-2/3}(2x-3)^2 + (x+6)^{1/3}(2)(2x-3)(2)$

$= \frac{1}{3}(x+6)^{-2/3}(2x-3)[(2x-3) + (3)(x+6)(4)] = \frac{1}{3}(x+6)^{-2/3}(2x-3)[2x-3+12x+72]$

$= \frac{1}{3}(x+6)^{-2/3}(2x-3)(14x+69)$

90. $\frac{1}{2}x^{-1/2}(3x+4)^{1/2} - \frac{3}{2}x^{1/2}(3x+4)^{-1/2} = \frac{1}{2}x^{-1/2}(3x+4)^{-1/2}[(3x+4) - 3x]$

$= \frac{1}{2}x^{-1/2}(3x+4)^{-1/2}(4) = 2x^{-1/2}(3x+4)^{-1/2}$

92. Difference of Cube:

$(A-B)(A^2+AB+B^2) = A^3 + A^2B + AB^2 - A^2B - AB^2 - B^3 = A^3 - B^3$

Sum of Cubes: $(A+B)(A^2-AB+B^2) = A^3 - A^2B + AB^2 + A^2B - AB^2 + B^3 = A^3 + B^3$

94. (a) *Moved portion = field − habitat.*

(b) Using the difference of squares we get:

$b^2 - (b-2x)^2 = [b-(b-2x)][b+(b-x)] = 2x(2b-2x) = 4x(b-x).$

96. (a) $A^4 - B^4 = (A^2 - B^2)(A^2 + B^2) = (A-B)(A+B)(A^2+B^2)$

$A^6 - B^6 = (A^3 - B^3)(A^3 + B^3)$ (the difference of squares)

$= (A-B)(A^2+AB+B^2)(A+B)(A^2-AB+B^2)$ (difference and sum of cubes)

(b) $12^4 - 7^4 = 20{,}736 - 2{,}401 = 18{,}335$

$12^6 - 7^6 = 2{,}985{,}984 - 117{,}649 = 2{,}868{,}335$

(c) $18{,}335 = 12^4 - 7^4 = (12-7)(12+7)(12^2+7^2) = 5(19)(144+49) = 5(19)(193)$

$2{,}868{,}335 = 12^6 - 7^6 = (12-7)(12+7)[12^2+12(7)+7^2][12^2-12(7)+7^2]$

$= 5(19)(144+84+49)(144-84+49) = 5(19)(277)(109)$

98. (a) $x^4 + x^2 - 2 = (x^2-1)(x^2+2) = (x-1)(x+1)(x^2+2)$

(b) $x^4 + 2x^2 + 9 = (x^4 + 6x^2 + 9) - 4x^2 = (x^2+3)^2 - (2x)^2 = [(x^2+3) - 2x][(x^2+3) + 2x]$

$= (x^2 - 2x + 3)(x^2 + 2x + 3)$

(c) $x^4 + 4x^2 + 16 = (x^4 + 8x^2 + 16) - 4x^2 = (x^2+4)^2 - (2x)^2$

$= [(x^2+4) - 2x][(x^2+4) + 2x] = (x^2 - 2x + 4)(x^2 + 2x + 4)$

(d) $x^4 + 2x^2 + 1 = (x^2+1)^2$

Exercises P.7

2. (a) When $x = -1$ we get $-(-1)^4 + (-1)^3 + 9(-1) = -1 - 1 - 9 = -11$

   (b) Domain: all real numbers.

4. (a) When $x = 1$ we get $\dfrac{2(1^2) - 5}{3(1) + 6} = \dfrac{-3}{9} = -\dfrac{1}{3}$.

   (b) Since $3x + 6 \neq 0$ we have $x \neq -2$. Domain: $\{x \mid x \neq -2\}$

6. (a) When $x = 5$ we get $\dfrac{1}{\sqrt{5 - 1}} = \dfrac{1}{\sqrt{4}} = \dfrac{1}{2}$.

   (b) Since $x - 1 > 0 \quad \Leftrightarrow \quad x > 1$. Domain; $\{x \mid x > 1\}$

8. $\dfrac{81x^3}{18x} = \dfrac{9x \cdot 9x^2}{9x \cdot 2} = \dfrac{9x^2}{2} = \tfrac{9}{2}x^2$

10. $\dfrac{14t^2 - t}{7t} = \dfrac{t \cdot (14t - 1)}{t \cdot 7} = \dfrac{14t - 1}{7} = 2t - \tfrac{1}{7}$

12. $\dfrac{4(x^2 - 1)}{12(x + 2)(x - 1)} = \dfrac{4(x + 1)(x - 1)}{12(x + 2)(x - 1)} = \dfrac{x + 1}{3(x + 2)}$

14. $\dfrac{x^2 - x - 2}{x^2 - 1} = \dfrac{(x - 2)(x + 1)}{(x - 1)(x + 1)} = \dfrac{x - 2}{x - 1}$

16. $\dfrac{x^2 - x - 12}{x^2 + 5x + 6} = \dfrac{(x - 4)(x + 3)}{(x + 2)(x + 3)} = \dfrac{x - 4}{x + 2}$

18. $\dfrac{y^2 - 3y - 18}{2y^2 + 5y + 3} = \dfrac{(y - 6)(y + 3)}{(2y + 3)(y + 1)}$

20. $\dfrac{1 - x^2}{x^3 - 1} = \dfrac{(1 - x)(1 + x)}{(x - 1)(x^2 + x + 1)} = \dfrac{-(x - 1)(1 + x)}{(x - 1)(x^2 + x + 1)} = \dfrac{-(x + 1)}{x^2 + x + 1}$

22. $\dfrac{x^2 - 25}{x^2 - 16} \cdot \dfrac{x + 4}{x + 5} = \dfrac{(x - 5)(x + 5)}{(x - 4)(x + 4)} \cdot \dfrac{x + 4}{x + 5} = \dfrac{x - 5}{x - 4}$

24. $\dfrac{x^2 + 2x - 3}{x^2 - 2x - 3} \cdot \dfrac{3 - x}{3 + x} = \dfrac{(x + 3)(x - 1)}{(x - 3)(x + 1)} \cdot \dfrac{-(x - 3)}{x + 3} = \dfrac{-(x - 1)}{x + 1} = \dfrac{-x + 1}{x + 1} = \dfrac{1 - x}{1 + x}$

26. $\dfrac{x^2 - x - 6}{x^2 + 2x} \cdot \dfrac{x^3 + x^2}{x^2 - 2x - 3} = \dfrac{(x - 3)(x + 2)}{x(x + 2)} \cdot \dfrac{x^2(x + 1)}{(x - 3)(x + 1)} = x$

28. $\dfrac{x^2 + 2xy + y^2}{x^2 - y^2} \cdot \dfrac{2x^2 - xy - y^2}{x^2 - xy - 2y^2} = \dfrac{(x + y)(x + y)}{(x - y)(x + y)} \cdot \dfrac{(x - y)(2x + y)}{(x - 2y)(x + y)} = \dfrac{2x + y}{x - 2y}$

30. $\dfrac{4y^2 - 9}{2y^2 + 9y - 18} \div \dfrac{2y^2 + y - 3}{y^2 + 5y - 6} = \dfrac{4y^2 - 9}{2y^2 + 9y - 18} \cdot \dfrac{y^2 + 5y - 6}{2y^2 + y - 3}$

$= \dfrac{(2y - 3)(2y + 3)}{(2y - 3)(y + 6)} \cdot \dfrac{(y - 1)(y + 6)}{(y - 1)(2y + 3)} = 1w$

32. $\dfrac{\dfrac{2x^2 - 3x - 2}{x^2 - 1}}{\dfrac{2x^2 + 5x + 2}{x^2 + x - 2}} = \dfrac{2x^2 - 3x - 2}{x^2 - 1} \cdot \dfrac{x^2 + x - 2}{2x^2 + 5x + 2} = \dfrac{(x - 2)(2x + 1)}{(x - 1)(x + 1)} \cdot \dfrac{(x - 1)(x + 2)}{(x + 2)(2x + 1)} = \dfrac{x - 2}{x + 1}$

34. $\dfrac{x}{y/z} = x \div \dfrac{y}{z} = \dfrac{x}{1} \cdot \dfrac{z}{y} = \dfrac{xz}{y}$

36. $\dfrac{2x - 1}{x + 4} - 1 = \dfrac{2x - 1}{x + 4} + \dfrac{-(x + 4)}{x + 4} = \dfrac{2x - 1 - x - 4}{x + 4} = \dfrac{x - 5}{x + 4}$

38. $\dfrac{1}{x + 1} + \dfrac{1}{x - 1} = \dfrac{x - 1}{(x + 1)(x - 1)} + \dfrac{x + 1}{(x + 1)(x - 1)} = \dfrac{x - 1 + x + 1}{(x + 1)(x - 1)} = \dfrac{2x}{(x + 1)(x - 1)}$

40. $\dfrac{x}{x - 4} - \dfrac{3}{x + 6} = \dfrac{x(x + 6)}{(x - 4)(x + 6)} + \dfrac{-3(x - 4)}{(x - 4)(x + 6)} = \dfrac{x^2 + 6x - 3x + 12}{(x - 4)(x + 6)} = \dfrac{x^2 + 3x + 12}{(x - 4)(x + 6)}$

42. $\dfrac{5}{2x - 3} - \dfrac{3}{(2x - 3)^2} = \dfrac{5(2x - 3)}{(2x - 3)^2} - \dfrac{3}{(2x - 3)^2} = \dfrac{10x - 15 - 3}{(2x - 3)^2} = \dfrac{10x - 18}{(2x - 3)^2} = \dfrac{2(5x - 9)}{(2x - 3)^2}$

44. $\dfrac{2}{a^2} - \dfrac{3}{ab} + \dfrac{4}{b^2} = \dfrac{2b^2}{a^2 b^2} - \dfrac{3ab}{a^2 b^2} + \dfrac{4a^2}{a^2 b^2} = \dfrac{2b^2 - 3ab + 4a^2}{a^2 b^2}$

46. $\dfrac{1}{x} + \dfrac{1}{x^2} + \dfrac{1}{x^3} = \dfrac{x^2}{x^3} + \dfrac{x}{x^3} + \dfrac{1}{x^3} = \dfrac{x^2 + x + 1}{x^3}$

48. $\dfrac{x}{x^2 - 4} + \dfrac{1}{x - 2} = \dfrac{x}{(x - 2)(x + 2)} + \dfrac{1}{x - 2} = \dfrac{x}{(x - 2)(x + 2)} + \dfrac{x + 2}{(x - 2)(x + 2)}$

$= \dfrac{2x + 2}{(x - 2)(x + 2)} = \dfrac{2(x + 1)}{(x - 2)(x + 2)}$

50. $\dfrac{x}{x^2 + x - 2} - \dfrac{2}{x^2 - 5x + 4} = \dfrac{x}{(x - 1)(x + 2)} + \dfrac{-2}{(x - 1)(x - 4)}$

$= \dfrac{x(x - 4)}{(x - 1)(x + 2)(x - 4)} + \dfrac{-2(x + 2)}{(x - 1)(x + 2)(x - 4)} = \dfrac{x^2 - 4x - 2x - 4}{(x - 1)(x + 2)(x - 4)}$

$= \dfrac{x^2 - 6x - 4}{(x - 1)(x + 2)(x - 4)}$

52. $\dfrac{x}{x^2 - x - 6} - \dfrac{1}{x + 2} - \dfrac{2}{x - 3} = \dfrac{x}{(x - 3)(x + 2)} + \dfrac{-1}{x + 2} + \dfrac{-2}{x - 3}$

$= \dfrac{x}{(x - 3)(x + 2)} + \dfrac{-1(x - 3)}{(x - 3)(x + 2)} + \dfrac{-2(x + 2)}{(x - 3)(x + 2)} = \dfrac{x - x + 3 - 2x - 4}{(x - 3)(x + 2)} = \dfrac{-2x - 1}{(x - 3)(x + 2)}$

54. $\dfrac{1}{x+1} - \dfrac{2}{(x+1)^2} + \dfrac{3}{x^2-1} = \dfrac{1}{x+1} + \dfrac{-2}{(x+1)^2} + \dfrac{3}{(x-1)(x+1)}$

$= \dfrac{(x+1)(x-1)}{(x-1)(x+1)^2} + \dfrac{-2(x-1)}{(x-1)(x+1)^2} + \dfrac{3(x+1)}{(x-1)(x+1)^2}$

$= \dfrac{x^2-1}{(x-1)(x+1)^2} + \dfrac{-2x+2}{(x-1)(x+1)^2} + \dfrac{3x+3}{(x-1)(x+1)^2} = \dfrac{x^2-1-2x+2+3x+3}{(x-1)(x+1)^2}$

$= \dfrac{x^2+x+4}{(x-1)(x+1)^2}$

56. $x - \dfrac{y}{\dfrac{x}{y}+\dfrac{y}{x}} = x - \dfrac{y}{\dfrac{x}{y}+\dfrac{y}{x}} \cdot \dfrac{xy}{xy} = x - \dfrac{xy^2}{x^2+y^2} = \dfrac{x(x^2+y^2)}{x^2+y^2} - \dfrac{xy^2}{x^2+y^2} = \dfrac{x^3+xy^2-xy^2}{x^2+y^2}$

$= \dfrac{x^3}{x^2+y^2}$

58. $1 + \dfrac{1}{1+\frac{1}{1+x}} = 1 + \dfrac{1}{1+\frac{1}{1+x}} \cdot \dfrac{1+x}{1+x} = 1 + \dfrac{1+x}{1+x+1} = 1 + \dfrac{1+x}{2+x} = \dfrac{2+x}{2+x} + \dfrac{1+x}{2+x} = \dfrac{3+2x}{2+x}$

60. $\dfrac{\dfrac{a-b}{a}-\dfrac{a+b}{b}}{\dfrac{a-b}{b}+\dfrac{a+b}{a}} = \dfrac{\dfrac{a-b}{a}-\dfrac{a+b}{b}}{\dfrac{a-b}{b}+\dfrac{a+b}{a}} \cdot \dfrac{ab}{ab} = \dfrac{(a-b)b-(a+b)a}{(a-b)a+(a+b)b} = \dfrac{ab-b^2-a^2-ab}{a^2-ab+ab+b^2}$

$= \dfrac{-a^2-b^2}{a^2+b^2} = \dfrac{-(a^2+b^2)}{a^2+b^2} = -1$

62. $\dfrac{x^{-1}+y^{-1}}{(x+y)^{-1}} = \dfrac{\dfrac{1}{x}+\dfrac{1}{y}}{\dfrac{1}{x+y}} = \dfrac{\dfrac{1}{x}+\dfrac{1}{y}}{\dfrac{1}{x+y}} \cdot \dfrac{xy(x+y)}{xy(x+y)} = \dfrac{y(x+y)+x(x+y)}{xy} = \dfrac{xy+y^2+x^2+xy}{xy}$

$= \dfrac{x^2+2xy+y^2}{xy} = \dfrac{(x+y)^2}{xy}$

64. $\dfrac{\left(a+\dfrac{1}{b}\right)^m\left(a-\dfrac{1}{b}\right)^n}{\left(b+\dfrac{1}{a}\right)^m\left(b-\dfrac{1}{a}\right)^n} = \dfrac{\left(a+\dfrac{1}{b}\right)^m}{\left(b+\dfrac{1}{a}\right)^m}\dfrac{\left(a-\dfrac{1}{b}\right)^n}{\left(b-\dfrac{1}{a}\right)^n} = \left(\dfrac{\frac{ab+1}{b}}{\frac{ab+1}{a}}\right)^m\left(\dfrac{\frac{ab-1}{b}}{\frac{ab-1}{a}}\right)^n = \left(\dfrac{a}{b}\right)^m\left(\dfrac{a}{b}\right)^n$

$= \left(\dfrac{a}{b}\right)^{m+n} = \dfrac{a^{m+n}}{b^{m+n}}$

66. $\dfrac{(x+h)^{-3}-x^{-3}}{h} = \dfrac{\frac{1}{(x+h)^3}-\frac{1}{x^3}}{h} = \left(\dfrac{1}{(x+h)^3}-\dfrac{1}{x^3}\right)\cdot\dfrac{1}{h} = \left(\dfrac{x^3}{(x+h)^3x^3}+\dfrac{-(x+h)^3}{(x+h)^3x^3}\right)\cdot\dfrac{1}{h}$

$= \dfrac{x^3-(x^3+3x^2h+3xh^2+h^3)}{(x+h)^3x^3}\cdot\dfrac{1}{h} = \dfrac{x^3-x^3-3x^2h-3xh^2-h^3}{h(x+h)^3x^3} = \dfrac{-3x^2h-3xh^2-h^3}{h(x+h)^3x^3}$

$= \dfrac{-h(3x^2+3xh+h^2)}{h(x+h)^3x^3} = \dfrac{-(3x^2+3xh+h^2)}{(x+h)^3x^3}$

68.
$$\frac{(x+h)^3 - 7(x+h) - (x^3 - 7x)}{h} = \frac{x^3 + 3x^2h + 3xh^2 + h^3 - 7x - 7h - x^3 + 7x}{h}$$

$$= \frac{3x^2h + 3xh^2 + h^3 - 7h}{h} = \frac{h(3x^2 + 3xh + h^2 - 7)}{h} = 3x^2 + 3xh + h^2 - 7$$

70.
$$\sqrt{1 + \left(x^3 - \frac{1}{4x^3}\right)^2} = \sqrt{1 + x^6 - \frac{2x^3}{4x^3} + \frac{1}{16x^6}} = \sqrt{1 + x^6 - \frac{1}{2} + \frac{1}{16x^6}}$$

$$= \sqrt{x^6 + \frac{1}{2} + \frac{1}{16x^6}} = \sqrt{\left(x^3 + \frac{1}{4x^3}\right)^2} = \left|x^3 + \frac{1}{4x^3}\right|$$

72.
$$\frac{2x(x+6)^4 - x^2(4)(x+6)^3}{(x+6)^8} = \frac{(x+6)^3[2x(x+6) - 4x^2]}{(x+6)^8} = \frac{2x^2 + 12x - 4x^2}{(x+6)^5} = \frac{12x - 2x^2}{(x+6)^5}$$

$$= \frac{2x(6-x)}{(x+6)^5}$$

74.
$$\frac{(1-x^2)^{1/2} + x^2(1-x^2)^{-1/2}}{1-x^2} = \frac{(1-x^2)^{-1/2}(1 - x^2 + x^2)}{1-x^2} = \frac{1}{(1-x^2)^{3/2}}$$

76.
$$\frac{(7-3x)^{1/2} + \frac{3}{2}x(7-3x)^{-1/2}}{7-3x} = \frac{(7-3x)^{-1/2}\left(7 - 3x + \frac{3}{2}x\right)}{7-3x} = \frac{7 - \frac{3}{2}x}{(7-3x)^{3/2}}$$

78.
$$\frac{2}{3-\sqrt{5}} = \frac{2}{3-\sqrt{5}} \cdot \frac{3+\sqrt{5}}{3+\sqrt{5}} = \frac{2\left(3+\sqrt{5}\right)}{9-5} = \frac{2\left(3+\sqrt{5}\right)}{4} = \frac{3+\sqrt{5}}{2}$$

80.
$$\frac{1}{\sqrt{x}+1} = \frac{1}{\sqrt{x}+1} \cdot \frac{\sqrt{x}-1}{\sqrt{x}-1} = \frac{\sqrt{x}-1}{x-1}$$

82.
$$\frac{2(x-y)}{\sqrt{x}-\sqrt{y}} = \frac{2(x-y)}{\sqrt{x}-\sqrt{y}} \cdot \frac{\sqrt{x}+\sqrt{y}}{\sqrt{x}+\sqrt{y}} = \frac{2(x-y)(\sqrt{x}+\sqrt{y})}{x-y} = 2(\sqrt{x}+\sqrt{y}) = 2\sqrt{x} + 2\sqrt{y}$$

84.
$$\frac{\sqrt{3}+\sqrt{5}}{2} = \frac{\sqrt{3}+\sqrt{5}}{2} \cdot \frac{\sqrt{3}-\sqrt{5}}{\sqrt{3}-\sqrt{5}} = \frac{3-5}{2\left(\sqrt{3}-\sqrt{5}\right)} = \frac{-2}{2\left(\sqrt{3}-\sqrt{5}\right)} = \frac{-1}{\sqrt{3}-\sqrt{5}}$$

86.
$$\frac{\sqrt{x} - \sqrt{x+h}}{h\sqrt{x}\sqrt{x+h}} = \frac{\sqrt{x} - \sqrt{x+h}}{h\sqrt{x}\sqrt{x+h}} \cdot \frac{\sqrt{x} + \sqrt{x+h}}{\sqrt{x} + \sqrt{x+h}} = \frac{x - (x+h)}{h\sqrt{x}\sqrt{x+h}\left(\sqrt{x} + \sqrt{x+h}\right)}$$

$$= \frac{-h}{h\sqrt{x}\sqrt{x+h}\left(\sqrt{x} + \sqrt{x+h}\right)} = \frac{-1}{\sqrt{x}\sqrt{x+h}\left(\sqrt{x} + \sqrt{x+h}\right)}$$

88.
$$\sqrt{x+1} - \sqrt{x} = \frac{\sqrt{x+1} - \sqrt{x}}{1} \cdot \frac{\sqrt{x+1} + \sqrt{x}}{\sqrt{x+1} + \sqrt{x}} = \frac{x+1-x}{\sqrt{x+1} + \sqrt{x}} = \frac{1}{\sqrt{x+1} + \sqrt{x}}$$

90. This statement is false. For example, take $b = 2$ and $c = 1$, then LHS $= \dfrac{b}{b-c} = \dfrac{2}{2-1} = 2$, while

RHS $= 1 - \dfrac{b}{c} = 1 - \dfrac{2}{1} = -1$, and $2 \neq -1$.

92. This statement is false. For example, take $x = 5$ and $y = 2$. Then substituting into the left hand side

we obtain LHS $= \dfrac{x+1}{y+1} = \dfrac{5+1}{2+1} = \dfrac{6}{3} = 2$, while the right hand side yields RHS $= \dfrac{x}{y} = \dfrac{5}{2}$, and

$2 \neq \dfrac{5}{2}$.

94. This statement is false. For example, take $x = 1$ and $y = 1$. Then substituting into the left hand side

we obtain LHS $= 2\left(\dfrac{a}{b}\right) = 2\left(\dfrac{1}{1}\right) = 2$, while the right hand side yields RHS $= \dfrac{2a}{2b} = \dfrac{2}{2} = 1$, and

$2 \neq 1$.

96. This statement is true: $\dfrac{1+x+x^2}{x} = \dfrac{1}{x} + \dfrac{x}{x} + \dfrac{x^2}{x} = \dfrac{1}{x} + 1 + x.$

98. (a) The average cost $A = \dfrac{Cost}{number\ of\ shirts} = \dfrac{500 + 6x + 0.01x^2}{x}$

(b)

x	10	20	50	100	200	500	1000
Average cost	$56.10	$31.20	$16.50	$12.00	$10.50	$12.00	$16.50

100. No, squaring $\dfrac{2}{\sqrt{x}}$ changes its value by a factor of $\dfrac{2}{\sqrt{x}}$.

102. (a) $A + B$ (b) AB (c) $A^{1/3}$ (d) A/B

Review Exercises for Chapter P

2. Commutative Property for multiplication.

4. Distributive Property.

6. $(0, 10]$ $\Leftrightarrow$ $0 < x \le 10$

8. $[-2, \infty)$ $\Leftrightarrow$ $-2 \le x$

10. $x < -3$ $\Leftrightarrow$ $(-\infty, -3)$

12. $0 \le x \le \frac{1}{2}$ $\Leftrightarrow$ $[0, \frac{1}{2}]$

14. $1 - \big|1 - |-1|\big| = 1 - |1 - 1| = 1 - |0| = 1$

16. $\sqrt[3]{-125} = \sqrt[3]{(-5)^3} = -5$

18. $64^{2/3} = (4^3)^{2/3} = 4^2 = 16$

20. $\sqrt[4]{4}\sqrt[4]{324} = \sqrt[4]{4}\sqrt[4]{4 \cdot 81} = \sqrt[4]{2^2}\sqrt[4]{2^2 \cdot 3^4} = 2 \cdot 3 = 6$

22. $\sqrt{2}\sqrt{50} = \sqrt{100} = 10$

24. $x\sqrt{x} = x^1 \cdot x^{1/2} = x^{3/2}$

26. $\big((x^m)^2\big)^n = (x^{2m})^n = x^{2mn}$

28. $\Big((x^a)^b\Big)^c = (x^{ab})^c = x^{abc}$

30. $\dfrac{(x^2)^n x^5}{x^n} = \dfrac{x^{2n} \cdot x^5}{x^n} = x^{2n+5-n} = x^{n+5}$

32. $(a^2)^{-3}(a^3 b)^2 (b^3)^4 = a^{-6} \cdot a^6 b^2 \cdot b^{12} = a^{-6+6} b^{2+12} = b^{14}$

34. $\left(\dfrac{r^2 s^{4/3}}{r^{1/3} s}\right)^6 = \dfrac{r^{12} s^8}{r^2 s^6} = r^{12-2} s^{8-6} = r^{10} s^2$

36. $\sqrt{x^2 y^4} = \sqrt{x^2} \cdot \sqrt{(y^2)^2} = |x| y^2$

38. $\dfrac{\sqrt{x}+1}{\sqrt{x}-1} = \dfrac{\sqrt{x}+1}{\sqrt{x}-1} \cdot \dfrac{\sqrt{x}+1}{\sqrt{x}+1} = \dfrac{(\sqrt{x})^2 + 2\sqrt{x} + 1}{x-1} = \dfrac{x + 2\sqrt{x} + 1}{x-1}$. Here simplify means to rationalize the denominator.

40. $\left(\dfrac{ab^2 c^{-3}}{2a^3 b^{-4}}\right)^{-2} = \dfrac{a^{-2} b^{-4} c^6}{2^{-2} a^{-6} b^8} = 2^2\, a^{-2-(-6)}\, b^{-4-8}\, c^6 = 4a^4\, b^{-12}\, c^6 = \dfrac{4a^4 c^6}{b^{12}}$

42. $2.08 \times 10^{-8} = 0.0000000208$

44. $80\,\dfrac{\text{times}}{\text{minute}} \cdot \dfrac{60 \text{ minutes}}{\text{hour}} \cdot \dfrac{24 \text{ hours}}{\text{day}} \cdot \dfrac{365 \text{ day}}{\text{years}} \cdot 90 \text{ years} \approx 3.8 \times 10^9 \text{ times}$

46. $12x^2 y^4 - 3xy^5 + 9x^3 y^2 = 3xy^2(4xy^2 - y^3 + 3x^2)$

48. $x^2 + 3x - 10 = (x+5)(x-2)$

50. $6x^2 + x - 12 = (3x - 4)(2x + 3)$

52. $x^4 - 2x^2 + 1 = (x^2 - 1)^2 = [(x - 1)(x + 1)]^2 = (x - 1)^2(x + 1)^2$

54. $2y^6 - 32y^2 = 2y^2(y^4 - 16) = 2y^2(y^2 + 4)(y^2 - 4) = 2y^2(y^2 + 4)(y + 2)(y - 2)$

56. $y^3 - 2y^2 - y + 2 = y^2(y - 2) - 1(y - 2) = (y - 2)(y^2 - 1) = (y - 2)(y - 1)(y + 1)$

58. $a^4b^2 + ab^5 = ab^2(a^3 + b^3) = ab^2(a + b)(a^2 - ab + b^2)$

60. $8x^3 + y^6 = (2x)^3 + (y^2)^3 = (2x + y^2)(4x^2 - 2xy^2 + y^4)$

62. $3x^3 - 2x^2 + 18x - 12 = x^2(3x - 2) + 6(3x - 2) = (3x - 2)(x^2 + 6)$

64. $ax^2 + bx^2 - a - b = x^2(a + b) - 1(a + b) = (a + b)(x^2 - 1) = (a + b)(x - 1)(x + 1)$

66. $(a + b)^2 + 2(a + b) - 15 = [(a + b) - 3][(a + b) + 5] = (a + b - 3)(a + b + 5)$

68. $(2y - 7)(2y + 7) = 4y^2 - 49$

70. $(1 + x)(2 - x) - (3 - x)(3 + x) = 2 + x - x^2 - (9 - x^2) = 2 + x - x^2 - 9 + x^2 = -7 + x$

72. $(2x + 1)^3 = (2x)^3 + 3(2x)^2(1) + 3(2x)(1)^2 + (1)^3 = 8x^3 + 12x^2 + 6x + 1$

74. $x^3(x - 6)^2 + x^4(x - 6) = x^3(x - 6)[(x - 6) + x] = x^3(x - 6)(2x - 6) = x^3[2x^2 - 18x + 36]$
      $= 2x^5 - 18x^4 + 36x^3$

76. $\dfrac{x^3 + 2x^2 + 3x}{x} = \dfrac{x(x^2 + 2x + 3)}{x} = x^2 + 2x + 3$

78. $\dfrac{t^3 - 1}{t^2 - 1} = \dfrac{(t - 1)(t^2 + t + 1)}{(t - 1)(t + 1)} = \dfrac{t^2 + t + 1}{t + 1}$

80. $\dfrac{x^3/(x - 1)}{x^2/(x^3 - 1)} = \dfrac{x^3}{x - 1} \cdot \dfrac{x^3 - 1}{x^2} = \dfrac{x^3}{x - 1} \cdot \dfrac{(x - 1)(x^2 + x + 1)}{x^2} = x(x^2 + x + 1) = x^3 + x^2 + x$

82. $x - \dfrac{1}{x + 1} = \dfrac{x(x + 1)}{x + 1} - \dfrac{1}{x + 1} = \dfrac{x^2 + x - 1}{x + 1}$

84. $\dfrac{2}{x} + \dfrac{1}{x - 2} + \dfrac{3}{(x - 2)^2} = \dfrac{2(x - 2)^2}{x(x - 2)^2} + \dfrac{x(x - 2)}{x(x - 2)^2} + \dfrac{3x}{x(x - 2)^2}$
      $= \dfrac{2(x^2 - 4x + 4) + x^2 - 2x + 3x}{x(x - 2)^2} = \dfrac{2x^2 - 8x + 8 + x^2 - 2x + 3x}{x(x - 2)^2} = \dfrac{3x^2 - 7x + 8}{x(x - 2)^2}$

86. $\dfrac{1}{x + 2} + \dfrac{1}{x^2 - 4} - \dfrac{2}{x^2 - x - 2} = \dfrac{1}{x + 2} + \dfrac{1}{(x - 2)(x + 2)} - \dfrac{2}{(x - 2)(x + 1)}$
      $= \dfrac{(x - 2)(x + 1)}{(x - 2)(x + 1)(x + 2)} + \dfrac{x + 1}{(x - 2)(x + 1)(x + 2)} - \dfrac{2(x + 2)}{(x - 2)(x + 1)(x + 2)}$

$$= \frac{x^2 - x - 2 + x + 1 - 2x - 4}{(x-2)(x+1)(x+2)} = \frac{x^2 - 2x - 5}{(x-2)(x+1)(x+2)}$$

88. $$\dfrac{\dfrac{1}{x} - \dfrac{1}{x+1}}{\dfrac{1}{x} + \dfrac{1}{x+1}} = \dfrac{\dfrac{1}{x} - \dfrac{1}{x+1}}{\dfrac{1}{x} + \dfrac{1}{x+1}} \cdot \dfrac{x(x+1)}{x(x+1)} = \dfrac{(x+1) - x}{(x+1) + x} = \dfrac{1}{2x+1}$$

90. $$\frac{\sqrt{x+h} - \sqrt{x}}{h} = \frac{\sqrt{x+h} - \sqrt{x}}{h} \cdot \frac{\sqrt{x+h} + \sqrt{x}}{\sqrt{x+h} + \sqrt{x}} = \frac{(x+h) - x}{h\left(\sqrt{x+h} + \sqrt{x}\right)}$$

$$= \frac{h}{h\left(\sqrt{x+h} + \sqrt{x}\right)} = \frac{1}{\sqrt{x+h} + \sqrt{x}}$$

92. This statement is true, for $a \neq 1$: $\dfrac{1 + \sqrt{a}}{1 - a} = \dfrac{1 + \sqrt{a}}{1 - a} \cdot \dfrac{1 - \sqrt{a}}{1 - \sqrt{a}} = \dfrac{1 - a}{(1-a)(1 - \sqrt{a})} = \dfrac{1}{1 - \sqrt{a}}.$

94. This statement is false. For example, take $a = 1$ and $b = 1$, then LHS $= \sqrt[3]{a+b} = \sqrt[3]{1+1}$
    $= \sqrt[3]{2}$, while RHS $= \sqrt[3]{a} + \sqrt[3]{b} = \sqrt[3]{1} + \sqrt[3]{1} = 1 + 1 = 2$, and $\sqrt[3]{2} \neq 2$.

96. This statement is false. For example, take $x = 1$, then LHS $= \dfrac{1}{x+4} = \dfrac{1}{1+4} = \dfrac{1}{5}$, while
    RHS $= \dfrac{1}{x} + \dfrac{1}{4} = \dfrac{1}{1} + \dfrac{1}{4} = \dfrac{5}{4}$, and $\dfrac{1}{5} \neq \dfrac{5}{4}$.

98. This statement is false. For example, take $x = 1$, the LHS $= \dfrac{x^2 + 1}{x^2 + 2x + 1} = \dfrac{1+1}{1+2+1} = \dfrac{2}{5}$, while
    RHS $= \dfrac{1}{2x+1} = \dfrac{1}{2+1} = \dfrac{1}{3}$, and $\dfrac{2}{5} \neq \dfrac{1}{3}$.

100. Substituting we obtain $\sqrt{1 + t^2} = \sqrt{1 + \left[\dfrac{1}{2}\left(x^3 - \dfrac{1}{x^3}\right)\right]^2} = \sqrt{1 + \dfrac{1}{4}\left(x^6 - 2\dfrac{x^3}{x^3} + \dfrac{1}{x^6}\right)}$

$$= \sqrt{1 + \frac{x^6}{4} - \frac{1}{2} + \frac{1}{4x^6}} = \sqrt{\frac{x^6}{4} + \frac{1}{2} + \frac{1}{4x^6}} = \sqrt{\left(\frac{x^3}{2} + \frac{1}{2x^3}\right)^2}. \text{ Since } x > 0, \sqrt{\left(\frac{x^3}{2} + \frac{1}{2x^3}\right)^2}$$

$$= \tfrac{x^3}{2} + \tfrac{1}{2x^3} = \tfrac{1}{2}\left(x^3 + \tfrac{1}{x^3}\right).$$

Principles of Problem Solving

2. Let r be the rate of the descent. We use the formula $time = \dfrac{distance}{rate}$; the ascent takes $\dfrac{1}{15}$ hr, the descent takes $\dfrac{1}{r}$ hr, and the total trip should take $\dfrac{2}{30} = \dfrac{1}{15}$ hr. Thus we have $\dfrac{1}{15} + \dfrac{1}{r} = \dfrac{1}{15}$ $\Leftrightarrow$ $\frac{1}{r} = 0$, which is impossible. So the car can not go fast enough to average 30 mi/h for the 2 mile trip.

4. Let us start with a given price P. After a discount of 40%, the price decreases to $0.6P$. After a discount of 20%, the price decreases to $0.8P$, and after another 20% discount, it becomes $0.8(0.8P) = 0.64P$. Since $0.6P < 0.64P$, a 40% discount is better.

6. By placing two amoebas into the vessel, we skip the first simple division which took 3 minutes. Thus when we place two amoebas into the vessel, it will take $60 - 3 = 57$ minutes for the vessel to be full of amoebas.

8. The statement is false. Here is one particular counterexample:

	Player A	Player B
First half	1 hit in 99 at-bats: average $= \frac{1}{99}$	0 hit in 1 at-bat: average $= \frac{0}{1}$
Second half	1 hit in 1 at-bat: average $= \frac{1}{1}$	98 hits in 99 at-bats: average $= \frac{98}{99}$
Entire season	2 hits in 100 at-bats: average $= \frac{2}{100}$	99 hits in 100 at-bats: average $= \frac{99}{100}$

10. It remains the same. The weight of the ice cube is the same as the weight of the volume of water it displaces. Thus as the ice cube melts, it becomes water that fits exactly into the space displaced by the floating ice cube.

12. We know that $\sqrt{2}$ is irrational. Now $\sqrt{2}^{\sqrt{2}}$ is either rational or irrational. If $\sqrt{2}^{\sqrt{2}}$ is rational, we would be done because we would have found an irrational number raised to an irrational number which is rational. If $\sqrt{2}^{\sqrt{2}}$ is not rational, then consider $\left(\sqrt{2}^{\sqrt{2}}\right)^{\sqrt{2}} = \sqrt{2}^{\sqrt{2}\cdot\sqrt{2}} = \sqrt{2}^{2} = 2$. So, again we would have found an irrational number (namely $\sqrt{2}^{\sqrt{2}}$) raised to an irrational number (namely $\sqrt{2}$) which is rational (namely 2).

14. Let n, $n + 1$, and $n + 2$ be three consecutive integers. Then the product of three consecutive integers plus the middle is $n(n + 1)(n + 2) + (n + 1) = n^3 + 3n^2 + 2n + n + 1$
    $= n^3 + 3n^2 + 3n + 1 = (n + 1)^3$.

16. Let us see what happens when we square similar numbers with fewer 9's:
    $$39^2 = 1521, \quad 399^2 = 159201, \quad 3999^2 = 15992001, \quad 39999^2 = 1599920001.$$
    The pattern is that the square always seems to start with 15 and end with 1, and if $39\cdots9$ has n 9's, then the 2 in the middle of its square is preceded by $(n - 1)$ 9's and followed by $(n - 1)$ 0's. From this pattern, we make the guess that
    $$3{,}999{,}999{,}999{,}999^2 = 15{,}999{,}999{,}999{,}992{,}000{,}000{,}000{,}001$$
    This can be verified by writing the number as follows:

$$3,999,999,999,999^2 = (4,000,000,000,000 - 1)^2$$
$$= 4,000,000,000,000^2 - 2 \cdot 4,000,000,000,000 + 1$$
$$= 16,000,000,000,000,000,000,000,000 - 8,000,000,000,000 + 1$$
$$= 15,999,999,999,992,000,000,000,001$$

18. $x^2 + y^2 = 4z + 3$. We consider all possible cases for x and y even or odd.

Case 1: x, y are both odd. Since the square of an odd number is odd, and the sum of two odd numbers is even, the left hand side is even but the right hand side is odd. Thus the left hand side cannot equal the right hand side.

Case 2: x, y are both even. Since the square of an even number is even, and the sum of two even numbers is even, the left hand side is even but the right hand side is odd. Thus the left hand side cannot equal the right hand side.

Case 3: x odd and y even. Let $x = 2n + 1$ and let $y = 2m$ for some integers n and m. Then
$(2n + 1)^2 + (2m)^2 = 4z + 3 \quad \Leftrightarrow \quad 4n^2 + 4n + 1 + 4m^2 = 4z + 3 \quad \Leftrightarrow$
$4(n^2 + n + m^2) + 1 = 4z + 3$. However, when the left hand side is divided by 4 the remainder is 1, but when the right hand side is divided by 4 the remainder is 3. Thus the left hand side cannot equal the right hand side.

Case 4: x even and y odd. This is handled in the same way as case 3.

Thus there are no integer solutions to $x^2 + y^2 = 4z + 3$.

20. Let h_1 be the height of the pyramid whose square base length is a. Then $h_1 - h$ is the height of the pyramid whose square base length is b. Thus the volume of the truncated pyramid is
$V = \frac{1}{3} h_1 a^2 - \frac{1}{3}(h_1 - h)b^2 = \frac{1}{3} h_1 a^2 - \frac{1}{3} h_1 b^2 + \frac{1}{3} h b^2 = \frac{1}{3} h_1 (a^2 - b^2) + \frac{1}{3} h b^2$. Next we must find a relationship between h_1 and the other variables. Using geometry, the ratio of the height to the base of the two pyramids must be the same. Thus $\dfrac{h_1}{a} = \dfrac{h_1 - h}{b} \quad \Leftrightarrow \quad bh_1 = ah_1 - ah \quad \Leftrightarrow$

$ah = ah_1 - bh_1 = (a - b)h_1 \quad \Leftrightarrow \quad \dfrac{ah}{a - b} = h_1$. Substituting for h_1 we have

$V = \frac{1}{3}\left(\dfrac{ah}{a-b}\right)(a^2 - b^2) + \frac{1}{3} h b^2 = \frac{1}{3}(ah)(a + b) + \frac{1}{3} h b^2 = \frac{1}{3} h a^2 + \frac{1}{3} h ab + \frac{1}{3} h b^2$
$= \frac{1}{3} h(a^2 + ab + b^2)$.

22. Label the figures as shown. The unshaded area is c^2 in the first figure and $a^2 + b^2$ in the second figure. These unshaded areas are equal, since in each case they equal the area of the large square minus the areas of the four shaded triangles. Thus $a^2 + b^2 = c^2$. Finally, note that a, b, and c are the legs and hypotenuse of the shaded triangle.

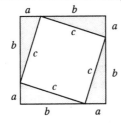 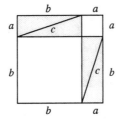

24. (a) $\left(\sqrt{3 + 2\sqrt{2}} - \sqrt{3 - 2\sqrt{2}}\right)^2 = 3 + 2\sqrt{2} - 2 \cdot \sqrt{\left(3 + 2\sqrt{2}\right)\left(3 - 2\sqrt{2}\right)} + 3 - 2\sqrt{2}$

$= 6 - 2 \cdot \sqrt{9-8} = 6 - 2 = 4.$ Therefore, $\sqrt{3 + 2\sqrt{2}} - \sqrt{3 - 2\sqrt{2}} = \sqrt{4} = 2.$

(b) $\dfrac{\sqrt{2} + \sqrt{6}}{\sqrt{2 + \sqrt{3}}} = \sqrt{\dfrac{\left(\sqrt{2} + \sqrt{6}\right)^2}{2 + \sqrt{3}}} = \sqrt{\dfrac{2 + 2\sqrt{12} + 6}{2 + \sqrt{3}}} = \sqrt{\dfrac{8 + 4\sqrt{3}}{2 + \sqrt{3}}} = \sqrt{\dfrac{4\left(2 + \sqrt{3}\right)}{2 + \sqrt{3}}} = 2$

Chapter One

Exercises 1.1

2. (a) When $x = -1$, LHS $= 2 - 5(-1) = 2 + 5 = 7$ and RHS $= 8 + (-1) = 7$. Since LHS $=$ RHS, $x = -1$ is a solution.

   (b) When $x = 1$, LHS $= 2 - 5(1) = 2 - 5 = -3$ and RHS $= 8 + (1) = 9$. Since LHS $\neq$ RHS, $x = 1$ is not a solution.

4. (a) When $x = 2$, LHS $= \dfrac{1}{2} - \dfrac{1}{2-4} = \dfrac{1}{2} - \dfrac{1}{-2} = \dfrac{1}{2} + \dfrac{1}{2} = 1$ and RHS $= 1$. Since LHS $=$ RHS, $x = 2$ is a solution.

   (b) When $x = 4$ the expression $\dfrac{1}{4-4}$ is not defined, so $x = 4$ is not a solution.

6. (a) When $x = 4$, LHS $= \dfrac{4^{3/2}}{4-6} = \dfrac{2^3}{-2} = \dfrac{8}{-2} = -4$ and RHS $= (4) - 8 = -4$. Since LHS $=$ RHS, $x = 4$ is a solution.

   (b) When $x = 8$, LHS $= \dfrac{8^{3/2}}{8-6} = \dfrac{(2^3)^{3/2}}{2} = \dfrac{2^{9/2}}{2} = 2^{7/2}$ and RHS $= (8) - 8 = 0$. Since LHS $\neq$ RHS, $x = 8$ is not a solution.

8. (a) When $x = b + \sqrt{b^2 - 2}$,

   $$\begin{aligned}
   \text{LHS} &= \tfrac{1}{2}(b + \sqrt{b^2 - 2})^2 - b(b + \sqrt{b^2 - 2}) + 1 \\
   &= \tfrac{1}{2}\left[b^2 + 2b\sqrt{b^2 - 2} + (b^2 - 2) \right] - b^2 - b\sqrt{b^2 - 2} + 1 \\
   &= \tfrac{1}{2}b^2 + b\sqrt{b^2 - 2} + \tfrac{1}{2}b^2 - 1 - b^2 - b\sqrt{b^2 - 2} + 1 \\
   &= 0 = \text{RHS}
   \end{aligned}$$

   Since LHS $=$ RHS, $x = b + \sqrt{b^2 - 2}$ is a solution.

   (b) When $x = b - \sqrt{b^2 - 2}\,1$,

   $$\begin{aligned}
   \text{LHS} &= \tfrac{1}{2}(b - \sqrt{b^2 - 2})^2 - b(b - \sqrt{b^2 - 2}) + 1 \\
   &= \tfrac{1}{2}\left[b^2 - 2b\sqrt{b^2 - 2} + (b^2 - 2) \right] - b^2 + b\sqrt{b^2 - 2} + 1 \\
   &= \tfrac{1}{2}b^2 - b\sqrt{b^2 - 2} + \tfrac{1}{2}b^2 - 1 - b^2 + b\sqrt{b^2 - 2} + 1 \\
   &= 0 = \text{RHS}
   \end{aligned}$$

   Since LHS $=$ RHS, $x = b - \sqrt{b^2 - 2}$ is a solution.

10. $5x - 3 = 4 \quad \Leftrightarrow \quad 5x = 7 \quad \Leftrightarrow \quad x = \tfrac{7}{5}$

12. $3 + \tfrac{1}{3}x = 5 \quad \Leftrightarrow \quad \tfrac{1}{3}x = 2 \quad \Leftrightarrow \quad x = 6$

14. $4x + 7 = 9x - 13 \quad \Leftrightarrow \quad 20 = 5x \quad \Leftrightarrow \quad x = 4$

16. $5t - 13 = 12 - 5t \iff 10t = 25 \iff t = \dfrac{5}{2}$

18. $\dfrac{z}{5} = \dfrac{3}{10}z + 7 \iff 2z = 3z + 70 \iff z = -70$

20. $5(x + 3) + 9 = -2(x - 2) - 1 \iff 5x + 15 + 9 = -2x + 4 - 1 \iff 5x + 24 = -2x + 3$
$\iff 7x = -21 \iff x = -3$

22. $\dfrac{2}{3}y + \dfrac{1}{2}(y - 3) = \dfrac{y + 1}{4} \iff 8y + 6(y - 3) = 3(y + 1) \iff 8y + 6y - 18 = 3y + 3 \iff$
$14y - 18 = 3y + 3 \iff 11y = 21 \iff y = \dfrac{21}{11}$

24. $\dfrac{2x - 1}{x + 2} = \dfrac{4}{5} \Rightarrow 5(2x - 1) = 4(x + 2) \iff 10x - 5 = 4x + 8 \iff 6x = 13 \iff$
$x = \frac{13}{6}$

26. $\dfrac{1}{t - 1} + \dfrac{t}{3t - 2} = \dfrac{1}{3} \Rightarrow 3(3t - 2) + 3t(t - 1) = (t - 1)(3t - 2) \iff$
$9t - 6 + 3t^2 - 3t = 3t^2 - 5t + 2 \iff 11t = 8 \iff t = \frac{8}{11}$

28. $(t - 4)^2 = (t + 4)^2 + 32 \iff t^2 - 8t + 16 = t^2 + 8t + 16 + 32 \iff -16t = 32 \iff$
$t = -2$

30. $\dfrac{2}{3}x - \dfrac{1}{4} = \dfrac{1}{6}x - \dfrac{1}{9} \iff 24x - 9 = 6x - 4 \iff 18x = 5 \iff x = \dfrac{5}{18}$

32. $\dfrac{6}{x - 3} = \dfrac{5}{x + 4} \Rightarrow 6(x + 4) = 5(x - 3) \iff 6x + 24 = 5x - 15 \iff x = -39$

34. $\dfrac{4}{x - 1} + \dfrac{2}{x + 1} = \dfrac{35}{x^2 - 1} \Rightarrow 4(x + 1) + 2(x - 1) = 35 \iff 4x + 4 + 2x - 2 = 35 \iff$
$6x + 2 = 35 \iff 6x = 33 \iff x = \dfrac{11}{2}$

36. $\dfrac{12x - 5}{6x + 3} = 2 - \dfrac{5}{x} \Rightarrow (12x - 5)x = 2x(6x + 3) - 5(6x + 3) \iff$
$12x^2 - 5x = 12x^2 + 6x - 30x - 15 \iff 12x^2 - 5x = 12x^2 - 24x - 15 \iff 19x = -15$
$\iff x = -\frac{15}{19}$

38. $2x - \dfrac{x}{2} + \dfrac{x + 1}{4} = 6x \iff 8x - 2x + x + 1 = 24x \iff 7x + 1 = 24x \iff 1 = 17x$
$\iff x = \frac{1}{17}$

40. $\dfrac{1}{1 - \dfrac{3}{2+w}} = 60 \iff \dfrac{2 + w}{2 + w - 3} = 60 \text{ (Multiplying the LHS by } \dfrac{2 + w}{2 + w}\text{). Thus } \dfrac{2 + w}{w - 1} = 60$
$\Rightarrow 2 + w = 60(w - 1) \iff 2 + w = 60w - 60 \iff 62 = 59w \iff w = \frac{62}{59}$

42. $\dfrac{1}{3-t} + \dfrac{4}{3+t} + \dfrac{16}{9-t^2} = 0 \quad \Rightarrow \quad (3+t) + 4(3-t) + 16 = 0 \quad \Leftrightarrow$

$3 + t + 12 - 4t + 16 = 0 \quad \Leftrightarrow \quad -3t + 31 = 0 \quad \Leftrightarrow \quad -3t = -31 \quad \Leftrightarrow \quad t = \dfrac{31}{3}$

44. $\dfrac{1}{x+3} + \dfrac{5}{x^2-9} = \dfrac{2}{x-3} \quad \Rightarrow \quad (x-3) + 5 = 2(x+3) \quad \Leftrightarrow \quad x + 2 = 2x + 6 \quad \Leftrightarrow$

$x = -4$

46. $\dfrac{1}{x} - \dfrac{2}{2x+1} = \dfrac{1}{2x^2+x} \quad \Rightarrow \quad (2x+1) - 2(x) = 1 \quad \Leftrightarrow \quad 1 = 1.$ This is an identity for $x \neq 0$

and $x \neq -\frac{1}{2}$, so the solutions are all real numbers except 0 and $-\frac{1}{2}$.

48. $x^2 = 18 \quad \Rightarrow \quad x = \pm\sqrt{18} = \pm 3\sqrt{2}$

50. $x^2 - 7 = 0 \quad \Leftrightarrow \quad x^2 = 7 \quad \Rightarrow \quad x = \pm\sqrt{7}$

52. $5x^2 - 125 = 0 \quad \Leftrightarrow \quad 5(x^2 - 25) = 0 \quad \Leftrightarrow \quad x^2 = 25 \quad \Rightarrow \quad x = \pm 5$

54. $6x^2 + 100 = 0 \quad \Leftrightarrow \quad 6x^2 = -100 \quad \Leftrightarrow \quad x^2 = -\frac{50}{3}$, which has no real solutions.

56. $3(x-5)^2 = 15 \quad \Leftrightarrow \quad (x-5)^2 = 5 \quad \Rightarrow \quad x - 5 = \pm\sqrt{5} \quad \Leftrightarrow \quad x = 5 \pm \sqrt{5}$

58. $x^5 + 32 = 0 \quad \Leftrightarrow \quad x^5 = -32 \quad \Leftrightarrow \quad x = -32^{1/5} = -2$

60. $64x^6 = 27 \quad \Leftrightarrow \quad x^6 = \frac{27}{64} \quad \Rightarrow \quad x = \pm\left(\frac{27}{64}\right)^{1/6} = \pm\frac{27^{1/6}}{64^{1/6}} = \pm\frac{\sqrt{3}}{2}$

62. $(x-1)^3 + 8 = 0 \quad \Leftrightarrow \quad (x-1)^3 = -8 \quad \Leftrightarrow \quad x - 1 = (-8)^{1/3} = -2 \quad \Leftrightarrow \quad x = -1.$

64. $(x+1)^4 + 16 = 0 \quad \Leftrightarrow \quad (x+1)^4 = -16$, which has no real solutions.

66. $4(x+2)^5 = 1 \quad \Leftrightarrow \quad (x+2)^5 = \frac{1}{4} \quad \Rightarrow \quad x + 2 = \sqrt[5]{\frac{1}{4}} \quad \Leftrightarrow \quad x = -2 + \sqrt[5]{\frac{1}{4}}$

68. $x^{4/3} - 16 = 0 \quad \Leftrightarrow \quad x^{4/3} = 16 = 2^4 \quad \Leftrightarrow \quad \left(x^{4/3}\right)^3 = (2^4)^3 = 2^{12} \quad \Leftrightarrow \quad x^4 = 2^{12} \quad \Leftrightarrow$

$x = \pm\left(2^{12}\right)^{1/4} = \pm 2^3 = \pm 8.$

70. $6x^{2/3} - 216 = 0 \quad \Leftrightarrow \quad 6x^{2/3} = 216 \quad \Leftrightarrow \quad x^{2/3} = 36 = (\pm 6)^2 \quad \Leftrightarrow \quad \left(x^{2/3}\right)^{3/2} = [(\pm 6)^2]^{3/2}$

$\Leftrightarrow \quad x = (\pm 6)^3 = \pm 216$

72. $8.36 - 0.95x = 9.97 \quad \Leftrightarrow \quad -0.95x = 1.61 \quad \Leftrightarrow \quad x = \dfrac{1.61}{-0.95} \approx -1.69$

74. $3.95 - x = 2.32x + 2.00 \quad \Leftrightarrow \quad 1.95 = 3.32x \quad \Leftrightarrow \quad x = \dfrac{1.95}{3.32} \approx 0.59$

76. $2.14(x - 4.06) = 2.27 - 0.11x \quad \Leftrightarrow \quad 2.14x - 8.6684 = 2.27 - 0.11x \quad \Leftrightarrow \quad 2.25x = 10.9584$

$\Leftrightarrow \quad x = 4.8704 \approx 4.87$

78. $\dfrac{1.73x}{2.12 + x} = 1.51 \quad \Leftrightarrow \quad 1.73x = 1.51(2.12 + x) \quad \Leftrightarrow \quad 1.73x = 3.20 + 1.51x \quad \Leftrightarrow$

$0.22x = 3.20 \quad \Leftrightarrow \quad x = \dfrac{3.20}{0.22} \approx 14.55$

80. $F = G\dfrac{mM}{r^2} \quad \Leftrightarrow \quad m = \dfrac{Fr^2}{GM}$

82. $P = 2l + 2w \quad \Leftrightarrow \quad 2w = P - 2l \quad \Leftrightarrow \quad w = \dfrac{P - 2l}{2}$

84. $a - 2[b - 3(c - x)] = 6 \quad \Leftrightarrow \quad a - 2(b - 3c + 3x) = 6 \quad \Leftrightarrow \quad a - 2b + 6c - 6x = 6 \quad \Leftrightarrow$

$-6x = 6 - a + 2b - 6c \quad \Leftrightarrow \quad x = \dfrac{6 - a + 2b - 6c}{-6} = -\dfrac{6 - a + 2b - 6c}{6}$

86. $\dfrac{a+1}{b} = \dfrac{a-1}{b} + \dfrac{b+1}{a} \quad \Leftrightarrow \quad a(a+1) = a(a-1) + b(b+1) \quad \Leftrightarrow$

$a^2 + a = a^2 - a + b^2 + b \quad \Leftrightarrow \quad 2a = b^2 + b \quad \Leftrightarrow \quad a = \tfrac{1}{2}(b^2 + b)$

88. $F = G\dfrac{mM}{r^2} \quad \Leftrightarrow \quad r^2 = G\dfrac{mM}{F} \quad \Rightarrow \quad r = \pm\sqrt{G\dfrac{mM}{F}}$

90. $A = P\left(1 + \dfrac{i}{100}\right)^2 \quad \Leftrightarrow \quad \dfrac{A}{P} = \left(1 + \dfrac{i}{100}\right)^2 \quad \Rightarrow \quad 1 + \dfrac{i}{100} = \pm\sqrt{\dfrac{A}{P}} \quad \Leftrightarrow$

$\dfrac{i}{100} = -1 \pm \sqrt{\dfrac{A}{P}} \quad \Leftrightarrow \quad i = -100 \pm 100\sqrt{\dfrac{A}{P}}$

92. $x^4 + y^4 + z^4 = 100 \quad \Leftrightarrow \quad x^4 = 100 - y^4 - z^4 \quad \Rightarrow \quad x = \pm\sqrt[4]{100 - y^4 - z^4}$

94. Substituting $C = 3600$ we get $3600 = 450 + 3.75x \quad \Leftrightarrow \quad 3150 = 3.75x \quad \Leftrightarrow \quad x = \frac{3150}{3.75} = 840$. So the toy manufacturer can manufacture 840 toy trucks.

96. Substituting $F = 300$ we get $300 = 0.3x^{3/4} \quad \Leftrightarrow \quad 1000 = 10^3 = x^{3/4} \quad \Leftrightarrow \quad x^{1/4} = 10 \quad \Leftrightarrow$ $x = 10^4 = 10{,}000$ lb.

98. When we multiplied by x, we introduced $x = 0$ as a solution. When we divided by $x - 1$, we are really dividing by 0, since $x = 1 \quad \Leftrightarrow \quad x - 1 = 0$.

Exercises 1.2

2. If n is the middle integer, then $n - 1$ is the first integer, and $n + 1$ is the third integer. So the sum of the three consecutive integers is $(n - 1) + n + (n + 1) = 3n$.

4. If q is the fourth quiz score, then since the other quiz scores are 8, 8, and 8, the average of the four quiz scores is $\dfrac{8 + 8 + 8 + q}{4} = \dfrac{24 + q}{4}$.

6. If n is the number of months the apartment is rented, and each month the rent is \$795, then the total rent paid is $795n$.

8. Since w is the width of the rectangle, the length is $w + 5$. The perimeter is
$2 \times length + 2 \times width = 2(w + 5) + 2(w) = 4w + 10$.

10. If d is the given distance, in miles, and $distance = rate \times time$, we have $time = \dfrac{distance}{rate} = \dfrac{d}{55}$.

12. If p is the number of pennies in the purse, then the number of nickels is $2p$, the number of dimes is $4 + 2p$, and the number of quarters is $(2p) + (4 + 2p) = 4p + 4$. Thus the value (in cents) of the change in the purse is $1 \cdot p + 5 \cdot 2p + 10 \cdot (4 + 2p) + 25 \cdot (4p + 4)$
$= p + 10p + 40 + 20p + 100p + 100 = 131p + 140$.

14. Let x be the first consecutive odd integer. Then $x + 2$, $x + 4$, and $x + 6$ are the next consecutive odd integers. So $x + (x + 2) + (x + 4) + (x + 6) = 272 \quad \Leftrightarrow \quad 4x + 12 = 272 \quad \Leftrightarrow \quad 4x = 260$
$\Leftrightarrow \quad x = 65$. Thus the consecutive odd integers are $65, 67, 69$, and 71.

16. Let m be amount invested at $5\frac{1}{2}\%$. Then $4000 + m$ is the total amount invested. Thus $4\frac{1}{2}\%$ of the total investment $=$ interest earned at 4% $+$ the interest earned at $5\frac{1}{2}\%$. So
$0.045(4000 + m) = 0.04(4000) + 0.055m \quad \Leftrightarrow \quad 180 + 0.045m = 160 + 0.055m \quad \Leftrightarrow$
$20 = 0.01m \quad \Leftrightarrow \quad m = \frac{20}{0.01} = 2000$. Thus \$2,000 needs to be invested at $5\frac{1}{2}\%$.

18. Let r be the interest rate at which Jack invests the \$1000. Then $r + 0.005$ is the interest rate at which Jack invest the \$2000. Since total interest $=$ (interest earned at r) $+$ (interest earned at $r + 0.005$), we have $190 = 1000r + 2000(r + 0.005) \quad \Leftrightarrow \quad 190 = 1000r + 2000r + 10 \quad \Leftrightarrow$
$180 = 3000r \quad \Leftrightarrow \quad r = \frac{180}{3000} = 0.06$. Thus Jack invests \$1,000 is invested at 6%.

20. Let s be the husband's annual salary. Then her annual salary is $1.15s$. Since
husband's annual salary $+$ wife's annual salary $=$ total annual income, we have
$s + 1.15s = 69{,}875 \quad \Leftrightarrow \quad 2.15s = 69{,}875 \quad \Leftrightarrow \quad s = 32{,}500$. Thus the husband's annual salary is \$32,500.

22. Let x be overtime hours Helen works. Since gross pay $=$ regular salary $+$ overtime pay, we obtain the equation $352.50 = 7.50 \times 35 + 7.50 \times 1.5 \times x \quad \Leftrightarrow \quad 352.50 = 262.50 + 11.25x \quad \Leftrightarrow$
$90 = 11.25x \quad \Leftrightarrow \quad x = \dfrac{90}{11.25} = 8$. Thus Helen worked 8 hours of overtime.

24. Let a be the daughter's age now. Then her father's age now is $4a$. In 6 years, her age will $a + 6$, and her father's age will be $3(a + 6)$. But her father's age in 6 years is also $4a + 6$. So $3(a + 6) = 4a + 6 \quad \Leftrightarrow \quad 3a + 18 = 4a + 6 \quad \Leftrightarrow \quad 12 = a$. Thus the daughter is 12 years old.

26. Let h be number of home runs Babe Ruth hit. Then $h + 31$ is the number of home runs that Hank Aaron hit. So $1459 = h + h + 31 \quad \Leftrightarrow \quad 1428 = 2h \quad \Leftrightarrow \quad h = 714$. Thus Babe Ruth hit 714 home runs.

28. Let q be the number of quarters. Then $2q$ is the number of dimes, and $2q + 5$ is the number of nickels. Thus $3.00 = \textit{value of the nickels} + \textit{value of the dimes} + \textit{value of the quarters}$. So $3.00 = 0.05(2q + 5) + 0.10(2q) + 0.25q \quad \Leftrightarrow \quad 3.00 = 0.10q + 0.25 + 0.20q + 0.25q \quad \Leftrightarrow$ $2.75 = 0.55q \quad \Leftrightarrow \quad q = \dfrac{2.75}{0.55} = 5$. Thus Mary has 5 quarters, $2(5) = 10$ dimes, and $2(5) + 5 = 15$ nickels.

30. Let w be the width of the pasture. Then the length of the pasture is $2w$. Since $\textit{area} = \textit{length} \times \textit{width}$ we have $115{,}200 = w(2w) = 2w^2 \quad \Leftrightarrow \quad w^2 = 57{,}600 \quad \Rightarrow \quad w = \pm 240$. Thus the width of the pasture is 240 feet.

32. Let w be the width of the building lot. Then the length of the building lot is $5w$. Since a half-acre is $\frac{1}{2} \cdot 43{,}560 = 21{,}780$ and $\textit{area} = \textit{length} \times \textit{width}$ we have $21{,}780 = w(5w) = 5w^2 \quad \Leftrightarrow$ $w^2 = 4{,}356 \quad \Rightarrow \quad w = \pm 66$. Thus the width of the building lot is 66 feet and the length of the building lot is $5(66) = 330$ feet.

34. The figure can be broken up into two rectangles. Since the total area equals the sum of the areas of the rectangles, we have $144 = 10x + 6x \quad \Leftrightarrow$ $144 = 16x \quad \Leftrightarrow \quad 9 = x$. Thus the length of x is 9 cm.

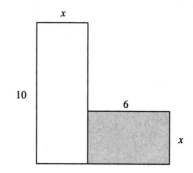

36. Let x be the amount in mL of 60% acid solution to be used. Then $(300 - x)$ mL of 30% solution would have to be used to yield a total of 300 mL of solution.

	60% acid	30% acid	mixture
mL	x	$300 - x$	300
rate (% acid)	0.60	0.30	0.50
value	$0.60x$	$0.30(300 - x)$	$0.50(300)$

Thus the total amount of pure acid used is $0.60x + 0.30(300 - x) = 0.50(300) \quad \Leftrightarrow$ $0.3x + 90 = 150 \quad \Leftrightarrow \quad x = \frac{60}{0.3} = 200$. So 200 mL of 60% acid solution must be mixed with 100 mL of 30% solution to get 300 mL of 50% acid solution.

38. Let x be the number of liters of water to be boiled off. The result will contain $6 - x$ liters.

	Original	Water	Final
liters	6	$-x$	$6 - x$
concentration	120	0	200
amount	120(6)	0	$200(6 - x)$

So $120(6) + 0 = 200(6 - x)$ $\Leftrightarrow$ $720 = 1200 - 200x$ $\Leftrightarrow$ $200x = 480$ $\Leftrightarrow$ $x = 2.4$.
Thus 2.4 liters need to be boiled off.

40. Let x be the number of gallons of 2% bleach removed from the tank. This is also the number of gallons of pure bleach added to make the 5% mixture.

	original 2%	pure bleach	5% mixture
gallons	$100 - x$	x	100
concentration	0.02	1	0.05
bleach	$0.02(100 - x)$	$1x$	$0.05 \cdot 100$

So $0.02(100 - x) + x = 0.05 \cdot 100$ $\Leftrightarrow$ $2 - 0.02x + x = 5$ $\Leftrightarrow$ $0.98x = 3$ $\Leftrightarrow$
$x = 3.06$. Thus 3.06 gallons need to removed and replaced with pure bleach.

42. Let x be the number of pounds of $3.00/lb tea Then $80 - x$ is the number of pounds of $2.75/lb tea.

	3.00 tea	2.75 tea	mixture
pounds	x	$80 - x$	80
rate (cost per pound)	3.00	2.75	2.90
value	$3.00x$	$2.75(80 - x)$	$2.90(80)$

So $3.00x + 2.75(80 - x) = 2.90(80)$ $\Leftrightarrow$ $3.00x + 220 - 2.75x = 232$ $\Leftrightarrow$ $0.25x = 12$
$\Leftrightarrow$ $x = 48$. The mixture uses 48 pounds of $3.00/lb tea and $80 - 48 = 32$ pounds of $2.75/lb tea.

44. Let x be the height of the tall tree. Here we use the property that corresponding sides in similar triangles are proportional. The base of the similar triangles starts at eye level of the woodcutter, 5 feet. Thus we obtain the proportion $\dfrac{x - 5}{15} = \dfrac{150}{25}$ $\Leftrightarrow$ $25(x - 5) = 15(150)$ $\Leftrightarrow$
$25x - 125 = 2250$ $\Leftrightarrow$ $25x = 2375$ $\Leftrightarrow$ $x = 95$. Thus the tree is 95 feet tall.

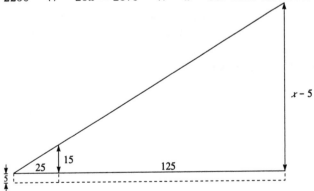

46. Let w be the largest weight that can be hung. In this exercise, the edge of the building acts as the fulcrum, so the 240 lb man is sitting 25 feet from the fulcrum. Then substituting the known values into the formula given in Exercise 51, we have $240(25) = 5w$ $\Leftrightarrow$ $6000 = 5w$ $\Leftrightarrow$
$w = 1200$. Therefore, 1200 pounds is the largest weight that can be hung.

48. Let t be the time it takes Hilda, in minutes, to mow the lawn. Since Hilda is twice as fast as Stan, it takes Stan $2t$ minutes to mow the lawn by himself. Thus $40 \cdot \dfrac{1}{t} + 40 \cdot \dfrac{1}{2t} = 1 \quad \Leftrightarrow \quad 40 + 20 = t \quad \Leftrightarrow \quad t = 60$. So it would take Stan $2(60) = 120$ minutes to mow the lawn.

50. Let h be the time, in hours, to fill the swimming pool using Jim's hose alone. Since Bob's hose takes 20% less time, it uses only 80% of the time, or $0.8h$. Thus $18 \cdot \dfrac{1}{h} + 18 \cdot \dfrac{1}{0.8h} = 1 \quad \Leftrightarrow \quad 18 \cdot 0.8 + 18 = 0.8h \quad \Leftrightarrow \quad 14.4 + 18 = 0.8h \quad \Leftrightarrow \quad 32.4 = 0.8h \quad \Leftrightarrow \quad h = 40.5$. Jim's hose takes 40.5 hours, and Bob's hose takes 32.4 hours to fill the pool alone.

52. Let r be the speed of the slower cyclist, in mi/h. Then the speed of the faster cyclist is $2r$.

	Rate	Time	Distance
Slower cyclist	r	2	$2r$
Faster cyclist	$2r$	2	$4r$

When they meet, they will have traveled a total of 90 miles, so $2r + 4r = 90 \quad \Leftrightarrow \quad 6r = 90 \quad \Leftrightarrow \quad r = 15$. The speed of the slower cyclist is 15 mi/h, while the speed of the faster cyclist is $2(15) = 30$ mi/h.

54. Let x be the speed of the car in mi/h. Since a mile contains 5280 ft and an hour contains 3600 s, $1 \text{ mi/h} = \dfrac{5280 \text{ ft}}{3600 \text{sec}} = \dfrac{22}{15}$ feet/second. The truck is traveling at $50 \cdot \dfrac{22}{15} = \dfrac{220}{3}$ ft/s. So in 6 seconds, the truck travels $6 \cdot \frac{220}{3} = 440$ feet. Thus the back end of the car must travel the length of the car, the length of the truck, and the 440 feet in 6 seconds, so its speed must be $\frac{14+30+440}{6} = \frac{242}{3}$ ft/s. Converting to mi/h, we have that the speed of the car is $\frac{242}{3} \cdot \frac{15}{22} = 55$ mi/h.

56. Let r be the radius of the running track. The running track consists of 2 semicircles and 2 straight sections 110 yards long, so we get the equation $2\pi r + 220 = 440 \quad \Leftrightarrow \quad 2\pi r = 220 \quad \Leftrightarrow \quad r = \dfrac{110}{\pi} = 35.03$. Thus the radius of the semicircle is about 35 yards.

58. Let h be the height of the break in feet. Then the portion of the bamboo above the break is $10 - h$. Applying the Pythagorean Theorem, we obtain $h^2 + 3^2 = (10 - h)^2 \quad \Leftrightarrow \quad h^2 + 9 = 100 - 20h + h^2 \quad \Leftrightarrow \quad -91 = -20h \quad \Leftrightarrow \quad h = \frac{91}{20} = 4.55$. Thus the break is 4.55 ft above the ground.

Exercises 1.3

2. $x^2 + 3x - 4 = 0$ $\Leftrightarrow$ $(x-1)(x+4) = 0$ $\Leftrightarrow$ $x - 1 = 0$ or $x + 4 = 0$. Thus $x = 1$ or $x = -4$.

4. $x^2 + 8x + 12 = 0$ $\Leftrightarrow$ $(x+2)(x+6) = 0$ $\Leftrightarrow$ $x + 2 = 0$ or $x + 6 = 0$. Thus $x = -2$ or $x = -6$.

6. $4x^2 - 4x - 15 = 0$ $\Leftrightarrow$ $(2x+3)(2x-5) = 0$ $\Leftrightarrow$ $2x + 3 = 0$ or $2x - 5 = 0$. Thus $x = -\frac{3}{2}$ or $x = \frac{5}{2}$.

8. $4w^2 = 4w + 3$ $\Leftrightarrow$ $4w^2 - 4w - 3 = 0$ $\Leftrightarrow$ $(2w+1)(2w-3) = 0$ $\Leftrightarrow$ $2w + 1 = 0$ or $2w - 3 = 0$. If $2w + 1 = 0$, then $w = -\frac{1}{2}$; if $2w - 3 = 0$, then $w = \frac{3}{2}$.

10. $3x^2 + 1 = 4x$ $\Leftrightarrow$ $3x^2 - 4x + 1 = 0$ $\Leftrightarrow$ $(3x-1)(x-1) = 0$ $\Leftrightarrow$ $3x - 1 = 0$ or $x - 1 = 0$. If $3x - 1 = 0$, then $x = \frac{1}{3}$; if $x - 1 = 0$, then $x = 1$.

12. $6x(x-1) = 21 - x$ $\Leftrightarrow$ $6x^2 - 6x = 21 - x$ $\Leftrightarrow$ $6x^2 - 5x - 21 = 0$ $\Leftrightarrow$ $(2x+3)(3x-7) = 0$ $\Leftrightarrow$ $2x + 3 = 0$ or $3x - 7 = 0$. If $2x + 3 = 0$, then $x = -\frac{3}{2}$; if $3x - 7 = 0$, then $x = \frac{7}{3}$.

14. $x^2 - 4x + 2 = 0$ $\Leftrightarrow$ $x^2 - 4x = -2$ $\Leftrightarrow$ $x^2 - 4x + 4 = -2 + 4$ $\Leftrightarrow$ $(x-2)^2 = 2$ $\Rightarrow$ $x - 2 = \pm\sqrt{2}$ $\Leftrightarrow$ $x = 2 \pm \sqrt{2}$.

16. $x^2 + 3x - \frac{7}{4} = 0$ $\Leftrightarrow$ $x^2 + 3x = \frac{7}{4}$ $\Leftrightarrow$ $x^2 + 3x + \frac{9}{4} = \frac{7}{4} + \frac{9}{4}$ $\Leftrightarrow$ $(x + \frac{3}{2})^2 = \frac{16}{4} = 4$ $\Rightarrow$ $x + \frac{3}{2} = \pm 2$ $\Leftrightarrow$ $x = -\frac{3}{2} \pm 2$ $\Leftrightarrow$ $x = \frac{1}{2}$ or $x = -\frac{7}{2}$.

18. $x^2 - 5x + 1 = 0$ $\Leftrightarrow$ $x^2 - 5x = -1$ $\Leftrightarrow$ $x^2 - 5x + \frac{25}{4} = -1 + \frac{25}{4}$ $\Leftrightarrow$ $(x - \frac{5}{2})^2 = \frac{21}{4}$ $\Rightarrow$ $x - \frac{5}{2} = \pm\sqrt{\frac{21}{4}} = \pm\frac{\sqrt{21}}{2}$ $\Leftrightarrow$ $x = \frac{5}{2} \pm \frac{\sqrt{21}}{2}$.

20. $x^2 - 18x = 19$ $\Leftrightarrow$ $x^2 - 18x + (-9)^2 = 19 + (-9)^2 = 19 + 81$ $\Leftrightarrow$ $(x-9)^2 = 100$ $\Rightarrow$ $x - 9 = \pm 10$ $\Leftrightarrow$ $x = 9 \pm 10$, so $x = -1$ or $x = 19$.

22. $3x^2 - 6x - 1 = 0$ $\Leftrightarrow$ $x^2 - 2x - \frac{1}{3} = 0$ $\Leftrightarrow$ $x^2 - 2x = \frac{1}{3}$ $\Leftrightarrow$ $x^2 - 2x + 1 = \frac{1}{3} + 1$ $\Leftrightarrow$ $(x-1)^2 = \frac{4}{3}$ $\Rightarrow$ $x - 1 = \pm\sqrt{\frac{4}{3}}$ $\Leftrightarrow$ $x = 1 \pm \frac{2\sqrt{3}}{3}$.

24. $x^2 = \frac{3}{4}x - \frac{1}{8}$ $\Leftrightarrow$ $x^2 - \frac{3}{4}x = -\frac{1}{8}$ $\Leftrightarrow$ $x^2 - \frac{3}{4}x + \frac{9}{64} = -\frac{1}{8} + \frac{9}{64}$ $\Leftrightarrow$ $(x - \frac{3}{8})^2 = \frac{1}{64}$ $\Rightarrow$ $x - \frac{3}{8} = \pm\frac{1}{8}$ $\Leftrightarrow$ $x = \frac{3}{8} \pm \frac{1}{8}$, so $x = \frac{1}{2}$ or $x = \frac{1}{4}$.

26. $x^2 + 5x - 6 = 0$ $\Leftrightarrow$ $(x-1)(x+6) = 0$ $\Leftrightarrow$ $x - 1 = 0$ or $x + 6 = 0$. Thus $x = 1$ or $x = -6$.

28. $x^2 + 30x + 200 = 0$ $\Leftrightarrow$ $(x+10)(x+20) = 0$ $\Leftrightarrow$ $x + 10 = 0$ or $x + 20 = 0$. Thus $x = -10$ or $x = -20$.

30. $3x^2 + 7x + 4 = 0$ $\Leftrightarrow$ $(3x + 4)(x + 1) = 0$ $\Leftrightarrow$ $3x + 4 = 0$ or $x + 1 = 0$. Thus $x = -\frac{4}{3}$ or $x = -1$.

32. $x^2 - 4x + 2 = 0$ $\Leftrightarrow$ $x^2 - 4x = -2$ $\Leftrightarrow$ $x^2 - 4x + 4 = -2 + 4$ $\Leftrightarrow$ $(x - 2)^2 = 2$
     $\Leftrightarrow$ $x - 2 = \pm\sqrt{2}$ $\Leftrightarrow$ $x = 2 \pm \sqrt{2}$.

34. $8x^2 - 6x - 9 = 0$ $\Leftrightarrow$ $(2x - 3)(4x + 3) = 0$ $\Leftrightarrow$ $2x - 3 = 0$ or $4x + 3 = 0$. If
     $2x - 3 = 0$, then $x = \frac{3}{2}$; if $4x + 3 = 0$, then $x = -\frac{3}{4}$.

36. $x^2 - 6x + 1 = 0$ $\Rightarrow$ $x = \dfrac{-b \pm \sqrt{b^2 - 4ac}}{2a} = \dfrac{-(-6) \pm \sqrt{(-6)^2 - 4(1)(1)}}{2(1)} = \dfrac{6 \pm \sqrt{36 - 4}}{2}$
     $= \dfrac{6 \pm \sqrt{32}}{2} = \dfrac{6 \pm 4\sqrt{2}}{2} = 3 \pm 2\sqrt{2}$.

38. $\theta^2 - \dfrac{3}{2}\theta + \dfrac{9}{16} = 0$ $\Leftrightarrow$ $\left(\theta - \dfrac{3}{4}\right)^2 = 0$ $\Leftrightarrow$ $\theta - \dfrac{3}{4} = 0$ $\Leftrightarrow$ $\theta = \dfrac{3}{4}$

40. $0 = x^2 - 4x + 1 = 0$ $\Rightarrow$ $x = \dfrac{-b \pm \sqrt{b^2 - 4ac}}{2a} = \dfrac{-(-4) \pm \sqrt{(-4)^2 - 4(1)(1)}}{2(1)} =$
     $\dfrac{4 \pm \sqrt{16 - 4}}{2} = \dfrac{4 \pm \sqrt{12}}{2} = \dfrac{4 \pm 2\sqrt{3}}{2} = 2 \pm \sqrt{3}$.

42. $w^2 = 3(w - 1)$ $\Leftrightarrow$ $w^2 - 3w + 3 = 0$ $\Rightarrow$
     $w = \frac{-(-3) \pm \sqrt{(-3)^2 - 4(1)(3)}}{2(1)} = \frac{3 \pm \sqrt{9-12}}{2} = \dfrac{3 \pm \sqrt{-3}}{2}$. Since the discriminant is less than 0, the
     equation has no real solutions.

44. (Method 1: place directly into the quadratic formula.) $\sqrt{6}x^2 + 2x - \sqrt{\dfrac{3}{2}} = 0$ $\Rightarrow$

     $x = \dfrac{-b \pm \sqrt{b^2 - 4ac}}{2a} = \dfrac{-(2) \pm \sqrt{(2)^2 - 4\left(\sqrt{6}\right)\left(-\sqrt{\frac{3}{2}}\right)}}{2\left(\sqrt{6}\right)} =$

     $\dfrac{-2 \pm \sqrt{4 + 12}}{2\sqrt{6}} = \dfrac{-2 \pm \sqrt{16}}{2\sqrt{6}} = \dfrac{-2 \pm 4}{2\sqrt{6}}$. So $x = -\dfrac{\sqrt{6}}{2}$ or $x = \dfrac{\sqrt{6}}{6}$.

     (Method 2: multiply by $\sqrt{6}$ and use the result in the quadratic formula.) $\sqrt{6}x^2 + 2x - \sqrt{\dfrac{3}{2}} = 0$
     $\Leftrightarrow$ $6x^2 + 2\sqrt{6}x - 3 = 0$ $\Rightarrow$

     $x = \dfrac{-b \pm \sqrt{b^2 - 4ac}}{2a} = \dfrac{-\left(2\sqrt{6}\right) \pm \sqrt{\left(2\sqrt{6}\right)^2 - 4(6)(-3)}}{2(6)} =$

     $\dfrac{-2\sqrt{6} \pm \sqrt{24 + 72}}{12} = \dfrac{-2\sqrt{6} \pm \sqrt{96}}{12} = \dfrac{-2\sqrt{6} \pm 4\sqrt{6}}{12}$. So $x = -\dfrac{\sqrt{6}}{2}$ or $x = \dfrac{\sqrt{6}}{6}$

46. $25x^2 + 70x + 49 = 0$ $\Leftrightarrow$ $(5x + 7)^2 = 0$ $\Leftrightarrow$ $5x + 7 = 0$ $\Leftrightarrow$ $5x = -7$ $\Leftrightarrow$
     $x = -\frac{7}{5}$.

48. $5x^2 - 7x + 5 \quad \Rightarrow \quad x = \dfrac{-b \pm \sqrt{b^2 - 4ac}}{2a} = \dfrac{-(-7) \pm \sqrt{(-7)^2 - 4(5)(5)}}{2(5)}$

$= \dfrac{7 \pm \sqrt{49 - 100}}{10} = \dfrac{7 \pm \sqrt{-51}}{10}$. Since the discriminant is less than 0, the equation has no real

solutions.

50. $x^2 - 2.450x + 1.500 = 0 \quad \Rightarrow \quad x = \dfrac{-(-2.450) \pm \sqrt{(-2.450)^2 - 4(1)(1.500)}}{2(1)} =$

$\dfrac{2.450 \pm \sqrt{6.0025 - 6}}{2} = \dfrac{2.450 \pm \sqrt{0.0025}}{2} = \dfrac{2.450 \pm 0.050}{2}$.

Thus $x = \dfrac{2.450 + 0.050}{2} = 1.250$ or $x = \dfrac{2.450 - 0.050}{2} = 1.200$.

52. $x^2 - 1.800x + 0.810 = 0 \quad \Rightarrow \quad x = \dfrac{-(-1.800) \pm \sqrt{(-1.800)^2 - 4(1)(0.810)}}{2(1)}$

$= \dfrac{1.800 \pm \sqrt{3.24 - 3.24}}{2} = \dfrac{1.800 \pm \sqrt{0}}{2} = 0.900$. Thus the only solution is $x = 0.900$.

54. $12.714x^2 + 7.103x = 0.987 \quad \Leftrightarrow \quad 12.714x^2 + 7.103x - 0.987 = 0 \quad \Rightarrow$

$x = \dfrac{-(7.103) \pm \sqrt{(7.103)^2 - 4(12.714)(-0.987)}}{2(12.714)} = \dfrac{-7.103 \pm \sqrt{50.4526 + 50.1949}}{25.428} =$

$\dfrac{-7.103 \pm \sqrt{100.6475}}{25.428} = \dfrac{-7.103 \pm 10.032}{25.428}$. Thus $x \approx \dfrac{-7.103 - 10.032}{25.428} = -0.674$ or

$x \approx \dfrac{-7.103 + 10.032}{25.428} = 0.115$.

56. $S = \dfrac{n(n+1)}{2} \quad \Leftrightarrow \quad 2S = n^2 + n \quad \Leftrightarrow \quad n^2 + n - 2S = 0$. Using the quadratic formula,

$n = \dfrac{-1 \pm \sqrt{(1)^2 - 4(1)(-2S)}}{2(1)} = \dfrac{-1 \pm \sqrt{1 + 8S}}{2}$.

58. $A = 2\pi r^2 + 2\pi rh \quad \Leftrightarrow \quad 2\pi r^2 + 2\pi rh - A = 0$. Using the quadratic formula,

$r = \dfrac{-(2\pi h) \pm \sqrt{(2\pi h)^2 - 4(2\pi)(-A)}}{2(2\pi)} = \dfrac{-2\pi h \pm \sqrt{4\pi^2 h^2 + 8\pi A}}{4\pi} = \dfrac{-\pi h \pm \sqrt{\pi^2 h^2 + 2\pi A}}{2\pi}$.

60. $\dfrac{1}{r} + \dfrac{2}{1 - r} = \dfrac{4}{r^2} \quad \Leftrightarrow \quad r^2(1 - r)\left(\dfrac{1}{r} + \dfrac{2}{1 - r}\right) = r^2(1 - r)\left(\dfrac{4}{r^2}\right) \quad \Leftrightarrow$

$r(1 - r) + 2r^2 = 4(1 - r) \quad \Leftrightarrow \quad r - r^2 + 2r^2 = 4 - 4r \quad \Leftrightarrow \quad r^2 + 5r - 4 = 0$. Using the

quadratic formula, $r = \dfrac{-(5) \pm \sqrt{(5)^2 - 4(1)(-4)}}{2(1)} = \dfrac{-5 \pm \sqrt{25 + 16}}{2} = \dfrac{-5 \pm \sqrt{41}}{2}$.

62. $x^2 = 6x - 9 \quad \Leftrightarrow \quad x^2 - 6x + 9$, so $D = b^2 - 4ac = (-6)^2 - 4(1)(9) = 36 - 36 = 0$. Since $D = 0$, this equation has one real solution.

64. $D = b^2 - 4ac = (2.21)^2 - 4(1)(1.21) = 4.8841 - 4.84 = 0.0441$. Since $D \neq 0$, this equation has two real solutions.

66. $D = b^2 - 4ac = (-4)^2 - 4(9)(\frac{4}{9}) = 16 - 16 = 0$. Since $D = 0$, this equation has one real solution.

68. $D = b^2 - 4ac = (-r)^2 - 4(1)(s) = r^2 - 4s > 0$, (since $r > 2\sqrt{s}$). Since D is positive, this equation has two real solutions.

70. $b^2 x^2 - 5bx + 4 = 0 \iff (bx - 4)(bx - 1) = 0 \iff bx - 4 = 0$ or $bx - 1 = 0$. If $bx - 4 = 0$, then $bx = 4 \iff x = \frac{4}{b}$. If $bx - 1 = 0$, then $bx = 1 \iff x = \frac{1}{b}$.

72. $bx^2 + 2x + \frac{1}{b} = 0 \iff b^2 x^2 + 2bx + 1 = 0 \iff (bx + 1)^2 = 0 \iff bx + 1 = 0 \iff$
$\iff x = -\frac{1}{b}$.

74. We want to find the values of k that make the discriminant 0. Thus $D = 36^2 - 4(k)(k) = 0 \iff 4k^2 = 36^2 \implies 2k = \pm 36 \iff k = \pm 18$.

76. Let n be one even number. Then the next even number is $n + 2$. Thus we get the equation $n^2 + (n + 2)^2 = 1252 \iff n^2 + n^2 + 4n + 4 = 1252 \iff 0 = 2n^2 + 4n - 1248 = 2(n^2 + 2n - 624) = 2(n - 24)(n + 26)$. So $n = 24$ or $n = -26$. Thus the consecutive even integers are 24 and 26 or -26 and -24.

78. Let w be the width of the bedroom. Then it's length is $w + 7$. Since $area = length \times width$, we have $228 = (w + 7)w = w^2 + 7w \iff w^2 + 7w - 228 = 0 \iff (w + 19)(w - 12) = 0 \iff w + 19 = 0$ or $w - 12 = 0$. Thus $w = -19$ or $w = 12$. Since the width must be positive, the width is 12 feet.

80. The shaded area can be broken down into three rectangles as shown in the on the next page. So $160 = 14x + x^2 + 13x \iff x^2 + 27x - 160 = 0 \iff (x - 5)(x + 32) = 0 \iff x - 5 = 0$ or $x + 32 = 0$. Thus $x = 5$ or $x = -32$. Since x represents a length, it must be positive, so x is 5 in.

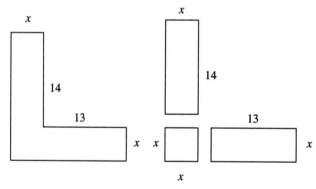

82. Setting $P = 1250$ and solving for x, we have $1250 = \frac{1}{10}x(300 - x) = 30x - \frac{1}{10}x^2 \iff$
$\frac{1}{10}x^2 - 30x + 1250 = 0$. Using quadratic formula $x = \dfrac{-(-30) \pm \sqrt{(-30)^2 - 4(\frac{1}{10})(1250)}}{2(\frac{1}{10})}$

$= \dfrac{30\pm\sqrt{900-500}}{0.2} = \dfrac{30\pm20}{0.2}$. Thus $x = \dfrac{30-20}{0.2} = 50$ or $x = \dfrac{30+20}{0.2} = 250$. Since he must have $0 \le x \le 200$, he should make 50 ovens per week.

84. Let r be the radius of the can. Now using the formula $V = \pi r^2 h$ with $V = 40\pi$ cm^3 and $h = 10$, we solve for r. Thus $40\pi = \pi r^2(10) \quad \Leftrightarrow \quad 4 = r^2 \quad \Rightarrow \quad r = \pm 2$. Since r represents radius, $r > 0$. Thus $r = 2$, and the diameter is 4 cm.

86. Let h be the height of the flagpole, in feet. Then the length of each guy wire is $h + 5$. Since the distance between the points where the wires are fixed to the ground is equal to one guy wire, the triangle is equilateral, and the flagpole is the perpendicular bisector of the base. Thus from the Pythagorean Theorem, we get $\left[\frac{1}{2}(h+5)\right]^2 + h^2 = (h+5)^2 \quad \Leftrightarrow$
$h^2 + 10h + 25 + 4h^2 = 4h^2 + 40h + 100 \quad \Leftrightarrow \quad h^2 - 30h - 75 = 0 \quad \Rightarrow$
$h = \dfrac{-(-30)\pm\sqrt{(-30)^2-4(1)(-75)}}{2(1)} = \dfrac{30\pm\sqrt{900+300}}{2} = \dfrac{30\pm\sqrt{1200}}{2} = \dfrac{30\pm20\sqrt{3}}{2}$. Since $h = \dfrac{30-20\sqrt{3}}{2} < 0$, we reject it. Thus the height is $h = \dfrac{30+20\sqrt{3}}{2} = 15 + 10\sqrt{3} \approx 32.32$ feet ≈ 32 feet 4 inches.

88. (a) Using $h_0 = 96$, half the distance is 48, so we solve the equation $48 = -16t^2 + 96 \quad \Leftrightarrow$
$-48 = -16t^2 \quad \Leftrightarrow \quad 3 = t^2 \quad \Rightarrow \quad t = \pm\sqrt{3}$. Since $t \ge 0$, it takes $\sqrt{3} \approx 1.732$ sec.

(b) The ball hits the ground when $h = 0$, so we solve the equation $0 = -16t^2 + 96 \quad \Leftrightarrow$
$16t^2 = 96 \quad \Leftrightarrow \quad t^2 = 6 \quad \Rightarrow \quad t = \pm\sqrt{6}$. Since $t \ge 0$, it takes $\sqrt{6} \approx 2.449$ sec.

90. If the maximum height is 100 feet, then the discriminant of the equation, $16t^2 - v_o t + 100 = 0$, must equal zero. So $0 = b^2 - 4ac = (-v_o)^2 - 4(16)(100) \quad \Leftrightarrow \quad v_o^2 = 6400 \quad \Rightarrow \quad v_o = \pm 80$. Since $v_o = -80$ does not make sense, we must have $v_o = 80$ ft/s.

92. Let y be the circumference of the circle, so $360 - y$ is the perimeter of the square. Use the circumference to find the radius, r, in terms of y: $y = 2\pi r \quad \Rightarrow \quad r = \dfrac{y}{2\pi}$. Thus the area of the circle is $\pi\left(\dfrac{y}{2\pi}\right)^2 = \dfrac{y^2}{4\pi}$. Now if the perimeter of the square is $360 - y$, the length of each side is $\frac{1}{4}(360 - y)$, and the area of the square is $\left(\dfrac{360-y}{4}\right)^2$. Setting these areas equal, we obtain
$\dfrac{y^2}{4\pi} = \left(\dfrac{360-y}{4}\right)^2 \quad \Leftrightarrow \quad \dfrac{y}{2\sqrt{\pi}} = \dfrac{360-y}{4} \quad \Leftrightarrow \quad 2y = 360\sqrt{\pi} - \sqrt{\pi}\,y \quad \Leftrightarrow$
$(2 + \sqrt{\pi})y = 360\sqrt{\pi}$. Therefore, $y = \dfrac{360\sqrt{\pi}}{2 + \sqrt{\pi}} \approx 169.1$. Thus one wire is 169.1 in long and the other is 190.9 in long.

94. Let x be the rate, in mi/h, at which Kiran drove from Tortula to Cactus.

Cities	Distance	Rate	Time
Tortula $\rightarrow$ Cactus	250	x	$\dfrac{250}{x}$
Cactus $\rightarrow$ Dry Junction	360	$x + 10$	$\dfrac{360}{x + 10}$

We have used *Time* = $\frac{Distance}{Rate}$ to fill in the "Time" column of the table. We are given that the sum of the times is 11 hours. Thus we get the equation $\dfrac{250}{x} + \dfrac{360}{x+10} = 11$ $\Leftrightarrow$

$250(x+10) + 360x = 11x(x+10)$ $\Leftrightarrow$ $250x + 2500 + 360x = 11x^2 + 110x$ $\Leftrightarrow$
$11x^2 - 500x - 2500 = 0$ $\Rightarrow$

$x = \dfrac{-(-500) \pm \sqrt{(-500)^2 - 4(11)(-2500)}}{2(11)} = \dfrac{500 \pm \sqrt{250000 + 110000}}{22} = \dfrac{500 \pm \sqrt{360000}}{22}$

$= \dfrac{500 \pm 600}{22}$. Hence, Kiran drove either -4.54 mi/h (impossible) or 50 mi/h between Tortula and Cactus.

96. Let r be the speed of the south-bound boat. Then $r + 3$ is the speed of the east-bound boat. In two hours the south-bound boat has traveled $2r$ miles and the east-bound boat has traveled $2(r+3) = 2r + 6$ miles. Since they are traveling is directions with are 90° apart, we can use the Pythagorean theorem to get $(2r)^2 + (2r+6)^2 = 30^2$ $\Leftrightarrow$ $4r^2 + 4r^2 + 24r + 36 = 900$ $\Leftrightarrow$ $8r^2 + 24r - 864 = 0$ $\Leftrightarrow$ $8(r^2 + 3r - 108) = 0$ $\Leftrightarrow$ $8(r+12)(r-9) = 0$. So $r = -12$ or $r = 9$. Since speed is positive, the speed of the south-bound boat is 9 mi/h.

98. Let h be the height the ladder reaches (in feet). Using the Pythagorean theorem we have $\left(7\frac{1}{2}\right)^2 + h^2 = \left(19\frac{1}{2}\right)^2$ $\Leftrightarrow$ $\left(\frac{15}{2}\right)^2 + h^2 = \left(\frac{39}{4}\right)^2$ $\Leftrightarrow$ $h^2 = \left(\frac{39}{4}\right)^2 - \left(\frac{15}{2}\right)^2 = \frac{1521}{4} - \frac{225}{4} = \frac{1296}{4} = 324$. So $h = \sqrt{324} = 18$.

100. Let t be the time, in hours, it takes Kay to deliver all the flyers alone. Then it takes Lynn $t + 1$ hours to deliver all the flyers alone, and it takes the group $0.4t$ hours to do it together. Thus $\frac{1}{4} + \frac{1}{t} + \frac{1}{t+1} = \frac{1}{0.4t}$ $\Leftrightarrow$ $\frac{1}{4}(0.4t) + \frac{1}{t}(0.4t) + \frac{1}{t+1}(0.4t) = 1$ $\Leftrightarrow$ $t + 4 + \frac{4t}{t+1} = 10$ $\Leftrightarrow$ $t(t+1) + 4(t+1) + 4t = 10(t+1)$ $\Leftrightarrow$ $t^2 + t + 4t + 4 + 4t = 10t + 10$ $\Leftrightarrow$ $t^2 - t - 6 = 0$ $\Leftrightarrow$ $(t-3)(t+2) = 0$. So $t = 3$ or $t = -2$. Since $t = -2$ is impossible, it takes Kay 3 hours to deliver all the flyers alone.

102. $x^2 - 9x + 20 = (x-4)(x-5) = 0$, so $x = 4$ or $x = 5$. The roots are 4 and 5. The product is $4 \cdot 5 = 20$, and the sum is $4 + 5 = 9$.

$x^2 - 2x - 8 = (x-4)(x+2) = 0$, so $x = 4$ or $x = -2$. The roots are 4 and -2. The product is $4 \cdot (-2) = -8$, and the sum is $4 + (-2) = 2$.

$x^2 + 4x + 2 = 0$, so using the quadratic formula, $x = \dfrac{-4 \pm \sqrt{4^2 - 4(1)(2)}}{2(1)} = \dfrac{-4 \pm \sqrt{8}}{2}$

$= \dfrac{-4 \pm 2\sqrt{2}}{2} = -2 \pm \sqrt{2}$. The roots are $-2 - \sqrt{2}$ and $-2 + \sqrt{2}$. The product is $(-2 - \sqrt{2}) \cdot (-2 + \sqrt{2}) = 4 - 2 = 2$, and the sum is $(-2 - \sqrt{2}) + (-2 + \sqrt{2}) = -4$.

In general, if $x = r_1$ and $x = r_2$ are roots, then $x^2 + bx + c = (x - r_1)(x - r_2) = x^2 - r_1x - r_2x + r_1r_2 = x^2 - (r_1 + r_2)x + r_1r_2$. Equating the coefficients, we get $c = r_1r_2$ and $b = -(r_1 + r_2)$.

Exercises 1.4

2. $-6 + 4i$: real part -6, imaginary part 4.

4. $\dfrac{4 + 7i}{2} = 2 + \frac{7}{2}i$: real part 2, imaginary part $\frac{7}{2}$.

6. $-\frac{1}{2}$: real part $-\frac{1}{2}$, imaginary part 0.

8. $i\sqrt{3}$: real part 0, imaginary part $\sqrt{3}$.

10. $2 - \sqrt{-5} = 2 - i\sqrt{5}$: real part 2, imaginary part $-\sqrt{5}$.

12. $(2 + 5i) + (4 - 6i) = (2 + 4) + (5 - 6)i = 6 - i$.

14. $(3 - 2i) + (-5 - \frac{1}{3}i) = (3 - 5) + (-2 - \frac{1}{3})i = -2 - \frac{7}{3}i$.

16. $(-4 + i) - (2 - 5i) = -4 + i - 2 + 5i = (-4 - 2) + (1 + 5)i = -6 + 6i$.

18. $6i - (4 - i) = 6i - 4 + i = (-4) + (6 + 1)i = -4 + 7i$.

20. $2i\left(\frac{1}{2} - i\right) = i - 2i^2 = 2 + i$.

22. $(5 - 3i)(1 + i) = 5 + 5i - 3i - 3i^2 = (5 + 3) + (5 - 3)i = 8 + 2i$.

24. $\left(\frac{2}{3} + 12i\right)\left(\frac{1}{6} + 24i\right) = \frac{1}{9} + 16i + 2i + 288i^2 = \left(\frac{1}{9} - 288\right) + (16 + 2)i = -\frac{2591}{9} + 18i$.

26. $(-2 + i)(3 - 7i) = -6 + 14i + 3i - 7i^2 = (-6 + 7) + (14 + 3)i = 1 + 17i$.

28. $\dfrac{1}{1 + i} = \dfrac{1}{1 + i} \cdot \dfrac{1 - i}{1 - i} = \dfrac{1 - i}{1 - i^2} = \dfrac{1 - i}{1 + 1} = \dfrac{1 - i}{2} = \frac{1}{2} - \frac{1}{2}i$.

30. $\dfrac{5 - i}{3 + 4i} = \dfrac{5 - i}{3 + 4i} \cdot \dfrac{3 - 4i}{3 - 4i} = \dfrac{15 - 20i - 3i + 4i^2}{9 - 16i^2} = \dfrac{(15 - 4) + (-20 - 3)i}{9 + 16} = \dfrac{11 - 23i}{25}$
　　$= \frac{11}{25} - \frac{23}{25}i$.

32. $\dfrac{25}{4 - 3i} = \dfrac{25}{4 - 3i} \cdot \dfrac{4 + 3i}{4 + 3i} = \dfrac{100 + 75i}{16 - 9i^2} = \dfrac{100 + 75i}{16 + 9} = \dfrac{100 + 75i}{25} = \dfrac{25(4 + 3i)}{25} = 4 + 3i$.

34. $(2 - 3i)^{-1} = \dfrac{1}{2 - 3i} = \dfrac{1}{2 - 3i} \cdot \dfrac{2 + 3i}{2 + 3i} = \dfrac{2 + 3i}{4 - 9i^2} = \dfrac{2 + 3i}{4 + 9} = \dfrac{2 + 3i}{13} = \frac{2}{13} + \frac{3}{13}i$.

36. $\dfrac{-3 + 5i}{15i} = \dfrac{-3 + 5i}{15i} \cdot \dfrac{i}{i} = \dfrac{-3i + 5i^2}{15i^2} = \dfrac{-5 - 3i}{-15} = \dfrac{-5}{-15} + \dfrac{-3}{-15}i = \frac{1}{3} + \frac{1}{5}i$.

38. $\dfrac{(1 + 2i)(3 - i)}{2 + i} = \dfrac{3 - i + 6i - 2i^2}{2 + i} = \dfrac{5 + 5i}{2 + i} \cdot \dfrac{2 - i}{2 - i} = \dfrac{10 - 5i + 10i - 5i^2}{4 - i^2} =$
　　$\dfrac{(10 + 5) + (-5 + 10)i}{5} = \dfrac{15 + 5i}{5} = \dfrac{15}{5} + \dfrac{5}{5}i = 3 + i$.

40. $(2i)^4 = 2^4 \cdot i^4 = 16 \cdot (1) = 16$.

42. $i^{1002} = (i^4)^{250} \cdot i^2 = (1)^{250} \cdot (-1) = -1$.

44. $\sqrt{-\frac{9}{4}} = \frac{3}{2}i$.

46. $\sqrt{\frac{1}{3}}\sqrt{-27} = \sqrt{\frac{1}{3}} \cdot 3\sqrt{3}\,i = 3\,i.$

48. $\dfrac{1-\sqrt{-1}}{1+\sqrt{-1}} = \dfrac{1-i}{1+i} = \dfrac{1-i}{1+i} \cdot \dfrac{1-i}{1-i} = \dfrac{1-i-i+i^2}{1-i^2} = \dfrac{(1-1)+(-1-1)i}{1+1} = \dfrac{-2i}{2} = -i.$

50. $\left(\sqrt{3}-\sqrt{-4}\right)\left(\sqrt{6}-\sqrt{-8}\right) = \left(\sqrt{3}-2i\right)\left(\sqrt{6}-2\sqrt{2}\,i\right)$

     $= \sqrt{18} - 2\sqrt{6}\,i - 2\sqrt{6}\,i + 4\sqrt{2}\,i^2 = \left(3\sqrt{2}-4\sqrt{2}\right) + \left(-2\sqrt{6}-2\sqrt{6}\right)i = -\sqrt{2} - 4\sqrt{6}\,i.$

52. $\dfrac{\sqrt{-7}\,\sqrt{-49}}{\sqrt{28}} = \dfrac{\left(\sqrt{7}\,i\right)(7i)}{2\sqrt{7}} = \dfrac{7\,i^2}{2} = \dfrac{-7}{2}.$

54. $9x^2 + 4 = 0 \quad\Leftrightarrow\quad 9x^2 = -4 \quad\Leftrightarrow\quad x^2 = -\dfrac{4}{9} \quad\Rightarrow\quad x = \pm\dfrac{2}{3}\,i.$

56. $x^2 + 2x + 2 = 0 \quad\Rightarrow\quad x = \dfrac{-(2) \pm \sqrt{(2)^2 - 4(1)(2)}}{2(1)} = \dfrac{-2 \pm \sqrt{4-8}}{2} = \dfrac{-2 \pm \sqrt{-4}}{2} =$

     $\dfrac{-2 \pm 2i}{2} = -1 \pm i.$

58. $x^2 - 3x + 3 = 0 \quad\Rightarrow\quad x = \dfrac{-(-3) \pm \sqrt{(-3)^2 - 4(1)(3)}}{2(1)} = \dfrac{3 \pm \sqrt{9-12}}{2} = \dfrac{3 \pm \sqrt{-3}}{2} =$

     $\dfrac{3 \pm \sqrt{3}\,i}{2} = \dfrac{3}{2} \pm \dfrac{\sqrt{3}}{2}\,i.$

60. $2x^2 + 3 = 2x \quad\Leftrightarrow\quad 2x^2 - 2x + 3 = 0 \quad\Rightarrow\quad x = \dfrac{-(-2) \pm \sqrt{(-2)^2 - 4(2)(3)}}{2(2)}$

     $= \dfrac{2 \pm \sqrt{4-24}}{4} = \dfrac{2 \pm \sqrt{-20}}{4} = \dfrac{2 \pm 2\sqrt{5}\,i}{4} = \dfrac{1}{2} \pm \dfrac{\sqrt{5}}{2}\,i.$

62. $z + 4 + \dfrac{12}{z} = 0 \quad\Leftrightarrow\quad z^2 + 4z + 12 = 0 \quad\Rightarrow\quad z = \dfrac{-(4) \pm \sqrt{(4)^2 - 4(1)(12)}}{2(1)}$

     $= \dfrac{-4 \pm \sqrt{16-48}}{2} = \dfrac{-4 \pm \sqrt{-32}}{2} = \dfrac{-4 \pm 4\sqrt{2}\,i}{2} = -2 \pm 2\sqrt{2}\,i.$

64. $4x^2 - 16x + 19 = 0 \quad\Rightarrow\quad x = \dfrac{-(-16) \pm \sqrt{(-16)^2 - 4(4)(19)}}{2(4)} = \dfrac{16 \pm \sqrt{256-304}}{8}$

     $= \dfrac{16 \pm \sqrt{-48}}{8} = \dfrac{16 \pm 4\sqrt{3}\,i}{8} = \dfrac{16}{8} \pm \dfrac{4\sqrt{3}}{8}\,i = 2 \pm \dfrac{\sqrt{3}}{2}\,i.$

66. $x^2 + \frac{1}{2}x + 1 = 0 \quad\Rightarrow\quad x = \dfrac{-\left(\frac{1}{2}\right) \pm \sqrt{\left(\frac{1}{2}\right)^2 - 4(1)(1)}}{2(1)} = \dfrac{-\frac{1}{2} \pm \sqrt{\frac{1}{4}-4}}{2} = \dfrac{-\frac{1}{2} \pm \sqrt{-\frac{15}{4}}}{2} =$

     $= \dfrac{-\frac{1}{2} \pm \frac{1}{2}\sqrt{15}\,i}{2} = -\dfrac{1}{4} \pm \dfrac{\sqrt{15}}{4}\,i.$

68. LHS $= \overline{zw} = \overline{(a+b\,i)(c+d\,i)} = \overline{ac + ad\,i + bc\,i + bd\,i^2} = \overline{(ac - bd) + (ad + bc)\,i} =$
     $(ac - bd) - (ad + bc)\,i.$

RHS $= \overline{z} \cdot \overline{w} = \overline{a+bi} \cdot \overline{c+di} = (a-bi)(c-di) = ac - adi - bci + bdi^2$
$= (ac - bd) - (ad + bc)i$.

Since LHS $=$ RHS, this proves the statement.

70. $\overline{\overline{z}} = \overline{\overline{a+bi}} = \overline{a-bi} = a+bi = z$.

72. $z - \overline{z} = (a+bi) - \overline{(a+bi)} = a+bi - (a-bi) = a+bi-a+bi = 2bi$, which is a pure imaginary number.

74. Suppose $z = \overline{z}$. Then we have $(a+bi) = \overline{(a+bi)} \quad \Rightarrow \quad a+bi = a-bi \quad \Rightarrow \quad 0 = -2bi$
     $\Rightarrow \quad b = 0$, so z is real. Now if z is real, then $z = a + 0i$ (where a is real). Since $\overline{z} = a - 0i$, we have $z = \overline{z}$.

76.
$i = i$	$i^5 = i^4 \cdot i = i$	$i^9 = i^8 \cdot i = i$
$i^2 = -1$	$i^6 = i^4 \cdot i^2 = -1$	$i^{10} = i^8 \cdot i^2 = -1$
$i^3 = -i$	$i^7 = i^4 \cdot i^3 = -i$	$i^{11} = i^8 \cdot i^3 = -i$
$i^4 = 1$	$i^8 = i^4 \cdot i^4 = 1$	$i^{12} = i^8 \cdot i^4 = 1$

Because $i^4 = 1$, we have $i^n = i^r$, where r is the remainder when n is divided by 4, that is,
$n = 4 \cdot k + r$, where k is an integer and $0 \le r < 4$.

Since $4446 = 4 \cdot 1111 + 2$, we must have $i^{4446} = i^2 = -1$.

Exercises 1.5

2. $x^5 = 27x^2$ $\Leftrightarrow$ $0 = x^5 - 27x^2 = x^2(x^3 - 27)$ $\Leftrightarrow$ $x^2 = 0$ or $x^3 - 27 = 0$. If $x^2 = 0$, then $x = 0$. If $x^3 - 27 = 0$, then $x^3 = 27$ $\Leftrightarrow$ $x = 3$. The solutions are 0 and 3.

4. $0 = x^5 - 16x = x(x^4 - 16) = x(x^2 + 4)(x^2 - 4) = x(x^2 + 4)(x - 2)(x + 2)$. Since $x^2 + 4 = 0$ has no real solution, thus either x $x = 0$, or $x = 2$, or $x = -2$. The solutions are 0, 2, and -2.

6. $0 = x^4 + 64x = x(x^3 + 64)$ $\Leftrightarrow$ $x = 0$ or $x^3 + 64 = 0$. If $x^3 + 64 = 0$, then $x^3 = -64$ $\Leftrightarrow$ $x = -4$. The solutions are 0 and -4.

8. $0 = x^4 - x^3 - 6x^2 = x^2(x^2 - x - 6) = x^2(x - 3)(x + 2)$. Thus either $x^2 = 0$, so $x = 0$,or $x = 3$, or $x = -2$. The solutions are $0, 3$, and -2.

10. $0 = (x - 2)^5 - 9(x - 2)^3 = (x - 2)^3[(x - 2)^2 - 9] = (x - 2)^3[(x - 2) - 3][(x - 2) + 3]$
    $= (x - 2)^3(x - 5)(x + 1)$. Thus either $x = 2$,or $x = 5$, or $x = -1$. The solutions are $2, 5$, and -1.

12. $0 = 2x^3 + x^2 - 18x - 9 = x^2(2x + 1) - 9(2x + 1) = (2x + 1)(x^2 - 9)$
    $= (2x + 1)(x - 3)(x + 3)$. The solutions are $-\frac{1}{2}$, 3, and -3.

14. $7x^3 - x + 1 = x^3 + 3x^2 + x$ $\Leftrightarrow$ $0 = 6x^3 - 3x^2 - 2x + 1 = 3x^2(2x - 1) - (2x - 1)$
    $= (2x - 1)(3x^2 - 1)$ $\Leftrightarrow$ $2x - 1 = 0$ or $3x^2 - 1 = 0$. If $2x - 1 = 0$, then $x = \frac{1}{2}$. If $3x^2 - 1 = 0$, then $3x^2 = 1$ $\Leftrightarrow$ $x^2 = \frac{1}{3}$ $\Rightarrow$ $x = \pm\sqrt{\frac{1}{3}}$. The solutions are $\frac{1}{2}$ and $\pm\sqrt{\frac{1}{3}}$.

16. $\dfrac{10}{x} - \dfrac{12}{x - 3} + 4 = 0$ $\Leftrightarrow$ $x(x - 3)\left(\dfrac{10}{x} - \dfrac{12}{x - 3} + 4\right) = 0$ $\Leftrightarrow$
    $(x - 3)10 - 12x + 4x(x - 3) = 0$ $\Leftrightarrow$ $10x - 30 - 12x + 4x^2 - 12x = 0$ $\Leftrightarrow$
    $4x^2 - 14x - 30 = 0$. Using the quadratic formula, we have
    $x = \dfrac{-(-14)\pm\sqrt{(-14)^2 - 4(4)(-30)}}{2(4)} = \dfrac{14\pm\sqrt{196 + 480}}{8} = \dfrac{14\pm\sqrt{676}}{8} = \dfrac{14\pm26}{8}$. So the solutions are 5 and $-\frac{3}{2}$.

18. $1 + \dfrac{2x}{(x + 3)(x + 4)} = \dfrac{2}{x + 3} + \dfrac{4}{x + 4}$ $\Rightarrow$ $(x + 3)(x + 4) + 2x = 2(x + 4) + 4(x + 3)$ $\Leftrightarrow$
    $x^2 + 7x + 12 + 2x = 2x + 8 + 4x + 12$ $\Leftrightarrow$ $x^2 + 3x - 8 = 0$. Using the quadratic formula, we
    have $x = \dfrac{-(3)\pm\sqrt{(3)^2 - 4(1)(-8)}}{2(1)} = \dfrac{-3\pm\sqrt{9 + 32}}{2} = \dfrac{-3\pm\sqrt{41}}{2}$. The solutions are $\dfrac{-3\pm\sqrt{41}}{2}$.

20. $\dfrac{x}{2x + 7} - \dfrac{x + 1}{x + 3} = 1$ $\Leftrightarrow$ $x(x + 3) - (x + 1)(2x + 7) = (2x + 7)(x + 3)$ $\Leftrightarrow$
    $x^2 + 3x - 2x^2 - 9x - 7 = 2x^2 + 13x + 21$ $\Leftrightarrow$ $3x^2 + 19x + 28 = 0$ $\Leftrightarrow$
    $(3x + 7)(x + 4) = 0$. Thus either $3x + 7 = 0$, so $x = -\frac{7}{3}$, or $x = -4$. The solutions are $-\frac{7}{3}$ and -4.

22. $\dfrac{x + \frac{2}{x}}{3 + \frac{4}{x}} = 5x$ $\Rightarrow$ $\left(\dfrac{x + \frac{2}{x}}{3 + \frac{4}{x}}\right) \cdot \dfrac{x}{x} = \dfrac{x^2 + 2}{3x + 4} = 5x$ $\Rightarrow$ $x^2 + 2 = 5x(3x + 4)$ $\Leftrightarrow$

    $x^2 + 2 = 15x^2 + 20x$ $\Leftrightarrow$ $0 = 14x^2 + 20x - 2$ $\Rightarrow$ $x = \dfrac{-(20)\pm\sqrt{(20)^2 - 4(14)(-2)}}{2(14)}$

$$= \frac{-20 \pm \sqrt{400+112}}{28} = \frac{-20 \pm \sqrt{512}}{28} = \frac{-20 \pm 16\sqrt{2}}{28} = \frac{-5 \pm 4\sqrt{2}}{7}. \text{ The solutions are } \frac{-5 \pm 4\sqrt{2}}{7}.$$

24. Let $w = \frac{x+1}{x}$. Then $0 = \left(\frac{x+1}{x}\right)^2 + 4\left(\frac{x+1}{x}\right) + 3$ becomes

$0 = w^2 + 4w + 3 = (w+1)(w+3)$. Now if $w+1 = 0$, then $\frac{x+1}{x} + 1 = 0 \quad \Leftrightarrow \quad \frac{x+1}{x} = -1$

$\Leftrightarrow \quad x + 1 = -x \quad \Leftrightarrow \quad x = -\frac{1}{2}$, and if $w + 3 = 0$, then $\frac{x+1}{x} + 3 = 0 \quad \Leftrightarrow \quad \frac{x+1}{x} = -3$

$\Leftrightarrow \quad x + 1 = -3x \quad \Leftrightarrow \quad x = -\frac{1}{4}$. The solutions are $-\frac{1}{2}$ and $-\frac{1}{4}$.

26. Let $w = \frac{x}{x+2}$. Then $\left(\frac{x}{x+2}\right)^2 = \frac{4x}{x+2} - 4$ becomes $w^2 = 4w - 4 \quad \Leftrightarrow$

$0 = w^2 - 4w + 4 = (w-2)^2$. Now if $w - 2 = 0$, then $\frac{x}{x+2} - 2 = 0 \quad \Leftrightarrow \quad \frac{x}{x+2} = 2 \quad \Leftrightarrow$

$x = 2x + 4 \quad \Leftrightarrow \quad x = -4$. The solution is -4.

28. $0 = x^4 - 5x^2 + 4 = (x^2 - 4)(x^2 - 1) = (x - 2)(x + 2)(x - 1)(x + 1)$. So $x = 2$, $x = -2$, or $x = 1$, or $x = -1$. The solutions are -2, 2, -1, and 1.

30. $0 = x^6 - 2x^3 - 3 = (x^3 - 3)(x^3 + 1)$. If $x^3 - 3 = 0$, then $x^3 = 3 \quad \Leftrightarrow \quad x = \sqrt[3]{3}$, or if $x^3 = -1$

$\Leftrightarrow \quad x = -1$. Thus $x = \sqrt[3]{3}$ or $x = -1$. The solutions are $\sqrt[3]{3}$ and -1.

32. $x^8 + 15x^4 = 16 \quad \Leftrightarrow \quad 0 = x^8 + 15x^4 - 16 = (x^4 + 16)(x^4 - 1)$. If $x^4 + 16 = 0$, then $x^4 = -16$ which is impossible (for real numbers). If $x^4 - 1 = 0 \quad \Leftrightarrow \quad x^4 = 1$, so $x = \pm 1$. The solutions are 1 and -1.

34. Let $u = \sqrt[4]{x}$; then $0 = \sqrt{x} - 3\sqrt[4]{x} - 4 = u^2 - 3u - 4 = (u - 4)(u + 1)$. So $u - 4 = \sqrt[4]{x} - 4 = 0 \quad \Leftrightarrow \quad \sqrt[4]{x} = 4 \quad \Rightarrow \quad x = 4^4 = 256$, or $u + 1 = \sqrt[4]{x} + 1 = 0 \quad \Leftrightarrow$ $\sqrt[4]{x} = -1$. However, $\sqrt[4]{x}$ is the positive fourth root, so this cannot equal -1. The only solution is 256.

36. Let $u = x - 4$; then $0 = 2(x - 4)^{7/3} - (x - 4)^{4/3} - (x - 4)^{1/3} = 2u^{7/3} - u^{4/3} - u^{1/3} = u^{1/3}(2u^2 - u - 1) = u^{1/3}(2u + 1)(u - 1)$. So $u^{1/3} = x - 4 = 0 \quad \Leftrightarrow \quad x = 4$, or $2u + 1 = 2(x - 4) + 1 = 2x - 7 = 0 \quad \Leftrightarrow \quad 2x = 7 \quad \Leftrightarrow \quad x = \frac{7}{2}$, or $u - 1 = (x - 4) - 1 = x - 5 = 0 \quad \Leftrightarrow \quad x = 5$. The solutions are 4, $\frac{7}{2}$, and 5.

38. $x^{1/2} + 3x^{-1/2} = 10x^{-3/2} \quad \Leftrightarrow \quad 0 = x^{1/2} + 3x^{-1/2} - 10x^{-3/2} = x^{-3/2}(x^2 + 3x - 10)$ $= x^{-3/2}(x - 2)(x + 5)$. Now $x^{-3/2}$ never equals 0, and no solution can be negative, because we cannot take the $\frac{1}{2}$ power of a negative number. Thus 2 is the only solution.

40. Let $u = \sqrt{x}$. Then $0 = x - 5\sqrt{x} + 6$ becomes $u^2 - 5u + 6 = (u - 3)(u - 2) = 0$. If $u - 3 = 0$, then $\sqrt{x} - 3 = 0 \quad \Leftrightarrow \quad \sqrt{x} = 3 \quad \Rightarrow \quad x = 9$. If $u - 2 = 0$, then $\sqrt{x} - 2 = 0 \quad \Leftrightarrow$ $\sqrt{x} = 2 \quad \Rightarrow \quad x = 4$. The solutions are 9 and 4.

42. $0 = 4x^{-4} - 16x^{-2} + 4$. Multiplying by $\frac{x^4}{4}$ we get, $0 = 1 - 4x^2 + x^4$. Substituting $u = x^2$, we get $0 = 1 - 4u + u^2$, and using the quadratic formula, we get $u = \frac{-(-4) \pm \sqrt{(-4)^2 - 4(1)(1)}}{2(1)}$

$$= \frac{4 \pm \sqrt{16-4}}{2} = \frac{4 \pm \sqrt{12}}{2} = \frac{4 \pm 2\sqrt{3}}{2} = 2 \pm \sqrt{3}.$$ Substituting back, we have $x^2 = 2 \pm \sqrt{3}$, and since $2 + \sqrt{3}$ and $2 - \sqrt{3}$ are both positive we have $x = \pm\sqrt{2+\sqrt{3}}$ or $x = \pm\sqrt{2-\sqrt{3}}$. Thus the solutions are $-\sqrt{2-\sqrt{3}}, \ \sqrt{2-\sqrt{3}}, \ -\sqrt{2+\sqrt{3}},$ and $\sqrt{2+\sqrt{3}}$.

44. $x - \sqrt{9-3x} = 0 \ \Leftrightarrow \ x = \sqrt{9-3x} \ \Rightarrow \ x^2 = 9 - 3x \ \Leftrightarrow \ 0 = x^2 + 3x - 9.$ Using the quadratic formula to find the potential solutions, we have $x = \frac{-3 \pm \sqrt{3^2 - 4(1)(-9)}}{2(1)} = \frac{-3 \pm \sqrt{45}}{2}$ $= \frac{-3 \pm 3\sqrt{5}}{2}$. Substituting each of these solutions into the original equation, we see that $x = \frac{-3+3\sqrt{5}}{2}$ is a solution , but $x = \frac{-3-3\sqrt{5}}{2}$ is not. Thus $x = \frac{-3+3\sqrt{5}}{2}$ is the only solution.

46. $2x + \sqrt{x+1} = 8 \ \Leftrightarrow \ \sqrt{x+1} = 8 - 2x \ \Rightarrow \ x + 1 = (8-2x)^2 \ \Leftrightarrow$ $x + 1 = 64 - 32x + 4x^2 \ \Leftrightarrow \ 0 = 4x^2 - 33x + 63 = (4x-21)(x-3).$ Potential solutions are $x = \frac{21}{4}$ and $x = 3$. Substituting each of these solutions into the original equation, we see that $x = 3$ is a solution , but $x = \frac{21}{4}$ is not. Thus 3 is the only solution.

48. $x + 2\sqrt{x-7} = 10 \ \Leftrightarrow \ 2\sqrt{x-7} = 10 - x \ \Rightarrow \ 4(x-7) = (10-x)^2 \ \Leftrightarrow$ $4x - 28 = 100 - 20x + x^2 \ \Leftrightarrow \ 0 = x^2 - 24x + 128 = (x-8)(x+16).$ Potential solutions are $x = 8$ and $x = 16$. Substituting each of these solutions into the original equation, we see that $x = 8$ is a solution , but $x = 16$ is not. Thus 8 is the only solution.

50. $\sqrt[3]{4x^2 - 4x} = x \ \Leftrightarrow \ 4x^2 - 4x = x^3 \ \Leftrightarrow$ $0 = x^3 - 4x^2 + 4x = x(x^2 - 4x + 4) = x(x-2)^2.$ So $x = 0$ or $x = 2$. The solutions are 0 and 2.

52. Let $u = \sqrt{11 - x^2}$. By definition of u we require it to be nonnegative. Now $\sqrt{11-x^2} - \dfrac{2}{\sqrt{11-x^2}} = 1 \ \Leftrightarrow \ u - \dfrac{2}{u} = 1.$ Multiplying both sides by u we obtain $u^2 - 2 = u \ \Leftrightarrow \ 0 = u^2 - u - 2 = (u-2)(u+1).$ So $u = 2$ or $u = -1$. But since u must be nonnegative, we only have $u = 2 \ \Leftrightarrow \ \sqrt{11-x^2} = 2 \ \Rightarrow \ 11 - x^2 = 4 \ \Leftrightarrow \ x^2 = 7$ $\Leftrightarrow \ x = \pm\sqrt{7}$. The solutions are $\pm\sqrt{7}$.

54. $\sqrt{1 + \sqrt{x + \sqrt{2x+1}}} = \sqrt{5 + \sqrt{x}}$. We square both sides to get $1 + \sqrt{x + \sqrt{2x+1}} = 5 + \sqrt{x}$ $\Rightarrow \ x + \sqrt{2x+1} = \left(4 + \sqrt{x}\right)^2 = 16 + 8\sqrt{x} + x \ \Leftrightarrow \ \sqrt{2x+1} = 16 + 8\sqrt{x}.$ Again, squaring both sides, we obtain $2x + 1 = \left(16 + 8\sqrt{x}\right)^2 = 256 + 256\sqrt{x} + 64x \ \Leftrightarrow$ $-62x - 255 = 256\sqrt{x}.$ We could continue squaring both sides until we found possible solutions; however, consider the last equation. Since we are working with real numbers, for $\sqrt{x}$ to be defined, we must have $x \geq 0$. Then $-62x - 255 < 0$ while $256\sqrt{x} \geq 0$, so there is no solution.

56. $0 = x^4 - 16 = (x^2 - 4)(x^2 + 4) = (x-2)(x+2)(x-2\,i)(x+2\,i).$ Setting each factor in turn equal to zero, we see that the solutions are ± 2 and $\pm 2\,i$.

58. $x^4 + x^3 + x^2 + x = 0 \ \Leftrightarrow$ $0 = x(x^3 + x^2 + x + 1) = x[x^2(x+1) + (x+1)] = x(x+1)(x^2+1) = x(x+1)(x-i)(x+i).$ Setting each factor in turn equal to zero, we see that the solutions are $0, -1,$ and $\pm i$.

60. $0 = x^3 + 3x^2 + 9x + 27 = x^2(x+3) + 9(x+3) = (x+3)(x^2+9) = (x+3)(x-3i)(x+3i)$.
Setting each factor in turn equal to zero, we see that the solutions are -3 and $\pm 3\,i$.

62. $0 = x^6 + 9x^4 - 4x^2 - 36 = x^4(x^2+9) - 4(x^2+9) = (x^2+9)(x^4-4)$
$= (x-3i)(x+3i)(x^2-2)(x^2+2)$
$= (x-3i)(x+3i)\left(x-\sqrt{2}\right)\left(x+\sqrt{2}\right)\left(x-\sqrt{2}\,i\right)\left(x-\sqrt{2}\,i\right)$. The six solutions are $\pm 3\,i$,
$\pm\sqrt{2}$, and $\pm\sqrt{2}\,i$.

64. $1 - \sqrt{x^2+7} = 6 - x^2 \quad\Leftrightarrow\quad x^2 - 5 = \sqrt{x^2+7}$. Squaring both sides, we get
$x^4 - 10x^2 + 25 = x^2 + 7 \quad\Leftrightarrow\quad 0 = x^4 - 11x^2 + 18 = (x^2-9)(x^2-2) =$
$(x-3)(x+3)(x-\sqrt{2})(x+\sqrt{2})$. The possible solutions are $x = \pm 3$ and $x = \pm\sqrt{2}$.
Checking $x = \pm 3$, we have LHS $= 1 - \sqrt{(\pm 3)^2+7} = 1 - \sqrt{16} = -3$;
RHS $= 6 - (\pm 3)^2 = -3$. Since LHS $=$ RHS these are solutions.
Checking $\pm\sqrt{2}$, we have LHS $= 1 - \sqrt{\left(\pm\sqrt{2}\right)^2+7} = 1 - \sqrt{9} = -2$;
RHS $= 6 - \left(\pm\sqrt{2}\right)^2 = 4$. Since LHS $\neq$ RHS these are not solutions. Thus the only solutions are
± 3.

66. $0 = a^3x^3 + b^3 = (ax+b)(a^2x^2 - abx + b^2)$. So $ax + b = 0 \quad\Leftrightarrow\quad ax = -b \quad\Leftrightarrow\quad x = -\dfrac{b}{a}$ or
$x = \dfrac{-(-ab) \pm \sqrt{(-ab)^2 - 4(a^2)(b^2)}}{2(a^2)} = \dfrac{ab \pm \sqrt{-3a^2b^2}}{2a^2} = \dfrac{ab \pm \sqrt{3}ab\,i}{2a^2} = \dfrac{b \pm \sqrt{3}b\,i}{2a}$. Thus the
three solutions are $-\dfrac{b}{a}$ and $\dfrac{b \pm \sqrt{3}b\,i}{2a}$.

68. Let $w = x^{1/6}$. Then $x^{1/3} = w^2$ and $x^{1/2} = w^3$, and so $0 = w^3 + aw^2 + bw + ab$
$= w^2(w+a) + b(w+a) = (w+a)(w^2+b) \quad\Leftrightarrow\quad \left(\sqrt[6]{x}+a\right)\left(\sqrt[3]{x}+b\right)$. So $\sqrt[6]{x} + a = 0$
$\Leftrightarrow\quad \sqrt[6]{x} = -a$; however, since $\sqrt[6]{x}$ is positive by definition is positive and $-a$ is negative, this is
impossible. Setting the other factor equal to zero, we have $\sqrt[3]{x} + b = 0 \quad\Leftrightarrow\quad \sqrt[3]{x} = -b \quad\Rightarrow$
$x = -b^3$. Checking $x = -b^3$, we have $\sqrt{-b^3} + a\sqrt[3]{-b^3} + b\sqrt[6]{-b^3} + ab$
$= b\sqrt{b}\,i + a(-b) + b\left(\sqrt{-b}\right) + ab = 2b\sqrt{b}\,i \neq 0$, so this is not a solution either. Therefore, there
are no solutions to this equation.

70. Let n be the number of people in the group, so each person now pays $\dfrac{120{,}000}{n}$. If one person joins
the group, then there would be $n+1$ members in the group, and each person would pay
$\dfrac{120{,}000}{n} - 6000$. So $(n+1)\left(\dfrac{120{,}000}{n} - 6000\right) = 120{,}000 \quad\Leftrightarrow$
$\left[\left(\dfrac{n}{6000}\right)\left(\dfrac{120{,}000}{n} - 6000\right)\right](n+1) = \left(\dfrac{n}{6000}\right)120{,}000 \quad\Leftrightarrow\quad (20-n)(n+1) = 20n \quad\Leftrightarrow$
$-n^2 + 19n + 20 = 20n \quad\Leftrightarrow\quad 0 = n^2 + n - 20 = (n-4)(n+5)$. Thus $n = 4$ or $n = -5$.
Since n must be positive, there are now 4 friends in the group.

72. Let d be the distance from the lens to the object. Then the distance from the lens to the image is
$d - 4$. So substituting $F = 4.8$, $x = d$, and $y = d - 4$, and then solving for x, we have

$\dfrac{1}{4.8} = \dfrac{1}{d} + \dfrac{1}{d-4}$. Now we multiply by the LCD, $4.8d(d-4)$, to get

$d(d-4) = 4.8(d-4) + 4.8d \quad \Leftrightarrow \quad d^2 - 4d = 9.6d - 19.2 \quad \Leftrightarrow \quad 0 = d^2 - 13.6d + 19.2$

$\Rightarrow \quad d = \dfrac{13.6 \pm 10.4}{2}$. So $d = 1.6$ or $d = 12$. Since $d - 4$ must also be positive, the object is 12 cm from the lens.

74. Let r be the radius of the larger sphere, in mm. Equating the volumes, we have

$\frac{4}{3}\pi r^3 = \frac{4}{3}\pi(2^3 + 3^3 + 4^3) \quad \Leftrightarrow \quad r^3 = 2^3 + 3^3 + 4^4 \quad \Leftrightarrow \quad r^3 = 99 \quad \Leftrightarrow \quad r = \sqrt[3]{99} \approx 4.63$.

Therefore, the radius of the larger sphere is about 4.63 mm.

76. Let x be the distance, in feet, that he goes on the boardwalk before veering off onto the sand. The distance along the boardwalk from where he started to the point on the boardwalk closest to the umbrella is $\sqrt{750^2 - 210^2} = 720$ ft. Thus the distance that he walks on the sand is

$\sqrt{(720-x)^2 + 210^2} = \sqrt{518400 - 1440x + x^2 + 44100} = \sqrt{x^2 - 1440x + 562500}$

	Distance	Rate	Time
Along Boardwalk	x	4	$\dfrac{x}{4}$
Across Sand	$\sqrt{x^2 - 1440x + 562500}$	2	$\dfrac{\sqrt{x^2 - 1440x + 562500}}{2}$

Since 4 minutes 45 seconds $= 285$ seconds, we equate the time it takes to walk along the Boardwalk

and across the sand to the total time to get $285 = \dfrac{x}{4} + \dfrac{\sqrt{x^2 - 1440x + 562500}}{2} \quad \Leftrightarrow$

$1140 - x = 2\sqrt{x^2 - 1440x + 562500}$. Squaring both sides, we get

$(1140 - x)^2 = 4(x^2 - 1440x + 562500) \quad \Leftrightarrow$

$1299600 - 2280x + x^2 = 4x^2 - 5760x + 2250000 \quad \Leftrightarrow \quad 0 = 3x^2 - 3480x + 950400$

$= 3(x^2 - 1160x + 316800) = 3(x - 720)(x - 440)$. So $x - 720 = 0 \quad \Leftrightarrow \quad x = 720$, and

$x - 440 = 0 \quad \Leftrightarrow \quad x = 440$. Checking $x = 720$, the distance across the sand is 210 feet. So

$\dfrac{720}{4} + \dfrac{210}{2} = 180 + 105 = 285$ seconds. Checking $x = 440$, the distance across the sand is

$\sqrt{(720 - 440)^2 + 210^2} = 350$ feet. So $\dfrac{440}{4} + \dfrac{350}{2} = 110 + 175 = 285$ seconds. Since both

solutions are less than or equal to 720 feet, we have two solutions: he walks 440 feet down the boardwalk and then heads towards his umbrella, or he walks 720 feet down the boardwalk and then heads toward his umbrella.

78. Let r be the radius of the tank, in feet. The volume of the spherical tank is $\dfrac{4}{3}\pi r^3$ and is also

$750 \times 0.1337 = 100.275$. So $\dfrac{4}{3}\pi r^3 = 100.275 \quad \Leftrightarrow \quad r^3 = 23.938 \quad \Leftrightarrow \quad r = 2.88$ feet.

80. Let h be the height of the screens in inches. Hence, the width of the smaller screen is $h + 5$ inches, and the width of the bigger screen is $1.8h$ inches. The diagonal measure of the smaller screen is $\sqrt{h^2 + (h+5)^2}$, and the diagonal measure of the larger screen is $\sqrt{h^2 + (1.8\,h)^2} = \sqrt{4.24\,h^2}$ $\approx 2.06h$. Thus $\sqrt{h^2 + (h+5)^2} + 14 = 2.06\,h \quad \Leftrightarrow \quad \sqrt{h^2 + (h+5)^2} = 2.06\,h - 14$. Squaring both sides yields $h^2 + h^2 + 10h + 25 = 4.24h^2 - 57.68h + 196 \quad \Leftrightarrow$ $0 = 2.24h^2 - 67.68h + 171$. Applying the quadratic formula, we obtain

$$h = \frac{67.68 \pm \sqrt{(-67.68)^2 - 4(2.24)(171)}}{2(2.24)} = \frac{67.68 \pm \sqrt{3048.4224}}{4.48} = \frac{67.68 \pm 55.21}{4.48}. \text{ So}$$

$h = 27.4$ or $h = 2.8$. Since high definition television monitors don't come 2.8 inch height, the height of the screens is 27.4 inches.

82. (a) <u>Method 1:</u> Let $u = \sqrt{x}$, so $u^2 = x$. Thus $x - \sqrt{x} - 2 = 0$ becomes $u^2 - u - 2 = 0 \quad \Leftrightarrow$
$(u - 2)(u + 1) = 0$. So $u = 2$ or $u = -1$. If $u = 2$, then $\sqrt{x} = 2 \quad \Rightarrow \quad x = 4$. If $u = -1$,
then $\sqrt{x} = -1 \quad \Rightarrow \quad x = 1$. So the possible solutions are 4 and 1. Checking $x = 4$ we have
$4 - \sqrt{4} - 2 = 4 - 2 - 2 = 0$. Checking $x = 1$ we have $1 - \sqrt{1} - 2 = 1 - 1 - 2 \neq 0$. The
only solution is 4.

<u>Method 2:</u> $x - \sqrt{x} - 2 = 0 \quad \Leftrightarrow \quad x - 2 = \sqrt{x} \quad \Rightarrow \quad x^2 - 4x + 4 = x \quad \Leftrightarrow$
$x^2 - 5x + 4 = 0 \quad \Leftrightarrow \quad (x - 4)(x - 1) = 0$. So the possible solutions are 4 and 1. Checking
will result in the same solution.

(b) <u>Method 1:</u> Let $u = \dfrac{1}{x - 3}$, so $u^2 = \dfrac{1}{(x - 3)^2}$. Thus $\dfrac{12}{(x - 3)^2} + \dfrac{10}{x - 3} + 1 = 0$ becomes

$12u^2 + 10u + 1 = 0$. Using the quadratic formula, we have

$u = \dfrac{-10 \pm \sqrt{10^2 - 4(12)(1)}}{2(12)} = \dfrac{-10 \pm \sqrt{52}}{24} = \dfrac{-10 \pm 2\sqrt{13}}{24} = \dfrac{-5 \pm \sqrt{13}}{12}$. If $u = \dfrac{-5 - \sqrt{13}}{12}$, then

$\dfrac{1}{x - 3} = \dfrac{-5 - \sqrt{13}}{12} \quad \Leftrightarrow \quad x - 3 = \dfrac{12}{-5 - \sqrt{13}} \cdot \dfrac{-5 + \sqrt{13}}{-5 + \sqrt{13}} = \dfrac{12(-5 + \sqrt{13})}{12} = -5 + \sqrt{13}$. So

$x = -2 + \sqrt{13}$.

If $u = \dfrac{-5 + \sqrt{13}}{12}$, then $\dfrac{1}{x - 3} = \dfrac{-5 + \sqrt{13}}{12} \quad \Leftrightarrow$

$x - 3 = \dfrac{12}{-5 + \sqrt{13}} \cdot \dfrac{-5 - \sqrt{13}}{-5 - \sqrt{13}} = \dfrac{12(-5 - \sqrt{13})}{12} = -5 - \sqrt{13}$. So $x = -2 - \sqrt{13}$.

The solutions are $-2 \pm \sqrt{13}$.

<u>Method 2:</u> Multiply by LCD, $(x - 3)^2$, we get

$(x - 3)^2 \left(\dfrac{12}{(x - 3)^2} + \dfrac{10}{x - 3} + 1 \right) = 0 \cdot (x - 3)^2 \quad \Leftrightarrow \quad 12 + 10(x - 3) + (x - 3)^2 = 0$

$\Leftrightarrow \quad 12 + 10x - 30 + x^2 - 6x + 9 = 0 \quad \Leftrightarrow \quad x^2 + 4x - 9 = 0$. Using the quadratic

formula, we have $u = \dfrac{-4 \pm \sqrt{4^2 - 4(1)(-9)}}{2} = \dfrac{-4 \pm \sqrt{52}}{2} = \dfrac{-4 \pm 2\sqrt{13}}{22} = -2 \pm \sqrt{13}$. The solutions are

$-2 \pm \sqrt{13}$.

Exercises 1.6

2. $x = -2$: $-2 + 1 \overset{?}{<} 2$. Yes, $-1 < 2$. $x = -1$: $-1 + 1 \overset{?}{<} 2$. Yes, $0 < 2$.

     $x = 0$: $0 + 1 \overset{?}{<} 2$. Yes, $1 < 2$. $x = \frac{1}{2}$: $\frac{1}{2} + 1 \overset{?}{<} 2$. Yes, $\frac{3}{2} < 2$.

     $x = 1$: $1 + 1 \overset{?}{<} 2$. No, $2 \not< 2$. $x = \sqrt{2}$: $\sqrt{2} + 1 \overset{?}{<} 2$. No, since $\sqrt{2} > 1$.

     $x = 2$: $2 + 1 \overset{?}{<} 2$. No, $3 \not< 2$. $x = 4$: $4 + 1 \overset{?}{<} 2$. No, $5 \not< 2$.

     The elements $-2, -1, 0,$ and $\frac{1}{2}$ all satisfies the inequality.

4. $x = -2$: $2(-2) - 1 \overset{?}{\geq} -2$. No, $-5 \not\geq -2$. $x = -1$: $2(-1) - 1 \overset{?}{\geq} \frac{1}{2}$. No, $6 - 1 \not\geq \frac{1}{2}$.

     $x = 0$: $2(0) - 1 \overset{?}{\geq} 0$. No, $-1 \not\geq 0$. $x = \frac{1}{2}$: $2(\frac{1}{2}) - 1 \overset{?}{\geq} \frac{1}{2}$. No, $0 \not\geq \frac{1}{2}$.

     $x = 1$: $2(1) - 1 \overset{?}{\geq} 1$. Yes since $1 \geq 1$.

     $x = \sqrt{2}$: $2(\sqrt{2}) - 1 \overset{?}{\geq} \sqrt{2}$. Yes, since $2\sqrt{2} - 1 \approx 1.8$ and $\sqrt{2} \approx 1.4$.

     $x = 2$: $2(2) - 1 \overset{?}{\geq} 2$. Yes, $3 \geq 2$. $x = 4$: $2(4) - 1 \overset{?}{\geq} 4$. Yes, $7 \geq 4$.

     The elements $1, \sqrt{2}, 2,$ and 4 all satisfies the inequality.

6. $x = -2$: $2 \overset{?}{\leq} 3 - (-2) \overset{?}{<} $. No, since $3 - (-2) = 5$ and $5 \not< 2$.

     $x = -1$: $-2 \overset{?}{\leq} 3 - (-1) \overset{?}{<} 2$. No, since $3 - (-1) = 4$ and $4 \not< 2$.

     $x = 0$: $-2 \overset{?}{\leq} 3 - 0 \overset{?}{<} 2$. No, since $3 - (0) = 3$ and $3 \not< 2$.

     $x = \frac{1}{2}$: $-2 \overset{?}{\leq} 3 - \frac{1}{2} \overset{?}{<} 2$. No, since $3 - (\frac{1}{2}) = \frac{5}{2}$ and $\frac{5}{2} \not< 2$.

     $x = 1$: $-2 \overset{?}{\leq} 3 - 1 \overset{?}{<} 2$ No, since $3 - (1) = 2$ and $2 \not< 2$.

     $x = \sqrt{2}$: $-2 \overset{?}{\leq} 3 - \sqrt{2} \overset{?}{<} 2$. Yes, since $3 - \sqrt{2} \approx 1.5$ and $-2 \leq 1.5 < 2$.

     $x = 2$: $-2 \overset{?}{\leq} 3 - 2 \overset{?}{<} 2$. Yes, since $3 - 2 = 1$ and $-2 \leq 1 < 2$.

     $x = 4$: $-2 \overset{?}{\leq} 3 - 4 \overset{?}{<} 2$. Yes, $3 - 4 = -1$ and $-2 \leq -1 < 2$.

     The elements $\sqrt{2}, 2,$ and 4 all satisfies the inequality.

8. $x = -2$: $(-2)^2 + 2 \overset{?}{<} 4$. No, since $(-2)^2 + 2 = 6$ and $6 \not< 4$.

     $x = -1$: $(-1)^2 + 2 \overset{?}{<} 4$. Yes, since $(-1)^2 + 2 = 3$ and $3 < 4$.

     $x = 0$: $0^2 + 2 \overset{?}{<} 4$. Yes, since $0^2 + 2 = 2$ and $2 < 4$.

     $x = \frac{1}{2}$: $\left(\frac{1}{2}\right)^2 + 2 \overset{?}{<} 4$. Yes, since $(\frac{1}{2})^2 + 2 = \frac{9}{4}$ and $\frac{9}{4} < 4$.

     $x = 1$: $1^2 + 2 \overset{?}{<} 4$. Yes, since $1^2 + 2 = 3$ and $3 < 4$.

     $x = \sqrt{2}$: $\left(\sqrt{2}\right)^2 + 2 \overset{?}{<} 4$. No, since $\left(\sqrt{2}\right)^2 + 2 = 4$ and $4 \not< 4$.

$x = 2$: $2^2 + 2 \overset{?}{<} 4$. No, since $2^2 + 2 = 6$ and $6 \not< 4$.

$x = 4$: $4^2 + 2 \overset{?}{<} 4$. No, $4^2 + 2 = 18$ and $18 \not< 4$.

The elements -1, 0, $\frac{1}{2}$, and 1 all satisfies the inequality.

10. $-4x \geq 10 \quad \Leftrightarrow \quad x \leq -\frac{5}{2}$.   Interval: $(-\infty, -\frac{5}{2}]$.

     Graph:

12. $3x + 11 < 5 \quad \Leftrightarrow \quad 3x < -6 \quad \Leftrightarrow \quad x < -2$.   Interval: $(-\infty, -2)$.

     Graph:

14. $5 - 3x \leq -16 \quad \Leftrightarrow \quad -3x \leq -21 \quad \Leftrightarrow \quad x \geq 7$.   Interval: $[7, \infty)$.

     Graph:

16. $0 < 5 - 2x \quad \Leftrightarrow \quad 2x < 5 \quad \Leftrightarrow \quad x < \frac{5}{2}$.   Interval: $\left(-\infty, \frac{5}{2}\right)$.

     Graph:

18. $6 - x \geq 2x + 9 \quad \Leftrightarrow \quad -3 \geq 3x \quad \Leftrightarrow \quad -1 \geq x$.   Interval: $(-\infty, -1]$.

     Graph:

20. $\frac{2}{5}x + 1 < \frac{1}{5} - 2x \quad \Leftrightarrow \quad \frac{12}{5}x < -\frac{4}{5} \quad \Leftrightarrow \quad x < -\frac{1}{3}$. Interval: $\left(-\infty, -\frac{1}{3}\right)$.

     Graph:

22. $\frac{2}{3} - \frac{1}{2}x \geq \frac{1}{6} + x$ (multiply both sides by 6) $\quad \Leftrightarrow \quad 4 - 3x \geq 1 + 6x \quad \Leftrightarrow \quad 3 \geq 9x \quad \Leftrightarrow$

     $\frac{1}{3} \geq x$.   Interval: $(-\infty, \frac{1}{3}]$.   Graph:

24. $2(7x - 3) \leq 12x + 16 \quad \Leftrightarrow \quad 14x - 6 \leq 12x + 16 \quad \Leftrightarrow \quad 2x \leq 22 \quad \Leftrightarrow \quad x \leq 11$.

     Interval: $(-\infty, 11]$.   Graph:

26. $5 \leq 3x - 4 \leq 14 \quad \Leftrightarrow \quad 9 \leq 3x \leq 18 \quad \Leftrightarrow \quad 3 \leq x \leq 6$.

     Interval: $[3, 6]$.   Graph:

28. $1 < 3x + 4 \leq 16 \quad \Leftrightarrow \quad -3 < 3x \leq 12 \quad \Leftrightarrow \quad -1 < x \leq 4$. Interval: $(-1, 4]$.

     Graph:

30. $-3 \leq 3x + 7 \leq \frac{1}{2} \quad \Leftrightarrow \quad -10 \leq 3x \leq -\frac{13}{2} \quad \Leftrightarrow \quad -\frac{10}{3} \leq x \leq -\frac{13}{6}$.

     Interval: $\left[-\frac{10}{3}, -\frac{13}{6}\right]$.   Graph:

32. $-\frac{1}{2} < \frac{4 - 3x}{5} \leq \frac{1}{4} \quad \Leftrightarrow \quad$ (multiply each expression by 20) $-10 < 4(4 - 3x) \leq 5 \quad \Leftrightarrow$

     $-10 < 16 - 12x \leq 5 \quad \Leftrightarrow \quad -26 < -12x \leq -11 \quad \Leftrightarrow \quad \frac{13}{6} > x \geq \frac{11}{12} \quad \Leftrightarrow \quad$ (expressing in

     standard form) $\frac{11}{12} \leq x < \frac{13}{6}$.   Interval: $\left[\frac{11}{12}, \frac{13}{6}\right)$ Graph:

34. $(x-5)(x+4) \geq 0$. The expression on the left of the inequality changes sign when $x = 5$ and $x = -4$. Thus we must check the intervals in the following table.

Interval	$(-\infty, -4)$	$(-4, 5)$	$(5, \infty)$
Sign of $x - 5$	$-$	$-$	$+$
Sign of $x + 4$	$-$	$+$	$+$
Sign of $(x-5)(x+4)$	$+$	$-$	$+$

From the table, the solution set is $\left\{ x \mid x \leq -\frac{1}{3} \text{ or } 1 \leq x \right\}$.

Solution: $(-\infty, -4] \cup [5, \infty)$.   Graph:

36. $x(2 - 3x) \leq 0$. The expression on the left of the inequality changes sign when $x = 0$ and $x = \frac{2}{3}$. Thus we must check the intervals in the following table.

Interval	$(-\infty, 0)$	$(0, \frac{2}{3})$	$(\frac{2}{3}, \infty)$
Sign of x	$-$	$+$	$+$
Sign of $2 - 3x$	$+$	$+$	$-$
Sign of $x(2 - 3x)$	$-$	$+$	$-$

From the table, the solution set is $\left\{ x \mid x \leq 0 \text{ or } \frac{2}{3} \leq x \right\}$.

Solution: $(-\infty, 0] \cup [\frac{2}{3}, \infty)$.   Graph:

38. $x^2 + 5x + 6 > 0 \quad \Leftrightarrow \quad (x+3)(x+2) > 0$. The expression on the left of the inequality changes sign when $x = -3$ and $x = -2$. Thus we must check the intervals in the following table.

Interval	$(-\infty, -3)$	$(-3, -2)$	$(-2, \infty)$
Sign of $x + 3$	$-$	$+$	$+$
Sign of $x + 2$	$-$	$-$	$+$
Sign of $(x+3)(x+2)$	$+$	$-$	$+$

From the table, the solution set is $\{ x \mid x < -3 \text{ or } -2 < x \}$.

Solution: $(-\infty, -3) \cup (-2, \infty)$.   Graph:

40. $x^2 < x + 2 \quad \Leftrightarrow \quad x^2 - x - 2 < 0 \quad \Leftrightarrow \quad (x+1)(x-2) < 0$. The expression on the left of the inequality changes sign when $x = -1$ and $x = 2$. Thus we must check the intervals in the following table.

Interval	$(-\infty, -1)$	$(-1, 2)$	$(2, \infty)$
Sign of $x + 1$	$-$	$+$	$+$
Sign of $x - 2$	$-$	$-$	$+$
Sign of $(x+1)(x-2)$	$+$	$-$	$+$

From the table, the solution set is $\{ x \mid -1 < x < 2 \}$.

Solution: $(-1, 2)$.   Graph:

42. $5x^2 + 3x \geq 3x^2 + 2 \quad \Leftrightarrow \quad 2x^2 + 3x - 2 \geq 0 \quad \Leftrightarrow \quad (2x - 1)(x + 2) \geq 0$. The expression on the left of the inequality changes sign when $x = \frac{1}{2}$ and $x = -2$. Thus we must check the intervals

in the following table.

Interval	$(-\infty, -2)$	$\left(-2, \frac{1}{2}\right)$	$\left(\frac{1}{2}, \infty\right)$
Sign of $2x - 1$	$-$	$-$	$+$
Sign of $x + 2$	$-$	$+$	$+$
Sign of $(2x - 1)(x + 2)$	$+$	$-$	$+$

From the table, the solution set is $\left\{x \mid x \leq -2 \text{ or } \frac{1}{2} \leq x\right\}$.

Solution: $(-\infty, -2] \cup \left[\frac{1}{2}, \infty\right)$.　　　Graph:

44. $x^2 + 2x > 3$　$\Leftrightarrow$　$x^2 + 2x - 3 > 0$　$\Leftrightarrow$　$(x + 3)(x - 1) > 0$. The expression on the left of the inequality changes sign when $x = -3$ and $x = 1$. Thus we must check the intervals in the following table.

Interval	$(-\infty, -3)$	$(-3, 1)$	$(1, \infty)$
Sign of $x + 3$	$-$	$+$	$+$
Sign of $x - 1$	$-$	$-$	$+$
Sign of $(x + 3)(x - 1)$	$+$	$-$	$+$

From the table, the solution set is $\{x \mid x < -3 \text{ or } 1 < x\}$.

Solution: $(-\infty, -3) \cup (1, \infty)$.　　　Graph:

46. $x^2 \geq 9$　$\Leftrightarrow$　$x^2 - 9 \geq 0$　$\Leftrightarrow$　$(x + 3)(x - 3) \geq 0$. The expression on the left of the inequality changes sign when $x = -3$ and $x = 3$. Thus we must check the intervals in the following table.

Interval	$(-\infty, -3)$	$(-3, 3)$	$(3, \infty)$
Sign of $x + 3$	$-$	$+$	$+$
Sign of $x - 3$	$-$	$-$	$+$
Sign of $(x + 3)(x - 3)$	$+$	$-$	$+$

From the table, the solution set is $\{x \mid x \leq -3 \text{ or } 3 \leq x\}$.

Solution: $(-\infty, -3] \cup [3, \infty)$.　　　Graph:

48. $(x + 2)(x - 1)(x - 3) \leq 0$. The expression on the left of the inequality changes sign when $x = -2$, $x = 1$, and $x = 3$. Thus we must check the intervals in the following table.

Interval	$(-\infty, -2)$	$(-2, 1)$	$(1, 3)$	$(3, \infty)$
Sign of $x + 2$	$-$	$+$	$+$	$+$
Sign of $x - 1$	$-$	$-$	$+$	$+$
Sign of $x - 3$	$-$	$-$	$-$	$+$
Sign of $(x + 2)(x - 1)(x - 3)$	$-$	$+$	$-$	$+$

From the table, the solution set is $\{x \mid x \leq -2 \text{ or } 1 \leq x \leq 3\}$.

Solution: $(-\infty, -2] \cup [1, 3]$.　　　Graph:

50. $16x \leq x^3$　$\Leftrightarrow$　$0 \leq x^3 - 16x = x(x^2 - 16) = x(x - 4)(x + 4)$. The expression on the left of the inequality changes sign when $x = -4$, $x = 0$, and $x = 4$. Thus we must check the intervals in the following table.

Interval	$(-\infty, -4)$	$(-4, 0)$	$(0, 4)$	$(4, \infty)$
Sign of $x + 4$	$-$	$+$	$+$	$+$
Sign of x	$-$	$-$	$+$	$+$
Sign of $x - 4$	$-$	$-$	$-$	$+$
Sign of $x(x + 4)(x - 4)$	$-$	$+$	$-$	$+$

From the table, the solution set is $\{x \mid -4 \leq x \leq 0 \text{ or } 4 \leq x\}$.

Solution: $[-4, 0] \cup [4, \infty)$.   Graph:

52. $\dfrac{2x + 6}{x - 2} < 0$. The expression on the left of the inequality changes sign when $x = -3$ and $x = 2$.
Thus we must check the intervals in the following table.

Interval	$(-\infty, -3)$	$(-3, 2)$	$(2, \infty)$
Sign of $2x + 6$	$-$	$+$	$+$
Sign of $x - 2$	$-$	$-$	$+$
Sign of $\dfrac{2x + 6}{x - 2}$	$+$	$-$	$+$

From the table, the solution set is $\{x \mid -3 < x < 2\}$.

Solution: $(-3, 2)$.   Graph:

54. $-2 < \dfrac{x + 1}{x - 3} \Leftrightarrow 0 < \dfrac{x + 1}{x - 3} + 2 \Leftrightarrow 0 < \dfrac{x + 1}{x - 3} + \dfrac{2(x - 3)}{x - 3} \Leftrightarrow 0 < \dfrac{3x - 5}{x - 3}$. The
expression on the left of the inequality changes sign when $x = \frac{5}{3}$ and $x = 3$. Thus we must check
the intervals in the following table.

Interval	$\left(-\infty, \frac{5}{3}\right)$	$\left(\frac{5}{3}, 3\right)$	$(3, \infty)$
Sign of $3x - 5$	$-$	$+$	$+$
Sign of $x - 3$	$-$	$-$	$+$
Sign of $\dfrac{3x - 5}{x - 3}$	$+$	$-$	$+$

From the table, the solution set is $\left\{x \mid x < \frac{5}{3} \text{ or } 3 < x < \infty\right\}$.

Solution: $\left(-\infty, \frac{5}{3}\right) \cup (3, \infty)$.   Graph:

56. $\dfrac{3 + x}{3 - x} \geq 1 \Leftrightarrow \dfrac{3 + x}{3 - x} - 1 \geq 0 \Leftrightarrow \dfrac{3 + x}{3 - x} - \dfrac{3 - x}{3 - x} \geq 0 \Leftrightarrow \dfrac{2x}{3 - x} \geq 0$. The expression
on the left of the inequality changes sign when $x = 0$ and $x = 3$. Thus we must check the intervals
in the following table.

Interval	$(-\infty, 0)$	$(0, 3)$	$(3, \infty)$
Sign of $3 - x$	$+$	$+$	$-$
Sign of $2x$	$-$	$+$	$+$
Sign of $\dfrac{2x}{3 - x}$	$-$	$+$	$-$

Since the denominator cannot equal 0, we must have $x \neq 3$. The solution set is $\{x \mid 0 \leq x < 3\}$.

Solution: $[0, 3)$.   Graph:

58. $\dfrac{x}{x+1} > 3x \quad\Leftrightarrow\quad \dfrac{x}{x+1} - 3x > 0 \quad\Leftrightarrow\quad \dfrac{x}{x+1} - \dfrac{3x(x+1)}{x+1} > 0 \quad\Leftrightarrow\quad \dfrac{-2x-3x^2}{x+1} > 0$

$\Leftrightarrow\quad \dfrac{-x(2+3x)}{x+1} > 0.$ The expression on the left of the inequality changes sign when $x = 0$,

$x = -\frac{2}{3}$, and $x = -1$. Thus we must check the intervals in the following table.

Interval	$(-\infty, -1)$	$(-1, -\frac{2}{3})$	$(-\frac{2}{3}, 0)$	$(0, \infty)$
Sign of $-x$	$+$	$+$	$+$	$-$
Sign of $2 + 3x$	$-$	$-$	$+$	$+$
Sign of $x + 1$	$-$	$+$	$+$	$+$
Sign of $\dfrac{(2-x)(2+x)}{x}$	$+$	$-$	$+$	$-$

From the table, the solution set is $\left\{x \mid x < -1 \text{ or } -\frac{2}{3} < x < 0\right\}$.

Solution: $(-\infty, -1) \cup (-\frac{2}{3}, 0)$.   Graph:

60. $\dfrac{3}{x-1} - \dfrac{4}{x} \geq 1 \quad\Leftrightarrow\quad \dfrac{3}{x-1} - \dfrac{4}{x} - 1 \geq 0 \quad\Leftrightarrow\quad \dfrac{3x}{x(x-1)} - \dfrac{4(x-1)}{x(x-1)} - \dfrac{x(x-1)}{x(x-1)} \geq 0 \quad\Leftrightarrow$

$\dfrac{3x - 4x + 4 - x^2 + x}{x(x-1)} \geq 0 \quad\Leftrightarrow\quad \dfrac{4 - x^2}{x(x-1)} \geq 0 \quad\Leftrightarrow\quad \dfrac{(2-x)(2+x)}{x(x-1)} \geq 0.$ The expression

on the left of the inequality changes sign when $x = 2$, $x = -2$, $x = 0$, and $x = 1$. Thus we must
check the intervals in the following table.

Interval	$(-\infty, -2)$	$(-2, 0)$	$(0, 1)$	$(1, 2)$	$(2, \infty)$
Sign of $2 - x$	$+$	$+$	$+$	$+$	$-$
Sign of $2 + x$	$-$	$+$	$+$	$+$	$+$
Sign of x	$-$	$-$	$+$	$+$	$+$
Sign of $x - 1$	$-$	$-$	$-$	$+$	$+$
Sign of $\dfrac{(2-x)(2+x)}{x(x-1)}$	$-$	$+$	$-$	$+$	$-$

Since $x = 0$ and $x = 1$ yield undefined expressions, we cannot include them in the solution. From
the table, the solution set is $\left\{x \mid -2 \leq x < 0 \text{ or } 1 < x \leq 2\right\}$.

Solution: $[-2, 0) \cup (1, 2]$.   Graph:

62. $\dfrac{x}{2} \geq \dfrac{5}{x+1} + 4 \quad\Leftrightarrow\quad \dfrac{x}{2} - \dfrac{5}{x+1} - 4 \geq 0 \quad\Leftrightarrow\quad \dfrac{x(x+1)}{2(x+1)} - \dfrac{2 \cdot 5}{2(x+1)} - \dfrac{4(2)(x+1)}{2(x+1)} \geq 0$

$\Leftrightarrow\quad \dfrac{x^2 + x - 10 - 8x - 8}{2(x+1)} \geq 0 \quad\Leftrightarrow\quad \dfrac{x^2 - 7x - 18}{2(x+1)} \geq 0 \quad\Leftrightarrow\quad \dfrac{(x-9)(x+2)}{2(x+1)} \geq 0.$ The

expression on the left of the inequality changes sign when $x = 9$, $x = -2$, and $x = -1$. Thus we
must check the intervals in the following table.

Interval	$(-\infty, -2)$	$(-2, -1)$	$(-1, 9)$	$(9, \infty)$
Sign of $x - 9$	$-$	$-$	$-$	$+$
Sign of $x + 2$	$-$	$+$	$+$	$+$
Sign of $x + 1$	$-$	$-$	$+$	$+$
Sign of $\dfrac{(x-9)(x+2)}{2(x+1)}$	$-$	$+$	$-$	$+$

From the table, the solution set is $\{x| \ -2 \le x < -1 \text{ or } 9 \le x\}$. The point $x = -1$ is excluded from the solution set because it makes the expression undefined.

Solution: $[-2, -1) \cup [9, \infty)$ Graph:

64. $\dfrac{1}{x+1} + \dfrac{1}{x+2} \le 0 \ \Leftrightarrow \ \dfrac{x+2}{(x+1)(x+2)} + \dfrac{x+1}{(x+1)(x+2)} \le 0 \ \Leftrightarrow \ \dfrac{x+2+x+1}{(x+1)(x+2)} \le 0$

$\Leftrightarrow \ \dfrac{2x+3}{(x+1)(x+2)} \le 0$. The expression on the left of the inequality changes sign when

$x = -\frac{3}{2}$, $x = -1$, and $x = -2$. Thus we must check the intervals in the following table.

Interval	$(-\infty, -2)$	$\left(-2, -\frac{3}{2}\right)$	$\left(-\frac{3}{2}, -1\right)$	$(-1, \infty)$
Sign of $2x + 3$	$-$	$-$	$+$	$+$
Sign of $x + 1$	$-$	$-$	$-$	$+$
Sign of $x + 2$	$-$	$+$	$+$	$+$
Sign of $\dfrac{2x+3}{(x+1)(x+2)}$	$-$	$+$	$-$	$+$

From the table, the solution set is $\left\{x| \ x < -2 \text{ or } -\frac{3}{2} \le x < -1\right\}$. The points $x = -2$ and $x = -1$ are excluded from the solution because the expression is undefined at those values.

Solution: $(-\infty, -2) \cup \left[-\frac{3}{2}, -1\right)$.   Graph:

66. $x^5 > x^2 \ \Leftrightarrow \ x^5 - x^2 > 0 \ \Leftrightarrow \ x^2(x^3 - 1) > 0 \ \Leftrightarrow \ x^2(x-1)(x^2 + x + 1) > 0$. The expression on the left of the inequality changes sign when $x = 0$ and $x = 1$. But the solution of $x^2 + x + 1 = 0$ are $x = \frac{-1 \pm \sqrt{(1)^2 - 4(1)(1)}}{2(1)} = \frac{-1 \pm \sqrt{-3}}{2}$. Since these are not real solutions. The expression $x^2 + x + 1$ does not changes signs, so we must check the intervals in the following table.

Interval	$(-\infty, 0)$	$(0, 1)$	$(1, \infty)$
Sign of x^2	$+$	$+$	$+$
Sign of $x - 1$	$-$	$-$	$+$
Sign of $x^2 + x + 1$	$+$	$+$	$+$
Sign of $x^2(x-1)(x^2 + x + 1)$	$-$	$-$	$+$

From the table, the solution set is $\{x| \ 1 < x\}$.

Solution: $(1, \infty)$.   Graph:

68. For $\sqrt{3x^2 - 5x + 2}$ to be defined as a real number, we must have $3x^2 - 5x + 2 \ge 0 \ \Leftrightarrow \ (3x - 2)(x - 1) \ge 0$. The expression on the left of the inequality changes sign when $x = \frac{2}{3}$ and $x = 1$. Thus we must check the intervals in the following table.

Interval	$\left(-\infty, \frac{2}{3}\right)$	$\left(\frac{2}{3}, 1\right)$	$(1, \infty)$
Sign of $3x - 2$	$-$	$+$	$+$
Sign of $x - 1$	$-$	$-$	$+$
Sign of $(3x - 2)(x - 1)$	$+$	$-$	$+$

Thus $x \le \frac{2}{3}$ or $1 \le x$.

70. For $\sqrt[4]{\dfrac{1-x}{2+x}}$ to be defined as a real number we must have $\dfrac{1-x}{2+x} \geq 0$. The expression on the left of the inequality changes sign when $x = 1$ and $x = -2$. Thus we must check the intervals in the following table.

Interval	$(-\infty, -2)$	$(-2, 1)$	$(1, \infty)$
Sign of $1 - x$	$+$	$+$	$-$
Sign of $2 + x$	$-$	$+$	$+$
Sign of $\dfrac{1-x}{2+x}$	$-$	$+$	$-$

Thus $-2 < x \leq 1$. Note that $x = -2$ has been excluded from the solution set because the expression is undefined at that value.

72. We are given $\dfrac{a}{b} < \dfrac{c}{d}$, where a, b, c, and d are all positive. So multiplying by d, we get $\dfrac{ad}{b} < c$ and so by Rule 1 for Inequalities, $a + \dfrac{ad}{b} < a + c \quad \Leftrightarrow \quad \dfrac{ab + ad}{b} < a + c \quad \Leftrightarrow \quad \dfrac{a}{b}(b + d) < a + c$

$\Leftrightarrow \quad \dfrac{a}{b} < \dfrac{a+c}{b+d}$, since $b + d$ is positive. Similarly, $\dfrac{a}{b} < \dfrac{c}{d} \quad \Leftrightarrow \quad a < \dfrac{bc}{d} \quad \Leftrightarrow$

$a + c < \dfrac{bc}{d} + c \quad \Leftrightarrow \quad a + c < \dfrac{bc}{d} + \dfrac{dc}{d} \quad \Leftrightarrow \quad a + c < \dfrac{c}{d}(b + d) \quad \Leftrightarrow \quad \dfrac{a+c}{b+d} < \dfrac{c}{d}$, since

$b + d > 0$. Combining the two inequalities, $\dfrac{a}{b} < \dfrac{a+c}{b+d} < \dfrac{c}{d}$.

74. Inserting the relationship $F = \frac{9}{5}C + 32$, we have $50 \leq F \leq 95 \quad \Leftrightarrow \quad 50 \leq \frac{9}{5}C + 32 \leq 95 \quad \Leftrightarrow$
$18 \leq \frac{9}{5}C \leq 63 \quad \Leftrightarrow \quad 10 \leq C \leq 35$.

76. Let m be the number of minutes of long-distance calls placed per month. Then under Plan A, the cost will be $25 + 0.05m$, and under Plan B, the cost will be $5 + 0.12m$. To determine when Plan B is advantageous, we must solve $25 + 0.05m > 5 + 0.12m \quad \Leftrightarrow \quad 20 > 0.07m \quad \Leftrightarrow \quad 285.7 > m$. So Plan B is advantageous if a person places less than 286 minutes of long-distance calls during a month.

78. (a) $T = 20 - \frac{h}{100}$, where T is the temperature in °C, and h is the height in meters.

(b) Solving the expression in part (a) for h, we get $h = 100(20 - T)$. So $0 \leq h \leq 5000 \quad \Leftrightarrow$
$0 \leq 100(20 - T) \leq 5000 \quad \Leftrightarrow \quad 0 \leq 20 - T \leq 50 \quad \Leftrightarrow \quad -20 \leq -T \leq 30 \quad \Leftrightarrow$
$20 \geq T \geq -30$. Thus the range of temperature is from 20°C down to -30°C.

80. If the customer buys x pounds of coffee at \$6.50 per pound, then his cost c will be $6.50x$. Thus $x = \frac{c}{6.5}$. Since the scale's accuracy is ± 0.03 lb, and the scale shows 3 lb, we have
$3 - 0.03 \leq x \leq 3 + 0.03 \quad \Leftrightarrow \quad 2.97 \leq \frac{c}{6.5} \leq 3.03 \quad \Leftrightarrow \quad (6.50)2.97 \leq c \leq (6.50)3.03 \quad \Leftrightarrow$
$19.305 \leq c \leq 19.695$. Since the customer paid \$19.50, he could have been over- or undercharged by as much as $19\frac{1}{2}$¢.

82. $\dfrac{600{,}000}{x^2 + 300} < 500 \quad \Leftrightarrow \quad 600{,}000 < 500(x^2 + 300)$ (Note that $x^2 + 300 \geq 300 > 0$, so we can multiply both sides by the denominator and not worry that we might be multiplying both sides by a negative number or by zero.) $1200 < x^2 + 300 \quad \Leftrightarrow \quad 0 < x^2 - 900 \quad \Leftrightarrow$
$0 < (x - 30)(x + 30)$. The expression in the inequality changes sign at $x = 30$ and $x = -30$.

However, since x represents distance, we must have $x > 0$.

Interval	$(0, 30)$	$(30, \infty)$
Sign of $x - 30$	$-$	$+$
Sign of $x + 30$	$+$	$+$
Sign of $(x - 30)(x + 30)$	$-$	$+$

So $x > 30$ and you must stand at least 30 meters from the center of the fire.

84.   Solve $30 \le 10 + 0.9v - 0.01v^2$ for $10 \le v \le 75$. We have $30 \le 10 + 0.9v - 0.01v^2$ $\Leftrightarrow$
    $0.01v^2 - 0.9v + 20 \le 0$ $\Leftrightarrow$ $(0.1v - 4)(0.1v - 5) \le 0$. The possible endpoints are
    $0.1v - 4 = 0$ $\Leftrightarrow$ $0.1v = 4$ $\Leftrightarrow$ $v = 40$ and $0.1v - 5 = 0$ $\Leftrightarrow$ $0.1v = 5$ $\Leftrightarrow$ $v = 50$.

Interval	$(10, 40)$	$(40,\ 50)$	$(50, 75)$
Sign of $0.1v - 4$	$-$	$+$	$+$
Sign of $0.1v - 5$	$-$	$-$	$+$
Sign of $(0.1v - 4)(0.1v - 5)$	$+$	$-$	$+$

Thus he must drive between 40 and 50 mi/h.

86.   Solve $2400 \le 20x - (2000 + 8x + 0.0025x^2)$ $\Leftrightarrow$ $2400 \le 20x - 2000 - 8x - 0.0025x^2$ $\Leftrightarrow$
    $0.0025x^2 - 12x + 4400 \le 0$ $\Leftrightarrow$ $(0.0025x - 1)(x - 4400) \le 0$. The expression on the left of
    the inequality changes sign when $x = 400$ and $x = 4400$. Since the manufacturer can only sell
    positive units, we check the intervals in the following table.

Interval	$(0, 400)$	$(400,\ 4400)$	$(4400, \infty)$
Sign of $0.0025x - 1$	$-$	$+$	$+$
Sign of $x - 4400$	$-$	$-$	$+$
Sign of $(0.0025x - 1)(x - 4400)$	$+$	$-$	$+$

So the manufacturer must sell between 400 and 4400 units to enjoy a profit of at least \$2400.

88.   Let x be the length of the garden. Using the fact that the perimeter is 120 ft, we must have
    $2x + 2width = 120$ $\Leftrightarrow$ $width = 60 - x$. Now since the area must be at least 800 ft^2, we have
    $800 < x(60 - x)$ $\Leftrightarrow$ $800 < 60x - x^2$ $\Leftrightarrow$ $x^2 - 60x + 800 < 0$ $\Leftrightarrow$
    $(x - 20)(x - 40) < 0$. The expression in the inequality changes sign at $x = 20$ and $x = 40$.
    However, since x represents length, we must have $x > 0$.

Interval	$(0, 20)$	$(20,\ 40)$	$(40, \infty)$
Sign of $x - 20$	$-$	$+$	$+$
Sign of $x - 40$	$-$	$-$	$+$
Sign of $(x - 20)(x - 40)$	$+$	$-$	$+$

The length of the garden should be between 20 and 40 feet.

90.   The rule we want to apply here is "$a < b \Rightarrow ac < bc$ if $c > 0$ and $a < b \Rightarrow ac > bc$ if $c < 0$". Thus
    we cannot simply multiply by x, since we don't yet know if x is positive or negative, so in solving
    $1 < \dfrac{3}{x}$, we must consider two cases

    Case 1: $x > 0$ Then multiplying all sides by x, we have $x < 3$. Together with our initial condition,
    we have $0 < x < 3$.

    Case 2: $x < 0$ Then multiplying all sides by x, we have $x > 3$. But $x < 0$ and $x > 3$ have no
    elements in common, so this gives no additional solutions.

    Hence, the only solutions are $0 < x < 3$.

Exercises 1.7

2. $|6x| = 15 \quad \Leftrightarrow \quad 6x = \pm 15 \quad \Leftrightarrow \quad x = \pm\frac{5}{2}$.

4. $\frac{1}{2}|x| - 7 = 2 \quad \Leftrightarrow \quad \frac{1}{2}|x| = 9 \quad \Leftrightarrow \quad |x| = 18 \quad \Leftrightarrow \quad x = \pm 18$.

6. $|2x - 3| = 7$ is equivalent to either $2x - 3 = 7 \quad \Leftrightarrow \quad 2x = 10 \quad \Leftrightarrow \quad x = 5$; or $2x - 3 = -7$
$\Leftrightarrow \quad 2x = -4 \quad \Leftrightarrow \quad x = -2$. The two solutions are $x = 5$ and $x = -2$.

8. $|x + 4| = -3$. Since the absolute value is always nonnegative, there is no solution.

10. $\left|\frac{1}{2}x - 2\right| = 1$ is equivalent to either $\frac{1}{2}x - 2 = 1 \quad \Leftrightarrow \quad \frac{1}{2}x = 3 \quad \Leftrightarrow \quad x = 6$; or $\frac{1}{2}x - 2 = -1$
$\Leftrightarrow \quad \frac{1}{2}x = 1 \quad \Leftrightarrow \quad x = 2$. The two solutions are $x = 6$ and $x = 2$.

12. $|5 - 2x| + 6 = 14 \quad \Leftrightarrow \quad |5 - 2x| = 8$ which is equivalent to either $5 - 2x = 8 \quad \Leftrightarrow \quad -2x = 3$
$\Leftrightarrow \quad x = -\frac{3}{2}$; or $5 - 2x = -8 \quad \Leftrightarrow \quad -2x = -13 \quad \Leftrightarrow \quad x = \frac{13}{2}$. The two solutions are
$x = -\frac{3}{2}$ and $x = \frac{13}{2}$.

14. $20 + |2x - 4| = 15 \quad \Leftrightarrow \quad |2x - 4| = -5$. Since the absolute value is always nonnegative, there is
no solution.

16. $\left|\frac{3}{5}x + 2\right| - \frac{1}{2} = 4 \quad \Leftrightarrow \quad \left|\frac{3}{5}x + 2\right| = \frac{9}{2}$ which is equivalent to either $\frac{3}{5}x + 2 = \frac{9}{2} \quad \Leftrightarrow \quad \frac{3}{5}x = \frac{5}{2}$
$\Leftrightarrow \quad x = \frac{25}{6}$; or $\frac{3}{5}x + 2 = -\frac{9}{2} \quad \Leftrightarrow \quad \frac{3}{5}x = -\frac{13}{2} \quad \Leftrightarrow \quad x = -\frac{65}{6}$. The two solutions are $x = \frac{25}{6}$
and $x = -\frac{65}{6}$.

18. $|x + 3| = |2x + 1|$ is equivalent to either $x + 3 = 2x + 1 \quad \Leftrightarrow \quad -x = -2 \quad \Leftrightarrow \quad x = 2$ or to
$x + 3 = -(2x + 1) \quad \Leftrightarrow \quad x + 3 = -2x - 1 \quad \Leftrightarrow \quad 3x = -4 \quad \Leftrightarrow \quad x = -\frac{4}{3}$. The two
solutions are $x = 2$ and $x = -\frac{4}{3}$.

20. $|3x| < 15 \quad \Leftrightarrow \quad -15 < 3x < 15 \quad \Leftrightarrow \quad -5 < x < 5$. Interval: $(-5, 5)$.

22. $\frac{1}{2}|x| \geq 1 \quad \Leftrightarrow \quad |x| \geq 2$ is equivalent to $x \geq 2$ or $x \leq -2$. Interval: $(-\infty, -2] \cup [2, \infty)$.

24. $|x - 9| > 9$ is equivalent to $x - 9 > 9 \quad \Leftrightarrow \quad x > 18$, or $x - 9 < -9 \quad \Leftrightarrow \quad x < 0$.
Interval: $(-\infty, 0) \cup (18, \infty)$.

26. $|x + 4| \leq 0$ is equivalent to $|x + 4| = 0 \quad \Leftrightarrow \quad x + 4 = 0 \quad \Leftrightarrow \quad x = -4$. The solution is just the
point $x = -4$.

28. $|x + 1| \geq 3$ is equivalent either to $x + 1 \geq 3 \quad \Leftrightarrow \quad x \geq 2$, or to $x + 1 \leq -3 \quad \Leftrightarrow \quad x \leq -4$.
Interval: $(-\infty, -4] \cup [2, \infty)$.

30. $|5x - 2| < 6 \quad \Leftrightarrow \quad -6 < 5x - 2 < 6 \quad \Leftrightarrow \quad -4 < 5x < 8 \quad \Leftrightarrow \quad -\frac{4}{5} < x < \frac{8}{5}$.
Interval: $\left(-\frac{4}{5}, \frac{8}{5}\right)$.

32. $\left|\dfrac{x+1}{2}\right| \geq 4 \iff \left|\dfrac{1}{2}(x+1)\right| \geq 4 \iff \dfrac{1}{2}|x+1| \geq 4 \iff |x+1| \geq 8$ which is equivalent either to $x+1 \geq 8 \iff x \geq 7$, or to $x+1 \leq -8 \iff x \leq -9$.

Interval: $(-\infty, -9] \cup [7, \infty)$.

34. $|x-a| < d \iff -d < x-a < d \iff a-d < x < a+d$. Interval: $(a-d,\, a+d)$.

36. $3 - |2x+4| \leq 1 \iff -|2x+4| \leq -2 \iff |2x+4| \geq 2$ which is equivalent to either $2x+4 \geq 2 \iff 2x \geq -2 \iff x \geq -1$; or $2x+4 \leq -2 \iff 2x \leq -6 \iff x \leq -3$. Interval: $(-\infty, -3] \cup [-1, \infty)$.

38. $7|x+2| + 5 > 4 \iff 7|x+2| > -1 \iff |x+2| > -\frac{1}{7}$. Since the absolute value is always nonnegative, the inequality is true for all real numbers. In interval notation, we have $(-\infty, \infty)$.

40. $2\left|\frac{1}{2}x + 3\right| + 3 \leq 51 \iff 2\left|\frac{1}{2}x + 3\right| \leq 48 \iff \left|\frac{1}{2}x + 3\right| \leq 24 \iff -24 < \frac{1}{2}x + 3 < 24$ $\iff -27 < \frac{1}{2}x < 21 \iff -54 < x < 42$. Interval: $(-54, 42)$.

42. $0 < |x - 5| \leq \frac{1}{2}$. For $x \neq 5$, this is equivalent to $-\frac{1}{2} \leq x - 5 \leq \frac{1}{2} \iff \frac{9}{2} \leq x \leq \frac{11}{2}$. Since $x = 5$ is excluded, the solution is $\left[\frac{9}{2}, 5\right) \cup \left(5, \frac{11}{2}\right]$.

44. $\dfrac{1}{|2x-3|} \leq 5 \iff \frac{1}{5} \leq |2x - 3|$, since $|2x-3| > 0$, provided $2x - 3 \neq 0 \iff x \neq \frac{3}{2}$. Now for $x \neq \frac{3}{2}$, we have $\frac{1}{5} \leq |2x - 3|$ is equivalent to $\frac{1}{5} \leq 2x - 3 \iff \frac{16}{5} \leq 2x \iff \frac{8}{5} \leq x$; or $2x - 3 \leq -\frac{1}{5} \iff 2x \leq \frac{14}{5} \iff x \leq \frac{7}{5}$. Interval: $\left(-\infty, \frac{7}{5}\right] \cup \left[\frac{8}{5}, \infty\right)$.

46. $|x| > 2$.

48. $|x - 2| \leq 4$

50. $|x| \geq 1$.

52. $|x| < 4$.

54. $\left|\dfrac{h - 68.2}{2.9}\right| \leq 2 \iff -2 \leq \dfrac{h - 68.2}{2.9} \leq 2 \iff -5.8 \leq h - 68.2 \leq 5.8 \iff$ $62.4 \leq h \leq 74.0$. Thus 95% of the adult males are between 62.4 in and 74.0 in.

Review Exercises for Chapter 1

2. $3 - x = 5 + 3x$ $\Leftrightarrow$ $-2 = 4x$ $\Leftrightarrow$ $x = -\frac{1}{2}$.

4. $5x - 7 = 42$ $\Leftrightarrow$ $5x = 49$ $\Leftrightarrow$ $x = \frac{49}{5}$.

6. $8 - 2x = 14 + x$ $\Leftrightarrow$ $-3x = 6$ $\Leftrightarrow$ $x = -2$.

8. $\frac{2}{3}x + \frac{3}{5} = \frac{1}{5} - 2x$ $\Leftrightarrow$ $10x + 9 = 3 - 30x$ $\Leftrightarrow$ $40x = -6$ $\Leftrightarrow$ $x = -\frac{6}{40} = -\frac{3}{20}$.

10. $\dfrac{x - 5}{2} - \dfrac{2x + 5}{3} = \dfrac{5}{6}$ $\Leftrightarrow$ $3(x - 5) - 2(2x + 5) = 5$ $\Leftrightarrow$ $3x - 15 - 4x - 10 = 5$ $\Leftrightarrow$
$-x = 30$ $\Leftrightarrow$ $x = -30$.

12. $\dfrac{x}{x + 2} - 3 = \dfrac{1}{x + 2}$ $\Leftrightarrow$ $x - 3(x + 2) = 1$ $\Leftrightarrow$ $x - 3x - 6 = 1$ $\Leftrightarrow$ $-2x = 7$ $\Leftrightarrow$
$x = -\frac{7}{2}$.

14. $4x^2 = 49$ $\Leftrightarrow$ $x^2 = \frac{49}{4}$ $\Rightarrow$ $x = \pm\frac{7}{2}$

16. $x^3 - 27 = 0$ $\Leftrightarrow$ $x^3 = 27$ $\Rightarrow$ $x = 3$.

18. $6x^4 + 15 = 0$ $\Leftrightarrow$ $6x^4 = -15$ $\Leftrightarrow$ $x^4 = -\frac{5}{2}$. Since x^4 must be nonnegative, there is no real solution.

20. $(x + 2)^2 - 2 = 0$ $\Leftrightarrow$ $(x + 2)^2 = 2$ $\Leftrightarrow$ $x + 2 = \pm\sqrt{2}$ $\Leftrightarrow$ $x = -2 \pm \sqrt{2}$.

22. $x^{2/3} - 4 = 0$ $\Leftrightarrow$ $\left(x^{1/3}\right)^2 = 4$ $\Rightarrow$ $x^{1/3} = \pm 2$ $\Leftrightarrow$ $x = \pm 8$.

24. $(x - 2)^{1/5} = 2$ $\Leftrightarrow$ $x - 2 = 2^5 = 32$ $\Leftrightarrow$ $x = 2 + 32 = 34$.

26. $(x + 2)^2 = (x - 4)^2$ $\Leftrightarrow$ $(x + 2)^2 - (x - 4)^2 = 0$ $\Leftrightarrow$
$[(x + 2) - (x - 4)][(x + 2) + (x - 4)] = 0$ $\Leftrightarrow$
$[x + 2 - x + 4][x + 2 + x - 4] = 6(2x - 2) = 0$ $\Leftrightarrow$ $2x - 2 = 0$ $\Leftrightarrow$ $x = 1$.

28. $x^2 + 24x + 144 = 0$ $\Leftrightarrow$ $(x + 12)^2 = 0$ $\Leftrightarrow$ $x + 12 = 0$ $\Leftrightarrow$ $x = -12$.

30. $3x^2 + 5x - 2 = 0$ $\Leftrightarrow$ $(3x - 1)(x + 2) = 0$ $\Leftrightarrow$ $x = \frac{1}{3}$ or $x = -2$.

32. $x^3 - 2x^2 - 5x + 10 = 0$ $\Leftrightarrow$ $x^2(x - 2) - 5(x - 2) = 0$ $\Leftrightarrow$ $(x - 2)(x^2 - 5) = 0$ $\Leftrightarrow$
$x = 2$ or $x = \pm\sqrt{5}$.

34. $x^2 - 3x + 9 = 0$ $\Rightarrow$ $x = \dfrac{-b \pm \sqrt{b^2 - 4ac}}{2a} = \dfrac{-(-3) \pm \sqrt{(-3)^2 - 4(1)(9)}}{2(1)} = \dfrac{3 \pm \sqrt{9 - 36}}{2}$
$= \dfrac{3 \pm \sqrt{-27}}{2}$, which are not real numbers. There are no real solutions.

36. $\dfrac{x}{x-2} + \dfrac{1}{x+2} = \dfrac{8}{x^2-4}$ $\Leftrightarrow$ $x(x+2) + (x-2) = 8$ $\Leftrightarrow$ $x^2 + 2x + x - 2 = 8$ $\Leftrightarrow$

     $x^2 + 3x - 10 = 0$ $\Leftrightarrow$ $(x-2)(x+5) = 0$ $\Leftrightarrow$ $x = 2$ or $x = -5$. However, since $x = 2$

     makes the expression undefined, we reject this solution. Hence the only solution is $x = -5$.

38. $x - 4\sqrt{x} = 32$. Let $u = \sqrt{x}$. Then $u^2 - 4u = 32$ $\Leftrightarrow$ $u^2 - 4u - 32 = 0$ $\Leftrightarrow$

     $(u-8)(u+4) = 0$. So $u - 8 = 0$ $\Rightarrow$ $\sqrt{x} - 8 = 0$ $\Leftrightarrow$ $\sqrt{x} = 8$ $\Rightarrow$ $x = 64$, or

     $u + 4 = 0$ $\Rightarrow$ $\sqrt{x} + 4 = 0$ $\Leftrightarrow$ $\sqrt{x} = -4$, which has no real solution. The only solution is

     $x = 64$.

40. $(1 + \sqrt{x})^2 - 2(1 + \sqrt{x}) - 15 = 0$. Let $u = 1 + \sqrt{x}$, then the equation becomes

     $u^2 - 2u - 15 = 0$ $\Leftrightarrow$ $(u-5)(u+3) = 0$. So $u - 5 = 1 + \sqrt{x} - 5 = 0$ $\Leftrightarrow$ $\sqrt{x} = 4$

     $\Rightarrow$ $x = 16$; or $u + 3 = 1 + \sqrt{x} + 3 = 0$ $\Leftrightarrow$ $\sqrt{x} = -4$, which has no real solution. The

     only solution is $x = 16$.

42. $|3x| = 18$ is equivalent to $3x = \pm 18$ $\Leftrightarrow$ $x = \pm 6$.

44. $4|3-x| + 3 = 15$ $\Leftrightarrow$ $4|3-x| = 12$ $\Leftrightarrow$ $|3-x| = 3$ $\Leftrightarrow$ $3 - x = \pm 3$. So $3 - x = 3$

     $\Leftrightarrow$ $-x = 0$ $\Leftrightarrow$ $x = 0$; or $3 - x = -3$ $\Leftrightarrow$ $-x = -6$ $\Leftrightarrow$ $x = 6$.   The solutions are

     $x = 0$ and $x = 6$.

46.   Let t be the number of hours that Anthony drives. Then Helen drives for $t - \frac{1}{4}$ hours.

	Rate	Time	Distance
Anthony	45	t	$45t$
Helen	40	$t - \dfrac{1}{4}$	$40\left(t - \dfrac{1}{4}\right)$

     When they pass each other, they will have traveled a total of 160 miles. So $45t + 40\left(t - \frac{1}{4}\right) = 160$

     $\Leftrightarrow$ $45t + 40t - 10 = 160$ $\Leftrightarrow$ $85t = 170$ $\Leftrightarrow$ $t = 2$. Since Anthony leaves at 2:00 P.M.

     and travels for 2 hours, they pass each other at 4:00 P.M.

48.   Substituting 75 for d, we have $75 = x + \dfrac{x^2}{20}$ $\Leftrightarrow$ $1500 = 20x + x^2$ $\Leftrightarrow$

     $x^2 + 20x - 1500 = 0$ $\Leftrightarrow$ $(x-30)(x+50) = 0$. So $x = 30$ or $x = -50$. Since x is the speed

     of the car, $x > 0$, the speed of the car must have been 30 mi/h.

50.   Let t be the time it would take Abbie to paint a living room if she works alone. It would take Beth

     $2t$ hours to paint the living room alone, and it would take $3t$ hours for Cathie to paint the living

     room. Thus Abbie does $\dfrac{1}{t}$ of the job per hour, Beth does $\dfrac{1}{2t}$ of the job per hour, and Cathie does $\dfrac{1}{3t}$

     of the job per hour. So $\dfrac{1}{t} + \dfrac{1}{2t} + \dfrac{1}{3t} = 1$ $\Leftrightarrow$ $6 + 3 + 2 = 6t$ $\Leftrightarrow$ $6t = 11$ $\Leftrightarrow$ $t = \dfrac{11}{6}$.

     So it would take Abbie 1 hour and 50 minutes to paint the living room alone.

52.   Let l be length of each garden plot. The width of each plot is then $\dfrac{80}{l}$ and the total amount of

     fencing material is $4(l) + 6\left(\dfrac{80}{l}\right) = 88$. Thus $4l + \dfrac{480}{l} = 88$ $\Leftrightarrow$ $4l^2 + 480 = 88l$ $\Leftrightarrow$

     $4l^2 - 88l + 480 = 0$ $\Leftrightarrow$ $4(l^2 - 22l + 120) = 0$ $\Leftrightarrow$ $4(l-10)(l-12) = 0$. So $l = 10$ or

$l = 12$. If $l = 10$ feet, then the width of each plot is $\frac{80}{10} = 8$ feet. If $l = 12$ feet, then the width of each plot is $\frac{80}{12} = 6.67$ feet. Both solutions are possible.

54. $(-2 + 3\,i) + (\frac{1}{2} - i) = -2 + 3\,i + \frac{1}{2} - i = -\frac{3}{2} + 2\,i.$

56. $3(5 - 2\,i)\dfrac{i}{5} = (15 - 6\,i)\dfrac{i}{5} = 3i - \dfrac{6}{5}i^2 = \dfrac{6}{5} + 3\,i.$

58. $\dfrac{2 + i}{4 - 3\,i} = \dfrac{2 + i}{4 - 3\,i} \cdot \dfrac{4 + 3\,i}{4 + 3\,i} = \dfrac{8 + 6\,i + 4\,i + 3\,i^2}{16 - 9\,i^2} = \dfrac{8 + 10\,i - 3}{16 + 9} = \dfrac{5 + 10\,i}{25} = \dfrac{1}{5} + \dfrac{2}{5}\,i.$

60. $(3 - i)^3 = (3)^3 - 3(3)^2(i) + 3(3)(i)^2 - (i)^3 = 27 - 27\,i + 9i^2 - i^3 = 27 - 27\,i - 9 + i$
     $= 18 - 26\,i.$

62. $\sqrt{-5} \cdot \sqrt{-20} = \sqrt{5}\,i \cdot 2\sqrt{5}\,i = 10\,i^2 = -10.$

64. $x^2 = -12 \quad \Rightarrow \quad x = \pm\sqrt{-12} = \pm 2\sqrt{3}\,i.$

66. $2x^2 - 3x + 2 = 0 \quad \Rightarrow \quad x = \dfrac{-(-3) \pm \sqrt{(-3)^2 - 4(2)(2)}}{2(2)} = \dfrac{3 \pm \sqrt{9 - 16}}{4} = \dfrac{3 \pm \sqrt{-7}}{4}$
     $= \dfrac{3 \pm \sqrt{7}\,i}{4}.$

68. $x^3 - 2x^2 + 4x - 8 = 0 \quad \Leftrightarrow \quad x^2(x - 2) + 4(x - 2) = 0 \quad \Leftrightarrow \quad (x - 2)(x^2 + 4) = 0 \quad \Leftrightarrow$
     $x = 2$, or $x^2 + 4 = 0 \quad \Leftrightarrow \quad x^2 = -4 \quad \Rightarrow \quad x = \pm\sqrt{-4} = \pm 2i$. The solutions are 2 and $\pm 2i$.

70. $x^3 = 125 \quad \Leftrightarrow \quad x^3 - 125 = 0 \quad \Leftrightarrow \quad (x - 5)(x^2 + 5x + 25) = 0.$ So $x = 5$, or
     $x = \dfrac{-5 \pm \sqrt{5^2 - 4(1)(25)}}{2(1)} = \dfrac{-5 \pm \sqrt{25 - 100}}{2} = \dfrac{-5 \pm \sqrt{-75}}{2} = \dfrac{-5 \pm 5\sqrt{3}\,i}{2}.$
     The solutions are 5 and $\dfrac{-5 \pm 5\sqrt{3}\,i}{2}.$

72. $12 - x \geq 7x \quad \Leftrightarrow \quad 12 \geq 8x \quad \Leftrightarrow \quad \frac{3}{2} \geq x.$ Interval: $\left(-\infty, \frac{3}{2}\right]$

Graph:

74. $3 - x \leq 2x - 7 \quad \Leftrightarrow \quad 10 \leq 3x \quad \Leftrightarrow \quad \frac{10}{3} \leq x.$ Interval: $\left[\frac{10}{3}, \infty\right)$

Graph:

76. $x^2 \leq 1 \quad \Leftrightarrow \quad x^2 - 1 \leq 0 \quad \Leftrightarrow \quad (x - 1)(x + 1) \leq 0.$ The expression on the left of the inequality changes sign when $x = -1$ and $x = 1$. Thus we must check the intervals in the following table.

Interval	$(-\infty, -1)$	$(-1,\, 1)$	$(1,\, \infty)$
Sign of $x - 1$	$-$	$-$	$+$
Sign of $x + 1$	$-$	$+$	$+$
Sign of $(x - 1)(x + 1)$	$+$	$-$	$+$

Interval: $[-1, 1]$ Graph:

78. $2x^2 \geq x + 3 \iff 2x^2 - x - 3 \geq 0 \iff (2x - 3)(x + 1) \geq 0.$ The expression on the left of the inequality changes sign when -1 and $\frac{3}{2}$. Thus we must check the intervals in the following table.

Interval	$(-\infty, -1)$	$\left(-1, \frac{3}{2}\right)$	$\left(\frac{3}{2}, \infty\right)$
Sign of $2x - 3$	$-$	$-$	$+$
Sign of $x + 1$	$-$	$+$	$+$
Sign of $(2x - 3)(x + 1)$	$+$	$-$	$+$

Interval: $(-\infty, -1] \cup \left[\frac{3}{2}, \infty\right)$ Graph:

80. $\dfrac{5}{x^3 - x^2 - 4x + 4} < 0 \iff \dfrac{5}{x^2(x - 1) - 4(x - 1)} < 0 \iff \dfrac{5}{(x - 1)(x^2 - 4)} < 0 \iff$

$\dfrac{5}{(x - 1)(x - 2)(x + 2)} < 0.$ The expression on the left of the inequality changes sign when $-2, 1,$ and 2. Thus we must check the intervals in the following table.

Interval	$(-\infty, -2)$	$(-2, 1)$	$(1, 2)$	$(2, \infty)$
Sign of $x - 1$	$-$	$-$	$+$	$+$
Sign of $x - 2$	$-$	$-$	$-$	$+$
Sign of $x + 2$	$-$	$+$	$+$	$+$
Sign of $\dfrac{5}{(x - 1)(x - 2)(x + 2)}$	$-$	$+$	$-$	$+$

Interval: $(-\infty, -2) \cup (1, 2)$ Graph:

82. $|x - 4| < 0.02 \iff -0.02 < x - 4 < 0.02 \iff 3.98 < x < 4.02.$ Interval: $(3.98, 4.02)$.
Graph:

84. $|x - 1|$ is the distance between x and 1 on the number line, and $|x - 3|$ is the distance between x and 3. We want those points that are closer to 1 than to 3. Since 2 is midway between 1 and 3, we get $x \in (-\infty, 2)$ as the solution.   Graph:

86. We have $8 \leq \frac{4}{3}\pi r^3 \leq 12 \iff \dfrac{6}{\pi} \leq r^3 \leq \dfrac{9}{\pi} \iff \sqrt[3]{\dfrac{6}{\pi}} \leq r \leq \sqrt[3]{\dfrac{9}{\pi}}.$ Thus $r \in \left[\sqrt[3]{\dfrac{6}{\pi}}, \sqrt[3]{\dfrac{9}{\pi}}\right].$

Focus on Modeling

2. (a) Plan 1: $Cost = 3 \cdot \left(\dfrac{daily}{cost}\right) + \left(\dfrac{cost\ per}{mile}\right) \times miles = 3 \cdot 65 + 0.15x = 195 + 0.15x$.

Plan 2: $Cost = 3 \cdot \left(\dfrac{daily}{cost}\right) = 3 \cdot 90 = 270$.

(b) When $x = 400$, Plan 1 cost $195 + 0.15(400) = \$255$ and Plan 2 cost \$270, so Plan 1 is cheaper.

When $x = 800$, Plan 1 cost $195 + 0.15(800) = \$315$ and Plan 2 cost \$270, so Plan 2 is cheaper.

(c) The cost will be the same when $195 + 0.15x = 270 \quad \Leftrightarrow \quad 0.15 = 75x \quad \Leftrightarrow \quad x = 500$. So both plans cost \$270 when the businessman drives 500 miles.

4. (a) Option 1: In this option the *width* does not change, thus *width* $= 100$. Let $x =$ the increase in the length. Then the additional area is given by the model

$Area = width \times \left(\dfrac{increase}{length}\right) = 100x$. The $cost = \left(\dfrac{cost\ of\ moving}{the\ old\ fence}\right) + \left(\dfrac{cost\ of\ installing}{new\ fence}\right)$.

The cost of moving the width is $\$6 \cdot 100 = \$600$ and the cost of installing the new fencing is $2 \cdot 10 \cdot x = 20x$, so the cost is $C = 20x + 600$. Solving for x we get $C = 20x + 600 \quad \Leftrightarrow$

$20x = C - 600 \quad \Leftrightarrow \quad x = \dfrac{C - 600}{20}$. Substituting in the area we have

$A_1 = 100\left(\dfrac{C - 600}{20}\right) = 5(C - 600) = 5C - 3{,}000$.

Option 2: In this option the *length* does not change, thus *length* $= 180$. Let $y =$ the increase in the width. Then the additional area is given by the model

$Area = length \times \left(\dfrac{increase}{width}\right) = 180y$. The $cost = \left(\dfrac{cost\ of\ moving}{the\ old\ fence}\right) + \left(\dfrac{cost\ of\ installing}{new\ fence}\right)$.

The cost of moving the width is $\$6 \cdot 180 = \$1080$ and the cost of installing the new fencing is $2 \cdot 10 \cdot y = 20x$, so the cost is $C = 20y + 1080$. Solving for x we get $C = 20y + 1080 \quad \Leftrightarrow$

$20y = C - 1080 \quad \Leftrightarrow \quad x = \dfrac{C - 1080}{20}$. Substituting in the area we have

$A_2 = 180\left(\dfrac{C - 1080}{20}\right) = 9(C - 1080) = 9C - 9{,}720$.

(b)

Cost C	Area gain (Option 1) A_1	Area gain (Option 2) A_2
\$1100	2,500 ft^2	180 ft^2
\$1200	3,000 ft^2	1,080 ft^2
\$1500	4,500 ft^2	3,780 ft^2
\$2000	7,000 ft^2	8,280 ft^2
\$2500	9,500 ft^2	12,780 ft^2
\$3000	12,000 ft^2	17,280 ft^2

(c) When the farmer has only $1200 Option 1 gives him the greatest gain. When the farmer has only $2000 Option 2 gives him the greatest gain.

6. (a) Plan 1: Tomatoes every year.
   $Profit = acres \times (Revenue - cost) = 100(1600 - 300) = 130{,}000$. Then for n years the profit is $P_1(n) = 130{,}000n$.

   (b) Plan 2: Soybeans followed by tomatoes. So the profit for two years is
   $$profit = acres \times \left[\left(\begin{matrix} revenue\ from \\ soybeans \end{matrix} \right) + \left(\begin{matrix} revenue\ from \\ tomatoes \end{matrix} \right) \right] = 100(1200 + 1600) = 280{,}000.$$
   (Remember *no* fertilizer *is* needed in this plan.) Then for $2k$ years the profit is $P_2(2k) = 280{,}000k$.

   (c) $P_1(10) = 130{,}000(10) = 1{,}300{,}000$. Since $2k = 10$ when $k = 5$. Thus $P_2(10) = 280{,}000(5) = 1{,}400{,}000$. So Plan B is more profitable.

8. (a) In this plan, Company A get $3.2 million and Company B get $3.2 million. Company A's investment was $1.4 million so they make a profit of $3.2 - 1.4 = $1.8 million. Company B's investment was $2.6 million so they make a profit of $3.2 - 2.6 = $0.6 million. So Company A makes three times the profit that Company B does.

   (b) The original investment is $1.4 + 2.6 = $4 million. So after giving the original investment back, they then share the profit of $2.4 million. So each gets an additional $1.2 million. So Company A gets $1.4 + 1.2 = $2.6 million and Company B gets $2.6 + 1.2 = $3.8 million. So even though Company B invests more, they make the same profit as Company A.

   (c) The original investment is $4 million. So Company A gets $\dfrac{1.4}{4} \cdot 6.4 = $2.24 million and Company B gets $\dfrac{2.6}{4} \cdot 6.4 = $4.16 million. This seems the fairest.

Chapter Two

Exercises 2.1

2. $A\,(5,\,1)$ $B\,(1,\,2)$ $C\,(-2,\,6)$ $D\,(-6,\,2)$
    $E\,(-4,\,-1)$ $F\,(-2,\,0)$ $G\,(-1,\,-3)$ $H\,(2,\,-2)$

4. The two points are $(-2,-1)$ and $(2,2)$.

    (a) $d = \sqrt{(-2-2)^2 + (-1-2)^2} = \sqrt{(-4)^2 + (-3)^2} = \sqrt{16+9} = \sqrt{25} = 5$

    (b)   midpoint: $\left(\frac{-2+2}{2}, \frac{-1+2}{2}\right) = \left(0, \frac{1}{2}\right)$

6. The two points are $(-2,-3)$ and $(4,-1)$.

    (a) $d = \sqrt{(-2-4)^2 + (-3-(-1))^2} = \sqrt{(-6)^2 + (-2)^2} = \sqrt{36+4} = \sqrt{50} = 2\sqrt{10}$

    (b)   midpoint: $\left(\frac{-3+5}{2}, \frac{-3+(-1)}{2}\right) = (1, -2)$

8. (a)

    (b) $d = \sqrt{(-2-10)^2 + (5-0)^2}$
        $= \sqrt{(-12)^2 + (5)^2} = \sqrt{169} = 13$

    (c)   midpoint: $\left(\frac{-2+10}{2}, \frac{5+0}{2}\right) = \left(4, \frac{5}{2}\right)$

10. (a)

    (b) $d = \sqrt{(-1-9)^2 + (-1-9)^2}$
        $= \sqrt{(-10)^2 + (-10)^2} =$
        $= \sqrt{200} = 10\sqrt{2}$

    (c)   midpoint: $\left(\frac{-1+9}{2}, \frac{3+2}{2}\right) = (4, 4)$

12. (a)

    (b) $d = \sqrt{(1-(-1))^2 + (-6-(-3))^2}$
        $= \sqrt{2^2 + (-3)^2} = \sqrt{4+9} = \sqrt{13}$

    (c)   midpoint: $\left(\frac{1+(-1)}{2}, \frac{-6+(-3)}{2}\right) = \left(0, -\frac{9}{2}\right)$

14. (a)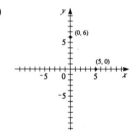

    (b) $d = \sqrt{(5-0)^2 + (0-6)^2}$
    $$= \sqrt{5^2 + (-6)^2} = \sqrt{25 + 36}$$
    $$= \sqrt{61}$$

    (c) midpoint: $\left(\frac{5+0}{2}, \frac{0+6}{2}\right) = \left(\frac{5}{2}, 3\right)$

16. The area of a parallelogram = *base · height*. Since two sides are parallel to the x-axis, we use the length of one of these as the *base*. Thus *base* is

    $$d(A, B) = \sqrt{(1-5)^2 + (2-2)^2} = \sqrt{(-4)^2} = 4.$$

    The height is the change in the y coordinates, thus, the *height* is $6 - 2 = 4$. So the area of the parallelogram is *base · height* $= 4 \cdot 4 = 16$.

    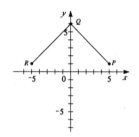

18. The point S must be located at $(0, -4)$. To find the area, we find the length of one side and square it. This gives

    $$d(Q, R) = \sqrt{(-5-0)^2 + (1-6)^2} = \sqrt{(-5)^2 + (-5)^2}$$
    $$= \sqrt{25 + 25} = \sqrt{50} = 5\sqrt{2}.$$

    So the area is $\left(5\sqrt{2}\right)^2 = 50.$

    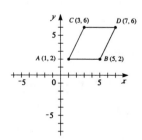

20.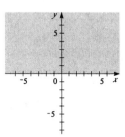

22.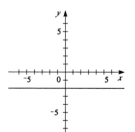

24.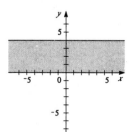

26. $xy > 0 \quad \Leftrightarrow \quad x < 0$ and $y < 0$ or $x > 0$ and $y > 0$.

    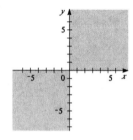

28.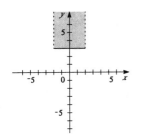

30.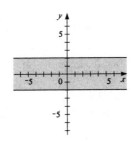

32.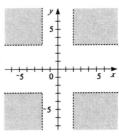

34. $d(E, C) = \sqrt{(-6 - (-2))^2 + (3 - 1)^2} = \sqrt{(-4)^2 + 2^2} = \sqrt{16 + 4} = \sqrt{20}$.
$d(E, D) = \sqrt{(3 - (-2))^2 + (0 - 1)^2} = \sqrt{5^2 + (-1)^2} = \sqrt{25 + 1} = \sqrt{26}$.
Thus point C is closer to point E.

36. (a) The distance from $(7, 3)$ to the origin is
$\sqrt{(7 - 0)^2 + (3 - 0)^2} = \sqrt{7^2 + 3^2} = \sqrt{49 + 9} = \sqrt{58}$. The distance from $(3, 7)$ to the
origin is $\sqrt{(3 - 0)^2 + (7 - 0)^2} = \sqrt{3^2 + 7^2} = \sqrt{9 + 49} = \sqrt{58}$. So the points are the same
distance from the origin.

(b) The distance from (a, b) to the origin is $\sqrt{(a - 0)^2 + (b - 0)^2} = \sqrt{a^2 + b^2}$. The distance
from (b, a) to the origin is $\sqrt{(b - 0)^2 + (a - 0)^2} = \sqrt{b^2 + a^2} = \sqrt{a^2 + b^2}$. So the points are
the same distance from the origin.

38. Since the side AB is parallel to the x-axis, we use this as the *base* in the formula
Area $= \frac{1}{2}(base \cdot height)$. The *height* is the change in the y coordinates. The *base* is $|-2 - 4| = 6$,
and the *height* is $|4 - 1| = 3$. So *Area* is $\frac{1}{2}(6 \cdot 3) = 9$.

40. $d(A, B) = \sqrt{(11 - 6)^2 + (-3 - (-7))^2} = \sqrt{5^2 + 4^2} = \sqrt{25 + 16} = \sqrt{41}$;
$d(A, C) = \sqrt{(2 - 6)^2 + (-2 - (-7))^2} = \sqrt{(-4)^2 + 5^2} = \sqrt{16 + 25} = \sqrt{41}$;
$d(B, C) = \sqrt{(2 - 11)^2 + (-2 - (-3))^2} = \sqrt{(-9)^2 + 1^2} = \sqrt{81 + 1} = \sqrt{82}$.
Since $[d(A, B)]^2 + [d(A, C)]^2 = [d(B, C)]^2$, we conclude that the triangle is a right triangle. The
Area is $\frac{1}{2}\left(\sqrt{41} \cdot \sqrt{41}\right) = \frac{41}{2}$.

42. $d(A, B) = \sqrt{(3 - (-1))^2 + (11 - 3)^2} = \sqrt{4^2 + 8^2} = \sqrt{16 + 64} = \sqrt{80} = 4\sqrt{5}$.
$d(B, C) = \sqrt{(5 - 3)^2 + (15 - 11)^2} = \sqrt{2^2 + 4^2} = \sqrt{4 + 16} = \sqrt{20} = 2\sqrt{5}$.
$d(A, C) = \sqrt{(5 - (-1))^2 + (15 - 3)^2} = \sqrt{6^2 + 12^2} = \sqrt{36 + 144} = \sqrt{180} = 6\sqrt{5}$. So
$d(A, B) + d(B, C) = d(A, C)$, and the points are collinear.

44. The midpoint of AB is $C' = \left(\dfrac{1 + 3}{2}, \dfrac{0 + 6}{2}\right) = (2, 3)$. So the length of the median CC' is
$d(C, C') = \sqrt{(2 - 8)^2 + (3 - 2)^2} = \sqrt{37}$. The midpoint of AC is $B' = \left(\dfrac{1 + 8}{2}, \dfrac{0 + 2}{2}\right)$
$= \left(\frac{9}{2}, 1\right)$. So the length of the median BB' is $d(B, B') = \sqrt{\left(\frac{9}{2} - 3\right)^2 + (1 - 6)^2} = \dfrac{\sqrt{109}}{2}$. The

midpoint of BC is $A' = \left(\dfrac{3+8}{2}, \dfrac{6+2}{2}\right) = \left(\dfrac{11}{2}, 4\right)$. So the length of the median AA' is

$d(A, A') = \sqrt{\left(\frac{11}{2} - 1\right)^2 + (4 - 0)^2} = \dfrac{\sqrt{145}}{2}$.

46. Points on a perpendicular bisector of PQ are the same distance from the points P and Q.

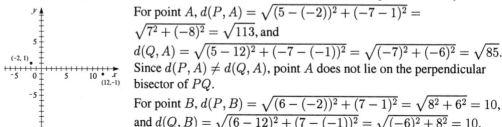

For point A, $d(P, A) = \sqrt{(5 - (-2))^2 + (-7 - 1)^2} = \sqrt{7^2 + (-8)^2} = \sqrt{113}$, and

$d(Q, A) = \sqrt{(5 - 12)^2 + (-7 - (-1))^2} = \sqrt{(-7)^2 + (-6)^2} = \sqrt{85}$. Since $d(P, A) \neq d(Q, A)$, point A does not lie on the perpendicular bisector of PQ.

For point B, $d(P, B) = \sqrt{(6 - (-2))^2 + (7 - 1)^2} = \sqrt{8^2 + 6^2} = 10$, and $d(Q, B) = \sqrt{(6 - 12)^2 + (7 - (-1))^2} = \sqrt{(-6)^2 + 8^2} = 10$.

Since $d(P, B) = d(Q, B)$, point $B(6, 7)$ lies on the perpendicular bisector of PQ.

48. We solve the equation $6 = \dfrac{2 + x}{2}$ to find the x coordinate of B. This gives $6 = \dfrac{2 + x}{2} \Leftrightarrow$

$12 = 2 + x \quad \Leftrightarrow \quad x = 10$. Likewise, $8 = \dfrac{3 + y}{2} \quad \Leftrightarrow \quad 16 = 3 + y \quad \Leftrightarrow \quad y = 13$. Thus, $B = (10, 13)$.

50. We have $M = \left(\dfrac{a + 0}{2}, \dfrac{b + 0}{2}\right) = \left(\dfrac{a}{2}, \dfrac{b}{2}\right)$. Thus,

$d(C, M) = \sqrt{\left(\dfrac{a}{2} - 0\right)^2 + \left(\dfrac{b}{2} - 0\right)^2} = \sqrt{\dfrac{a^2}{4} + \dfrac{b^2}{4}} = \dfrac{\sqrt{a^2 + b^2}}{2}$;

$d(A, M) = \sqrt{\left(\dfrac{a}{2} - a\right)^2 + \left(\dfrac{b}{2} - 0\right)^2} = \sqrt{\left(-\dfrac{a}{2}\right)^2 + \left(\dfrac{b}{2}\right)^2} = \sqrt{\dfrac{a^2}{4} + \dfrac{b^2}{4}} = \dfrac{\sqrt{a^2 + b^2}}{2}$;

$d(B, M) = \sqrt{\left(\dfrac{a}{2} - 0\right)^2 + \left(\dfrac{b}{2} - b\right)^2} = \sqrt{\left(\dfrac{a}{2}\right)^2 + \left(-\dfrac{b}{2}\right)^2} = \sqrt{\dfrac{a^2}{4} + \dfrac{b^2}{4}} = \dfrac{\sqrt{a^2 + b^2}}{2}$.

52. (a) The midpoint is at $\left(\frac{3+27}{2}, \frac{7+17}{2}\right) = (15, 12)$, which is at the intersection of 15th Street and 12th Avenue.

(b) They each must walk $|15 - 3| + |12 - 7| = 12 + 5 = 17$ blocks.

54. (a) The point $(5, 3)$ is shifted to $(5 + 3, 3 + 2) = (8, 5)$.

(b) The point (a, b) is shifted to $(a + 3, b + 2)$.

(c) Let (x, y) be the point that is shifted to $(3, 4)$. Then $(x + 3, y + 2) = (3, 4)$. Setting the x-coordinates equal, we get $x + 3 = 3 \quad \Leftrightarrow \quad x = 0$. Setting the y-coordinates equal, we get $y + 2 = 4 \quad \Leftrightarrow \quad y = 2$. So the point is $(0, 2)$.

(d) $A = (-5, -1)$ so $A' = (-5 + 3, -1 + 2) = (-2, 1)$;
$B = (-3, 2)$ so $B' = (-3 + 3, 2 + 2) = (0, 4)$;
$C = (2, 1)$ so $C' = (2 + 3, 1 + 2) = (5, 3)$.

56. We solve the equation $6 = \dfrac{2+x}{2}$ to find the x coordinate of B: $\quad 6 = \dfrac{2+x}{2} \quad \Leftrightarrow \quad 12 = 2+x$ $\quad \Leftrightarrow \quad x = 10$. Likewise, for the y coordinate of B, we have $8 = \dfrac{3+y}{2} \quad \Leftrightarrow \quad 16 = 3+y \quad \Leftrightarrow$ $y = 13$. Thus $B = (10,\ 13)$.

Exercises 2.2

2. $(1,0)$: $0 \overset{?}{=} \sqrt{(1)+1}$ $\Leftrightarrow$ $0 \overset{?}{=} \sqrt{2}$ No.
    $(0,1)$: $1 \overset{?}{=} \sqrt{(0)+1}$ $\Leftrightarrow$ $1 \overset{?}{=} \sqrt{1}$ Yes.
    $(3,2)$: $2 \overset{?}{=} \sqrt{(3)+1}$ $\Leftrightarrow$ $2 \overset{?}{=} \sqrt{4}$ Yes.
    So $(0,1)$ and $(3,2)$ are points on the graph of this equation.

4. $(1,1)$: $(1)[(1)^2 + 1] \overset{?}{=} 1$ $\Leftrightarrow$ $1(2) \overset{?}{=} 1$ No.
    $\left(1, \frac{1}{2}\right)$: $\left(\frac{1}{2}\right)[(1)^2 + 1] \overset{?}{=} 1$ $\Leftrightarrow$ $\frac{1}{2}(2) \overset{?}{=} 1$ Yes.
    $\left(-1, \frac{1}{2}\right)$: $\left(\frac{1}{2}\right)[(-1)^2 + 1] \overset{?}{=} 1$ $\Leftrightarrow$ $\frac{1}{2}(2) \overset{?}{=} 1$ Yes.
    So both $\left(1, \frac{1}{2}\right)$ and $\left(-1, \frac{1}{2}\right)$ are points on the graph of this equation.

6. $(0,1)$: $(0)^2 + (1)^2 - 1 \overset{?}{=} 0$ $\Leftrightarrow$ $0 + 1 - 1 \overset{?}{=} 0$. Yes.
    $\left(\frac{1}{\sqrt{2}}, \frac{1}{\sqrt{2}}\right)$: $\left(\frac{1}{\sqrt{2}}\right)^2 + \left(\frac{1}{\sqrt{2}}\right)^2 - 1 \overset{?}{=} 0$ $\Leftrightarrow$ $\frac{1}{2} + \frac{1}{2} - 1 \overset{?}{=} 0$. Yes.
    $\left(\frac{3}{\sqrt{2}}, \frac{1}{2}\right)$: $\left(\frac{3}{\sqrt{2}}\right)^2 + \left(\frac{1}{2}\right)^2 - 1 \overset{?}{=} 0$ $\Leftrightarrow$ $\frac{3}{4} + \frac{1}{4} - 1 \overset{?}{=} 0$. Yes.
    So $(0,1)$, $\left(\frac{1}{\sqrt{2}}, \frac{1}{\sqrt{2}}\right)$ and $\left(\frac{3}{\sqrt{2}}, \frac{1}{2}\right)$, are all points on the graph of this equation.

8. To find x-intercepts, set $y = 0$. This gives $\dfrac{x^2}{9} + \dfrac{0^2}{4} = 1$ $\Leftrightarrow$ $\dfrac{x^2}{9} = 1$ $\Leftrightarrow$ $x^2 = 9$ $\Leftrightarrow$
    $x = \pm 3$, so the x-intercept are -3 and 3.
    To find y-intercepts, set $x = 0$. This gives $\dfrac{0^2}{9} + \dfrac{y^2}{4} = 1$ $\Leftrightarrow$ $\dfrac{y^2}{4} = 1$ $\Leftrightarrow$ $y^2 = 4$ $\Leftrightarrow$
    $x = \pm 2$, so the y-intercept are -2 and 2.

10. To find x-intercepts, set $y = 0$. This gives $x^2 + 0^3 - x^2(0)^2 = 64$ $\Leftrightarrow$ $x^2 = 64$ $\Leftrightarrow$
    $x = \pm 8$. So the x-intercept are -8 and 8.
    To find y-intercepts, set $x = 0$. This gives $0^2 + y^3 - (0)^2 y^2 = 64$ $\Leftrightarrow$ $y^3 = 64$ $\Leftrightarrow$ $y = 4$.
    So the y-intercept is 4.

12. To find x-intercepts, set $y = 0$. This gives $0 = x^2 - 5x + 6$ $\Leftrightarrow 0 = (x-2)(x-3)$. So
    $x - 2 = 0$ and $x = 2$ or $x - 3 = 0$ and $x = 3$, and the x-intercepts are at 2 and 3. To find
    y-intercepts, set $x = 0$. This gives $y = 0^2 - 0 + 6$ $\Leftrightarrow$ $y = 6$, so the y-intercept is 6.

14. To find x-intercepts, set $y = 0$. This gives $0 - 2x(0) + 2x = 1$ $\Leftrightarrow$ $2x = 1$ $\Leftrightarrow$ $x = \frac{1}{2}$, so
    the x-intercept is $\frac{1}{2}$. To find y-intercepts, set $x = 0$. This gives $y - 2(0)y + 2(0) = 1$ $\Leftrightarrow$
    $y = 1$, so the y-intercept is 1.

16. To find x-intercepts, set $y = 0$. This gives $0 = \sqrt{x+1}$ $\Leftrightarrow$ $0 = x + 1$ $\Leftrightarrow$ $x = -1$, so the
    x-intercept is -1. To find y-intercepts, set $x = 0$. This gives $y = \sqrt{0+1}$ $\Leftrightarrow$ $y = 1$, so the
    y-intercept is 1.

18. To find x-intercepts, set $y = 0$. This gives $x^2 - x(0) + (0) = 1$ $\Leftrightarrow$ $x^2 = 1$ $\Rightarrow$ $x = \pm 1$, so
    the x-intercepts are -1 and 1. To find y-intercepts, set $x = 0$. This gives $y = (0)^2 - (0)y + y = 1$
    $\Leftrightarrow$ $y = 1$, so the y-intercept is 1.

20. $y = 2x$

x	y
-4	4
-2	2
0	0
1	-1
2	-2
3	-3
4	-4

$y = 0 \iff 2x = 0$. So the x-intercept is 0, and the y-intercept is also 0.

x-axis symmetry: $(-y) = 2x \iff -y = 2x$, which is not the same, so not symmetric with respect to the x-axis.

y-axis symmetry: $y = 2(-x) \iff y = -2x$, which is not the same, so not symmetric with respect to the y-axis.

Origin symmetry: $(-y) = 2(-x)$ is the same as $y = 2x$, so it is symmetric with respect to the origin.

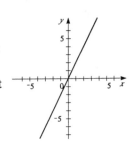

22. $y = -3x + 3$

x	y
-3	-1
-2	1
-1	3
0	5
1	7
2	9
3	11

$y = 0 \Rightarrow 0 = -3x + 3 \iff 3x = 3 \iff x = 1$, so the x-intercept is 1, and $x = 0 \Rightarrow y = -3(0) + 3 = 3$, so the y-intercept is 3.

x-axis symmetry: $(-y) = -3x + 3 \iff -y = -3x + 3$, which is not the same as $y = -3x + 3$, so not symmetric with respect to the x-axis.

y-axis symmetry: $y = -3(-x) + 3 \iff y = 3x + 3$, which is not the same as $y = -3x + 3$, so not symmetric with respect to the y-axis.

Origin symmetry: $(-y) = -3(-x) + 3 \iff -y = 3x + 3$, which is not the same as $y = -3x + 3$, so not symmetric with respect to the origin.

24. Solve for y: $x + y = 3 \iff y = -x + 3$.

x	y
-2	5
-1	4
0	3
1	2
2	1
3	0
4	-1

$y = 0 \Rightarrow 0 = -x + 3 \iff x = 3$, so the x-intercept is 3, and $x = 0 \Rightarrow y = -(0) + 3 = 3$, so the y-intercept is 3.

x-axis symmetry: $x + (-y) = 3 \iff x - y = 3$, which is not the same as $x + y = 3$, so not symmetric with respect to the x-axis.

y-axis symmetry: $(-x) + y = 5 \iff -x + y = 3$, which is not the same as $x + y = 3$, so not symmetric with respect to the y-axis.

Origin symmetry: $(-x) + (-y) = 3 \iff -x - y = 3$, which is not the same as $x + y = 3$, so not symmetric with respect to the origin.

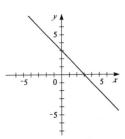

26. $y = x^2 + 2$

x	y
-3	11
-2	6
-1	3
0	2
1	3
2	6
3	11

$y = 0 \Rightarrow 0 = x^2 + 2 \iff -2 = x^2$, since $x^2 \geq 0$, there are no x-intercepts, and $x = 0 \Rightarrow y = (0)^2 + 2 = 2$, so the y-intercept is 2.

x-axis symmetry: $(-y) = x^2 + 2 \iff -y = x^2 + 2$, which is not the same, so not symmetric with respect to the x-axis.

y-axis symmetry: $y = (-x)^2 + 2 = x^2 + 2$, so it is symmetric with respect to the y-axis.

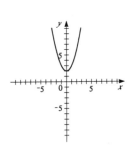

Origin symmetry: $(-y) = (-x)^2 + 2$ $\Leftrightarrow$ $-y = x^2 + 2$, which is not the same, so not symmetric with respect to the origin.

28. $8y = x^3$ $\Leftrightarrow$ $y = \frac{1}{8}x^3$

x	y
-6	-27
-4	-8
-2	-1
0	0
2	1
4	8
3	27

$y = 0$ $\Rightarrow$ $0 = \frac{1}{8}x^3$ $\Leftrightarrow$ $x^3 = 0$ $\Rightarrow$ $x = 0$, so the x-intercept is 0, and $x = 0$ $\Rightarrow$ $y = \frac{1}{8}(0)^3 = 0$, so the y-intercept is 0.

x-axis symmetry: $(-y) = \frac{1}{8}x^3$, which is not the same as $y = \frac{1}{8}x^3$, so not symmetric with respect to the x-axis.

y-axis symmetry: $y = \frac{1}{8}(-x)^3 = -\frac{1}{8}x^3$, which is not the same as $y = \frac{1}{8}x^3$, so not symmetric with respect to the y-axis.

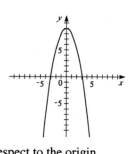

Origin symmetry: $(-y) = \frac{1}{8}(-x)^3$ $\Leftrightarrow$ $-y = -\frac{1}{8}x^3$, which is the same as $y = \frac{1}{8}x^3$, so it is symmetric with respect to the origin.

30. $y = 9 - x^2$

x	y
-4	-7
-3	0
-2	5
-1	8
0	9
1	8
2	5
3	0
4	-7

$y = 0$ $\Rightarrow$ $0 = 9 - x^2$ $\Leftrightarrow$ $x^2 = 9$ $\Rightarrow$ $x = \pm 3$, so the x-intercepts are -3 and 3, and $x = 0$ $\Rightarrow$ $y = 9 - (0)^2 = 9$, so the y-intercept is 9.

x-axis symmetry: $(-y) = 9 - x^2$, which is not the same as $y = 9 - x^2$, so not symmetric with respect to the x-axis.

y-axis symmetry: $y = 9 - (-x)^2 = 9 - x^2$, so it is symmetric with respect to the y-axis.

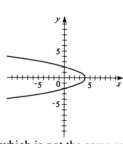

Origin symmetry: $(-y) = 9 - (-x)^2$ $\Leftrightarrow$ $-y = 9 - x^2$, which is not the same as $y = 9 - x^2$, so not symmetric with respect to the origin.

32. Solve for x in terms of y: $x + y^2 = 4$ $\Leftrightarrow$ $x = 4 - y^2$

x	y
-12	-4
-5	-3
0	-2
3	-1
4	0
3	1
0	2
-5	3
-12	4

$y = 0$ $\Rightarrow$ $x + 0^2 = 4$ $\Leftrightarrow$ $x = 4$, so the x-intercept is 4, and $x = 0$ $\Rightarrow$ $0 + y^2 = 4$ $\Rightarrow$ $y = \pm 2$, so the y-intercepts are -2 and 2.

x-axis symmetry: $x + (-y)^2 = 4$ $\Leftrightarrow$ $x + y^2 = 4$, so it is symmetric with respect to the x-axis.

y-axis symmetry: $(-x) + y^2 = 4$ $\Leftrightarrow$ $-x + y^2 = 4$, which is not the same, so not symmetric with respect to the y-axis.

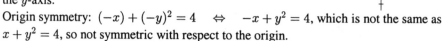

Origin symmetry: $(-x) + (-y)^2 = 4$ $\Leftrightarrow$ $-x + y^2 = 4$, which is not the same as $x + y^2 = 4$, so not symmetric with respect to the origin.

34. $x^2 + y^2 = 9$ $y = 0$ $\Rightarrow$ $x^2 + 0^2 = 9$ $\Leftrightarrow$ $x = \pm 3$, so the

x	y
-3	0
-2	$\pm\sqrt{5}$
-1	$\pm 2\sqrt{2}$
0	± 3
1	$\pm 2\sqrt{2}$
2	$\pm\sqrt{5}$
3	0

x-intercepts are -3 and 3, and $x = 0$ $\Rightarrow$ $0^2 + y^2 = 9$
$\Leftrightarrow$ $y = \pm 3$, so the y-intercepts are -3 and 3.
x-axis symmetry: $x^2 + (-y)^2 = 9$ $\Leftrightarrow$
$x^2 + y^2 = 9$, so it is symmetric with respect to the x-axis.
y-axis symmetry: $(-x)^2 + y^2 = 9$ $\Leftrightarrow$
$x^2 + y^2 = 9$, so it is symmetric with respect to the y-axis.
Origin symmetry: $(-x)^2 + (-y)^2 = 9$ $\Leftrightarrow$
$x^2 + y^2 = 9$, so it is symmetric with respect to the origin.

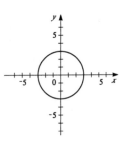

36. Since the radicand (the inside of the square root) cannot be negative, we must have $4 - x^2 \geq 0$
$\Leftrightarrow$ $x^2 \leq 4$ $\Leftrightarrow$ $|x| \leq 2$. So $-2 \leq x \leq 2$ is the only portion of the x-axis where this equation
is defined.

x	y
-2	0
-1	$-\sqrt{3}$
0	-4
1	$-\sqrt{3}$
2	0

$y = 0$ $\Rightarrow$ $0 = -\sqrt{4 - x^2}$ $\Leftrightarrow$ $4 - x^2 = 0$ $\Leftrightarrow$
$x^2 = 4$ $\Rightarrow$ $x = \pm 2$, so the x-intercepts are -2 and 2,
and $x = 0$ $\Rightarrow$ $y = -\sqrt{4 - (0)^2} = -\sqrt{4} = -2$, so the
y-intercept is -2.
x-axis symmetry: Since $y \leq 0$, this graph is not symmetric
with respect to the x-axis.

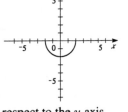

y-axis symmetry: $y = -\sqrt{4 - (-x)^2} = -\sqrt{4 - x^2}$, so it is symmetric with respect to the y-axis.
Origin symmetry: Since $y \leq 0$, the graph is not symmetric with respect to the origin.

38. $x = |y|$. Here we insert values of y and find the corresponding value of x.

x	y
3	-3
2	-2
1	-1
0	0
1	1
2	2
3	3

$y = 0$ $\Rightarrow$ $x = |0| = 0$, so the x-intercept is 0, and $x = 0$
$\Rightarrow$ $0 = |y|$ $\Leftrightarrow$ $y = 0$, so the y-intercept is 0.
x-axis symmetry: $x = |-y| = |y|$, so the graph is symmetric
with respect to the x-axis.
y-axis symmetry: Since $x \geq 0$, the graph is not symmetric
with respect to the y-axis.
Origin symmetry: Since $x \geq 0$, the graph is not symmetric
with respect to the origin.

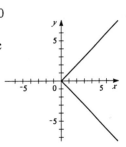

40. $y = |4 - x|$

x	y
-6	10
-4	8
-2	6
0	4
2	2
4	0
6	2
8	4
10	6

$y = 0$ $\Rightarrow$ $0 = |4 - x|$ $\Leftrightarrow$ $4 - x = 0$ $\Rightarrow$
$x = 4$, so the x-intercept is 4, and $x = 0$ $\Rightarrow$
$y = |4 - 0| = |4| = 4$, so the y-intercept is 4.
x-axis symmetry: $y = |4 - (-x)| = |4 + x|$, which is not the
same as $y = |4 - x|$, so not symmetric with respect to the
x-axis.
y-axis symmetry: $(-y) = |4 - x|$ $\Leftrightarrow$ $-y = |4 - x|$,
which is not the same as $y = |4 - x|$, so not symmetric with
respect to the y-axis.

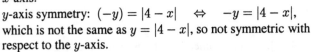

Origin symmetry: $(-y) = |4 - (-x)|$ $\Leftrightarrow$ $-y = |4 + x|$, which is not the same as
$y = |4 - x|$, so not symmetric with respect to the origin.

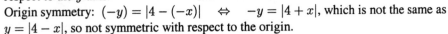

42. $y = x^3 - 1$

x	y
-3	-28
-2	-9
-1	-2
0	-1
1	1
2	7
3	26

$y = 0 \Rightarrow 0 = x^3 - 1 = (x-1)(x^2 + x + 1)$. So
$x - 1 = 0 \Leftrightarrow x = 1$, while $x^2 + x + 1 = 0$ has no real
solution, so the x-intercept is 1. And $x = 0 \Rightarrow$
$y = (0)^3 - 1 = -1$, so the y-intercept is -1.
x-axis symmetry: $-y = x^3 - 1$, which is not the same as
$y = x^3 - 1$, so not symmetric with respect to the x-axis.
y-axis symmetry: $y = (-x)^3 - 1 = -x^3 - 1$, which is not the
same as $y = x^3 - 1$, so not symmetric with respect to the y-axis.

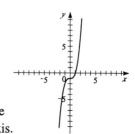

Origin symmetry: $(-y) = (-x)^3 - 1 \Leftrightarrow -y = -x^3 - 1 \Leftrightarrow y = x^3 + 1$ which is not the
same. Not symmetric with respect to the origin.

44. $y = 16 - x^4$

x	y
-3	-65
-2	0
-1	15
0	16
1	15
2	0
3	-65

$y = 0 \Rightarrow 0 = 16 - x^4 \Rightarrow x^4 = 16 \Rightarrow x^2 = 4$
$\Rightarrow x = \pm 2$, so the x-intercepts are ± 2, and so $x = 0 \Rightarrow$
$y = 16 - 0^4 = 16$, so the y-intercept is 16.
x-axis symmetry: $(-y) = 16 - x^4 \Leftrightarrow y = -16 + x^4$,
which is not the same as $y = 16 - x^4$, so not symmetric with
respect to the x-axis.
y-axis symmetry: $y = 16 - (-x)^4 = 16 - x^4$, so it is
symmetric with respect to the y-axis.

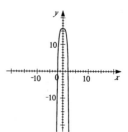

Origin symmetry: $(-y) = 16 - (-x)^4 \Leftrightarrow -y = 16 - x^4$, which is not the same as
$y = 16 - x^4$, so not symmetric with respect to the origin.

46. x-axis symmetry: $x = (-y)^4 - (-y)^2 = y^4 - y^2$, so it is symmetric with respect to the x-axis.
y-axis symmetry: $(-x) = y^4 - y^2$, which is not the same as $x = y^4 - y^2$, so not symmetric with
respect to the y-axis.
Origin symmetry: $(-x) = (-y)^4 - (-y)^2 \Leftrightarrow -x = y^4 - y^2$, which is not the same as
$x = y^4 - y^2$, so not symmetric with respect to the origin.

48. x-axis symmetry: $x^4(-y)^4 + x^2(-y)^2 = 1 \Leftrightarrow x^4 y^4 + x^2 y^2 = 1$, so it is symmetric with
respect to the x-axis.
y-axis symmetry: $(-x)^4 y^4 + (-x)^2 y^2 = 1 \Leftrightarrow x^4 y^4 + x^2 y^2 = 1$, so it is symmetric with respect
to the y-axis.
Origin symmetry: $(-x)^4(-y)^4 + (-x)^2(-y)^2 = 1 \Leftrightarrow x^4 y^4 + x^2 y^2 = 1$, so it is symmetric with
respect to the origin.

50. x-axis symmetry: $(-y) = x^2 + |x| \Leftrightarrow y = -x^2 - |x|$, which is not the same as $y = x^2 + |x|$,
so not symmetric with respect to the x-axis.
y-axis symmetry: $y = (-x)^2 + |-x| \Leftrightarrow y = x^2 + |x|$, so it is symmetric with respect to the
y-axis. Note: $|-x| = |x|$.
Origin symmetry: $(-y) = (-x)^2 + |-x| \Leftrightarrow -y = x^2 + |x| \Leftrightarrow y = -x^2 - |x|$, which is
not the same as $y = x^2 + |x|$, so not symmetric with respect to the origin.

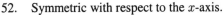

52. Symmetric with respect to the x-axis.

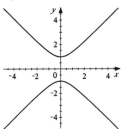

54. Symmetric with respect to the origin.

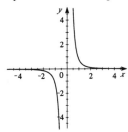

56. Using $h = -1$, $k = -4$, and $r = 8$ we get $(x - (-1))^2 + (y - (-4))^2 = 8^2$ $\Leftrightarrow$ $(x + 1)^2 + (y + 4)^2 = 64$.

58. Using $h = -1$ and $k = 5$ we get $(x - (-1))^2 + (y - 5)^2 = r^2$ $\Leftrightarrow$ $(x + 1)^2 + (y - 5)^2 = r^2$. Next, using the point $(-4, -6)$, we solve for r^2. This gives $(-4 + 1)^2 + (-6 - 5)^2 = r^2$ $\Leftrightarrow$ $130 = r^2$. Thus, the equation of the circle is $(x + 1)^2 + (y - 5)^2 = 130$.

60. The center is at the midpoint of the line segment, which is $\left(\frac{-1+7}{2}, \frac{3+-5}{2}\right) = (3, -1)$. The radius is one half the diameter, so $r = \frac{1}{2}\sqrt{(-1 - 7)^2 + (3 - (-5))^2} = 4\sqrt{2}$. Thus, the equation of the circle is $(x - 3)^2 + (y + 1)^2 = 32$.

62. Since the circle with $r = 5$ lies in the first quadrant and is tangent to both the x-axis and the y-axis, the center of the circle is at $(5, 5)$. Therefore, the equation of the circle is $(x - 5)^2 + (y - 5)^2 = 25$.

64. From the figure, the center of the circle is at $(-1, 1)$. The radius is the distance from the center to the point $(2, 0)$. Thus $r = \sqrt{(-1 - 2)^2 + (1 - 0)^2} = \sqrt{9 + 1} = \sqrt{10}$, and the equation of the circle is $(x + 1)^2 + (y - 1)^2 = 10$.

66. Completing the square gives $x^2 + y^2 - 2x - 2y = 2$ $\Leftrightarrow$ $x^2 - 2x + \underline{} + y^2 - 2y + \underline{} = 2$ $\Leftrightarrow$ $x^2 - 2x + \left(\frac{-2}{2}\right)^2 + y^2 - 2y + \left(\frac{-2}{2}\right)^2 = 2 + \left(\frac{-2}{2}\right)^2 + \left(\frac{-2}{2}\right)^2$ $\Leftrightarrow$ $x^2 - 2x + 1 + y^2 - 2y + 1 = 2 + 1 + 1$ $\Leftrightarrow$ $(x - 1)^2 + (y - 1)^2 = 4$. Thus, the center is at $(1, 1)$, and the radius is 2.

68. Completing the square gives $x^2 + y^2 + 6y + 2 = 0$ $\Leftrightarrow$ $x^2 + y^2 + 6y + \underline{} = -2$ $\Leftrightarrow$ $x^2 + y^2 + 6y + \left(\frac{6}{2}\right)^2 = -2 + \left(\frac{6}{2}\right)^2$ $\Leftrightarrow$ $x^2 + y^2 + 6y + 9 = -2 + 9$ $\Leftrightarrow$ $x^2 + (y + 3)^2 = 7$. Thus, the center is at $(0, -3)$, and the radius is $\sqrt{7}$.

70. Completing the square gives $x^2 + y^2 + 2x + y + 1 = 0$ $\Leftrightarrow$ $x^2 + 2x + \underline{} + y^2 + y = -1$ $\Leftrightarrow$ $x^2 + 2x + \left(\frac{2}{2}\right)^2 + y^2 + y + \left(\frac{1}{2}\right)^2 = -1 + 1 + \left(\frac{1}{2}\right)^2$ $\Leftrightarrow$ $x^2 + 2x + 1 + y^2 + y + \frac{1}{4} = \frac{1}{4}$ $\Leftrightarrow$ $(x + 1)^2 + \left(y + \frac{1}{2}\right)^2 = \frac{1}{4}$. Thus, the center is at $\left(-1, -\frac{1}{2}\right)$, and the radius is $\frac{1}{2}$.

72. Completing the square gives $x^2 + y^2 + \frac{1}{2}x + 2y + \frac{1}{16} = 0$ $\Leftrightarrow$ $x^2 + \frac{1}{2}x + \underline{} + y^2 + 2y + \underline{} = -\frac{1}{16}$ $\Leftrightarrow$ $x^2 + \frac{1}{2}x + \left(\frac{1/2}{2}\right)^2 + y^2 + 2y + \left(\frac{2}{2}\right)^2 = -\frac{1}{16} + \left(\frac{1/2}{2}\right)^2 + \left(\frac{2}{2}\right)^2$ $\Leftrightarrow$ $\left(x + \frac{1}{4}\right)^2 + (y + 1)^2 = 1$. Thus, the center is at $\left(-\frac{1}{4}, -1\right)$, and the radius is 1.

74. First divide by 4, then complete the square. This gives
$$4x^2 + 4y^2 + 2x = 0 \quad \Leftrightarrow \quad x^2 + y^2 + \tfrac{1}{2}x = 0 \quad \Leftrightarrow$$
$$x^2 + \tfrac{1}{2}x + \underline{\quad} + y^2 = 0 \quad \Leftrightarrow \quad x^2 + \tfrac{1}{2}x + \left(\tfrac{1/2}{2}\right)^2 + y^2 = \left(\tfrac{1/2}{2}\right)^2 \quad \Leftrightarrow$$
$$\left(x + \tfrac{1}{4}\right)^2 + y^2 = \tfrac{1}{16}.$$
Thus, the center is at $\left(-\tfrac{1}{4}, 0\right)$, and the radius is $\tfrac{1}{4}$.

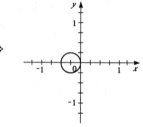

76. $x^2 + y^2 - 16x + 12y + 200 = 0 \quad \Leftrightarrow \quad x^2 - 16x + \underline{\quad} + y^2 + 12y + \underline{\quad} = -200 \quad \Leftrightarrow$
$x^2 - 16x + \left(\tfrac{-16}{2}\right)^2 + y^2 + 12y + \left(\tfrac{12}{2}\right)^2 = -200 + \left(\tfrac{-16}{2}\right)^2 + \left(\tfrac{12}{2}\right)^2 \quad \Leftrightarrow$
$(x - 8)^2 + (y + 6)^2 = -200 + 64 + 36 = -100.$ Since completing the square gives $r^2 = -100$,
this is not the equation of a circle. There is no graph.

78. $\{(x, y)|\ x^2 + y^2 > 4\}.$ This is the set of points outside the circle
$x^2 + y^2 = 4.$

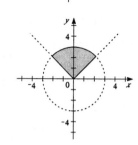

80. $\{(x, y)|\ 2x < x^2 + y^2 \le 4\}.$ Completing the square gives
$2x < x^2 + y^2 \quad \Leftrightarrow \quad 0 < x^2 - 2x + \underline{\quad} + y^2 \quad \Leftrightarrow$
$1 < x^2 + 2x + 1 + y^2 \quad \Leftrightarrow \quad 1 < (x + 1)^2 + y^2.$ Thus, this is the set
of points outside the circle $(x + 1)^2 + y^2 = 1$ and inside the circle
$x^2 + y^2 = 4.$

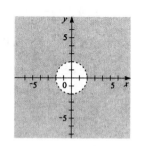

82. This is the top quarter of the circle of radius 3. Thus, the area is
$\tfrac{1}{4}(9\pi) = \tfrac{9\pi}{4}.$

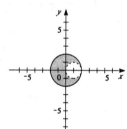

84. (a) Closest: 2 Mm. Farthest: 8 Mm.

(b) When $y = 2$ we have $\dfrac{(x - 3)^2}{25} + \dfrac{2^2}{16} = 1 \quad \Leftrightarrow \quad \dfrac{(x - 3)^2}{25} + \dfrac{1}{4} = 1 \quad \Leftrightarrow \quad \dfrac{(x - 3)^2}{25} = \dfrac{3}{4}$

$\Leftrightarrow \quad (x - 3)^2 = \dfrac{75}{4}.$ Taking the square root of both sides we get $x - 3 = \pm\sqrt{\dfrac{75}{4}} = \pm\dfrac{5\sqrt{3}}{2}$

$\Leftrightarrow \quad x = 3 \pm \dfrac{5\sqrt{3}}{2}.$ So $x = 3 - \dfrac{5\sqrt{3}}{2} \approx -1.33$ or $x = 3 + \dfrac{5\sqrt{3}}{2} \approx 7.33.$

The distance from $(-1.33, 2)$ to the center $(0, 0)$ is $d = \sqrt{(-1.33 - 0)^2 + (2 - 0)^2} = \sqrt{5.7689} \approx 2.40$.

The distance from $(7.33, 2)$ to the center $(0, 0)$ is $d = \sqrt{(7.33 - 0)^2 + (2 - 0)^2} = \sqrt{57.7307} \approx 7.60$.

86. (a) (i) $(x - 2)^2 + (y - 1)^2 = 9$, the center is at $(2, 1)$, and the radius is 3.
$(x - 6)^2 + (y - 4)^2 = 16$, the center is at $(6, 4)$, and the radius is 4.
The distance between centers is
$$\sqrt{(2 - 6)^2 + (1 - 4)^2} = \sqrt{(-4)^2 + (-3)^2} = \sqrt{16 + 9} = \sqrt{25} = 5.$$
Since $5 < 3 + 4$, these circles intersect.

(ii) $x^2 + (y - 2)^2 = 4$, the center is at $(0, 2)$, and the radius is 2.
$(x - 5)^2 + (y - 14)^2 = 9$, the center is at $(5, 1\,4)$, and the radius is 3.
The distance between centers is
$$\sqrt{(0 - 5)^2 + (2 - 14)^2} = \sqrt{(-5)^2 + (-12)^2} = \sqrt{25 + 144} = \sqrt{169} = 13.$$
Since $13 > 2 + 3$, these circles do not intersect.

(iii) $(x - 3)^2 + (y + 1)^2 = 1$, the center is at $(3, -1)$, and the radius is 1.
$(x - 2)^2 + (y - 2)^2 = 25$, the center is at $(2, 2)$, and the radius is 5.
The distance between centers is
$$\sqrt{(3 - 2)^2 + (-1 - 2)^2} = \sqrt{1^2 + (-3)^2} = \sqrt{1 + 9} = \sqrt{10}.$$
Since $\sqrt{10} < 1 + 5$, these circles intersect.

(b) As shown in the diagram, if two circles intersect, then the centers of the circles and one point of intersection form a triangle. So because in any triangle each side has length less than the sum of the other two, the two circles will intersect only if the distance between their centers, d, is less than or equal to the sum of the radii, r_1 and r_2. That is, the circles will intersect if $d \leq r_1 + r_2$.

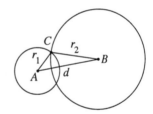

88. (a) *In geometry:* The line runs through the center of the circle and intersects it at the points A and B.
In coordinate geometry: The line $y = \frac{4}{3}x$ runs through the center of the circle $x^2 + y^2 = 25$ and intersects it at the points $A(-3, -4)$ and $B(3, 4)$.

(b)

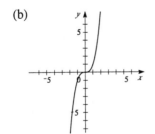

It is difficult to attach geometric significance to points on this curve, unlike the circle where the points on it are a fixed distance from the center.

(c) In coordinate geometry we can use equations to describe curves. In geometry without coordinates we are restricted to geometric descriptions (involving, for example, distances between points) to describe curves.

Exercises 2.3

2. $y = x^2 + 7x + 6$

   (a) $[-5, 5]$ by $[-5, 5]$

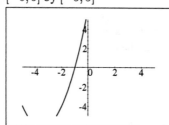

   (b) $[0, 10]$ by $[-20, 100]$

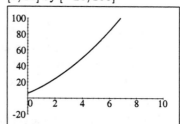

   (c) $[-15, 8]$ by $[-20, 100]$

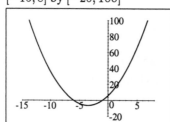

   (d) $[-10, 3]$ by $[-100, 20]$

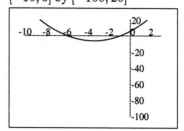

The viewing rectangle in part (c) produces the most appropriate graph of the equation.

4. $y = 2x^2 - 1000$

   (a) $[-10, 10]$ by $[-10, 10]$

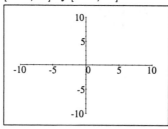

   (b) $[-10, 10]$ by $[-100, 100]$

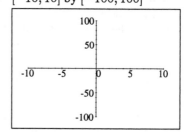

   (c) $[-10, 10]$ by $[-1000, 1000]$

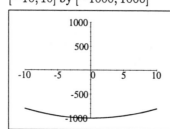

   (d) $[-25, 25]$ by $[-1200, 200]$

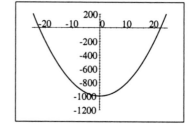

The viewing rectangle in part (d) produces the most appropriate graph of the equation.

6. $y = \sqrt{8x - x^2}$

(a) $[-4, 4]$ by $[-4, 4]$

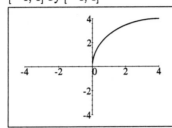

(b) $[-5, 5]$ by $[0, 100]$

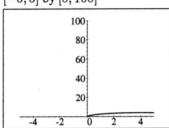

(c) $[-10, 10]$ by $[-10, 40]$

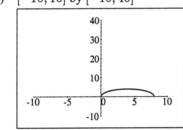

(d) $[-2, 10]$ by $[-2, 6]$

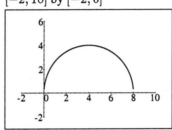

From the graphs we see that the viewing rectangle in (d) produces the most appropriate graph of the equation. Note: Squaring both sides yields the equation $y^2 = 8x - x^2$ $\Leftrightarrow$ $(x - 4)^2 + y^2 = 16$. Since this gives a circle, the original equation represents the top half of a circle.

8. $y = -100x^2$

$[-5, 5]$ by $[-1000, 100]$

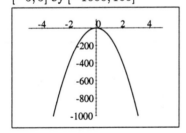

10. $y = 0.3x^2 + 1.7x - 3$

$[-10, 5]$ by $[-10, 20]$

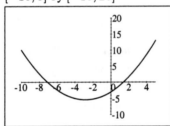

12. $y = \sqrt{12x - 17}$

$[0, 10]$ by $[0, 20]$

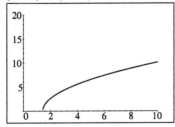

14. $y = x(x + 6)(x - 9)$

$[-10, 10]$ by $[-250, 150]$

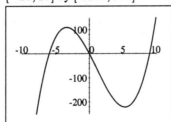

16. $y = \dfrac{x}{x^2 + 25}$

$[-10, 10]$ by $[-0.2, 0.2]$

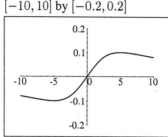

18. $y = 2x - |x^2 - 5|$

$[-10, 10]$ by $[-10, 10]$

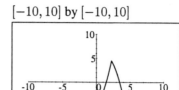

20. $(y - 1)^2 + x^2 = 1 \quad \Leftrightarrow \quad (y - 1)^2 = 1 - x^2 \quad \Rightarrow$
$y - 1 = \pm\sqrt{1 - x^2} \quad \Leftrightarrow \quad y = 1 \pm \sqrt{1 - x^2}$.
So we graph the functions $y_1 = 1 + \sqrt{1 - x^2}$ and
$y_2 = 1 - \sqrt{1 - x^2}$ in the viewing rectangle
$[-2, 2]$ by $[0, 3]$.

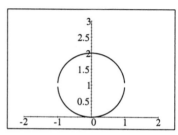

22. $y^2 - 9x^2 = 1 \quad \Leftrightarrow \quad y^2 = 1 + 9x^2 \quad \Rightarrow$
$y = \pm\sqrt{1 + 9x^2}$. So we graph the functions
$y_1 = \sqrt{1 + 9x^2}$ and $y_2 = -\sqrt{1 + 9x^2}$ in the
viewing rectangle $[-5, 5]$ by $[-5, 5]$.

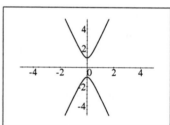

24. Although the graphs of $y = \sqrt{49 - x^2}$ and $y = \frac{1}{5}(41 - 3x)$
appear to intersect in the viewing rectangle $[-8, 8]$ by
$[-1, 8]$, there are no points of intersection. You can
verify that this is not an intersection by zooming in.

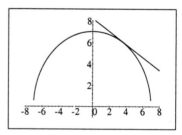

26. The graphs of $y = x^3 - 4x$ and $y = x + 5$ appears to
have one point of intersection in the viewing rectangle
$[-4, 4]$ by $[-15, 15]$. The solution is $x \approx 2.627$.

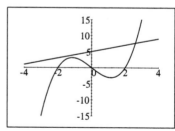

28. Algebraically: $\frac{1}{2}x - 3 = 6 + 2x \quad \Leftrightarrow \quad -9 = \frac{3}{2}x \quad \Leftrightarrow$

$x = -6$.

Graphically: We graph the two equations $y_1 = \frac{1}{2}x - 3$ and

$y_2 = 6 + 2x$ in the viewing rectangle $[-10, 5]$ by $[-10, 5]$.

Zooming in we see that solution is $x = -6$.

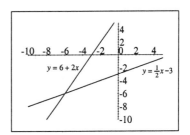

30. Algebraically: $\dfrac{4}{x+2} - \dfrac{6}{2x} = \dfrac{5}{2x+4} \quad \Leftrightarrow \quad 2x(x+2)\left(\dfrac{4}{x+2} - \dfrac{6}{2x}\right) = 2x(x+2)\left(\dfrac{5}{2x+4}\right)$

$\Leftrightarrow \quad 2x(4) - (x+2)(6) = x(5) \quad \Leftrightarrow \quad 8x - 6x - 12 = 5x \quad \Leftrightarrow \quad -12 = 3x \quad \Leftrightarrow$

$-4 = x$.

Graphically: We graph the two equations $y_1 = \dfrac{4}{x+2} - \dfrac{6}{2x}$

and $y_2 = \dfrac{5}{2x+4}$ in the viewing rectangle $[-5, 5]$ by $[-10, 10]$.

Zooming in we see that there is only one solution at $x = -4$.

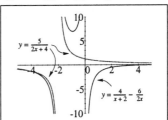

32. Algebraically: $x^3 + 16 = 0 \quad \Leftrightarrow \quad x^3 = -16 \quad \Leftrightarrow \quad x = -2\sqrt[3]{2}$.

Graphically: We graph the equation $y = x^3 + 16$ and

determine where this curve intersects the x-axis. We use

the viewing rectangle $[-5, 5]$ by $[-5, 5]$.

Zooming in, we see that solution is $x \approx -2.52$.

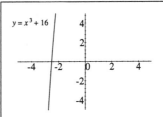

34. Algebraically: $2x^5 - 243 = 0 \quad \Leftrightarrow \quad 2x^5 = 243 \quad \Leftrightarrow \quad x^5 = \frac{243}{2} \quad \Leftrightarrow \quad x = \sqrt[5]{\frac{243}{2}} = \frac{3}{2}\sqrt[5]{16}$.

Graphically: We graph the equation $y = 2x^5 - 243$

and determine where this curve intersects the x-axis. We

use the viewing rectangle $[-5, 10]$ by $[-5, 5]$.

Zooming in, we see that solution is $x \approx 2.61$.

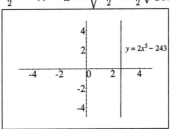

36. Algebraically: $6(x+2)^5 = 64 \quad \Leftrightarrow \quad (x+2)^5 = \frac{64}{6} = \frac{32}{3}$

$\Leftrightarrow \quad x + 2 = \sqrt[5]{\frac{32}{3}} = \frac{2}{3}\sqrt[5]{81} \quad \Leftrightarrow \quad x = -2 + \frac{2}{3}\sqrt[5]{81}$.

Graphically: We graph the two equations $y_1 = 6(x+2)^5$ and

$y_2 = 64$ in the viewing rectangle $[-5, 5]$ by $[50, 70]$.

Zooming in, we see that solution is $x \approx -0.39$.

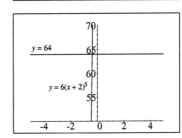

38. We graph $y = x^2 - 0.75x + 0.125$ in the
     viewing rectangle $[-2, 2]$ by $[-0.1, 0.1]$. The
     solutions are $x = 0.25$ and $x = 0.50$.

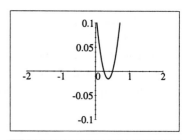

40. Since $16x^3 + 16x^2 = x + 1$ $\Leftrightarrow$ $16x^3 + 16x^2 - x - 1 = 0$,
     we graph $y = 16x^3 + 16x^2 - x - 1$ in the viewing rectangle
     $[-2, 2]$ by $[-0.1, 0.1]$. The solutions are: $x = -1.00$,
     $x = -0.25$, and $x = 0.25$.

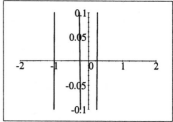

42. $1 + \sqrt{x} = \sqrt{1 + x^2}$ $\Leftrightarrow$ $1 + \sqrt{x} - \sqrt{1 + x^2} = 0$. Since
     $\sqrt{x}$ is only defined for $x \geq 0$, we start with the viewing
     rectangle $[-1, 5]$ by $[-1, 1]$. In this rectangle, we see a solution
     at $x = 0.00$ and another one between $x = 2$ and $x = 2.5$.

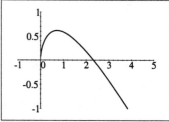

     We then use the viewing rectangle $[2.3, 2.35]$ by $[-.01, .01]$,
     and isolate the second solution as $x \approx 2.314$.
     Thus the solutions are $x = 0.00$ and $x \approx 2.31$.

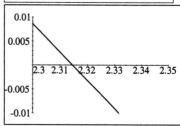

44. Since $x^{1/2}$ is only defined for $x \geq 0$, we start by graphing
     $y = x^{1/2} + x^{1/3} - x$ in the viewing rectangle $[-1, 5]$ by
     $[-1, 1]$. We see a solution at $x = 0.00$ and another one
     between $x = 3$ and $x = 3.5$.

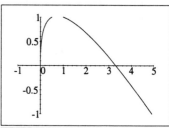

     We then use the viewing rectangle $[3.3, 3.4]$ by $[-.01, .01]$,
     and isolate the second solution as $x \approx 3.31$.
     Thus, the solutions are $x = 0$ and $x \approx 3.31$.

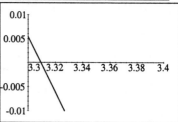

46. $x^4 - 8x^2 + 2 = 0$. We start by graphing the function
     $y = x^4 - 8x^2 + 2$ in the viewing rectangle $[-10, 10]$ by
     $[-10, 10]$. There appears to be four solutions between
     $x = -3$ and $x = 3$.

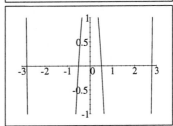

     We then use the viewing rectangle $[-1, 5]$ by $[-1, 1]$,
     and zoom to find the four solutions $x \approx -2.78$, $x \approx -0.51$,
     $x \approx 0.51$, and $x \approx 2.78$.

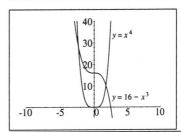

48. $x^4 = 16 - x^3$. We start by graphing the functions $y_1 = x^4$
     and $y_2 = 16 - x^3$ in the viewing rectangle $[-10, 10]$ by
     $[-5, 40]$. There appears to be two solutions, one near
     $x = -2$ and another one near $x = 2$.

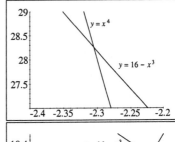

     We then use the viewing rectangle $[-2.4, -2.2]$ by $[27, 29]$,
     and zoom in to find the solution at $x \approx -2.31$.

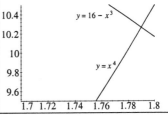

     We then use the viewing rectangle $[1.7, 1.8]$ by $[9.5, 10.5]$,
     and zoom in to find the solution at $x \approx 1.79$.

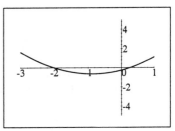

50. Since $0.5x^2 + 0.875x \le 0.25$ $\Leftrightarrow$
     $0.5x^2 + 0.875x - 0.25 \le 0$, we graph
     $y = 0.5x^2 + 0.875x - 0.25$ in the viewing rectangle $[-3, 1]$
     by $[-5, 5]$. Thus the solution to the inequality is
     $-2 \le x \le 0.25$.

52. Since $16x^3 + 24x^2 > -9x - 1$ $\Leftrightarrow$
     $16x^3 + 24x^2 + 9x + 1 > 0$, we graph
     $y = 16x^3 + 24x^2 + 9x + 1$ in the viewing rectangle $[-3, 1]$
     by $[-5, 5]$. From this rectangle, we see that $x = -1$ is an
     x-intercept, but it·is unclear what is occurring between
     $x = -0.5$ and $x = 0$.

     We then use the viewing rectangle $[-1, 0]$ by
     $[-0.01, 0.01]$. It shows $y = 0$ at $x = -0.25$. Thus
     in interval notation, the solution is
     $(-1, -0.25) \cup (-0.25, \infty)$.

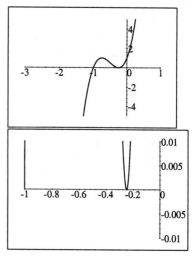

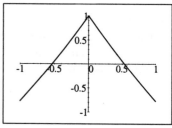

54. Since $\sqrt{0.5x^2 + 1} \le 2|x|$ $\Leftrightarrow$ $\sqrt{0.5x^2 + 1} - 2|x| \le 0$,
     we graph $y = \sqrt{0.5x^2 + 1} - 2|x|$ in the viewing
     rectangle $[-1, 1]$ by $[-1, 1]$. We locate the x-intercepts
     at $x \approx \pm 0.535$. Thus in interval notation, the solution
     is $(-\infty, -0.535] \cup [0.535, \infty)$.

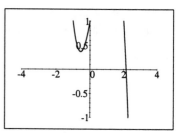

56. Since $(x + 1)^2 \le x^3$ $\Leftrightarrow$ $(x + 1)^2 - x^3 \le 0$, we
     graph $y = (x + 1)^2 - x^3$ in the viewing rectangle
     $[-4, 4]$ by $[-1, 1]$.
     The x-intercept is close to $x = 2$.
     Using TRACE we obtain $x \approx 2.148$.
     Thus the solution is $[2.148, \infty)$.

58. (a)

![graph with axis labels 20000, 15000, 10000, 5000, 0, -5000 and x-axis 100, 200, 300, 400]

     (b) From the graph it appears to be 100 cooktops. Verifying it by substituting $x = 100$ in
          $y = 10x + 0.5x^2 - 0.001x^3 - 5000$ we get $y = 0$.

     (c) Using the ZOOM or TRACE on the calculator we find that the company's profits are greater
          than \$15,000 for $280 < x < 400$.

60. Answers may vary. Possible answers are given.

   (a) Absolute value must be chosen from a menu or written as "abs()".

   (b) Some calculators or computer graphing programs will not take the odd root of a negative number, even though it is defined. Since $\sqrt[5]{-x} = -\sqrt[5]{x}$, you may have to enter two functions: $Y_1 = x^{\wedge}(1/5)$ and $Y_2 = -(-x)^{\wedge}(1/5)$.

   (c) Here you must remember to use parentheses and enter $x/(x-1)$.

   (d) As in part b, some graphing devices my not give the entire graph. When this is the case, you will need to determine where the radicand (inside of the radical) is negative. Here you might need to define 2 functions: $Y_1 = x^{\wedge}3 + (x+2)^{\wedge}(1/3)$ and $Y_2 = x^{\wedge}3 + -(-x-2)^{\wedge}(1/3)$.

62. Answers will vary.

Exercises 2.4

2. $m = \dfrac{y_2 - y_1}{x_2 - x_1} = \dfrac{0 - (-6)}{0 - (2)} = \dfrac{6}{-2} = -3.$

4. $m = \dfrac{y_2 - y_1}{x_2 - x_1} = \dfrac{2 - (3)}{1 - (3)} = \dfrac{-1}{-2} = \dfrac{1}{2}.$

6. $m = \dfrac{y_2 - y_1}{x_2 - x_1} = \dfrac{3 - (-5)}{(-4) - (2)} = \dfrac{8}{-6} = -\dfrac{4}{3}.$

8. $m = \dfrac{y_2 - y_1}{x_2 - x_1} = \dfrac{0 - (-4)}{6 - (-1)} = \dfrac{4}{7}.$

10. (a) 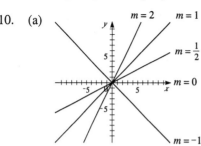   (b)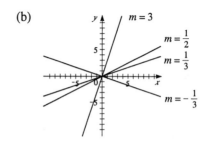

12. We find two points on the graph, $(0, 4)$ and $(-2, 0)$. So the slope is $m = \dfrac{0 - 4}{-2 - 0} = 2$. Since the y-intercept is 4, the equation of the line is $y = mx + b = 2x + 4$, so $y = 2x + 4$ $\Leftrightarrow$ $2x - y + 4 = 0$.

14. We choose the two intercepts, $(0, -4)$ and $(-3, 0)$. So the slope is $m = \dfrac{0 - (-4)}{-3 - 0} = -\dfrac{4}{3}$. Since the y-intercept is -4, the equation of the line is $y = mx + b = -\tfrac{4}{3}x - 4$ $\Leftrightarrow$ $4x + 3y + 12 = 0$.

16. Using the equation $y - y_1 = m(x - x_1)$, we get $y - 4 = -1(x - (-2))$ $\Leftrightarrow$ $y - 4 = -x - 2$ $\Leftrightarrow$ $x + y - 2 = 0$.

18. Using the equation $y - y_1 = m(x - x_1)$, we get $y - (-5) = -\tfrac{7}{2}(x - (-3))$ $\Leftrightarrow$ $2y + 10 = -7x - 21$ $\Leftrightarrow$ $7x + 2y + 31 = 0$.

20. First we find the slope, which is $m = \dfrac{y_2 - y_1}{x_2 - x_1} = \dfrac{3 - (-2)}{4 - (-1)} = \dfrac{5}{5} = 1$. Substituting into $y - y_1 = m(x - x_1)$, we get $y - 3 = 1(x - 4)$ $\Leftrightarrow$ $y - 3 = x - 4$ $\Leftrightarrow$ $x - y - 1 = 0$.

22. Using $y = mx + b$, we have $y = \tfrac{2}{5}x + 4$ $\Leftrightarrow$ $2x - 5y + 20 = 0$.

24. We are given two points, $(-8, 0)$ and $(0, 6)$. Thus, the slope is $m = \dfrac{y_2 - y_1}{x_2 - x_1} = \dfrac{6 - 0}{0 - (-8)} = \tfrac{6}{8} = \tfrac{3}{4}$. Using the y-intercept we have $y = \tfrac{3}{4}x + 6$ $\Leftrightarrow$ $3x - 4y + 24 = 0$.

26.   Any line parallel to the y-axis will have undefined slope and be of the form $x = a$. Since the graph
of the line passes through the point $(4, 5)$, the equation of the line is $x = 4$.

28.   Since $2x + 3y + 4 = 0$ $\Leftrightarrow$ $3y = -2x - 4$ $\Leftrightarrow$ $y = -\frac{2}{3}x - \frac{4}{3}$, the slope of this line is
$m = -\frac{2}{3}$. Substituting $m = -\frac{2}{3}$ and $b = 6$ into the slope intercept formula, the line we seek is given
by $y = -\frac{2}{3}x + 6$ $\Leftrightarrow$ $2x + 3y - 18 = 0$.

30.   Any line perpendicular to $y = 1$ has undefined slope and is of the form $x = a$. Since the graph of
the line passes through the point $(2, 6)$, the equation of the line is $x = 2$.

32.   First find the slope of the line $4x - 8y = 1$. This gives $4x - 8y = 1$ $\Leftrightarrow$ $-8y = -4x + 1$ $\Leftrightarrow$
$y = \frac{1}{2}x - \frac{1}{8}$. So the slope of the line that is perpendicular to $4x - 8y = 1$ is $m = -\frac{1}{1/2} = -2$. The
equation of the line we seek is $y - \left(-\frac{2}{3}\right) = -2\left(x - \frac{1}{2}\right)$ $\Leftrightarrow$ $y + \frac{2}{3} = -2x + 1$ $\Leftrightarrow$
$6x + 3y - 1 = 0$.

34.   First find the slope of the line passing through $(1, 1)$ and $(5, -1)$. This gives
$m = \dfrac{-1 - 1}{5 - 1} = \dfrac{-2}{4} = -\dfrac{1}{2}$, and so the slope of the line that is perpendicular is $m = -\dfrac{1}{-1/2} = 2$.
Thus the equation of the line we seek is $y + 11 = 2(x + 2)$ $\Leftrightarrow$ $2x - y - 7 = 0$.

36.   (a)

(b) $y - (-1) = -2(x - 4)$ $\Leftrightarrow$
$y + 1 = -2x + 8$ $\Leftrightarrow$
$2x + y - 7 = 0$.

38.

$y = mx - 3$, $m = 0$, ± 0.25, ± 0.75, ± 1.5.
Each of the lines contains the point $(0, -3)$ because
the point $(0, -3)$ satisfies each equation $y = mx - 3$.
Since $(0, -3)$ is on the y-axis, we could also say that
they all have the same y-intercept.

40.

$y = 2 + m(x + 3)$, $m = 0$, ± 0.25, ± 0.75, ± 1.5.
Each of the lines contains the point $(-3, 2)$ because
the point $(-3, 2)$ satisfies each equation
$y = 2 + m(x + 3)$.

42. $3x - 2y = 12 \quad \Leftrightarrow \quad -2y = -3x + 12 \quad \Leftrightarrow$
$y = \frac{3}{2}x - 6$. So the slope is $\frac{3}{2}$, and the y-intercept
is -6.

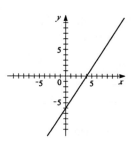

44. $2x - 5y = 0 \quad \Leftrightarrow \quad -5y = -2x \quad \Leftrightarrow \quad y = \frac{2}{5}x$.
So the slope is $\frac{2}{5}$, and the y-intercept is 0.

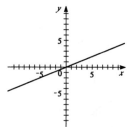

46. $-3x - 5y + 30 = 0 \quad \Leftrightarrow \quad -5y = 3x - 30 \quad \Leftrightarrow$
$y = -\frac{3}{5}x + 6$. So the slope is $-\frac{3}{5}$, and the y-intercept is 6

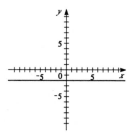

48. $4y + 8 = 0 \quad \Leftrightarrow \quad 4y = -8 \quad \Leftrightarrow \quad y = -2$,
which can also be expressed as $y = 0x - 2$.
So the slope is 0, and the y-intercept is -2.

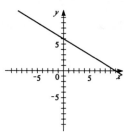

50. $x = -5$ cannot be expressed in the form $y = mx + b$.
So the slope is undefined, and there is no y-intercept.
This is a vertical line.

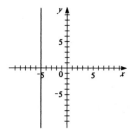

52. $4x + 5y = 10 \iff 5y = -4x + 10 \iff$
$y = -\frac{4}{5}x + 2$. So the slope is $-\frac{4}{5}$, and the
y-intercept is 2.

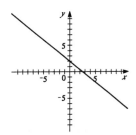

54. We first plot the points to determine the perpendicular sides.
Next find the slopes of the sides.

The slope of AB is $\frac{3-(-1)}{3-(-3)} = \frac{4}{6} = \frac{2}{3}$, and the slope of AC is

$\frac{8-(-1)}{-9-(-3)} = \frac{9}{-6} = -\frac{3}{2}$. Since the

(slope of AB) $\times$ (slope of AC) = $\left(\frac{2}{3}\right)\left(-\frac{3}{2}\right) = -1$, the sides

are perpendicular, and ABC is a right triangle.

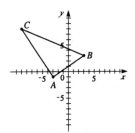

56. (a) The slope of the line passing through $(1, 1)$ and $(3, 9)$ is $\frac{9-1}{3-1} = \frac{8}{2} = 4$. The slope of the line
passing through $(1, 1)$ and $(6, 21)$ is $\frac{21-1}{6-1} = \frac{20}{5} = 4$. Since the slopes are equal, the points are
collinear.

(b) The slope of the line passing through $(-1, 3)$ and $(1, 7)$ is $\frac{7-3}{1-(-1)} = \frac{4}{2} = 2$. The slope of the
line passing through $(-1, 3)$ and $(4, 15)$ is $\frac{15-3}{4-(-1)} = \frac{12}{5}$. Since the slopes are not equal, the
points are not collinear.

58. We find the intercepts (the length of the sides). When $x = 0$, we have $2y + 3(0) - 6 = 0 \iff$
$2y = 6 \iff y = 3$, and when $y = 0$, we have $2(0) + 3x - 6 = 0 \iff 3x = 6 \iff x = 2$.
Thus, the area of the triangle is $\frac{1}{2}(3)(2) = 3$.

60. (a) The line tangent at $(3, -4)$ will be perpendicular to the line passing through the points $(0, 0)$
and $(3, -4)$. The slope of this line is $\frac{-4-0}{3-0} = -\frac{4}{3}$. Thus, the slope of the tangent line will be
$-\frac{1}{(-4/3)} = \frac{3}{4}$. Then the equation of the tangent line is $y - (-4) = \frac{3}{4}(x - 3) \iff$
$4(y + 4) = 3(x - 3) \iff 3x - 4y - 25 = 0$.

(b) Since diametrically opposite points on the circle have parallel tangent lines, the other point is
$(-3, 4)$.

62. (a) The slope represents the increase in the average surface temperature in degrees in °C per year.
The T-intercept is the average surface temperature in 1900, or 8.5°C.

(b) In 2100, $T = 2100 - 1900 = 200$, so $T = 0.02(200) + 8.50 = 12.5$°C.

64. (a)

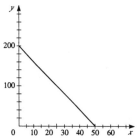

(b) The slope, -4, represents the decline in number of spaces sold for each $1 increase in rent. The y-intercept is the number of spaces at the flea market, 200, and the x-intercept is the cost per space when the manager rents no spaces, $50.

66. (a)

C	$-30°$	$-20°$	$-10°$	$0°$	$10°$	$20°$	$30°$
F	$-22°$	$-4°$	$14°$	$32°$	$50°$	$68°$	$86°$

(b) Substituting a for both F and C, we have $a = \frac{9}{5}a + 32$ $\Leftrightarrow$ $-\frac{4}{5}a = 32$ $\Leftrightarrow$ $a = -40°$. Thus both scales agree at $-40°$.

68. (a) Using t in place of x and V in place of y, we find the slope of the line using the points $(0, 4000)$ and $(4, 200)$. Thus, the slope is $m = \frac{200-4000}{4-0} = \frac{-3800}{4} = -950$. Using the V-intercept, the linear equation is $V = -950t + 4000$.

(b)

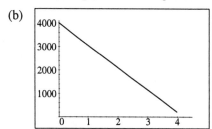

(c) The slope represents the rate of depreciation of the computer. The V-intercept represents the cost of the computer.

(d) When $t = 3$, the value of the computer is given by $V = -950(3) + 4000 = 1150$.

70. (a) Using $rate = \frac{distance}{time}$, we have $rate = \dfrac{40}{50/60} = 48$ mi/h. So $d = 48t$.

(b)

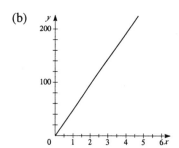

(c) The slope of the line is 48 and it represents the speed in mi/h.

72. (a) Using the points $(100, 2200)$ and $(300, 4800)$, we
find that the slope is $\frac{4800-2200}{300-100} = \frac{2600}{200} = 13$. So
$y - 2200 = 13(x - 100) \quad \Leftrightarrow \quad y = 13x + 900$.

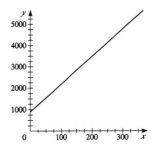

(b) The slope of the line in part (a) is 13, and it
represents the cost of producing each additional chair.

(c) The y-intercept is 900, and it represents the fixed daily
costs of operating the factory.

74. We label the three points A, B, and C.

If the slope of the line segment $\overline{AB}$ is equal to the slope of the line segment $\overline{BC}$, then the points A,
B, and C are collinear.

Using the distance formula, we find the distance between A and B, between B and C, and between
A and C. If the sum of the two smaller distances equals the largest distance, the points A, B, and C
are collinear.

Another method: Find an equation for the line through A and B. Then check if C satisfies the
equation. If so, the points are collinear.

Exercises 2.5

2. $P = kw$, where k is constant.   4. $w = kmn$, where k is constant.

6. $P = \dfrac{k}{T}$, where k is constant.   8. $A = k\dfrac{t^2}{x^3}$, where k is constant.

10. $S = kr^2\theta^2$, where k is constant.   12. $A = k\sqrt{xy}$, where k is constant.

14. $z = \dfrac{k}{t}$. Since $z = 5$ when $t = 3$, we have $5 = \dfrac{k}{3}$ $\Leftrightarrow$ $k = 15$, so $z = \dfrac{15}{t}$.

16. $S = kpq$. Since $S = 180$ when $p = 4$ and $q = 5$, we have $180 = k(4)(5)$ $\Leftrightarrow$ $180 = 20k$ $\Leftrightarrow$ $k = 9$. So $S = 9pq$.

18. $t = k\dfrac{xy}{r}$. Since $t = 25$ when $x = 2$, $y = 3$, and $r = 12$, we have $25 = k\dfrac{(2)(3)}{12}$ $\Leftrightarrow$ $k = 50$. So $t = 50\dfrac{xy}{r}$.

20. $H = kl^2w^2$. Since $H = 36$ when $l = 2$ and $w = \frac{1}{3}$, we have $36 = k(2)^2\left(\frac{1}{3}\right)^2$ $\Leftrightarrow$ $36 = \frac{4}{9}k$ $\Leftrightarrow$ $k = 81$. So $H = 81l^2w^2$.

22. $M = k\dfrac{abc}{d}$. Since $M = 128$ when $a = d$ and $b = c = 2$, we have $128 = k\dfrac{a(2)(2)}{a} = 4k$ $\Leftrightarrow$ $k = 32$. So $M = 32\dfrac{abc}{d}$.

24. (a) Let T and l be the period and the length of the pendulum, respectively. Then $T = k\sqrt{l}$.

    (b) $T = k\sqrt{l}$ $\Rightarrow$ $T^2 = k^2l$ $\Leftrightarrow$ $l = \dfrac{T^2}{k^2}$. If the period is doubled, the new length is $\dfrac{(2T)^2}{k^2} = 4\dfrac{T^2}{k^2} = 4l$. So we would quadruple the length l to double the period T.

26. (a) $P = \dfrac{kT}{V}$.

    (b) Substituting $P = 33.2$, $T = 400$, and $V = 100$, we get $33.2 = \dfrac{k(400)}{100}$ $\Leftrightarrow$ $k = 8.3$. Thus $k = 8.3$ and the equation is $P = \dfrac{8.3T}{V}$.

    (b) Substituting $T = 500$ and $V = 80$, we have $P = \dfrac{8.3(500)}{80} = 51.875$ kPa. Hence the pressure of the sample of gas is about 51.9 kPa.

28. $P = ks^3$. Since $P = 80$ when $s = 10$ we have $80 = k(10)^3$ so $k = 0.08$. Thus $P = 0.08s^3$. When $s = 15$ we have $P = 0.08(15)^3 = 270$. So 270 hp is needed to power the boat at 15 knots.

30. $D = ks^2$. Since $D = 240$ when $s = 50$ we have $240 = k(50)^2$ so $k = 0.096$. Thus $D = 0.096s^2$. When $D = 160$ then $160 = 0.096s^2$ $\Leftrightarrow$ $s^2 = 1666.7$ so $s \approx 40$ mi/h (for safety reasons we round down).

32. $L = ks^2 A$. Since $L = 1700$ when $s = 50$ and $A = 500$, we have $1700 = k(50^2)(500)$ $\Leftrightarrow$ $k = 0.00136$. Thus $L = 0.00136s^2 A$. When $A = 600$ and $s = 40$ we get the lift is $L = 0.00136(40^2)(600) = 1305.6$ lb.

34. (a) $F = k\dfrac{ws^2}{r}$.

   (b) For the first car we have $w_1 = 1600$ and $s_1 = 60$ and for the second car we have $w_2 = 2500$. Since the forces are equal we have $k\dfrac{1600 \cdot 60^2}{r} = k\dfrac{2500 \cdot s_2^2}{r}$ $\Leftrightarrow$ $\dfrac{16 \cdot 60^2}{25} = s_2^2$ so $s_2 = 48$ mi/h.

36. (a) $T^2 = kd^3$.

   (b) Substituting $T = 365$ and $d = 93 \times 10^6$, we get $365^2 = k \cdot (93 \times 10^6)^3$ $\Leftrightarrow$ $k = 1.66 \times 10^{-19}$.

   (c) $T^2 = 1.66 \times 10^{-19}(2.79 \times 10^9)^3 = 3.60 \times 10^9$ $\Rightarrow$ $T = 6.00 \times 10^4$. Hence the period of Neptune is 6.00×10^4 days ≈ 164 years.

38. Let V be the value of a building lot on Galiano Island, A the area of the lot, and q the quantity of the water produced. Since V is jointly proportional to the area and water quantity, we have $V = kAq$. When $A = 200 \cdot 300 = 60,000$ and $q = 10$, we have $V = \$48,000$, so $48,000 = k(60,000)(10)$ $\Leftrightarrow$ $k = 0.08$. Thus $V = 0.08Aq$. Now when $A = 400 \cdot 400 = 160,000$ and $q = 4$, the value is $V = 0.08(160,000)(4) = \$51,200$.

40. Let H be the heat experienced by a hiker at a campfire, let A be the amount of wood, and let d be the distance from campfire. So $H = k\dfrac{A}{d^3}$. When the hiker is 20 feet from the fire, the heat experienced is $H = k\dfrac{A}{20^3}$, and when the amount of wood is doubled, the heat experienced is $H = k\dfrac{2A}{d^3}$. So $k\dfrac{A}{8,000} = k\dfrac{2A}{d^3}$ $\Leftrightarrow$ $d^3 = 16,000$ $\Leftrightarrow$ $d = 20\sqrt[3]{2} \approx 25.2$ feet.

42. (a) Since r is jointly proportional to x and $P - x$, we have $r = kx(P - x)$, where k is a positive constant.

   (b) When 10 people are infected the rate is $r = k10(5000 - 10) = 49{,}000k$. When 1000 people are infected the rate is $r = k \cdot 1000 \cdot (5000 - 1000) = 4{,}000{,}000k$. So the rate is much higher when 1000 people are infected. Comparing these rates we get $\dfrac{1000 \text{ people infected}}{10 \text{ people infected}} = \dfrac{4{,}00{,}000k}{49{,}000k} \approx 82$. So the infection rate when 1000 people are infected is about 82 times larger than when 10 people are infected.

   (c) When the entire population is infected the rate is $r = k(5000)(5000 - 5000) = 0$. This makes sense since there are no more people who can be infected.

Review Exercises for Chapter 2

2.　(a)

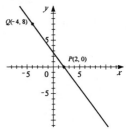

(b)　The distance from P to Q is

$$d(P,Q) = \sqrt{[2-(-4)]^2 + (0-8)^2}$$
$$= \sqrt{36+64} = \sqrt{100} = 10.$$

(c)　The midpoint is $\left(\frac{2+(-4)}{2}, \frac{0+8}{2}\right) = (-1, 4).$

(d)　The line has slope $m = \frac{0-8}{2-(-4)} = \frac{-8}{6} = -\frac{4}{3}$, and has equation $y - 0 = -\frac{4}{3}(x-2) \quad \Leftrightarrow$
$y = -\frac{4}{3}x + \frac{8}{3}.$

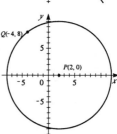

(e)　The radius of this circle was found in part (b). It is $r = d(P,Q) = 10.$ So the equation is
$(x-2)^2 + (y-0)^2 = (10)^2 \quad \Leftrightarrow$
$(x-2)^2 + y^2 = 100.$

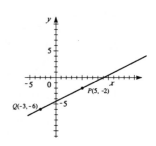

4.　(a)

(b)　The distance from P to Q is

$$d(P,Q) = \sqrt{[5-(-3)]^2 + [-2-(-6)]^2}$$
$$= \sqrt{64+16} = \sqrt{80} = 4\sqrt{5}.$$

(c)　The midpoint is $\left(\frac{5+(-3)}{2}, \frac{-2+(-6)}{2}\right) = (1, -4).$

(d)　The line has slope $m = \frac{-2-(-6)}{5-(-3)} = \frac{4}{8} = \frac{1}{2}$, and has
equation $y - (-2) = \frac{1}{2}(x-5) \quad \Leftrightarrow$
$y + 2 = \frac{1}{2}x - \frac{5}{2} \quad \Leftrightarrow \quad y = \frac{1}{2}x - \frac{9}{2}.$

(e) The radius of this circle was found in part (b).
It is $r = d(P, Q) = 4\sqrt{5}$. So the equation is

$(x - 5)^2 + [y - (-2)]^2 = \left(4\sqrt{5}\right)^2 \quad \Leftrightarrow$

$(x - 5)^2 + (y + 2)^2 = 80.$

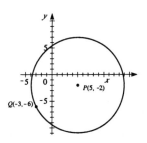

6. $\{(x, y)| \, x \geq 4 \text{ or } y \geq 2\}$

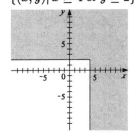

8. The circle with center at $(2, -5)$ and radius $\sqrt{2}$ has equation $(x - 2)^2 + (y + 5)^2 = \left(\sqrt{2}\right)^2 \quad \Leftrightarrow$
$(x - 2)^2 + (y + 5)^2 = 2.$

10. The midpoint of segment PQ is $\left(\frac{2-1}{2}, \frac{3+8}{2}\right) = \left(\frac{1}{2}, \frac{11}{2}\right)$, and the radius is $\frac{1}{2}$ of the distance from P to
Q, or $r = \frac{1}{2} \cdot d(P, Q) = \frac{1}{2}\sqrt{(2 - (-1))^2 + (3 - 8)^2} = \frac{1}{2}\sqrt{(2+1)^2 + (3 - 8)^2} \quad \Leftrightarrow$
$r = \frac{1}{2}\sqrt{34}$. Thus the equation is $(x - \frac{1}{2})^2 + (y - \frac{11}{2})^2 = \frac{17}{2}.$

12. $2x^2 + 2y^2 - 2x + 8y = \frac{1}{2} \quad \Leftrightarrow \quad x^2 - x + y^2 + 4y = \frac{1}{4} \quad \Leftrightarrow$
$\left(x^2 - x + \frac{1}{4}\right) + (y^2 + 4y + 4) = \frac{1}{4} + \frac{1}{4} + 4 \quad \Leftrightarrow \quad \left(x - \frac{1}{2}\right)^2 + (y + 2)^2 = \frac{9}{2}.$ This is the equation
of a circle whose center is $\left(\frac{1}{2}, -2\right)$ and radius is $\frac{3}{\sqrt{2}}$.

14. $x^2 + y^2 - 6x - 10y + 34 = 0 \quad \Leftrightarrow \quad x^2 - 6x + y^2 - 10y = -34 \quad \Leftrightarrow$
$(x^2 - 6x + 9) + (y^2 - 10y + 25) = -34 + 9 + 25 \quad \Leftrightarrow \quad (x - 3)^2 + (y - 5)^2 = 0.$ This is the
equation of the point $(3, 5)$.

16. $2x - y + 1 = 0 \quad \Leftrightarrow \quad y = 2x + 1$
x-axis symmetry: $(-y) = 2x + 1 \quad \Leftrightarrow \quad y = -2x - 1$, which is not the same as the original
equation, so not symmetric with respect to the x-axis.

y-axis symmetry: $y = 2(-x) + 1 \quad \Leftrightarrow \quad y = -2x + 1$, which is not the
same as the original equation, so not symmetric with respect to the y-axis.

Origin symmetry: $(-y) = 2(-x) + 1 \quad \Leftrightarrow$
$-y = -2x + 1 \quad \Leftrightarrow \quad y = 2x - 1$, which is not the
same as the original equation, so not symmetric with
respect to the origin.
Hence the graph has no symmetry.

x	y
-2	-3
0	1
$-\frac{1}{2}$	0

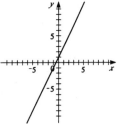

18. $x = 2y + 12$　$\Leftrightarrow$　$2y = x - 12$　$\Leftrightarrow$　$y = \frac{1}{2}x - 6$

x-axis symmetry: $(-y) = \frac{1}{2}x - 6$　$\Leftrightarrow$　$y = -\frac{1}{2}x + 6$, which is not the same as the original equation, so not symmetric with respect to the x-axis.

y-axis symmetry: $y = \frac{1}{2}(-x) - 6$　$\Leftrightarrow$　$y = -\frac{1}{2}x - 6$, which is not the same as the original equation, so not symmetric with respect to the y-axis.

Origin symmetry: $(-y) = \frac{1}{2}(-x) - 6$　$\Leftrightarrow$　$y = \frac{1}{2}x + 6$, which is not the same as the original equation, so not symmetric with respect to the origin.

x	y
-4	-8
0	-6
12	0

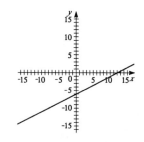

20. $\dfrac{x}{4} + \dfrac{y}{5} = 0$　$\Leftrightarrow$　$5x + 4y = 0$

x-axis symmetry: $5x + 4(-y) = 0$　$\Leftrightarrow$　$5x - 4y = 0$, which is not the same as the original equation, so not symmetric with respect to the x-axis.

y-axis symmetry: $5(-x) + 4y = 0$　$\Leftrightarrow$　$-5x + 4y = 0$, which is not the same as the original equation, so not symmetric with respect to the y-axis.

Origin symmetry: $5(-x) + 4(-y) = 0$　$\Leftrightarrow$　$-5x - 4y = 0$　$\Leftrightarrow$　$5x + 4y = 0$, which is the original equation, so it is symmetric with respect to the origin.

x	y
-4	5
0	0
4	-5

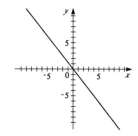

22. $8x + y^2 = 0$　$\Leftrightarrow$　$y^2 = -8x$

x-axis symmetry: $(-y)^2 = -8x$　$\Leftrightarrow$　$y^2 = -8x$, which is the same as the original equation, so it is symmetric with respect to the x-axis.

y-axis symmetry: $y^2 = -8(-x)$　$\Leftrightarrow$　$y^2 = 8x$, which is not the same as the original equation, so not symmetric with respect to the y-axis.

Origin symmetry: $(-y)^2 = -8(-x)$　$\Leftrightarrow$　$y^2 = 8x$, which is not the same as the original equation, so not symmetric with respect to the origin.

x	y
-8	± 8
-2	± 4
0	0

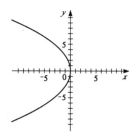

24. $y = -\sqrt{1 - x^2}$

x-axis symmetry: $(-y) = -\sqrt{1 - x^2}$ $\Leftrightarrow$ $y = \sqrt{1 - x^2}$, which is not the same as the original equation, so not symmetric with respect to the x-axis.

y-axis symmetry: $y = -\sqrt{1 - (-x)^2}$ $\Leftrightarrow$ $y = -\sqrt{1 - x^2}$, which is the same as the original equation, so it is symmetric with respect to the y-axis.

Origin symmetry: $(-y) = -\sqrt{1 - (-x)^2}$ $\Leftrightarrow$

$y = \sqrt{1 - x^2}$, which is not the same as the original equation, so not symmetric with respect to the origin.

x	y
-1	0
$\frac{1}{2}$	$-\frac{\sqrt{3}}{2}$
0	-1
1	0

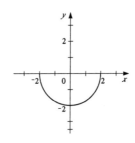

26. $y = \sqrt{5 - x}$;
Viewing rectangle $[-10, 6]$ by $[-1, 5]$.

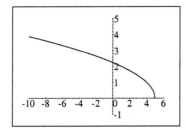

28. $\dfrac{x^2}{4} + y^2 = 1$ $\Leftrightarrow$ $y^2 = 1 - \dfrac{x^2}{4}$ $\Rightarrow$

$$y = \pm\sqrt{1 + \dfrac{x^2}{4}}$$

Viewing rectangle $[-3, 3]$ by $[-2, 2]$.

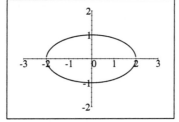

30. $\sqrt{x + 4} = x^2 - 5$. We graph the equations $y_1 = \sqrt{x + 4}$ and $y_2 = x^2 - 5$ in the viewing rectangle $[-4, 5]$ by $[0, 10]$. Using Zoom and/or Trace, we get the solutions $x \approx -2.50$ and $x \approx 2.76$.

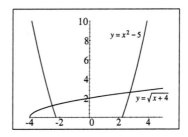

32. $\big||x + 3| - 5\big| = 2$. We graph the equations $y_1 = \big||x + 3| - 5\big|$ and $y_2 = 2$ in the viewing rectangle $[-20, 20]$ by $[0, 10]$. Using Zoom and/or Trace, we get the solutions $x = -10$, $x = -6$, $x = 0$, and $x = 4$.

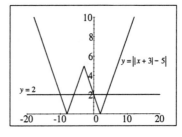

34. $x^3 - 4x^2 - 5x > 2$. We graph the equations
$y_1 = x^3 - 4x^2 - 5x$ and $y_2 = 2$ in the viewing rectangle
$[-10, 10]$ by $[-5, 5]$. We find that the point of intersection
is at $x \approx 5.07$. Since we want $x^3 - 4x^2 - 5x > 2$,
the solution is the interval $(5.07, \infty)$.

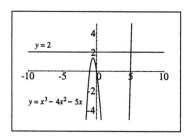

36. $|x^2 - 16| - 10 \geq 0$. We graph the equation
$y = |x^2 - 16| - 10$ in the viewing rectangle $[-10, 10]$
by $[-10, 10]$. Using Zoom and/or Trace, we find that
the points of intersection are at $x \approx \pm 5.10$ and $x \approx \pm 2.45$.
Since we want $|x^2 - 16| - 10 \geq 0$, the solution is
$(-\infty, -5.10] \cup [-2.45, 2.45] \cup [5.10, \infty)$.

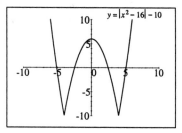

38. Using the point slope formula, the equation is $y + 3 = -\frac{1}{2}(x - 6)$ $\Leftrightarrow$ $y + 3 = -\frac{1}{2}x + 3$ $\Leftrightarrow$
$y = -\frac{1}{2}x$ or $x + 2y = 0$.

40. $x - 3y + 16 = 0$ $\Leftrightarrow$ $-3y = -x - 16$ $\Leftrightarrow$ $y = \frac{1}{3}x + \frac{16}{3}$. So the slope of the perpendicular
line we seek is $m = -\dfrac{1}{(1/3)} = -3$. Then the equation of the perpendicular line passing through the
point $(1, 7)$ is $y - 7 = -3(x - 1)$ $\Leftrightarrow$ $y = -3x + 10$ or $3x + y - 10 = 0$.

42. The line passing through the points $(-1, -3)$ and $(3, 2)$ has slope $m = \frac{2+3}{3+1} = \frac{5}{4}$. Therefore the
equation of the parallel line passing through the point $(5, 2)$ is $y - 2 = \frac{5}{4}(x - 5)$ $\Leftrightarrow$
$4y - 8 = 5x - 25$ $\Leftrightarrow$ $5x - 4y - 17 = 0$.

44. (a) We use the information to find two points, $(0, 60{,}000)$ and $(3, 70{,}500)$. Then the slope is
$m = \frac{70500 - 60000}{3 - 0} = \frac{10500}{3} = 3{,}500$. So $s = 3{,}500t + 60{,}000$.

     (b) The slope represents her annual increase, \$3,500, and the S-intercept represents her initial
salary, \$60,000.

     (c) When $t = 12$, her salary will be $S = 3{,}500(12) + 60{,}000 = 42{,}000 + 60{,}000 = \$102{,}000$.

46. Since z is inversely proportional to y, we have $z = \dfrac{k}{y}$. Substituting $z = 12$ when $y = 16$, we find
$12 = \frac{k}{16}$ $\Leftrightarrow$ $k = 192$. Therefore $z = \dfrac{192}{y}$.

48. Let f be the frequency of the string and l be the length of the string. Since the frequency is inversely
proportional to the length, we have $f = \dfrac{k}{l}$. Substituting $l = 12$ when $k = 440$, we find $440 = \dfrac{k}{12}$
$\Leftrightarrow$ $k = 5280$. Therefore $f = \dfrac{5280}{l}$. For $f = 660$, we must have $660 = \dfrac{5280}{l}$ $\Leftrightarrow$
$l = \frac{5280}{660} = 8$. So the string needs to be shortened to 8 inches.

50. Let r be the maximum range of the baseball and v be the velocity of the baseball. Since the maximum range is directly proportional to the square of the velocity, we have $r = lv^2$. Substituting $v = 60$ and $r = 242$, we find $242 = k(60)^2 \iff k \approx 0.0672$. If $v = 70$, then we have a maximum range of $r = 0.0672(70)^2 = 329.4$ feet.

52. Since the circle is tangent to the x-axis at the point $(5, 0)$ and tangent to the y-axis at the point $(0, 5)$, the center is at $(5, 5)$ and the radius is 5. Thus the equation is $(x - 5)^2 + (y - 5)^2 = 5^2 \iff (x - 5)^2 + (y - 5)^2 = 25$. The slope of the line passing through the points $(8, 1)$ and $(5, 5)$ is $m = \frac{5-1}{5-8} = \frac{4}{-3} = -\frac{4}{3}$, so the equation of the line we seek is $y - 1 = -\frac{4}{3}(x - 8) \iff 4x + 3y - 35 = 0$.

Principles of Modeling

2.　(a)

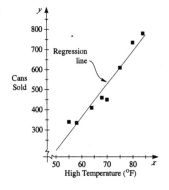

(b)　Using a graphing calculator, we obtain the regression line $y = 16.4163x - 621.83$.

(c)　Using $x = 95$ in the equation $y = 16.4163x - 621.83$, we get $y = 16.4163(95) - 621.83 \approx 938$ cans.

4.　(a)

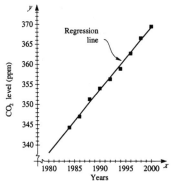

(b)　Using a graphing calculator, we obtain the regression line $y = 1.555x - 2740.8$.

(c)　Using $x = 2001$ in the equation $y = 1.555x - 2740.8$, we get $y = 1.555(2001) - 2740.8 \approx 370.8$ ppm CO_2.

6.　(a)

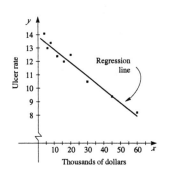

(b)　Using a graphing calculator, we obtain the regression line $y = -0.0995x + 13.9$, where x is measured in thousands of dollars.

(c)　Using the regression line equation, $y = -0.0995x + 13.9$, we get an estimated $y = 11.4$ ulcers per 100 population when $x = \$25,000$.

(d)　Again using $y = -0.0995x + 13.9$, we get an estimated $y = 5.8$ ulcers per 100 population when $x = \$80,000$.

8. (a)

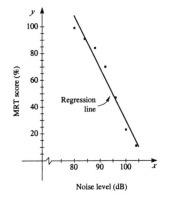

(b) Using a graphing calculator, we obtain
     $y = -3.9018x + 419.7$.

(c) The correlation coefficient is $r = -0.98$,
     so linear model is appropriate for x
     between 80 dB and 104 dB.

(d) Substituting $x = 94$ into the regression
     equation, we get
     $y = -3.9018(94) + 419.7 \approx 53$. So
     the intelligibility is about 53%.

10. (a)

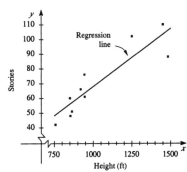

(b) Using a graphing calculator, we obtain
     the regression line $y = 0.0789x - 11.0$.

(c) The slope is 0.0789, it indicates the fraction
     of a story represented by one foot of height
     in the building. Since $\dfrac{1}{0.0789} \approx 12.67$, each
     story is about 12.7 feet.

12. (a) Using a graphing calculator, we obtain the regression line $y = 0.7804x + 15.5$.

    (b) Answers will vary.

14. Results will depend on student surveys in each class.

Chapter Three

Exercises 3.1

2. $f(x) = \dfrac{x+14}{7}$

4. $f(x) = \sqrt{x^2 + 9}$

6. Add 7, then divide by 2.

8. Multiply by 3, subtract 2, then take the square root.

10. Machine diagram for $f(x) = \dfrac{2}{x}$.

12. $g(x) = |2x - 3|$

x	$g(x)$
-2	$|2(-2) - 3| = 7$
0	$|2(0) - 3| = 3$
1	$|2(1) - 3| = 1$
3	$|2(3) - 3| = 3$
5	$|2(5) - 3| = 7$

14. $f(0) = 0^2 + 2(0) = 0;\quad f(3) = 3^2 + 2(3) = 9 + 6 = 15;\quad f(-3) = (-3)^2 + 2(-3) = 9 - 6 = 3;$
$f(a) = a^2 + 2(a) = a^2 + 2a;\quad f(-x) = (-x)^2 + 2(-x) = x^2 - 2x;$
$f\left(\dfrac{1}{a}\right) = \left(\dfrac{1}{a}\right)^2 + 2\left(\dfrac{1}{a}\right) = \dfrac{1}{a^2} + \dfrac{2}{a}.$

16. $h(1) = (1) + \dfrac{1}{(1)} = 2;\quad h(-1) = (-1) + \dfrac{1}{-1} = -1 - 1 = -2;\quad h(2) = 2 + \dfrac{1}{2} = \dfrac{5}{2};$
$h(\tfrac{1}{2}) = \dfrac{1}{2} + \dfrac{1}{\frac{1}{2}} = \dfrac{1}{2} + 2 = \dfrac{5}{2};\quad h(x) = x + \dfrac{1}{x};\quad h\left(\dfrac{1}{x}\right) = \dfrac{1}{x} + \dfrac{1}{\frac{1}{x}} = \dfrac{1}{x} + x.$

18. $f(0) = 0^3 - 4(0)^2 = 0 + 0 = 0;\quad f(1) = 1^3 - 4(1)^2 = 1 - 4 = -3;$
$f(-1) = (-1)^3 - 4(-1)^2 = -1 - 4 = -5;\quad f(\tfrac{3}{2}) = \left(\tfrac{3}{2}\right)^3 - 4\left(\tfrac{3}{2}\right)^2 = \dfrac{27}{8} - 9 = -\dfrac{45}{8};$
$f\left(\dfrac{x}{2}\right) = \left(\dfrac{x}{2}\right)^3 - 4\left(\dfrac{x}{2}\right)^2 = \dfrac{x^3}{8} - x^2;\quad f(x^2) = (x^2)^3 - 4(x^2)^2 = x^6 - 4x^4.$

20. $f(-2) = \dfrac{|-2|}{-2} = \dfrac{2}{-2} = -1;\quad f(-1) = \dfrac{|-1|}{-1} = \dfrac{1}{-1} = -1;\quad f(x)$ is not defined at $x = 0$;

$f(5) = \dfrac{|5|}{5} = \dfrac{5}{5} = 1;\quad f(x^2) = \dfrac{|x^2|}{x^2} = \dfrac{x^2}{x^2} = 1$ since $x^2 > 0,\ x \neq 0;\quad f\left(\dfrac{1}{x}\right) = \dfrac{\left|\frac{1}{x}\right|}{\frac{1}{x}} = \dfrac{x}{|x|}.$

22. Since $-3 \le 2$, we have $f(-3) = 5$. Since $0 \le 2$, we have $f(0) = 5$. Since $2 \le 2$, we have

$f(2) = 5$. Since $3 > 2$, we have $f(3) = 2(3) - 3 = 3$. Since $5 > 2$, we have $f(5) = 2(5) - 3 = 7$.

24. Since $-5 < 0$, we have $f(-5) = 3(-5) = -15$. Since $0 \le 0 \le 2$, we have $f(0) = 0 + 1 = 1$.

Since $0 \le 1 \le 2$, we have $f(1) = 1 + 1 = 2$. Since $0 \le 2 \le 2$, we have $f(2) = 2 + 1 = 3$. Since

$5 > 2$, we have $f(5) = (5 - 2)^2 = 9$.

26. $f(2x) = 3(2x) - 1 = 6x - 1;\quad 2f(x) = 2(3x - 1) = 6x - 2.$

28. $f\left(\dfrac{x}{3}\right) = 6\left(\dfrac{x}{3}\right) - 18 = 2x - 18;\quad \dfrac{f(x)}{3} = \dfrac{6x - 18}{3} = \dfrac{3(2x - 6)}{3} = 2x - 6.$

30. $f(a) = (a)^2 + 1 = a^2 + 1;\quad f(a+h) = (a+h)^2 + 1 = a^2 + 2ah + h^2 + 1;$

$\dfrac{f(a+h) - f(a)}{h} = \dfrac{(a^2 + 2ah + h^2 + 1) - (a^2 + 1)}{h} = \dfrac{a^2 + 2ah + h^2 + 1 - a^2 - 1}{h}$

$= \dfrac{2ah + h^2}{h} = \dfrac{h(2a + h)}{h} = 2a + h.$

32. $f(a) = \dfrac{1}{a + 1};\quad f(a+h) = \dfrac{1}{a + h + 1};$

$\dfrac{f(a+h) - f(a)}{h} = \dfrac{\dfrac{1}{a + h + 1} - \dfrac{1}{a + 1}}{h} = \dfrac{\dfrac{a + 1}{(a + 1)(a + h + 1)} - \dfrac{a + h + 1}{(a + 1)(a + h + 1)}}{h}$

$= \dfrac{\dfrac{-h}{(a + 1)(a + h + 1)}}{h} = \dfrac{-1}{(a + 1)(a + h + 1)}.$

34. $f(a) = a^3;\quad f(a+h) = (a+h)^3 = a^3 + 3a^2 h + 3ah^2 + h^3;$

$\dfrac{f(a+h) - f(a)}{h} = \dfrac{(a^3 + 3a^2 h + 3ah^2 + h^3) - (a^3)}{h}$

$= \dfrac{3a^2 h + 3ah^2 + h^3}{h} = \dfrac{h(3a^2 + 3ah + h^2)}{h} = 3a^2 + 3ah + h^2.$

36. $f(x) = x^2 + 1$. Since there is no restrictions, the domain is all real numbers, $(-\infty, \infty)$.

38. $f(x) = x^2 + 1$. The domain is restricted by the exercise to $[0, 5]$.

40. $f(x) = \dfrac{1}{3x - 6}$. Since the denominator cannot equal 0, we have $3x - 6 \neq 0 \quad \Leftrightarrow \quad 3x \neq 6 \quad \Leftrightarrow$

$x \neq 2$. In interval notation, the domain is $(-\infty, 2) \cup (2, \infty)$.

42. $f(x) = \dfrac{x^4}{x^2 + x - 6}$. Since the denominator cannot equal 0, $x^2 + x - 6 \neq 0 \quad \Leftrightarrow$

$(x + 3)(x - 2) \neq 0 \quad \Rightarrow \quad x \neq -3$ or $x \neq 2$. In interval notation, the domain is
$(-\infty, -3) \cup (-3, 2) \cup (2, \infty)$.

44. $f(x) = \sqrt[4]{x + 9}$. Since even roots are only defined for nonnegative numbers, we must have
$x + 9 \geq 0 \quad \Leftrightarrow \quad x \geq -9$, so the domain is $[-9, \infty)$.

46. $g(x) = \sqrt{7 - 3x}$. For the square root to be defined, we must have $7 - 3x \geq 0 \quad \Leftrightarrow \quad 7 \geq 3x$
$\Leftrightarrow \quad \frac{7}{3} \geq x$. Thus the domain is $(-\infty, \frac{7}{3}]$.

48. $G(x) = \sqrt{x^2 - 9}$. We must have $x^2 - 9 \geq 0 \quad \Leftrightarrow \quad x^2 \geq 9 \quad \Leftrightarrow \quad |x| \geq 3 \quad \Rightarrow \quad x \geq 3$ or
$x \leq -3$. Thus the domain is $(-\infty, -3] \cup [3, \infty)$.

50. $g(x) = \dfrac{\sqrt{x}}{2x^2 + x - 1}$. We must have $x \geq 0$ for the numerator and $2x^2 + x - 1 \neq 0$ for the

denominator. So $2x^2 + x - 1 \neq 0 \quad \Leftrightarrow \quad (2x - 1)(x + 1) \neq 0 \quad \Rightarrow \quad 2x - 1 \neq 0$ or $x + 1 \neq 0$
$\Leftrightarrow \quad x \neq \frac{1}{2}$ or $x \neq -1$. Thus the domain is $[0, \frac{1}{2}) \cup (\frac{1}{2}, \infty)$.

52. $g(x) = \sqrt{x^2 - 2x - 8}$. We must have $x^2 - 2x - 8 \geq 0 \quad \Leftrightarrow \quad (x - 4)(x + 2) \geq 0$. Using the
methods from earlier sections, we have

	$(-\infty, -2)$	$(-2, 4)$	$(4, \infty)$
Sign of $x - 4$	$-$	$-$	$+$
Sign of $x + 2$	$-$	$+$	$+$
Sign of $(x - 4)(x + 2)$	$+$	$-$	$+$

Thus the domain is $(-\infty, -2] \cup [4, \infty)$.

54. $f(x) = \dfrac{x^2}{\sqrt{6 - x}}$. Since the input to an even root must be nonnegative and the denominator cannot

equal 0, we have $6 - x > 0 \quad \Leftrightarrow \quad 6 > x$. Thus the domain is $(-\infty, 6)$.

56. $f(x) = \dfrac{x}{\sqrt[4]{9 - x^2}}$. Since the input to an even root must be nonnegative and the denominator cannot

equal 0, we have $9 - x^2 > 0 \quad \Leftrightarrow \quad (3 - x)(3 + x) > 0$. Using the methods from Chapter 1, we
have

	$(-\infty, -3)$	$(-3, 3)$	$(3, \infty)$
Sign of $3 - x$	$+$	$+$	$-$
Sign of $3 + x$	$-$	$+$	$+$
Sign of $(x - 4)(x + 2)$	$-$	$+$	$-$

Thus the domain is $(-3, 3)$.

58. (a) $S(2) = 4\pi(2)^2 = 16\pi \approx 50.27$, $S(3) = 4\pi(3)^2 = 36\pi \approx 113.10$.

    (b) $S(2)$ represents the surface area of a sphere of radius 2, and $S(3)$ represents the surface area of
    a sphere of radius 3.

60. (a) $V(0) = 50\left(1 - \frac{0}{20}\right)^2 = 50$ and $V(20) = 50\left(1 - \frac{20}{20}\right)^2 = 0$.

(b) $V(0) = 50$ represents the volume of the full tank at time $t = 0$, and $V(20) = 0$ represents the volume of the empty tank twenty minutes later.

(c)

x	$V(x)$
0	50
5	28.125
10	12.5
15	3.125
20	0

62. (a) $R(1) = \sqrt{\dfrac{13 + 7(1)^{0.4}}{1 + 4(1)^{0.4}}} = \sqrt{\dfrac{20}{5}} = 2$, $R(10) = \sqrt{\dfrac{13 + 7(10)^{0.4}}{1 + 4(10)^{0.4}}} \approx 1.66$, and

$R(100) = \sqrt{\dfrac{13 + 7(100)^{0.4}}{1 + 4(100)^{0.4}}} \approx 1.48$.

(b)

x	$R(x)$
1	2
10	1.66
100	1.48
200	1.44
500	1.41
1000	1.39

64. (a) Since $0 \le 5{,}000 \le 10{,}000$ we have $T(5{,}000) = 0$. Since $10{,}000 < 12{,}000 \le 20{,}000$ we have $T(12{,}000) = 0.08(12{,}000) = 960$. Since $20{,}000 < 25{,}000$ we have $T(25{,}000) = 1600 + 0.15(25{,}000) = 1{,}975$.

(b) There is no tax on $5,000, a tax of $960 on $12,000 income, and a tax of $5,350 on $25,000.

66. (a) $T(x) = \begin{cases} 75x & \text{if } 0 \le x \le 2 \\ 50x + 50 & \text{if } x > 2 \end{cases}$.

(b) $T(2) = 75(2) = 150$; $T(3) = 50(3) + 50 = 200$; and $T(5) = 5(50) + 50 = 300$.

(c) The total cost of the lodgings.

68.

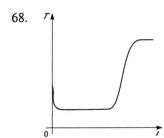

70. The sales of Christmas cards will increase sharply
    approximately from October to December and
    decrease sharply from January to September.

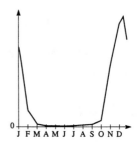

72.

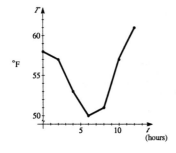

74. Answers will vary.

Exercises 3.2

2.

x	$f(x) = -3$
-4	-3
-2	-3
0	-3
2	-3
4	-3
6	-3

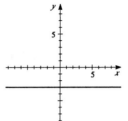

4.

x	$f(x) = 6 - 3x$
-2	12
-1	9
0	6
1	3
2	0
3	-3

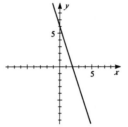

6.

x	$f(x) = \dfrac{x-3}{2}, 0 \le x \le 5$
0	-1.5
1	-1
2	-0.5
3	0
4	0.5
5	1

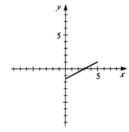

8.

x	$f(x) = x^2 - 4$
± 5	21
± 4	12
± 3	5
± 2	0
± 1	-3
0	-4

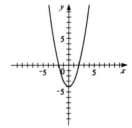

10.

x	$g(x) = 4x^2 - x^4$
± 4	-64
± 3	-45
± 2	0
± 1	3
0	0

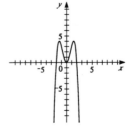

12.

x	$g(x) = \sqrt{-x}$
-9	3
-5	2.236
-4	2
-2	1.414
-1	1
0	0

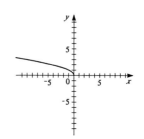

14.

x	$F(x) = \dfrac{1}{x+4}$
-6	-0.5
-5	-1
-4.5	-2
-4.1	-10
-3.9	10
-3.5	2
-3	1
-2	0.5

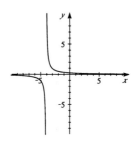

16.

x	$H(x) = |x+1|$
-5	4
-4	3
-3	2
-2	1
-1	0
0	1
1	2

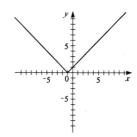

18.

x	$G(x) = |x| - x$
-5	10
-2	4
-1	2
0	0
1	0
3	0

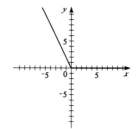

20.

x	$f(x) = \dfrac{x}{|x|}$
-3	-1
-2	-1
-1	-1
0	undefined
1	0
2	0
3	0

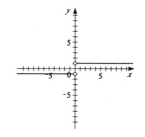

22.

x	$g(x) = \dfrac{\lvert x \rvert}{x^2}$
± 4	0.25
± 2	0.5
± 1	1
± 0.5	2
± 0.25	4
± 0.10	10

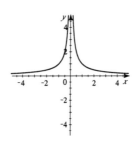

24. (a) $g(-4) = 3$; $g(-2) = 2$; $g(0) = -2$; $g(2) = 1$; $g(4) = 0$.

(b) Domain: $[-4, 4]$. Range: $[-2, 3]$.

26. (a) $f(0.5) \approx 1.2$

(b) $f(3) \approx 2.1$

(c) $x = 0.4$ and $x = 3.6$ are the only solutions to the equation $f(x) = 1$.

28. (a)

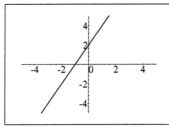

(b) Domain: $(-\infty, \infty)$
Range: $(-\infty, \infty)$

30. (a)

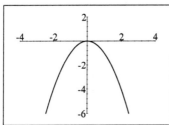

(b) Domain: $(-\infty, \infty)$
Range: $(-\infty, 0]$

32. (a)

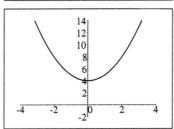

(b) Domain: $(-\infty, \infty)$
Range: $[4, \infty)]$

34. (a)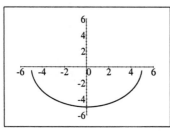

(b) Domain: $[-5, 5]$
    Range: $[-5, 0]$

36. (a)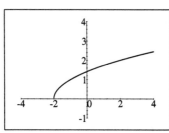

(b) Domain: $[-2, \infty)$
    Range: $[0, \infty)$

38. $f(x) = \begin{cases} 1 & \text{if } x \leq 1 \\ x+1 & \text{if } x > 1 \end{cases}$

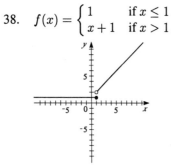

40. $f(x) = \begin{cases} 1-x & \text{if } x < -2 \\ 5 & \text{if } x \geq -2 \end{cases}$

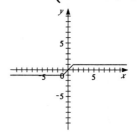

42. $f(x) = \begin{cases} 2x+3 & \text{if } x < -1 \\ 3-x & \text{if } x \geq -1 \end{cases}$

44. $f(x) = \begin{cases} -1 & \text{if } x < -1 \\ x & \text{if } -1 \leq x \leq 1 \\ 1 & \text{if } x > 1 \end{cases}$

46. $f(x) = \begin{cases} 1 - x^2 & \text{if } x \leq 2 \\ x & \text{if } x > 2 \end{cases}$

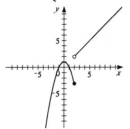

48. $f(x) = \begin{cases} x^2 & \text{if } |x| \leq 1 \\ 1 & \text{if } |x| > 1 \end{cases}$

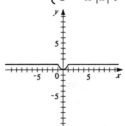

50. $f(x) = \begin{cases} -x & \text{if } x \leq 0 \\ 9 - x^2 & \text{if } 0 < x \leq 3 \\ x - 3 & \text{if } x > 3 \end{cases}$

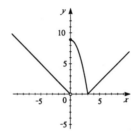

52. $f(x) = \begin{cases} 2x - x^2 & \text{if } x > 1 \\ (x - 1)^3 & \text{if } x \leq 1 \end{cases}$

The viewing rectangle below shows the output of a typical graphing device. However, the graph of this function is also shown below, and its difference from the graphing device's version should be noted.

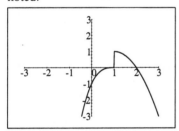

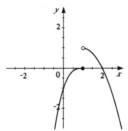

54. The curves in parts (b) and (c) are graphs of functions of x.

56. No, the given curve is not the graph of a function of x.

58. The given curve is the graph of a function of x. Domain: $[-3, 2]$. Range: $\{-2\} \cup (0, 3]$.

60. Solving for y in terms of x gives: $3x + 7y = 21$ $\Leftrightarrow$ $7y = -3x + 21$ $\Leftrightarrow$ $y = -\frac{3}{7}x + 3$. This defines y as a function of x.

62. Solving for y in terms of x gives: $x^2 + (y - 1)^2 = 4$ $\Leftrightarrow$ $(y - 1)^2 = 4 - x^2$ $\Leftrightarrow$ $y - 1 = \pm\sqrt{4 - x^2}$ $\Leftrightarrow$ $y = 1 \pm \sqrt{4 - x^2}$. The last equation gives two values of y for a given value of x. Thus, this equation does not define y as a function of x.

64. Solving for y in terms of x gives: $x^2 + y = 9$ $\Leftrightarrow$ $y = 9 - x^2$. This defines y as a function of x.

66. Solving for y in terms of x gives: $\sqrt{x} + y = 12 \quad \Leftrightarrow \quad y = 12 - \sqrt{x}$. This defines y as a function of x.

68. Solving for y in terms of x gives: $2x + |y| = 0 \quad \Leftrightarrow \quad |y| = -2x$. Since $|a| = |-a|$, the last equation gives two values of y for a given value of x. Thus, this equation does not define y as a function of x.

70. Solving for y in terms of x gives: $x = y^4 \quad \Leftrightarrow \quad y = \pm\sqrt[4]{x}$. The last equation gives two values of y for a given value of x. Thus, this equation does not define y as a function of x.

72. (a) $f(x) = (x - c)^2$, for $c = 0, 1, 2,$ and 3.   (b) $f(x) = (x - c)^2$, for $c = 0, -1, -2,$ and -3.

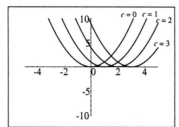   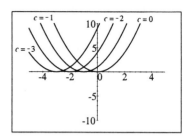

(c) The graphs in part (a) are obtained by shifting the graph of $y = x^2$ to the right 1, 2, and 3 units, while the graphs in part (b) are obtained by shifting the graph of $y = x^2$ to the left 1, 2, and 3 units.

74. (a) $f(x) = cx^2$, for $c = 1, \frac{1}{2}, 2,$ and 4.   (b) $f(x) = cx^2$, for $c = 1, -1, -\frac{1}{2},$ and -2.

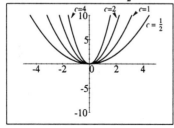   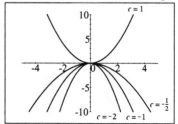

(c) As $|c|$ increases, the graph of $f(x) = cx^2$ is stretched vertically. As $|c|$ decreases, the graph of f is flattened. When $c < 0$, the graph is reflected about the x-axis.

76. (a) $f(x) = \dfrac{1}{x^n}$, for $n = 1$ and 3.   (b) $f(x) = \dfrac{1}{x^n}$, for $n = 2$ and 4.

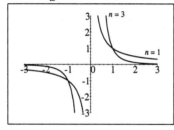   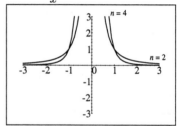

(c) As n increases, the graphs of $y = \dfrac{1}{x^n}$ go to zero faster for x large. Also, as n increases and x goes to 0, the graphs of $y = \dfrac{1}{x^n}$ go to infinity faster. The graphs of $y = \dfrac{1}{x^n}$ for n odd are similar to each other. Likewise, the graphs for n even are similar to each other.

78. The slope of the line containing the points $(-3, -2)$ and $(6, 3)$ is $m = \frac{-2-3}{-3-6} = \frac{-5}{-9} = \frac{5}{9}$. Using the point-slope equation of the line, we have $y - 3 = \frac{5}{9}(x - 6) \iff y = \frac{5}{9}x - \frac{10}{3} + 3 = \frac{5}{9}x - \frac{1}{3}$. Thus the function is $f(x) = \frac{5}{9}x - \frac{1}{3}$, for $-3 \le x \le 6$.

80. First solve the circle for y: $x^2 + y^2 = 9 \iff y^2 = 9 - x^2 \implies y = \pm\sqrt{9 - x^2}$. Since we seek the bottom half of the circle, we choose $y = -\sqrt{9 - x^2}$. So the function is $f(x) = -\sqrt{9 - x^2}$, $-3 \le x \le 3$.

82. The salesman travels away from home and stops to make a sales call between 9 a.m. and 10 a.m., and then travels further from home for a sales call between 12 noon and 1 p.m. Next he travels along a route that takes him closer to home before taking him further away from home. He then makes a final sales call between 5 p.m. and 6 p.m. and then returns home.

84. (a) At 6 A.M. the graph shows that the power consumption is about 500 megawatts. Since $t = 18$ represents 6 P.M., the graph shows that the power consumption at 6 P.M. is about 725 megawatts.

(b) The power consumption is lowest between 3 A.M. and 4 A.M..

(c) The power consumption is highest just before 12 noon.

86. $E(x) = \begin{cases} 6.00 + 0.10x & \text{if } 0 \le x \le 300 \\ 36.00 + 0.06(x - 300) & \text{if } 300 < x \end{cases}$

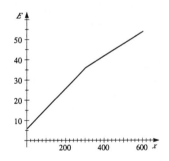

88. $P(x) = \begin{cases} 0.37 & \text{if } 0 < x \le 1 \\ 0.60 & \text{if } 1 < x \le 2 \\ 0.83 & \text{if } 2 < x \le 3 \\ \vdots & \\ 2.90 & \text{if } 11 < x \le 12 \end{cases}$

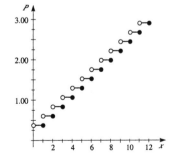

90. Answers vary. Some examples are almost anything we purchase based on weight, volume, length, or time, for example gasoline. Although the amount delivered by the pump is continuous, the amount we pay is rounded to the penny. An example involving time would be the cost of a telephone call.

92. (a) The graphs of $f(x) = x^2 + x - 6$ and $g(x) = |x^2 + x - 6|$ are shown in the viewing rectangle $[-10, 10]$ by $[-10, 10]$.

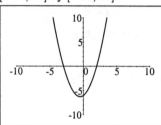 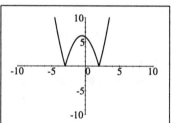

For those values of x where $f(x) \geq 0$, the graphs of f and g coincide, and for those values of x where $f(x) < 0$, the graph of g is obtained from that of f by reflecting the part below the x-axis about the x-axis.

(b) The graphs of $f(x) = x^4 - 6x^2$ and $g(x) = |x^4 - 6x^2|$ are shown in the viewing rectangle $[-5, 5]$ by $[-10, 15]$.

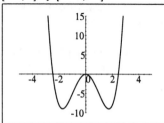 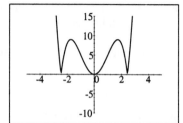

For those values of x where $f(x) \geq 0$, the graphs of f and g coincide, and for those values of x where $f(x) < 0$, the graph of g is obtained from that of f by reflecting the part below the x-axis above the x-axis.

(c) In general, if $g(x) = |f(x)|$, then for those values of x where $f(x) \geq 0$, the graphs of f and g coincide, and for those values of x where $f(x) < 0$, the graph of g is obtained from that of f by reflecting the part below the x-axis above the x-axis.

$y = f(x)$ $y = g(x)$

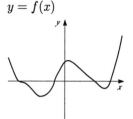 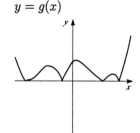

Exercises 3.3

2. The function is increasing on $[0, 1]$. It is decreasing on $[-2, 0]$ and $[1, 3]$.

4. The function is increasing on $[-1, 1]$. It is decreasing on $[-2, -1]$ and $[1, 2]$.

6. (a) $f(x) = 4 - x^{2/3}$ is graphed in the viewing rectangle $[-10, 10]$ by $[-10, 10]$.

    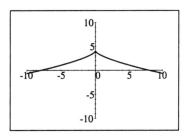

    (b) The function is increasing on $(-\infty, 0]$. It is decreasing on $[0, \infty)$.

8. (a) $f(x) = x^3 - 4x$ is graphed in the viewing rectangle $[-10, 10]$ by $[-10, 10]$.

    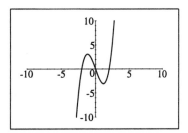

    (b) The function is increasing on $(-\infty, -1.15]$ and $[1.15, \infty)$. It is decreasing on $[-1.15, 1.15]$.

10. (a) $f(x) = x^4 - 16x^2$ is graphed in the viewing rectangle $[-10, 10]$ by $[-70, 10]$.

    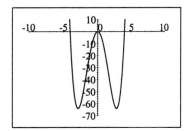

    (b) The function is increasing on $[-2.83, 0]$ and $[2.83, \infty)$. It is decreasing on $(-\infty, -2.83]$ and $[0, 2.83]$.

12. (a) $f(x) = x^4 - 4x^3 + 2x^2 + 4x - 3$ is graphed in the viewing rectangle $[-3, 5]$ by $[-5, 5]$.

    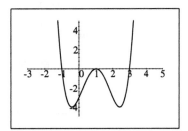

    (b) The function is increasing on $(-0.4, 1]$ and $[2.4, \infty)$. It is decreasing on $[-\infty, -0.4]$ and $[1, 2.4]$.

14. We use the points $(1, 4)$ and $(5, 2)$, so the average rate of change $= \frac{2-4}{5-1} = \frac{-2}{4} = -\frac{1}{2}$.

16. We use the points $(-1, 0)$ and $(5, 4)$, so the average rate of change $= \frac{4-0}{5-(-1)} = \frac{4}{6} = \frac{2}{3}$.

18. Average rate of change $= \frac{g(5) - g(1)}{5 - 1} = \frac{\left[5 + \frac{1}{2}(5)\right] - \left[5 + \frac{1}{2}(1)\right]}{4} = \frac{\frac{15}{2} - \frac{11}{2}}{4} = \frac{2}{4} = \frac{1}{2}$.

20. Average rate of change $= \frac{f(0) - f(-2)}{0 - (-2)} = \frac{[1 - 3(0)^2] - [1 - 3(-2)^2]}{2} = \frac{1 - (-11)}{2} = \frac{12}{2} = 6$.

22. Average rate of change $= \frac{f(3) - f(-1)}{3 - (-1)} = \frac{[3 + 3^4] - [(-1) + (-1)^4]}{4} = \frac{3 + 81}{4} = \frac{84}{4} = 21$.

24. Average rate of change $= \frac{f(1 + h) - f(1)}{(1 + h) - 1} = \frac{[4 - (1 + h)^2] - [4 - 1^2]}{h} = \frac{3 - 2h - h^2 - 3}{h}$

$$= \frac{-2h - h^2}{h} = \frac{h(-2 - h)}{h} = -2 - h.$$

26. Average rate of change $= \frac{g(h) - g(0)}{h - 0} = \frac{\frac{2}{h+1} - \frac{2}{0+1}}{h} \cdot \frac{(h+1)}{(h+1)} = \frac{2 - 2(h+1)}{h(h+1)}$

$$= \frac{-2h}{h(h+1)} = \frac{-2}{h+1}.$$

28. Average rate of change $= \frac{f(a + h) - f(a)}{(a + h) - a} = \frac{\sqrt{a+h} - \sqrt{a}}{h} \cdot \frac{\sqrt{a+h} + \sqrt{a}}{\sqrt{a+h} + \sqrt{a}}$

$$= \frac{(a + h) - a}{h\left(\sqrt{a+h} + \sqrt{a}\right)} = \frac{h}{h\left(\sqrt{a+h} + \sqrt{a}\right)} = \frac{1}{\sqrt{a+h} + \sqrt{a}}.$$

30. (a) Average rate of change $= \frac{g(a + h) - g(a)}{(a + h) - a} = \frac{[-4(a + h) + 2] - [-4a + 2]}{h}$

$$= \frac{-4a - 4h + 2 + 4a - 2}{h} = \frac{-4h}{h} = -4.$$

(b) The slope of the line $g(x) = -4x + 2$ is -4, which is also the average rate of change.

32. (a) The function P is increasing on $[0, 25]$ and decreasing on $[25, 50]$.

(b) Average rate of change $= \frac{P(40) - P(20)}{40 - 20} = \frac{40 - 40}{40 - 20} = \frac{0}{20} = 0$.

(c) The population increased and decreased the same amount during the 20 years.

34. (a) Average speed $= \frac{800 - 400}{152 - 68} = \frac{400}{84} = \frac{100}{21} \approx 4.76$ m/s.

(b) Average speed $= \frac{1,600 - 1,200}{412 - 263} = \frac{400}{149} \approx 2.68$ m/s.

(c)

Lap	Length of time to run lap	Average speed of lap.
1	32	6.25 m/s
2	36	5.56 m/s
3	40	5.00 m/s
4	44	4.55 m/s
5	51	3.92 m/s
6	60	3.33 m/s
7	72	2.78 m/s
8	77	2.60 m/s

The man is slowing down throughout the run.

36.

Year	Number of books
1980	420
1981	460
1982	500
1985	620
1990	820
1992	900
1995	1020
1997	1100
1998	1140
1999	1180
2000	1220

38. (a)

Time (s)	Average speed (ft/s)
0-2	17
2-4	18
4-6	63
6-8	147
8-10	237

The average speed is increasing, so the car is accelerating. From the shape of the graph, we see that the slope continues to get steeper.

(b)

Time (s)	Average speed (ft/s)
30-32	263
32-34	144
34-36	91
36-38	74
38-40	48

The average speed is decreasing, so the car is decelerating. From the shape of the graph, we see that the slope continues to get flatter.

Exercises 3.4

2. (a) Shift the graph of $y = f(x)$ to the left 4 units.

   (b) Shift the graph of $y = f(x)$ upward 4 units.

4. (a) Reflect the graph of $y = f(x)$ about the x-axis.

   (b) Reflect the graph of $y = f(x)$ about the y-axis.

6. (a) Reflect the graph of $y = f(x)$ about the x-axis, and then shift upward 5 units.

   (b) Stretch the graph of $y = f(x)$ vertically by a factor of 3, and then shift downward 5 units.

8. (a) Shift the graph of $y = f(x)$ to the left 2 units, stretch vertically by a factor of 2, and then shift downward 2 units.

   (b) Shift the graph of $y = f(x)$ to the right 2 units, stretch vertically by a factor of 2, and then shift upward 2 units.

10. (a) Shrink the graph of $y = f(x)$ horizontally by a factor of $\frac{1}{2}$, and then reflect the graph of $y = f(x)$ about the x-axis.

    (b) Shrink the graph of $y = f(x)$ horizontally by a factor of $\frac{1}{2}$ and then shift downward 1 unit.

12. $g(x) = f(x) + 3 = x^3 + 3$.

14. $g(x) = 2f(x) = 2|x|$.

16. $g(x) = -f(x - 2) + 1 = -(x - 2)^2 + 1 = -x^2 + 4x - 3$

18. (a) $y = \frac{1}{3}f(x)$ is graph #2.   (b) $y = -f(x + 4)$ is graph #3.

    (c) $y = f(x - 5) + 3$ is graph #1.

20. (a) $y = g(x + 1)$ (b) $y = -g(x + 1)$ (c) $y = g(x - 2)$

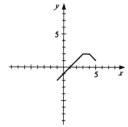

(d) $y = g(x) - 2$

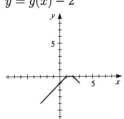

(e) $y = -g(x) + 2$

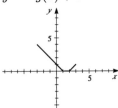

(f) $y = 2g(x)$

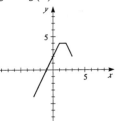

22. (a) $f(x) = \sqrt[3]{x}$

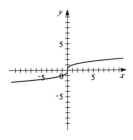

(b) (i) $\sqrt[3]{x} - 2$. Shift the graph of $f(x) = \sqrt[3]{x}$ to the right 2 units.

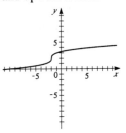

(ii) $y = \sqrt[3]{x + 2} + 2$. Shift the graph of $f(x) = \sqrt[3]{x}$ to the left 2 units and upward 2 units.

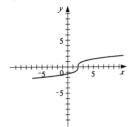

(iii) $y = 1 - \sqrt[3]{x}$. Reflect the graph of $f(x) = \sqrt[3]{x}$ about the x-axis, and then shift it upward 1 unit.

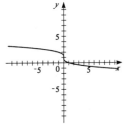

24. (a) The graph of $g(x) = (x - 4)^3$ is obtained by shifting the graph of $f(x)$ to the right 4 units.

(b) The graph of $g(x) = x^3 - 4$ is obtained by shifting the graph of $f(x)$ downward 4 units.

26. (a) The graph of $g(x) = 3|x| + 1$ is obtained by stretching the graph of $f(x)$ vertically by a factor of 3, then shifting the graph upward 1 unit.

(b) The graph of $g(x) = -|x+1|$ is obtained by shifting the graph of $f(x)$ left 1 unit, then reflecting the graph about the x-axis.

28. $y = f(x+4) - 1$. When $f(x) = x^3$, we get $y = (x+4)^3 - 1 = x^3 + 12x^2 + 48x + 64 - 1$
$= x^3 + 12x^2 + 48x + 63$.

30. $y = \frac{1}{2}f(-x) + \frac{3}{5}$. When $f(x) = \sqrt[3]{x}$, we get $y = \frac{1}{2}\sqrt[3]{-x} + \frac{3}{5}$.

32. $y = 3f(x+1) + 10$. When $f(x) = |x|$, we get $y = 3|x+1| + 10$

34. $f(x) = (x+7)^2$. Shift the graph of $y = x^2$ to the left 7 units.

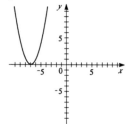

36. $f(x) = 1 - x^2$. Reflect the graph of $y = x^2$ about the x-axis, then shift it upward 1 unit.

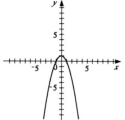

38. $f(x) = -x^3$. Reflect the graph of $y = x^3$ about the x-axis.

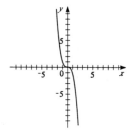

40. $y = 2 - \sqrt{x+1}$. Shift the graph of $y = \sqrt{x}$ to the left 1 unit, reflect it about the x-axis, and finally shift it upward 2 units.

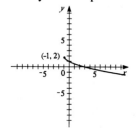

42. $y = 3 - 2(x-1)^2$. Shift the graph of $y = x^2$ to the right 1 unit, reflect it about the x-axis, stretch it vertically by a factor of 2, and then shift it upward 3 units.

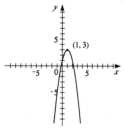

44. $y = \frac{1}{3}x^3 - 1$. Shrink the graph of $y = x^3$ vertically by a factor of $\frac{1}{3}$, then shift it downward 1 unit.

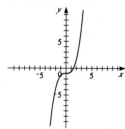

46. $y = |x - 1|$. Shift the graph of $y = |x|$ to the right 1 unit.

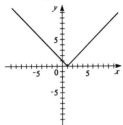

48. $y = 2 - |x|$. Reflect the graph of $y = |x|$ about the x-axis, and then shift it upward 2 units.

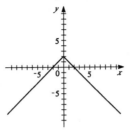

50.

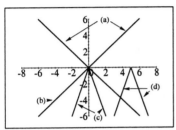

For (b), reflect the graph in (a) about the x-axis; for (c), stretch the graph in (a) vertically by a factor of 3 and reflect about the x-axis; for (d), shift the graph in (a) to the right 5 units, stretch it vertically by a factor of 3, and reflect it about the x-axis. The order in which each operation is applied to the graph in (a) is not important to obtain the graphs in part (c) and (d).

52.

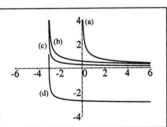

For (b), shift the graph in (a) to the left 3 units; for (c), shift the graph in (a) to the left 3 units and shrink it vertically by a factor of $\frac{1}{2}$; for (d), shift the graph in (a) to the left 3 units, shrink it vertically by a factor of $\frac{1}{2}$, and then shift it downward 3 units. The order in which each operation is applied to the graph in (a) is not important to sketch (c), while it is important in (d).

54. (a) $y = h(3x)$

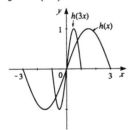

(b) $y = h\left(\frac{1}{3}x\right)$

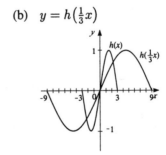

56. (a) Even

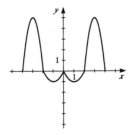

(b) Odd

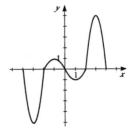

58. $y = \left[\!\left[\frac{1}{4}x\right]\!\right]$

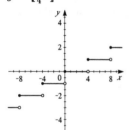

60. (a) $y = f(x) = \sqrt{2x - x^2}$ (b) $y = f(-x) = \sqrt{2(-x) - (-x)^2}$
                                                          $= \sqrt{-2x - x^2}$

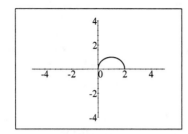   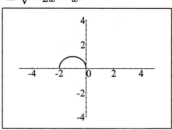

(c) $y = -f(-x) = -\sqrt{2(-x) - (-x)^2}$ (d) $y = f(-2x) = \sqrt{2(-2x) - (-2x)^2}$
        $= -\sqrt{-2x - x^2}$ $= \sqrt{-4x - 4x^2}$

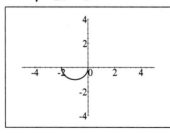   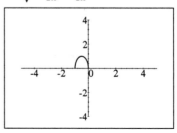

(e) $y = f(-\frac{1}{2}x) = \sqrt{2(-\frac{1}{2}x) - (-\frac{1}{2}x)^2}$ The graph in part (b) is obtained by reflecting the graph
        $= \sqrt{-x - \frac{1}{4}x^2}$ in part (a) about the y-axis. The graph in part (c) is
                                                                          obtained by reflecting the graph in part (a) about the
                                                                          origin. The graph in part (d) is obtained by reflecting
                                                                          the graph in part (a) about the y-axis and then horizontally
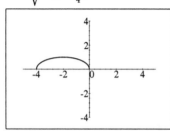   shrinking the graph by a factor of 2 (so the graph is $\frac{1}{2}$ as
                                                                          wide). The graph in part (e) is obtained by reflecting the
                                                                          graph is part (a) about the y-axis and then horizontally
                                                                          stretching the graph by a factor of 2 (so the graph is twice
                                                                          as wide).

62. $f(x) = x^{-3}$. $f(-x) = (-x)^{-3}$
     $= -x^{-3} = -f(x)$. Thus $f(x)$ is odd.

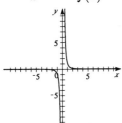

64. $f(x) = x^4 - 4x^2$. $f(-x) = (-x)^4 - 4(-x)^2$
     $= x^4 - 4x^2 = f(x)$. Thus $f(x)$ is even.

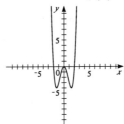

66. $f(x) = 3x^3 + 2x^2 + 1$.
     $f(-x) = 3(-x)^3 + 2(-x)^2 + 1$
     $= -3x^3 + 2x^2 + 1$. Thus $f(-x) \neq f(x)$.
     Also $f(-x) \neq -f(x)$, so $f(x)$ is neither odd
     nor even.

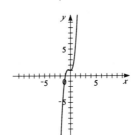

68. $f(x) = x + \dfrac{1}{x}$.

     $f(-x) = (-x) + \dfrac{1}{(-x)} = -x - \dfrac{1}{x}$

     $= -\left(x + \dfrac{1}{x}\right) = -f(x)$. Thus $f(x)$ is odd.

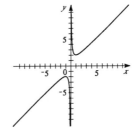

70. $g(x) = |x^4 - 4x^2|$

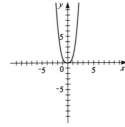

72. (a) $f(x) = x^3$

(b) $g(x) = |x^3|$

74. (a) The function $y = t^2$ must be shrunk vertically by a factor of $\frac{1}{2}$ and shifted up 2 units to obtain the function $y = C(t)$.

(b) The function $y = C(t)$ must be stretched vertically by a factor of $\frac{9}{5}$ and shifted up 32 units to obtain the function $y = F(t)$. So $F(t) = \frac{9}{5}C(t) + 32 = \frac{9}{5}(\frac{1}{2}t^2 + 2) + 32 = \frac{9}{10}t^2 + \frac{178}{5}$.

76. f even implies $f(-x) = f(x)$; g even implies $g(-x) = g(x)$; f odd implies $f(-x) = -f(x)$; and g odd implies $g(-x) = -g(x)$.

If f and g are both even, then $(fg)(-x) = f(-x) \cdot g(-x) = f(x) \cdot g(x) = (fg)(x)$. Thus fg is even.

If f and g are both odd, then $(fg)(-x) = f(-x) \cdot g(-x) = -f(x) \cdot (-g(x)) = f(x) \cdot g(x)$ $= (fg)(x)$. Thus fg is even

If f if odd and g is even, then $(fg)(-x) = f(-x) \cdot g(-x) = f(x) \cdot (-g(x)) = -f(x) \cdot g(x)$ $= -(fg)(x)$. Thus fg is odd.

Exercises 3.5

2. (a) Vertex: $(-2, 8)$ (b) Maximum value of f: 8.

4. (a) Vertex: $(-1, -4)$ (b) Minimum value of f: -4.

6. $y = x^2 + 8x$
Vertex: $y = x^2 + 8x = x^2 + 8x + 16 - 16 = (x + 4)^2 - 16$.
So the vertex is at $(-4, -16)$.
x-intercepts: $y = 0 \Rightarrow 0 = x^2 + 8x = x(x + 8)$. So $x = 0$ or
$x = -8$. The x-intercepts are at $x = 0$ and $x = -8$.
y-intercepts: $x = 0 \Rightarrow y = 0$. The y-intercept is at $y = 0$.

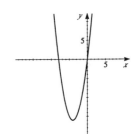

8. $y = -x^2 + 10x$
Vertex: $y = -x^2 + 10x = -(x^2 - 10x + 25) + 25 = -(x - 5)^2 + 25$.
Vertex is at $(5, 25)$.
x-intercepts: $y = 0 \Rightarrow 0 = -x^2 + 10x = -x(x - 10) = 0 \Rightarrow$
$x = 0$ or $x = 10$. The x-intercepts are at $x = 0$ and $x = 10$.
y-intercepts: $x = 0 \Rightarrow y = 0$. The y-intercept is at $y = 0$.

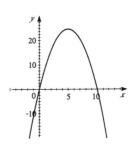

10. $y = x^2 - 2x + 2$
Vertex: $y = x^2 - 2x + 2 = x^2 - 2x + 1 + 1 = (x - 1)^2 + 1$. Vertex
s at $(1, 1)$.
x-intercepts: $y = 0 \Rightarrow (x - 1)^2 + 1 = 0 \Leftrightarrow (x - 1)^2 = -1$.
Since this last equation has no real solution, there is no x-intercept.
y-intercepts: $x = 0 \Rightarrow y = 2$. The y-intercept is at $y = 2$.

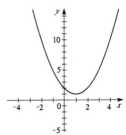

12. $y = -x^2 - 4x + 4$
Vertex: $y = -x^2 - 4x + 4 = -(x^2 + 4x) + 4$
$= -(x^2 + 4x + 4) + 4 + 4 = -(x + 2)^2 + 8$. Vertex is at $(-2, 8)$.
x-intercepts: $y = 0 \Rightarrow 0 = -x^2 - 4x + 4 \Leftrightarrow$
$0 = x^2 + 4x - 4$. Using the quadratic formula,
$x = \frac{-4 \pm \sqrt{(4)^2 - 4(1)(-4)}}{2(1)} = \frac{-4 \pm \sqrt{32}}{2} = \frac{2(-2 \pm 2\sqrt{2})}{2} = -2 \pm 2\sqrt{2}$.
The x-intercepts are at $x = -2 + 2\sqrt{2}$ and $x = -2 - 2\sqrt{2}$.
y-intercepts: $x = 0 \Rightarrow y = 4$. The y-intercept is at $y = 4$.

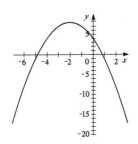

14. $y = -3x^2 + 6x - 2$

Vertex: $y = -3x^2 + 6x - 2 = -3(x^2 - 2x) - 2$
$= -3(x^2 - 2x + 1) - 2 + 3 = -3(x-1)^2 + 1$. Vertex is at $(1, 1)$.

x-intercepts: $y = 0 \Rightarrow 0 = -3(x-1)^2 + 1 = 0 \Leftrightarrow$
$(x-1)^2 = \frac{1}{3} \Rightarrow x - 1 = \pm\sqrt{\frac{1}{3}} \Leftrightarrow x = 1 \pm \sqrt{\frac{1}{3}}$. The

x-intercepts are at are at $x = 1 + \sqrt{\frac{1}{3}}$ and $x = 1 - \sqrt{\frac{1}{3}}$.

y-intercepts: $x = 0 \Rightarrow y = -2$. The y-intercept is at $y = -2$.

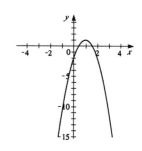

16. $y = 2x^2 + x - 6$

Vertex: $y = 22x^2 + x - 6 = 2(x^2 + \frac{1}{2}x) - 6$
$= 2(x^2 + \frac{1}{2}x + \frac{1}{16}) - 6 - \frac{1}{8} = 2(x + \frac{1}{4})^2 - \frac{49}{8}$. Vertex is
at $(-\frac{1}{4}, -\frac{49}{8})$.

x-intercepts: $y = 0 \Rightarrow 0 = 2x^2 + x - 6 = (2x - 3)(x + 2) \Rightarrow$
$x = \frac{3}{2}$ or $x = -2$. The x-intercepts are at $x = \frac{3}{2}$ and at $x = -2$

y-intercepts: $x = 0 \Rightarrow y = -6$. The y-intercept is at $y = -6$.

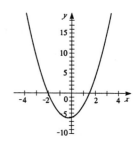

18. $y = 6x^2 + 12x - 5$

Vertex: $y = 6x^2 + 12x - 5 = 6(x^2 + 2x) - 5$
$= 6(x^2 + 2x + 1) - 5 - 6 = 6(x + 1)^2 - 11$. Vertex is at $(-1, -11)$

x-intercepts: $y = 0 \Rightarrow 0 = 6x^2 + 12x - 5$. Using the quadratic

formula, $x = \frac{-12 \pm \sqrt{(12)^2 - 4(6)(-5)}}{2(6)} = \frac{-12 \pm \sqrt{264}}{12} = \frac{-12 \pm 2\sqrt{66}}{12}$
$= \frac{2(-6 \pm \sqrt{66})}{12} = \frac{-6 \pm \sqrt{66}}{6}$. The x-intercepts are at $x = \frac{-6 \pm \sqrt{66}}{6}$.

y-intercepts: $x = 0 \Rightarrow y = -5$. The y-intercept is at $y = -5$.

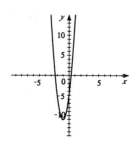

20.   (a) $f(x) = x + x^2 = (x^2 + x + \frac{1}{4}) - \frac{1}{4}$
            $= (x + \frac{1}{2})^2 - \frac{1}{4}$

      (b)
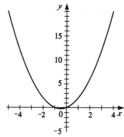

      (c) Therefore, the minimum value is $f(-\frac{1}{2}) = -\frac{1}{4}$.

22.   (a) $f(x) = x^2 - 8x + 8$
            $= (x^2 - 8x + 16) + 8 - 16$
            $= (x - 4)^2 - 8$

      (b)
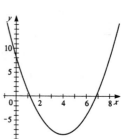

      (c) Therefore, the minimum value is $f(4) = -8$.

24. (a) $f(x) = 1 - 6x - x^2 = -(x^2 + 6x) + 1$ (b)
$$= -(x^2 + 6x + 9) + 1 + 9$$
$$= -(x+3)^2 + 10$$

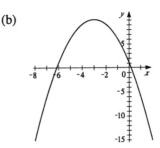

(c) Therefore, the maximum value is $f(-3) = 10$.

26. (a) $g(x) = 2x^2 + 8x + 11 = 2(x^2 + 4x) + 11$ (b)
$$= 2(x^2 + 4x + 4) + 11 - 8$$
$$= 2(x+2)^2 + 3$$

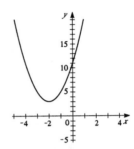

(c) Therefore, the minimum value is $g(-2) = 3$.

28. (a) $h(x) = 3 - 4x - 4x^2 = -4(x^2 + x) + 3$ (b)
$$= -4\left(x^2 + x + \tfrac{1}{4}\right) + 3 + 1$$
$$= -4\left(x + \tfrac{1}{2}\right)^2 + 4$$

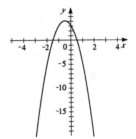

(c) Therefore, the maximum value is $h\left(-\tfrac{1}{2}\right) = 4$.

30. $f(x) = 1 + 3x - x^2 = -(x^2 - 3x) + 1 = -\left(x^2 - 3x + \tfrac{9}{4}\right) + 1 + \tfrac{9}{4} = -\left(x - \tfrac{3}{2}\right)^2 + \tfrac{13}{4}$.
Therefore, the maximum value is $f\left(\tfrac{3}{2}\right) = \tfrac{13}{4}$.

32. $f(t) = 10t^2 + 40t + 113 = 10(t^2 + 4t) + 113 = 10(t^2 + 4t + 4) + 113 - 40 = 10(t+2)^2 + 73$.
Therefore, the minimum value is $f(-2) = 73$.

34. $g(x) = 100x^2 - 1500x = 100(x^2 - 15x) = 100\left(x^2 - 15x + \tfrac{225}{4}\right) - 5625 = 10\left(x - \tfrac{15}{2}\right)^2 - 5625$.
Therefore, the minimum value is $g\left(\tfrac{15}{2}\right) = -5625$.

36. $f(x) = -\dfrac{x^2}{3} + 2x + 7 = -\tfrac{1}{3}(x^2 + 6x) + 7 = -\tfrac{1}{3}(x^2 + 6x + 9) + 7 + 3 = -\tfrac{1}{3}(x+3)^2 + 10$.

Therefore, the maximum value is $f(-3) = 10$.

38. $g(x) = 2x(x - 4) + 7 = 2x^2 - 8x + 7 = 2(x^2 - 4x) + 7 = 2(x^2 - 4x + 4) + 7 - 8$
$= 2(x - 2)^2 - 1$. Therefore, the minimum value is $g(2) = -1$.

40. Since the vertex is $(3, 4)$, the function is of the form $y = a(x - 3)^2 + 4$. Since the parabola passes through the point $(1, -8)$, it must satisfy $-8 = a(1 - 3)^2 + 4$ $\Leftrightarrow$ $-8 = 4a + 4$ $\Leftrightarrow$ $4a = -12$ $\Leftrightarrow$ $a = -3$. So the function is $y = -3(x - 3)^2 + 4 = -3x^2 + 18x - 23$.

42. $f(x) = x^2 - 2x - 3 = (x^2 - 2x + 1) - 3 - 1 = (x - 1)^2 - 4$. Then the domain of the function is all real numbers, and since the minimum value of the function is $f(1) = -4$, the range of the function is $[-4, \infty)$.

44. (a) $f(x) = 1 + x - \sqrt{2}\,x^2$ is shown in the viewing rectangle on the right. The maximum value is $f(x) \approx 1.18$.

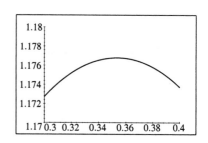

(b) $f(x) = 1 + x - \sqrt{2}\,x^2 = -\sqrt{2}\left[x^2 - \frac{\sqrt{2}}{2}x\right] + 1$

$= -\sqrt{2}\left[x^2 - \frac{\sqrt{2}}{2}x + \left(\frac{\sqrt{2}}{4}\right)^2\right] + 1 + \frac{\sqrt{2}}{8}$

$= -\sqrt{2}\left(x - \frac{\sqrt{2}}{4}\right)^2 + \frac{8+\sqrt{2}}{8}$

Therefore, the exact maximum of $f(x)$ is $\frac{8+\sqrt{2}}{8}$.

46. Local maximum: 2 at $x = -2$ and $x = 1$ at $x = 2$. Local minimum: -1 at $x = 0$.

48. Local maximum: 3 at $x = -2$ and 2 at $x = 1$. Local minimum: 0 at $x = -1$ and -1 at $x = 2$.

50. In the viewing rectangle on the right, we see that $f(x) = 3 + x + x^2 - x^3$ has a local minimum and a local maximum. Smaller x and y ranges (shown in the viewing rectangles below) shows that $f(x)$ has a local maximum of ≈ 4.00 when $x \approx 1.00$ and a local minimum of ≈ 2.81 when $x \approx -0.33$.

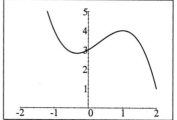

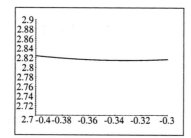

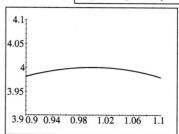

52. In the viewing rectangle on the right, we see that $g(x) = x^5 - 8x^3 + 20x$ has two local minimums and two local maximums. The local maximums are $g(x) \approx 13.02$ when $x = 1.04$ and $g(x) \approx -7.87$ when $x \approx -1.93$. Smaller x and y ranges (shown in the viewing rectangles below) shows that local minimums are $g(x) \approx -13.02$ when $x = -1.04$ and $g(x) \approx 7.87$ when $x \approx 1.93$. Notice that since $g(x)$ is odd, the local maximums and minimums are related.

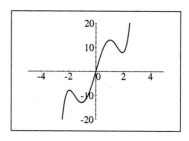

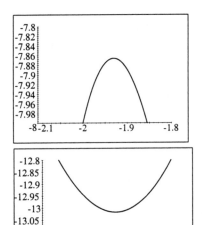

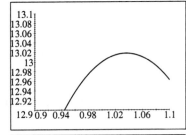

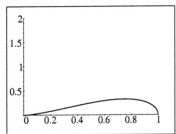

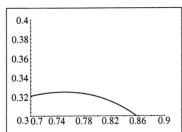

54. In the first viewing rectangle below, we see that $U(x) = x\sqrt{x - x^2}$ has only a local maximum.
     Smaller x and y ranges in the second viewing rectangle below show that $U(x)$ has a local maximum
     of ≈ 0.32 when $x \approx 0.75$.

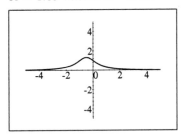

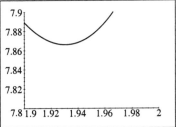

56. In the first viewing rectangle below, we see that $V(x) = \dfrac{1}{x^2 + x + 1}$ only has a local maximum.
     Smaller x and y ranges in the second viewing rectangle below show that $V(x)$ has a local maximum
     of ≈ 1.33 when $x \approx -0.50$.

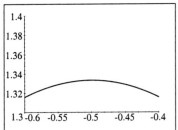

58. (a) $y = -0.005x^2 + x + 5 = -0.005(x^2 - 200x) + 5$
         $= -0.005(x^2 - 200x + 10,000) + 5 + 50$
         $= -0.005(x - 100)^2 + 55$
     Thus the maximum height attained by the football is 55 ft.

(b) We solve for $y = 0$: $0 = -0.005x^2 + x + 5$. Using the quadratic formula, we have
$x = \frac{-1 \pm \sqrt{1^2 - (4)(-0.005)(5)}}{2(-0.005)} = \frac{-1 \pm \sqrt{1.1}}{-0.01} = 100 \pm 100\sqrt{1.1}$. Since he throws the football down
field, we take the positive root, so $x = 100 + 100\sqrt{1.1} \approx 204.9$ feet.

60. $P(x) = -0.001x^2 + 3x - 1800 = -0.001(x^2 - 3000x) - 1800$
$= -0.001(x^2 - 3000x + 2,250,000) - 1800 + 2250$
$= -0.001(x - 1500)^2 + 450$
The vendor's maximum profit occurs when he sells 1500 cans and realizes a profit of $450.

62. $C(t) = 0.06t - 0.0002t^2 = -0.0002(t^2 - 300t) = -0.0002(t^2 - 300t + 22,500) + 4.5$
$= -0.0002(t - 150)^2 + 4.5$.
The maximum concentration occurs after 150 minutes, and the maximum concentration is 4.5 mg/L.

64. In the first viewing rectangle below, we see the general location of the minimum of
$E(v) = 2.73v^3 \dfrac{10}{v - 5}$. In the second viewing rectangle, we isolate the minimum, and from this
graph, we see that energy is minimized when $v \approx 7.5$ mi/h.

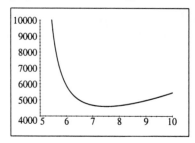   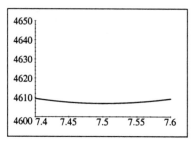

66. In the first viewing rectangle on the next page, we see the general location of the minimum of
$V = 999.87 - 0.06426T + 0.0085043T^2 - 0.0000679T^3$ is around $T = 4$. In the second viewing
rectangle, we isolate the minimum, and from this graph, we see that the minimum volume of 1 kg of
water occurs at $T \approx 3.96°$ C.

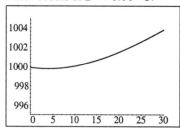   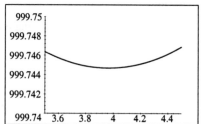

68. Numerous answers are possible.

Exercises 3.6

2. $f(x) = x^3 + 2x^2$ has domain $(-\infty, \infty)$. $g(x) = 3x^2 - 1$ has domain $(-\infty, \infty)$. The intersection of the domains of f and g is $(-\infty, \infty)$.

    $(f + g)(x) = (x^3 + 2x^2) + (3x^2 - 1) = x^3 + 5x^2 - 1$, and the domain is $(-\infty, \infty)$.

    $(f - g)(x) = (x^3 + 2x^2) - (3x^2 - 1) = x^3 - x^2 + 1$, and the domain is $(-\infty, \infty)$.

    $(fg)(x) = (x^3 + 2x^2)(3x^2 - 1) = 3x^5 + 6x^4 - x^3 - 2x^2$, and the domain is $(-\infty, \infty)$.

    $\left(\dfrac{f}{g}\right)(x) = \dfrac{x^3 + 2x^2}{3x^2 - 1}$, $\ 3x^2 - 1 \neq 0 \ \Leftrightarrow \ x^2 \neq \dfrac{1}{3}$, and the domain is $\left\{ x \mid x \neq \pm \dfrac{1}{\sqrt{3}} \right\}$.

4. $f(x) = \sqrt{9 - x^2}$ has domain $[-3, 3]$. $g(x) = \sqrt{x^2 - 1}$ has domain $(-\infty, -1] \cup [1, \infty)$. The intersection of the domains of f and g is $[-3, -1] \cup [1, 3]$.

    $(f + g)(x) = \sqrt{9 - x^2} + \sqrt{x^2 - 1}$, and the domain is $[-3, -1] \cup [1, 3]$.

    $(f - g)(x) = \sqrt{9 - x^2} - \sqrt{x^2 - 1}$, and the domain is $[-3, -1] \cup [1, 3]$.

    $(fg)(x) = \sqrt{9 - x^2} \cdot \sqrt{x^2 - 1} = \sqrt{-x^4 + 10x^2 - 9}$, and the domain is $[-3, -1] \cup [1, 3]$.

    $\left(\dfrac{f}{g}\right)(x) = \dfrac{\sqrt{9 - x^2}}{\sqrt{x^2 - 1}} = \sqrt{\dfrac{9 - x^2}{x^2 - 1}}$, and the domain is $[-3, -1) \cup (1, 3]$.

6. $f(x) = \dfrac{1}{x + 1}$ has domain $x \neq -1$. $g(x) = \dfrac{x}{x + 1}$ has domain $x \neq -1$. The intersection of the domains of f and g is $x \neq -1$, in interval notation, this is $(-\infty, -1) \cup (-1, \infty)$.

    $(f + g)(x) = \dfrac{1}{x + 1} + \dfrac{x}{x + 1} = \dfrac{x + 1}{x + 1} = 1$, and the domain is $(-\infty, -1) \cup (-1, \infty)$.

    $(f - g)(x) = \dfrac{1}{x + 1} - \dfrac{x}{x + 1} = \dfrac{1 - x}{x + 1}$, and the domain is $(-\infty, -1) \cup (-1, \infty)$.

    $(fg)(x) = \dfrac{1}{x + 1} \cdot \dfrac{x}{x + 1} = \dfrac{x}{(x + 1)^2}$, and the domain is $(-\infty, -1) \cup (-1, \infty)$.

    $\left(\dfrac{f}{g}\right)(x) = \dfrac{\frac{1}{x+1}}{\frac{x}{x+1}} = \dfrac{1}{x}$, so $x \neq 0$ as well. Thus the domain is $(-\infty, -1) \cup (-1, 0) \cup (0, \infty)$.

8. $g(x) = \sqrt{x + 1} + \dfrac{1}{x}$. The domain of $\sqrt{x + 1}$ is $[-1, \infty)$, and the domain of $\dfrac{1}{x}$ is $x \neq 0$. Since $x \neq 0$ is $(-\infty, 0) \cup (0, \infty)$, the domain is $[-1, \infty) \cap \{(-\infty, 0) \cup (0, \infty)\} = [-1, 0) \cup (0, \infty)$.

10. $k(x) = \dfrac{\sqrt{x + 3}}{x - 1}$. The domain of $\sqrt{x + 3}$ is $[-3, \infty)$, and the domain of $\dfrac{1}{x - 1}$ is $x \neq 1$. Since $x \neq 1$ is $(-\infty, 1) \cup (1, \infty)$, the domain is $[-3, \infty) \cap \{(-\infty, 1) \cup (1, \infty)\} = [-3, 1) \cup (1, \infty)$.

12.

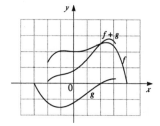

14.

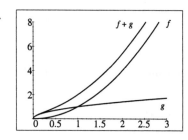

16.

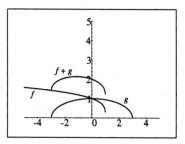

18. (a) $f(f(4)) = f(3(4) - 5) = f(7) = 3(7) - 5 = 16$

(b) $g(g(3)) = g(2 - (3)^2) = g(-7) = 2 - (-7)^2 = -47$

20. (a) $(f \circ f)(-1) = f(f(-1)) = f(3(-1) - 5) = f(-8) = 3(-8) - 5 = -29$

(b) $(g \circ g)(2) = g(g(2)) = g(2 - (2)^2) = g(-2) = 2 - (-2)^2 = -2$

22. (a) $(f \circ f)(x) = f(f(x)) = f(3x - 5) = 3(3x - 5) - 5 = 9x - 15 - 5 = 9x - 20$

(b) $(g \circ g)(x) = g(g(x)) = g(2 - x^2) = 2 - (2 - x^2)^2 = 2 - (4 - 4x^2 + x^4) = -x^4 + 4x^2 - 2$

24. $f(0) = 0$, so $g(f(0)) = g(0) = 3$.

26. $g(0) = 3$, so $(f \circ g)(0) = f(3) = 0$.

28. $f(4) = 2$, so $(f \circ f)(4) = f(2) = -2$.

30. $f(x) = 6x - 5$ has domain $(-\infty, \infty)$. $g(x) = \dfrac{x}{2}$ has domain $(-\infty, \infty)$.

$(f \circ g)(x) = f\left(\dfrac{x}{2}\right) = 6\left(\dfrac{x}{2}\right) - 5 = 3x - 5$, and the domain is $(-\infty, \infty)$.

$(g \circ f)(x) = g(6x - 5) = \dfrac{6x - 5}{2} = 3x - \dfrac{5}{2}$, and the domain is $(-\infty, \infty)$.

$(f \circ f)(x) = f(6x - 5) = 6(6x - 5) - 5 = 36x - 35$, and the domain is $(-\infty, \infty)$.

$(g \circ g)(x) = g\left(\dfrac{x}{2}\right) = \dfrac{\frac{x}{2}}{2} = \dfrac{x}{4}$, and the domain is $(-\infty, \infty)$.

32. $f(x) = x^3 + 2$ has domain $(-\infty, \infty)$. $g(x) = \sqrt[3]{x}$ has domain $(-\infty, \infty)$.

$(f \circ g)(x) = f(\sqrt[3]{x}) = (\sqrt[3]{x})^3 + 2 = x + 2$, and the domain is $(-\infty, \infty)$.

$(g \circ f)(x) = g(x^3 + 2) = \sqrt[3]{x^3 + 2}$ and the domain is $(-\infty, \infty)$.

$(f \circ f)(x) = f(x^3 + 2) = (x^3 + 2)^3 + 2 = x^9 + 6x^6 + 12x^3 + 8 + 2 = x^9 + 6x^6 + 12x^3 + 10$, and the domain is $(-\infty, \infty)$.

$(g \circ g)(x) = g(\sqrt[3]{x}) = \sqrt[3]{\sqrt[3]{x}} = (x^{1/3})^{1/3} = x^{1/9}$, and the domain is $(-\infty, \infty)$.

34. $f(x) = x^2$ has domain $(-\infty, \infty)$. $g(x) = \sqrt{x - 3}$ has domain $[3, \infty)$.

$(f \circ g)(x) = f\left(\sqrt{x - 3}\right) = \left(\sqrt{x - 3}\right)^2 = x - 3$, and the domain is $[3, \infty)$.

$(g \circ f)(x) = g(x^2) = \sqrt{x^2 - 3}$. For the domain we must have $x^2 \geq 3 \quad \Rightarrow \quad x \leq -\sqrt{3}$ or $x \geq \sqrt{3}$. Thus the domain is $(-\infty, -\sqrt{3}] \cup [\sqrt{3}, \infty)$.

$(f \circ f)(x) = f(x^2) = (x^2)^2 = x^4$, and the domain is $(-\infty, \infty)$.

$(g \circ g)(x) = g\left(\sqrt{x - 3}\right) = \sqrt{\sqrt{x - 3} - 3}$. For the domain we must have $\sqrt{x - 3} \geq 3 \quad \Rightarrow \quad x - 3 \geq 9 \quad \Rightarrow \quad x \geq 12$, so the domain is $[12, \infty)$.

36. $f(x) = x - 4$ has domain $(-\infty, \infty)$. $g(x) = |x + 4|$ has domain $(-\infty, \infty)$.

$(f \circ g)(x) = f(|x + 4|) = |x + 4| - 4$, and the domain is $(-\infty, \infty)$.

$(g \circ f)(x) = g(x - 4) = |(x - 4) + 4| = |x|$, and the domain is $(-\infty, \infty)$.

$(f \circ f)(x) = f(x - 4) = (x - 4) - 4 = x - 8$, and the domain is $(-\infty, \infty)$.

$(g \circ g)(x) = g(|x + 4|) = ||x + 4| + 4| = |x + 4| + 4$ ($|x + 4| + 4$ is always positive). The domain is $(-\infty, \infty)$.

38. $f(x) = \dfrac{1}{\sqrt{x}}$ has domain $\{x \mid x > 0\}$; $g(x) = x^2 - 4x$ has domain $(-\infty, \infty)$.

$(f \circ g)(x) = f(x^2 - 4x) = \dfrac{1}{\sqrt{x^2 - 4x}}$. $(f \circ g)(x)$ is defined whenever $0 < x^2 - 4x = x(x - 4)$.

The product of two numbers is positive either when both numbers are negative or when both numbers are positive. So the domain of $f \circ g$ is $\{x \mid x < 0 \text{ and } x < 4\} \cup \{x \mid x > 0 \text{ and } x > 4\}$ which is $(-\infty, 0) \cup (4, \infty)$.

$(g \circ f)(x) = g\left(\dfrac{1}{\sqrt{x}}\right) = \left(\dfrac{1}{\sqrt{x}}\right)^2 - 4\left(\dfrac{1}{\sqrt{x}}\right) = \dfrac{1}{x} - \dfrac{4}{\sqrt{x}}$. $(g \circ f)(x)$ is defined whenever both $f(x)$ and $g(f(x))$ are defined, that is, whenever $x > 0$. So the domain of $g \circ f$ is $(0, \infty)$.

$(f \circ f)(x) = f\left(\dfrac{1}{\sqrt{x}}\right) = \dfrac{1}{\sqrt{\dfrac{1}{\sqrt{x}}}} = x^{1/4}$. $(f \circ f)(x)$ is defined whenever both $f(x)$ and $f(f(x))$ are defined, that is, whenever $x > 0$. So the domain of $f \circ f$ is $(0, \infty)$.

$(g \circ g)(x) = g(x^2 - 4x) = (x^2 - 4x)^2 - 4(x^2 - 4x) = x^4 - 8x^3 + 16x^2 - 4x^2 + 16x = x^4 - 8x^3 + 12x^2 + 16x$, and the domain is $(-\infty, \infty)$.

40. $f(x) = \dfrac{2}{x}$ has domain $\{x \mid x \neq 0\}$; $g(x) = \dfrac{x}{x + 2}$ has domain $\{x \mid x \neq -2\}$.

$(f \circ g)(x) = f\left(\dfrac{x}{x+2}\right) = \dfrac{2}{\frac{x}{x+2}} = \dfrac{2x+4}{x}$. $(f \circ g)(x)$ is defined whenever both $g(x)$ and $f(g(x))$ are defined; that is, whenever $x \neq 0$ and $x \neq -2$. So the domain is $\{x \mid x \neq 0, -2\}$.

$(g \circ f)(x) = g\left(\dfrac{2}{x}\right) = \dfrac{\frac{2}{x}}{\frac{2}{x}+2} = \dfrac{2}{2+2x} = \dfrac{1}{1+x}$. $(g \circ f)(x)$ is defined whenever both $f(x)$ and $g(f(x))$ are defined; that is, whenever $x \neq 0$ and $x \neq -1$. So the domain is $\{x \mid x \neq 0, -1\}$.

$(f \circ f)(x) = f\left(\dfrac{2}{x}\right) = \dfrac{2}{\frac{2}{x}} = x$. $(f \circ f)(x)$ is defined whenever both $f(x)$ and $f(f(x))$ are defined; that is, whenever $x \neq 0$. So the domain is $\{x \mid x \neq 0\}$.

$(g \circ g)(x) = g\left(\dfrac{x}{x+2}\right) = \dfrac{\frac{x}{x+2}}{\frac{x}{x+2}+2} = \dfrac{x}{x+2(x+2)} = \dfrac{x}{3x+4}$. $(g \circ g)(x)$ is defined whenever both $g(x)$ and $g(g(x))$ are defined; that is, whenever $x \neq -2$ and $x \neq -\frac{4}{3}$. So the domain is $\{x \mid x \neq -2, -\frac{4}{3}\}$.

42. $(g \circ h)(x) = g(x^2 + 2) = (x^2 + 2)^3 = x^6 + 6x^4 + 12x^2 + 8$.

     $(f \circ g \circ h)(x) = f(x^6 + 6x^4 + 12x^2 + 8) = \dfrac{1}{x^6 + 6x^4 + 12x^2 + 8}$.

44. $(g \circ h)(x) = g(\sqrt[3]{x}) = \dfrac{\sqrt[3]{x}}{\sqrt[3]{x}-1}$. $(f \circ g \circ h)(x) = f\left(\dfrac{\sqrt[3]{x}}{\sqrt[3]{x}-1}\right) = \sqrt{\dfrac{\sqrt[3]{x}}{\sqrt[3]{x}-1}}$.

46. $F(x) = \sqrt{x} + 1$. If $f(x) = x + 1$ and $g(x) = \sqrt{x}$, then $F(x) = (f \circ g)(x)$.

48. $G(x) = \dfrac{1}{x+3}$. If $f(x) = \dfrac{1}{x}$ and $g(x) = x + 3$, then $G(x) = (f \circ g)(x)$.

50. $H(x) = \sqrt{1 + \sqrt{x}}$. If $f(x) = \sqrt{1+x}$ and $g(x) = \sqrt{x}$, then $H(x) = (f \circ g)(x)$.

For Exercises 52 and 54 there are several possible solutions only one of which is shown.

52. $F(x) = \sqrt[3]{\sqrt{x} - 1}$. If $g(x) = x - 1$ and $h(x) = \sqrt{x}$, then $(g \circ h)(x) = \sqrt{x} - 1$, and if $f(x) = \sqrt[3]{x}$, then $F(x) = (f \circ g \circ h)(x)$.

54. $G(x) = \dfrac{2}{(3 + \sqrt{x})^2}$. If $g(x) = 3 + x$ and $h(x) = \sqrt{x}$, then $(g \circ h)(x) = 3 + \sqrt{x}$, and if $f(x) = \dfrac{2}{x^2}$, then $G(x) = (f \circ g \circ h)(x)$.

56.   (a)   Let $f(t)$ be the radius of the spherical balloon in centimeters. Since the radius is increasing at a rate of 1 cm/s, the radius is $f(t) = t$ after t seconds.

(b)   The volume of the balloon can be written as $g(r) = \frac{4}{3}\pi r^3$.

(c) $g \circ f = \frac{4}{3}\pi(t)^3 = \frac{4}{3}\pi t^3$. $g \circ f$ represents the volume as a function of time.

58.   (a) $f(x) = 0.80x$

     (b) $g(x) = x - 50$

     (c) $f \circ g = f(x - 50) = 0.80(x - 50) = 0.80x - 40$. $f \circ g$ represents applying the \$50 coupon, then the 20% discount.

        $g \circ f = g(0.80x) = 0.80x - 50$. $g \circ f$ represents applying the 20% discount, then the \$50 coupon.

        So applying the 20% discount, then the \$50 coupon gives the lower price.

60.   Let t be the time since the plane flew over the radar station.

     (a)   Let s be the distance in miles between the plane and the radar station, and let d be the horizontal distance that the plane has flown.   Using the Pythagorean theorem,

        $s = f(d) = \sqrt{1 + d^2}$.

     (b)   Since *distance* = *rate* × *time* we have $d = g(t) = 350t$.

     (c) $s(t) = (f \circ g)(t) = f(350t) = \sqrt{1 + (350t)^2} = \sqrt{1 + 122{,}500t^2}$.

62.   Yes. If $f(x) = m_1 x + b_1$ and $g(x) = m_2 x + b_2$, then

     $(f \circ g)(x) = f(m_2 x + b_2) = m_1(m_2 x + b_2) + b_1 = m_1 m_2 x + m_1 b_2 + b_1$, which is a linear function, because it is of the form $y = mx + b$. The slope is $m_1 m_2$.

64.   If $g(x)$ is even, then $h(-x) = f(g(-x)) = f(g(x)) = h(x)$. So yes, h is always an even function.

If $g(x)$ is odd, then h is not necessarily an odd function. For example, if we let $f(x) = x - 1$ and $g(x) = x^3$, g is an odd function, but $h(x) = (f \circ g)(x) = f(x^3) = x^3 - 1$ is not an odd function.

If $g(x)$ is odd and f is also odd, then $h(-x) = (f \circ g)(-x) = f(g(-x)) = f(-g(x)) = -f(g(x))$ $= -(f \circ g)(x) = -h(x)$. So in this case, h is also an odd function.

If $g(x)$ is odd and f is even, then $h(-x) = (f \circ g)(-x) = f(g(-x)) = f(-g(x)) = f(g(x))$ $= (f \circ g)(x) = h(x)$, so in this case, h is an even function.

Exercises 3.7

2. By the Horizontal Line Test, f is one-to-one.

4. By the Horizontal Line Test, f is not one-to-one.

6. By the Horizontal Line Test, f is one-to-one.

8. $f(x) = -2x + 5$. If $x_1 \neq x_2$, then $-2x_1 \neq -2x_2$ and $-2x_1 + 5 \neq -2x_2 + 5$. So f is a one-to-one function.

10. $g(x) = |x|$. Since every number and its negative have the same absolute value, that is, $|-1| = 1 = |1|$, g is not a one-to-one function.

12. $h(x) = x^3 + 8$. If $x_1 \neq x_2$, then $x_1^3 \neq x_2^3$ and $x_1^3 + 8 \neq x_2^3 + 8$. So f is a one-to-one function.

14. $f(x) = x^4 + 5$, $0 \leq x \leq 2$. If $x_1 \neq x_2$, then $x_1^4 \neq x_2^4$ because two different <u>positive</u> numbers cannot have the same fourth power. Thus, $x_1^4 + 5 \neq x_2^4 + 5$. So f is a one-to-one function.

16. $f(x) = \dfrac{1}{x}$. If $x_1 \neq x_2$, then $\dfrac{1}{x_1} \neq \dfrac{1}{x_2}$. So f is a one-to-one function.

18. (a) $f(5) = 18$. Since f is one-to-one, $f^{-1}(18) = 5$.

    (b) $f^{-1}(4) = 2$. Since f is one-to-one, $f(2) = 4$.

20. To find $g^{-1}(5)$, we find the x value such that $g(x) = 5$; that is, we solve the equation
    $g(x) = x^2 + 4x = 5$. Now $x^2 + 4x = 5 \quad \Leftrightarrow \quad x^2 + 4x - 5 = 0 \quad \Leftrightarrow \quad (x - 1)(x + 5) = 0$
    $\Leftrightarrow \quad x = 1$ or $x = -5$. Since the domain of g is $[-2, \infty)$, $x = 1$ is the only value where $g(x) = 5$.
    Therefore, $g^{-1}(5) = 1$.

22. $f(g(x)) = f\left(\dfrac{x}{2}\right) = 2\left(\dfrac{x}{2}\right) = x$, for all x.

    $g(f(x)) = g(2x) = \dfrac{2x}{2} = x$, for all x. Thus f and g are inverses of each other.

24. $f(g(x)) = f(3 - 4x) = \dfrac{3 - (3 - 4x)}{4} = \dfrac{3 - 3 + 4x}{4} = x$, for all x.

    $g(f(x)) = g\left(\dfrac{3-x}{4}\right) = 3 - 4\left(\dfrac{3-x}{4}\right) = 3 - 3 + x = x$, for all x. Thus f and g are inverses of each other.

26. $f(g(x)) = f(\sqrt[5]{x}) = (\sqrt[5]{x})^5 = x$, for all x.

    $g(f(x)) = g(x^5) = \sqrt[5]{x^5} = x$, for all x. Thus f and g are inverses of each other.

28. $f(g(x)) = f((x - 1)^{1/3}) = ((x - 1)^{1/3})^3 + 1 = x - 1 + 1 = x$, for all x.

    $g(f(x)) = g(x^3 + 1) = \left[(x - 1)^{1/3}\right]^3 + 1 = x - 1 + 1 = x$, for all x. Thus f and g are inverses of each other.

30. $f(g(x)) = f\left(\sqrt{4-x^2}\right) = \sqrt{4 - \left(\sqrt{4-x^2}\right)^2} = \sqrt{4-4+x^2} = \sqrt{x^2} = x$, for all $0 \le x \le 2$.

(Note that the last equality is possible since $x \ge 0$.)

$g(f(x)) = g\left(\sqrt{4-x^2}\right) = \sqrt{4 - \left(\sqrt{4-x^2}\right)^2} = \sqrt{4-4+x^2} = \sqrt{x^2} = x$, for all $0 \le x \le 2$.

(Again, the last equality is possible since $x \ge 0$.) Thus f and g are inverses of each other.

32. $f(x) = 6 - x$. $y = 6 - x$ $\Leftrightarrow$ $x = 6 - y$. So $f^{-1}(x) = 6 - x$.

34. $f(x) = 3 - 5x$. $y = 3 - 5x$ $\Leftrightarrow$ $-5x = y - 3$ $\Leftrightarrow$ $x = -\frac{1}{5}(y-3) = \frac{1}{5}(3-y)$. So $f^{-1}(x) = \frac{1}{5}(3-x)$.

36. $f(x) = \dfrac{1}{x^2}$, $x > 0$. Since the function f is restricted to positive values, f is one-to-one. So $y = \dfrac{1}{x^2}$

$\Leftrightarrow$ $\sqrt{y} = \dfrac{1}{x}$ $\Leftrightarrow$ $x = \dfrac{1}{\sqrt{y}}$. Thus $f^{-1}(x) = \dfrac{1}{\sqrt{x}}$, $x > 0$.

38. $f(x) = \dfrac{x-2}{x+2}$. $y = \dfrac{x-2}{x+2}$ $\Leftrightarrow$ $y(x+2) = x-2$ $\Leftrightarrow$ $xy + 2y = x - 2$ $\Leftrightarrow$

$xy - x = -2 - 2y$ $\Leftrightarrow$ $x(y-1) = -2(y+1)$ $\Leftrightarrow$ $x = \dfrac{-2(y+1)}{y-1}$. So

$f^{-1}(x) = \dfrac{-2(x+1)}{x-1}$.

40. $f(x) = 5 - 4x^3$. $y = 5 - 4x^3$ $\Leftrightarrow$ $4x^3 = 5 - y$ $\Leftrightarrow$ $x^3 = \frac{1}{4}(5-y)$ $\Leftrightarrow$ $x = \sqrt[3]{\frac{1}{4}(5-y)}$. So $f^{-1}(x) = \sqrt[3]{\frac{1}{4}(5-x)}$.

42. $f(x) = x^2 + x = (x^2 + x + \frac{1}{4}) - \frac{1}{4} = (x+\frac{1}{2})^2 - \frac{1}{4}$, $x \ge -\frac{1}{2}$. $y = (x+\frac{1}{2})^2 - \frac{1}{4}$ $\Leftrightarrow$ $(x+\frac{1}{2})^2 = y + \frac{1}{4}$ $\Leftrightarrow$ $x + \frac{1}{2} = \sqrt{y + \frac{1}{4}}$ $\Leftrightarrow$ $x = \sqrt{y+\frac{1}{4}} - \frac{1}{2}$, $y \ge -\frac{1}{4}$. So

$f^{-1}(x) = \sqrt{x + \frac{1}{4}} - \frac{1}{2}$, $x \ge -\frac{1}{4}$. (Note that $x \ge -\frac{1}{2}$, so that $x + \frac{1}{2} \ge 0$, and hence

$(x+\frac{1}{2})^2 = y + \frac{1}{4}$ $\Leftrightarrow$ $x + \frac{1}{2} = \sqrt{y+\frac{1}{4}}$. Also, since $x \ge -\frac{1}{2}$, $y = (x+\frac{1}{2})^2 - \frac{1}{4} \ge -\frac{1}{4}$ so that

$y + \frac{1}{4} \ge 0$, and hence $\sqrt{y+\frac{1}{4}}$ is defined.)

44. $f(x) = \sqrt{2x-1}$. $y = \sqrt{2x-1}$ $\Leftrightarrow$ $2x - 1 = y^2$ $\Leftrightarrow$ $x = \frac{1}{2}(y^2 + 1)$. Since the range of f is $f(x) \ge 0$ so $f^{-1}(x) = \frac{1}{2}(x^2 + 1)$ for $x \ge 0$.

46. $f(x) = (2 - x^3)^5$. $y = (2 - x^3)^5$ $\Leftrightarrow$ $2 - x^3 = \sqrt[5]{y}$ $\Leftrightarrow$ $x^3 = 2 - \sqrt[5]{y}$ $\Leftrightarrow$ $x = \sqrt[3]{2 - \sqrt[5]{y}}$. So $f^{-1}(x) = \sqrt[3]{2 - \sqrt[5]{x}}$.

48. $f(x) = \sqrt{9 - x^2}$, $0 \le x \le 3$. $y = \sqrt{9 - x^2}$ $\Leftrightarrow$ $y^2 = 9 - x^2$ $\Leftrightarrow$ $x^2 = 9 - y^2$ $\Rightarrow$ $x = \sqrt{9 - y^2}$ (since we must have $x \ge 0$). So $f^{-1}(x) = \sqrt{9 - x^2}$, $0 \le x \le 3$.

50. $f(x) = 1 - x^3$. $y = 1 - x^3$ $\Leftrightarrow$ $x^3 = 1 - y$ $\Leftrightarrow$ $x = \sqrt[3]{1-y}$. So $f^{-1}(x) = \sqrt[3]{1-x}$.

52. (a) $f(x) = 16 - x^2$, $x \geq 0$ (b)

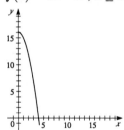   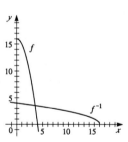

(c) $f(x) = 16 - x^2$, $x \geq 0$. $y = 16 - x^2$ $\Leftrightarrow$ $x^2 = 16 - y$ $\Leftrightarrow$ $x = \sqrt{16 - y}$. So
$f^{-1}(x) = \sqrt{16 - x}$, $x \leq 16$. (Note: $x \geq 0$ $\Rightarrow$ $f(x) = 16 - x^2 \leq 16$.)

54. (a) $f(x) = x^3 - 1$ (b)

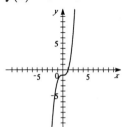   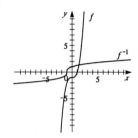

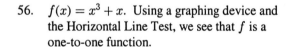

(c) $f(x) = x^3 - 1$ $\Leftrightarrow$ $y = x^3 - 1$ $\Leftrightarrow$ $x^3 = y + 1$ $\Leftrightarrow$ $x = \sqrt[3]{y + 1}$. So
$f^{-1}(x) = \sqrt[3]{x + 1}$.

56. $f(x) = x^3 + x$. Using a graphing device and
the Horizontal Line Test, we see that f is a
one-to-one function.

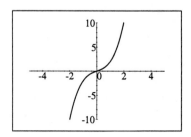

58. $f(x) = \sqrt{x^3 - 4x + 1}$. Using a graphing
device and the Horizontal Line Test, we see
that f is not a one-to-one function. For
example, $f(0) = 1 = f(2)$.

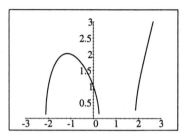

60. $f(x) = x \cdot |x|$. Using a graphing device and
     the Horizontal Line Test, we see that f is a
     one-to-one function.

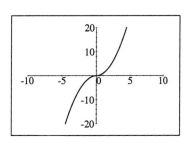

62. If we restrict the domain of $g(x)$ to $[1, \infty)$, then $y = (x - 1)^2 \;\Rightarrow\; x - 1 = \sqrt{y}$ (since $x \geq 1$ we
     take the positive square root) $\;\Leftrightarrow\; x = 1 + \sqrt{y}$. So $g^{-1}(x) = 1 + \sqrt{x}$.
     If we restrict the domain of $g(x)$ to $(-\infty, 1]$, then $y = (x - 1)^2 \;\Rightarrow\; x - 1 = -\sqrt{y}$ (since $x \leq 1$
     we take the negative square root) $\;\Leftrightarrow\; x = 1 - \sqrt{y}$. So $g^{-1}(x) = 1 - \sqrt{x}$.

64. $k(x) = |x - 3| = \begin{cases} -(x - 3) & \text{if } x - 3 < 0 \;\Leftrightarrow\; x < 3 \\ x - 3 & \text{if } x - 3 \geq 0 \;\Leftrightarrow\; x \geq 3 \end{cases}$

     If we restrict the domain of $k(x)$ to $[3, \infty)$, then $y = x - 3 \;\Leftrightarrow\; x = 3 + y$. So $k^{-1}(x) = 3 + x$.
     If we restrict the domain of $k(x)$ to $(-\infty, 3]$, then $y = -(x - 3) \;\Leftrightarrow\; y = -x + 3 \;\Leftrightarrow\;$
     $x = 3 - y$. So $k^{-1}(x) = 3 - x$.

66.

67.

68. (a) $V(t) = 100\left(1 - \dfrac{t}{40}\right)^2,\, 0 \leq t \leq 40.$ $y = 100\left(1 - \dfrac{t}{40}\right)^2 \;\Leftrightarrow\; \dfrac{y}{100} = \left(1 - \dfrac{t}{40}\right)^2 \;\Rightarrow\;$

     $1 - \dfrac{t}{40} = \pm\sqrt{\dfrac{y}{100}} \;\Leftrightarrow\; \dfrac{t}{40} = 1 \pm \dfrac{\sqrt{y}}{10} \;\Leftrightarrow\; t = 40 \pm 4\sqrt{y}.$ Since $t \leq 40$ we must

     have $V^{-1}(t) = 40 - 4\sqrt{t}$. V^{-1} represents time that has elapsed since the tank started to leak.

     (b) $V^{-1}(15) = 40 - 4\sqrt{15} \approx 24.5$ minutes. In 24.5 minutes the tank has drained to just 15
          gallons of water.

70. (a) $D(p) = -3p + 150.$ $y = -3p + 150 \;\Leftrightarrow\; 3p = 150 - y \;\Leftrightarrow\; p = 50 - \frac{1}{3}y.$ So
          $D^{-1}(p) = 50 - \frac{1}{3}p.$ D^{-1} represents the price that is associated with demand D.

     (b) $D^{-1}(30) = 50 - \frac{1}{3}(30) = 40.$ So when the demand is 30 units the price per unit is \$40.

72. (a) $f(x) = 0.7335x.$

     (b) $f(x) = 0.7335x.$ $y = 0.7335x \;\Leftrightarrow\; x = 1.3633y.$ So $f^{-1}(x) = 1.3633x.$ f^{-1} represents
          the exchange rate from US dollars to Canadian dollars.

(c) $f^{-1}(12{,}250) = 1.3633(12{,}250) = 16{,}700.75$. So $12,250 in US currency is worth $16,700.75 in Canadian currency.

74. (a) $f(x) = 0.85x$.

(b) $g(x) = x - 1000$.

(c) $H = f \circ g = f(x - 1000) = 0.85(x - 1000) = 0.85x - 850$.

(d) $H(x) = 0.85x - 850$. $y = 0.85x - 850$ $\Leftrightarrow$ $0.85x = y + 850$ $\Leftrightarrow$
$x = 1.176y + 1000$. So $H^{-1}(x) = 1.176x + 1000$. The function H^{-1} represents the original sticker price for a given discounted price.

(e) $H^{-1}(13{,}000) = 1.176(13{,}000) + 1000 = 16{,}266$. So the original price of the car is $16,288 when the discounted price ($1000 rebate, then 15% off) is $13,000.

76. $f(x) = mx + b$. Notice that $f(x_1) = f(x_2)$ $\Leftrightarrow$ $mx_1 + b = mx_2 + b$ $\Leftrightarrow$ $mx_1 = mx_2$.
We can conclude that $x_1 = x_2$ if and only if $m \neq 0$. Therefore f is one-to-one if and only if $m \neq 0$.
If $m \neq 0$, $f(x) = mx + b$ $\Leftrightarrow$ $y = mx + b$ $\Leftrightarrow$ $mx = y - b$ $\Leftrightarrow$ $x = \dfrac{y - b}{m}$. So,
$f^{-1}(x) = \dfrac{x - b}{m}$.

78. $f(I(x)) = f(x)$; therefore $f \circ I = f$. $I(f(x)) = f(x)$; therefore $I \circ f = f$.
By definition, $f \circ f^{-1}(x) = x = I(x)$; therefore $f \circ f^{-1} = I$. Similarly, $f^{-1} \circ f(x) = x = I(x)$;
therefore $f^{-1} \circ f = I$.

Review Exercises for Chapter 3

2. $f(x) = 2 - \sqrt{2x - 6}$; $f(5) = 2 - \sqrt{10 - 6} = 0$; $f(9) = 2 - \sqrt{18 - 6} = 2 - \sqrt{12} = 2 - 2\sqrt{3}$;
   $f(a + 3) = 2 - \sqrt{2a + 6 - 6} = 2 - \sqrt{2a}$; $f(-x) = 2 - \sqrt{2(-x) - 6} = 2 - \sqrt{-2x - 6}$;
   $f(x^2) = 2 - \sqrt{2x^2 - 6}$.
   $[f(x)]^2 = \left(2 - \sqrt{2x - 6}\right)^2 = 4 - 4\sqrt{2x - 6} + 2x - 6 = 2x - 4\sqrt{2x - 6} - 2$.

4. By the Vertical Line Test, figures (b) and (c) are graphs of functions. By the Horizontal Line Test, figure (c) is the graph of a one-to-one function.

6. $F(t) = t^2 + 2t + 5 = (t^2 + 2t + 1) + 5 - 1 = (t + 1)^2 + 4$. Therefore $F(t) \geq 4$ for all t. Since there are no restrictions on t, the domain of F is $(-\infty, \infty)$, and the range is $[4, \infty)$.

8. $f(x) = \dfrac{2x + 1}{2x - 1}$. Then $2x - 1 \neq 0 \quad \Leftrightarrow \quad x \neq \frac{1}{2}$. So the domain of f is $\{x \mid x \neq \frac{1}{2}\}$.

10. $f(x) = 3x - \dfrac{2}{\sqrt{x + 1}}$. The domain of f is the set of x where $x + 1 > 0 \quad \Leftrightarrow \quad x > -1$. So the domain is $(-1, \infty)$.

12. $g(x) = \dfrac{2x^2 + 5x + 3}{2x^2 - 5x - 3} = \dfrac{2x^2 + 5x + 3}{(2x + 1)(x - 3)}$. The domain of g is the set of all x where the denominator is not 0. So the domain is $\{x \mid 2x + 1 \neq 0 \text{ and } x - 3 \neq 0\} = \{x \mid x \neq -\frac{1}{2} \text{ and } x \neq 3\}$.

14. $f(x) = \dfrac{\sqrt[3]{2x + 1}}{\sqrt[3]{2x + 2}}$. Since we have an odd root, the domain is the set of all x where the denominator is not 0. Now $\sqrt[3]{2x + 2} \neq 0 \quad \Leftrightarrow \quad \sqrt[3]{2x} \neq -2 \quad \Leftrightarrow \quad 2x \neq -8 \quad \Leftrightarrow \quad x \neq -4$. Thus the domain of f is $\{x \mid x \neq -4\}$.

16. $f(x) = \frac{1}{3}(x - 5)$, $2 \leq x \leq 8$

18. $g(t) = t^2 - 2t = (t - 1)^2 - 1$

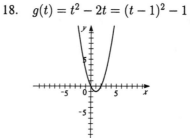

20. $f(x) = 3 - 8x - 2x^2 = -2(x - 2)^2 + 11$ 22. $y = -|x|$

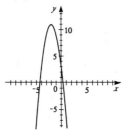   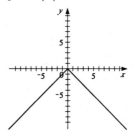

24. $y = \sqrt{x + 3}$ 26. $H(x) = x^3 - 3x^2 = x^2(x - 3)$

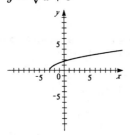   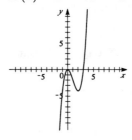

28. $G(x) = \dfrac{1}{(x - 3)^2}$ 30. $f(x) = \begin{cases} 1 - 2x & \text{if } x \le 0 \\ 2x - 1 & \text{if } x > 0 \end{cases}$

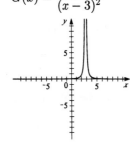   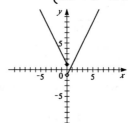

32. $f(x) = \begin{cases} -x & \text{if } x < 0 \\ x^2 & \text{if } 0 \le x < 2 \\ 1 & \text{if } x \ge 2 \end{cases}$

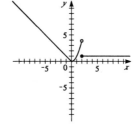

34. $f(x) = \sqrt{100 - x^3}$

   (i) $[-4, 4]$ by $[-4, 4]$ (ii) $[-10, 10]$ by $[-10, 10]$

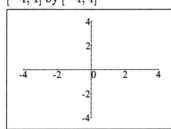   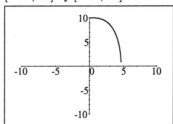

   (iii) $[-10, 10]$ by $[-10, 40]$ (iv) $[-100, 100]$ by $[-100, 100]$

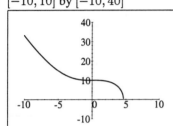   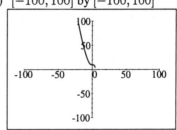

From the graphs, we see that the viewing rectangle in (iii) produces the most appropriate graph of f.

36. $f(x) = 1.1x^3 - 9.6x^2 - 1.4x + 3.2$. Here we experiment to find an appropriate viewing rectangle. The viewing rectangle shown is $[-10, 10]$ by $[-125, 10]$.

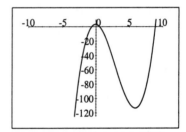

38. $y = |x(x + 2)(x + 4)|$. Here, the interesting parts of the function occur when $y = 0$. Thus we choose a viewing rectangle that includes the points $x = -4, -2$, and 0. So we choose the viewing rectangle $[-5, 5]$ by $[-1, 15]$.

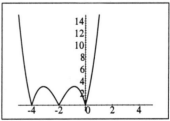

40. From the viewing rectangles, we see that the range of $f(x) = x^4 - x^3 + x^2 + 3x - 6$ is about $[-7.10, \infty)$.

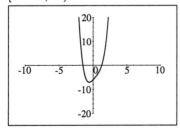   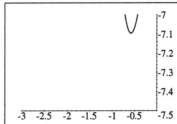

42. Average rate of change $= \dfrac{f(8) - f(4)}{8 - 4} = \dfrac{\left[\frac{1}{8-2}\right] - \left[\frac{1}{4-2}\right]}{4} = \dfrac{\frac{1}{6} - \frac{1}{2}}{4} \cdot \dfrac{6}{6} = \dfrac{1 - 3}{24} = -\dfrac{1}{12}.$

44. Average rate of change $= \dfrac{f(a+h) - f(a)}{(a+h) - a} = \dfrac{(a+h+1)^2 - (a+1)^2}{h}$

$= \dfrac{a^2 + 2ah + h^2 + 2a + 2h + 1 - a^2 - 2a - 1}{h} = \dfrac{2ah + h^2 + 2h}{h}$

$= \dfrac{h(2a + h + 2)}{h} = 2a + h + 2.$

46. $f(x) = |x^4 - 16|$ is graphed in the viewing rectangle $[-5, 5]$ by $[-5, 20]$. $f(x)$ is increasing on $[-2, 0]$ and $[2, \infty)$. It is decreasing on $(-\infty, -2]$ and $[0, 2]$.

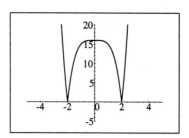

48. (a) $y = f(x - 2)$

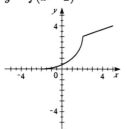

(b) $y = -f(x)$

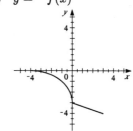

(c) $y = 3 - f(x)$

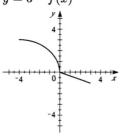

(d) $y = \frac{1}{2}f(x) - 1$

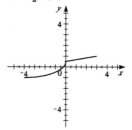

(e) $y = f^{-1}(x)$

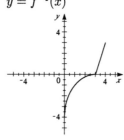

(f) $y = f(-x)$

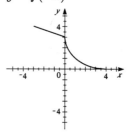

50. (a) $f(x)$ is odd;   (b) $g(x)$ is neither odd nor even;   (c) $h(x)$ is even.

52. $f(x) = -2x^2 + 12x + 12 = -2(x^2 - 6x) + 12 = -2(x^2 - 6x + 9) + 12 + 18$
     $= -2(x - 3)^2 + 30$

54. $f(x) = 1 - x - x^2 = -(x^2 + x) + 1 = -\left(x^2 + x + \frac{1}{4}\right) + 1 + \frac{1}{4} = -\left(x + \frac{1}{2}\right)^2 + \frac{5}{4}$. So the
     maximum value of f is $\frac{5}{4}$.

56. $P(x) = -1500 + 12x - 0.0004x^2 = -0.0004(x^2 - 30{,}000x) - 1500$
     $\qquad\quad = -0.0004(x^2 - 30{,}000x + 225{,}000{,}000) - 1500 + 90{,}000$
     $\qquad\quad = -0.0004(x - 15{,}000)^2 + 88{,}500$
     The maximum profit occurs when 15,000 units are sold, and the maximum profit is \$88,500.

58. $f(x) = x^{2/3}(6 - x)^{1/3}$. In the first viewing rectangle, $[-10, 10]$ by $[-10, 10]$, we see that $f(x)$ has a
     local maximum and a local minimum. The local minimum is 0 at $x = 0$ (and is easily verified). In
     the next viewing rectangle, $[3.95, 4.05]$ by $[3.16, 3.18]$, we isolate the local maximum value as
     approximately 3.175 when $x \approx 4.00$.

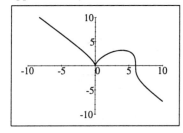
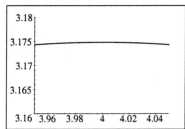

60. $f(x) = 1 + x^2$ and $g(x) = \sqrt{x - 1}$. (Remember that the proper domains must apply.)

   (a) $(f \circ g)(x) = f\left(\sqrt{x - 1}\right) = 1 + \left(\sqrt{x - 1}\right)^2 = 1 + x - 1 = x$

   (b) $(g \circ f)(x) = g(1 + x^2) = \sqrt{(1 + x^2) - 1} = \sqrt{x^2} = |x|$

   (c) $(f \circ g)(2) = f(g(2)) = f\left(\sqrt{(2) - 1}\right) = f(1) = 1 + (1)^2 = 2.$

   (d) $(f \circ f)(2) = f(f2)) = f(1 + (2)^2) = f(5) = 1 + (5)^2 = 26.$

   (e) $(f \circ g \circ f)(x) = f((g \circ f)(x)) = f(|x|) = 1 + (|x|)^2 = 1 + x^2.$ (Note $(g \circ f)(x) = |x|$ by
        part (b).)

   (f) $(g \circ f \circ g)(x) = g(\,(f \circ g)(x)) = g(x) = \sqrt{x - 1}.$ (Note $(f \circ g)(x) = x$ by part (a).)

62. $f(x) = \sqrt{x}$, has domain $\{x \mid x \ge 0\}$. $g(x) = \dfrac{2}{x - 4}$, has domain $\{x \mid x \ne 4\}$.

     $(f \circ g)(x) = f\left(\dfrac{2}{x - 4}\right) = \sqrt{\dfrac{2}{x - 4}}.$ $(f \circ g)(x)$ is defined whenever both $g(x)$ and $f(g(x))$ are

     defined; that is, whenever $x \ne 4$ and $\dfrac{2}{x - 4} \ge 0$. Now $\dfrac{2}{x - 4} \ge 0 \quad \Leftrightarrow \quad x - 4 > 0 \quad \Leftrightarrow$

     $x > 4$. So the domain of $f \circ g$ is $(4, \infty)$.

$(g \circ f)(x) = g(\sqrt{x}) = \dfrac{2}{\sqrt{x} - 4}$. $(g \circ f)(x)$ is defined whenever both $f(x)$ and $g(f(x))$ are

defined; that is, whenever $x \geq 0$ and $\sqrt{x} - 4 \neq 0$. Now $\sqrt{x} - 4 \neq 0$ $\Leftrightarrow$ $x \neq 16$. So the
domain of $g \circ f$ is $[0, 16) \cup (16, \infty)$.

$(f \circ f)(x) = f(\sqrt{x}) = \sqrt{\sqrt{x}} = x^{1/4}$. $(f \circ f)(x)$ is defined whenever both $f(x)$ and $f(f(x))$ are
defined; that is, whenever $x \geq 0$. So the domain of $f \circ f$ is $[0, \infty)$.

$(g \circ g)(x) = g\left(\dfrac{2}{x - 4}\right) = \dfrac{2}{\dfrac{2}{x - 4} - 4} = \dfrac{2(x - 4)}{2 - 4(x - 4)} = \dfrac{x - 4}{9 - 2x}$. $(g \circ g)(x)$ is defined whenever

both $g(x)$ and $g(g(x))$ are defined; that is, whenever $x \neq 4$ and $9 - 2x \neq 0$. Now $9 - 2x \neq 0$ $\Leftrightarrow$
$2x \neq 9$ $\Leftrightarrow$ $x \neq \frac{9}{2}$. So the domain of $g \circ g$ is $\{x \mid x \neq \frac{9}{2}, 4\}$.

64. If $h(x) = \sqrt{x}$ and $g(x) = 1 + x$, then $(g \circ h)(x) = g(\sqrt{x}) = 1 + \sqrt{x}$. If $f(x) = \dfrac{1}{\sqrt{x}}$, then

$(f \circ g \circ h)(x) = f(1 + \sqrt{x}) = \dfrac{1}{\sqrt{1 + \sqrt{x}}} = T(x).$

66. $g(x) = 2 - 2x + x^2 = (x^2 - 2x + 1) + 1 = (x - 1)^2 + 1$. Since $g(0) = 2 = g(2)$, as is true for all
pairs of numbers equidistant from 1, g is not a one-to-one function.

68. $r(x) = 2 + \sqrt{x + 3}$. If $x_1 \neq x_2$, then $x_1 + 3 \neq x_2 + 3$, so $\sqrt{x_1 + 3} \neq \sqrt{x_2 + 3}$ and
$2 + \sqrt{x_1 + 3} \neq 2 + \sqrt{x_2 + 3}$. Thus r is one-to-one.

70. $q(x) = 3.3 + 1.6x + 2.5x^3$. Using a graphing
device and the Horizontal Line Test, we see
that q is a one-to-one function.

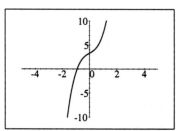

72. $f(x) = \dfrac{2x + 1}{3}$. $y = \dfrac{2x + 1}{3}$ $\Leftrightarrow$ $2x + 1 = 3y$ $\Leftrightarrow$ $2x = 3y - 1$ $\Leftrightarrow$ $x = \frac{1}{2}(3y - 1)$. So
$f^{-1}(x) = \frac{1}{2}(3x - 1)$.

74. $f(x) = 1 + \sqrt[5]{x - 2}$. $y = 1 + \sqrt[5]{x - 2}$ $\Leftrightarrow$ $y - 1 = \sqrt[5]{x - 2}$ $\Leftrightarrow$ $x - 2 = (y - 1)^5$ $\Leftrightarrow$
$x = 2 + (y - 1)^5$. So $f^{-1}(x) = 2 + (x - 1)^5$.

76. $f(x) = 1 + \sqrt[4]{x}$

(a) If $x_1 \neq x_2$, then $\sqrt[4]{x_1} \neq \sqrt[4]{x_2}$, and so $1 + \sqrt[4]{x_1} \neq 1 + \sqrt[4]{x_2}$. Therefore, f is a one-to-one
function.

(b)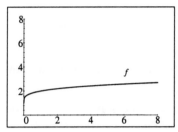

(c)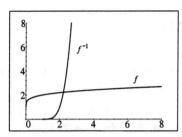

(d) $f(x) = 1 + \sqrt[4]{x}$. $y = 1 + \sqrt[4]{x}$ $\Leftrightarrow$ $\sqrt[4]{x} = y - 1$ $\Leftrightarrow$ $x = (y-1)^4$. So $f^{-1}(x) = (x-1)^4$, $x \geq 1$. Note that the domain of f is $[0, \infty)$, so $y = 1 + \sqrt[4]{x} \geq 1$. Hence, the domain of f^{-1} is $[1, \infty)$.

Focus on Modeling

2. Let w be the width of the poster. Then the length of the poster is $w + 10$. So the area of the poster is $A(w) = w(w + 10) = w^2 + 10w$.

4. Let r be the radius of the cylinder. Then the height of the cylinder is $4r$. Since for a cylinder $V = \pi r^2 h$, the volume of the cylinder is given by the function $V(r) = \pi r^2 (4r) = 4\pi r^3$.

6. Let A be the area and y be the length of the other side. Then $A = xy = 16$ $\Leftrightarrow$ $y = \dfrac{16}{x}$.

    Substituting into $P = 2x + 2y$ gives $P = 2x + 2 \cdot \dfrac{16}{x} = 2x + \dfrac{32}{x}$, where $x > 0$.

8. Let d represent the length of any side of a cube. Then the surface area is $S = 6d^2$, and the volume is $V = d^3$ $\Leftrightarrow$ $d = \sqrt[3]{V}$. Substituting for d gives $S(V) = 6\left(\sqrt[3]{V}\right)^2 = 6V^{2/3}$, $V > 0$.

10. Let r be the radius of a circle. Then the area is $A = \pi r^2$, and the circumference is $C = 2\pi r$ $\Leftrightarrow$ $r = \dfrac{C}{2\pi}$. Substituting for r gives $A(C) = \pi\left(\dfrac{C}{2\pi}\right)^2 = \dfrac{C^2}{4\pi}$, $C > 0$.

12. By similar triangles, $\dfrac{5}{L} = \dfrac{12}{L + d}$ $\Leftrightarrow$ $5(L + d) = 12L$ $\Leftrightarrow$ $5d = 7L$ $\Leftrightarrow$ $L = \dfrac{5d}{7}$. The model is $L(d) = \frac{5}{7}d$.

14. Let n be one of the numbers. Then the other number is $60 - n$, so the product is given by the function $P(n) = n(60 - n) = 60n - n^2$.

16. Let x be the length of the shorter leg of the right triangle. Then the length of the other triangle is $2x$. Since it is a right triangle, the length of the hypotenuse is $\sqrt{x^2 + (2x)^2} = \sqrt{5x^2} = \sqrt{5}\,x$ (since $x \geq 0$). Thus the perimeter of the triangle is $P(x) = x + 2x + \sqrt{5}\,x = (3 + \sqrt{5})x$.

18. Using the formula for the volume of a cone, $V = \frac{1}{3}\pi r^2 h$, we substitute $V = 100$ and solve for h. Thus $100 = \frac{1}{3}\pi r^2 h$ $\Leftrightarrow$ $h(r) = \dfrac{300}{\pi r^2}$.

20. Let the positive numbers be x and y. Since their sum is 100, we have $x + y = 100$ $\Leftrightarrow$ $y = 100 - x$. We wish to minimize the sum of squares, which is $S = x^2 + y^2 = x^2 + (100 - x)^2$. So $S(x) = x^2 + (100 - x)^2 = x^2 + 10000 - 200x + x^2 = 2x^2 - 200x + 10000$ $= 2(x^2 - 100x) + 10000 = 2(x^2 - 100x + 2500) + 10000 - 5000 = 2(x - 50)^2 + 5000$. Thus the minimum sum of squares occurs when $x = 50$. Then $y = 100 - 50 = 50$. Therefore both numbers are 50.

22. Let w and l be the width and the length of the rectangle in feet. We want all rectangles with perimeter equal to 20, so we have $2w + 2l = 20$ $\Leftrightarrow$ $l = 10 - w$. The area of a rectangle is given by $A(w) = l \cdot w = (10 - w)w = 10w - w^2 = -(w^2 - 10w) = -(w^2 - 10w + 25) + 25$ $= -(w - 5)^2 + 25$. So the area is maximized when $w = 5$, and hence the largest rectangle is a square where the dimension of each side is 5 feet.

24. (a) Let w be the width of the rectangular area (in feet) and l be the length of the field (in feet). Since the farmer has 750 feet of fencing, we must have $5w + 2l = 750$ $\Leftrightarrow$ $2l = 750 - 5w$

$\Leftrightarrow$ $l = \frac{5}{2}(150 - w)$. Thus the total area of the four pens is
$A(w) = l \cdot w = \frac{5}{2}w(150 - w) = -\frac{5}{2}(w^2 - 150w)$.

(b) We complete the square to get
$A(w) = -\frac{5}{2}(w^2 - 150w) = -\frac{5}{2}(w^2 - 150w + 75^2) + \left(\frac{5}{2}\right) \cdot 75^2 = -\frac{5}{2}(w - 75)^2 + 14062.5$.
Therefore, the largest possible total area of the four pens is 14,062.5 square feet.

26. (a) Let x be the length of wire in cm that is bent into a square. So $10 - x$ is the length of wire in cm that is bent into the second square. The width of each square is $\frac{x}{4}$ and $\frac{10 - x}{4}$, and the area of each square is $\left(\frac{x}{4}\right)^2 = \frac{x^2}{16}$ and $\left(\frac{10 - x}{4}\right)^2 = \frac{100 - 20x + x^2}{16}$. Thus the sum of the areas is $A(x) = \frac{x^2}{16} + \frac{100 - 20x + x^2}{16} = \frac{100 - 20x + 2x^2}{16} = \frac{1}{8}x^2 - \frac{5}{4}x + \frac{25}{4}$.

(b) We complete the square.
$A(x) = \frac{1}{8}x^2 - \frac{5}{4}x + \frac{25}{4} = \frac{1}{8}(x^2 - 10x) + \frac{25}{4} = \frac{1}{8}(x^2 - 10x + 25) + \frac{25}{4} - \frac{25}{8} = \frac{1}{8}(x - 5)^2 + \frac{25}{8}$
So the minimum area is $\frac{25}{8}$ cm^2 when each piece is 5 cm long.

28. (a) Let x be the number of one dollar increases in the price of a bird feeder. So the selling price will be $10 + x$ dollars, and the number of bird feeders sold will be $20 - 2x$. The revenue from the sales will be $(10 + x)(20 - 2x)$, and the cost will be $6(20 - 2x)$. The profits will be $P(x) = (10 + x)(20 - 2x) - 6(20 - 2x) = (4 + x)(20 - 2x) = 80 + 12x - 2x^2$.

(b) Completing the square we get $P(x) = 80 + 12x - 2x^2 = -2(x^2 - 6x) + 80$
$= -2(x^2 - 6x + 9) + 80 + 18 = -2(x - 3)^2 + 98$. Thus the profit would be maximized at $98 when $x = 3$. So the bird society should set the selling price at $10 + x = \$13$.

30. (a) The height of the box is x, the width of the box is $12 - 2x$, and the length of the box is $20 - 2x$. Therefore, the volume of the box is
$V(x) = x(12 - 2x)(20 - 2x) = 4x^3 - 64x^2 + 240x$, $0 < x < 6$.

(b) We graph the function $y = V(x)$ in the viewing rectangle $[0, 6] \times [200, 270]$. From the calculator we get that the volume o the box is greater than in^3 for $1.174 \le x \le 3.898$ (accurate to 3 decimal places).

(c) Also from the viewing rectangle, the volume of the box with the largest volume is 262.682 in^3 when $x \approx 2.427$.

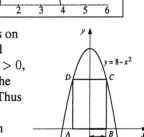

32. Let A, B, C, D be the vertices of the rectangle such that its base AB is on the x-axis, and its other two vertices, C and D, are above the x-axis and lying on the parabola $y = 8 - x^2$. Let C have the coordinates (x, y), $x > 0$, by symmetry, the coordinates of D must be $(-x, y)$. See the graph on the left. So the width of the rectangle is $2x$, and the length is $y = 8 - x^2$. Thus the area of the rectangle is $A(x) = length \cdot width = 2x(8 - x^2)$
$= 16x - 2x^3$. The graphs below show that the area is maximized when $x \approx 1.63$. Hence the maximum area occurs when the width is 3.27 and the length is 5.33.

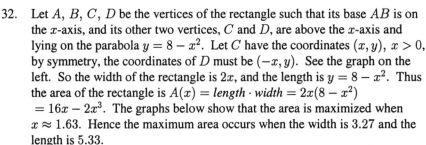

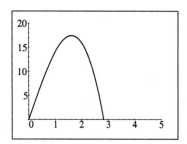

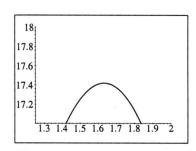

34. (a) Let t_1 represent the time, in hours, spent walking, and let t_2 represent the time spent rowing. Since the distance walked is x and the walking speed is 5 mi/h, the time spent walking is $t_1 = \frac{1}{5}x$. By the Pythagorean Theorem, the distance rowed is $d = \sqrt{2^2 + (7 - x)^2} = \sqrt{x^2 - 14x + 53}$, and so the time spent rowing is $t_2 = \frac{1}{2} \cdot \sqrt{x^2 - 14x + 53}$. Thus the total time is $T(x) = \frac{1}{2}\sqrt{x^2 - 14x + 53} + \frac{1}{5}x$.

    (b) We graph $y = T(x)$ in the viewing rectangle below. Using Zoom, we see that T is minimized when $x \approx 6.13$. He should land at a point 6.13 miles from point B.

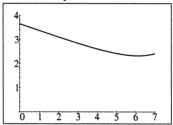

36. (a) Using the Pythagorean Theorem, we have that the height of the upper triangles is $\sqrt{25 - x^2}$ and the height of the lower triangles is $\sqrt{144 - x^2}$. So the area of the each of the upper triangles is $\frac{1}{2}x\sqrt{25 - x^2}$, and the area of the each of the lower triangles is $\frac{1}{2}x\sqrt{144 - x^2}$. Since there are two upper triangles and two lower triangles, we get that the total area is
$$A(x) = 2 \cdot \left[\frac{1}{2}x\sqrt{25 - x^2}\right] + 2 \cdot \left[\frac{1}{2}x\sqrt{144 - x^2}\right] = x\left(\sqrt{25 - x^2} + \sqrt{144 - x^2}\right).$$

    (b) The function $y = A(x) = x\left(\sqrt{25 - x^2} + \sqrt{144 - x^2}\right)$ is shown in the first viewing rectangle below. In the second viewing rectangle, we isolate the maximum, and we see that the area of the kite is maximized when $x \approx 4.615$. So the length of the horizontal crosspiece must be $2 \cdot 4.615 = 9.23$. The length of the vertical crosspiece is $\sqrt{5^2 - (4.615)^2} + \sqrt{12^2 - (4.615)^2} \approx 13.00$.

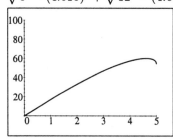

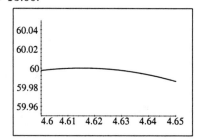

Chapter Four

Exercises 4.1

2. $P(x) = -x^3 + 8$

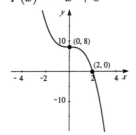

4. $P(x) = 2(x-1)^3$

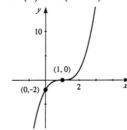

6. $P(x) = 3x^4 - 27$

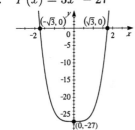

8. $P(x) = (x+2)^4 - 1$

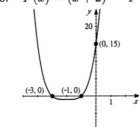

10. $P(x) = -\frac{1}{2}(x+3)^5 - 64$

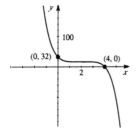

12. I

14. II

16. IV

18. $P(x) = (x-1)(x+1)(x-2)$

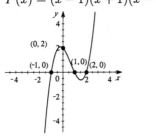

20. $P(x) = (2x-1)(x+1)(x+3)$

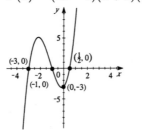

22. $P(x) = \frac{1}{5}x(x-5)^2$

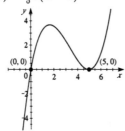

24. $P(x) = \frac{1}{4}(x+1)^3(x-3)$

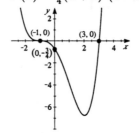

26. $P(x) = (x-1)^2(x+2)^3$

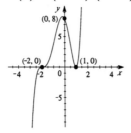

28. $P(x) = (x-3)^2(x+1)^2$

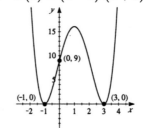

30. $P(x) = x^3 + 2x^2 - 8x$
     $= x(x-2)(x+4)$

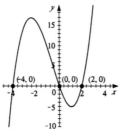

32. $P(x) = -2x^3 - x^2 + x = -x(2x^2 + x - 1)$
     $= -x(2x-1)(x+1)$

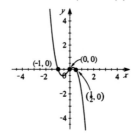

34. $P(x) = x^5 - 9x^3$
     $= x^3(x+3)(x-3)$

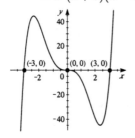

36. $P(x) = x^3 + 3x^2 - 4x - 12$
     $= (x+3)(x-2)(x+2)$

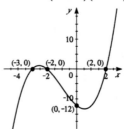

38. $P(x) = \frac{1}{8}(2x^4 + 3x^3 - 16x - 24)^2$
$= \frac{1}{8}(x-2)^2(2x+3)^2(x^2+2x+4)^2$

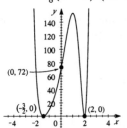

40. $P(x) = x^4 - 2x^3 + 8x - 16$
$= (x+2)(x-2)(x^2-2x+4)$

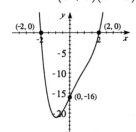

42. $P(x) = x^6 - 2x^3 + 1 = (x^3 - 1)^2$
$= (x-1)^2(x^2+x+1)^2$

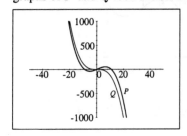

44. $P(x) = -\frac{1}{8}x^3 + \frac{1}{4}x^2 + 12x; \quad Q(x) = -\frac{1}{8}x^3$

Since P has odd degree and negative leading coefficient, it has the following end behavior:

$y \to -\infty$ as $x \to \infty$ and $y \to \infty$ as $x \to -\infty$.

On the large viewing rectangle, the graphs of P and Q look almost the same.

On the small viewing rectangle, the graphs of P and Q look very different and seem to have different end behavior.

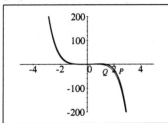

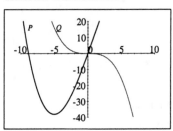

46. $P(x) = -x^5 + 2x^2 + x; \quad Q(x) = -x^5$

Since P has odd degree and negative leading coefficient, it has the following end behavior:

$y \to -\infty$ as $x \to \infty$ and $y \to \infty$ as $x \to -\infty$.

On the large viewing rectangle, the graphs of P and Q look almost the same.

On the small viewing rectangle, the graphs of P and Q look very different.

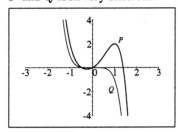

48. $P(x) = 2x^2 - x^{12}$; $Q(x) = -x^{12}$

Since P has even degree and negative leading coefficient, it has the following end behavior:

$y \rightarrow -\infty$ as $x \rightarrow \infty$ and $y \rightarrow -\infty$ as $x \rightarrow -\infty$.

On the large viewing rectangle, the graphs of P and Q look almost the same.

On the small viewing rectangle, the graphs of P and Q look very different.

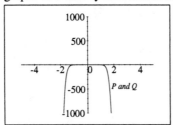 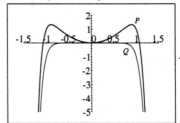

50.   (a) x-intercepts at 0 and 4.5. y-intercept at 0.

      (b)   Local maximum at $(0, 0)$ and local minimum at $(3, -3)$.

52.   (a) x-intercepts at 0 and 4. y-intercept at 0.

      (b)   No local maximum. Local minimum at $(3, -3)$.

54. $y = x^3 - 3x^2$, $[-2, 5]$ by $[-10, 10]$

Local minimum at $(2, -4)$.

Local maximum at $(0, 0)$.

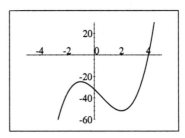

56. $y = 2x^3 - 3x^2 - 12x - 32$, $[-5, 5]$

by $[-60, 30]$

Local minimum at $(2, -52)$.

Local maximum at $(-1, -25)$.

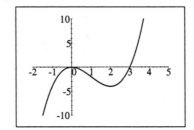

58. $y = x^4 - 18x^2 + 32$, $[-5, 5]$ by $[-100, 100]$

Local minima at $(-3, -49)$ and $(3, -49)$.

Local maximum at $(0, 32)$.

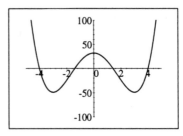

60. $y = x^5 - 5x^2 + 6$, $[-3, 3]$ by $[-5, 10]$
Local minimum at $(1.26, 1.24)$.
Local maximum at $(0, 6)$.

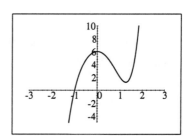

62. $y = x^3 + 12x$ has no local extremum.

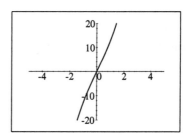

64. $y = 6x^3 + 3x + 1$ has no local extremum.

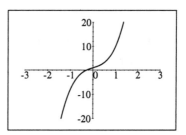

66. $y = 1.2x^5 + 3.75x^4 - 7x^3 - 15x^2 + 18x$
Two local maxima occur at $(0.50, 4.65)$ and
$(-2.97, 12.10)$, and two local minima occur
at $(-1.40, -27.44)$ and $(1.40, -2.54)$.

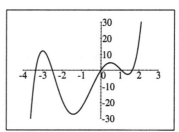

68. $y = (x^2 - 2)^3$ has one local minimum at
$(0, -8)$.

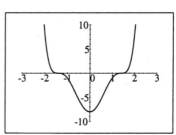

70. $y = \frac{1}{3}x^7 - 17x^2 + 7$.

One local maximum at $(0, 7)$ and one
local minimum at $(1.71, -28.46)$.

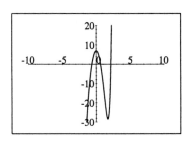

72. $P(x) = (x - c)^4$

Increasing the value of c shifts the graph
to the right.

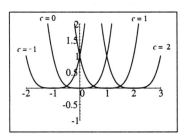

74. $P(x) = x^3 + cx$

Increasing the value of c makes the "bumps"
in the graph flatter.

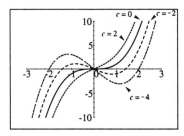

76. $P(x) = x^c$

The larger c gets, the flatter the graph is
near the origin, and the steeper it is away
from the origin.

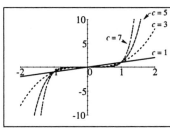

78. Graph 1 belongs to $y = x^4$. Graph 2 belongs to $y = x^2$. Graph 3 belongs to $y = x^6$. Graph 4 belongs to $y = x^3$. Graph 5 belongs to $y = x^5$.

80. (a) $P(x) = (x - 1)(x - 3)(x - 4)$.
Local maximum at $(1.8, 2.1)$.
Local minimum at $(3.6, -0.6)$.

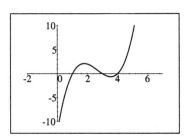

(b) Since $Q(x) = P(x) + 5$, each point on the
graph of Q has y-coordinate 5 units more
than the corresponding point on the graph
of P. Thus Q has:

Local maximum: $(1.8, 7.1)$

Local minimum: $(3.5, 4.4)$

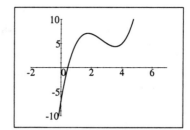

82. (a) From the graph,

$P(x) = x^3 - 4x = x(x - 2)(x + 2)$ has
three x-intercepts and two local extrema.

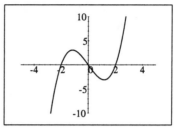

(b) From the graph,

$Q(x) = x^3 + 4x = x(x^2 + 4)$ has
one x-intercepts and no local extrema.

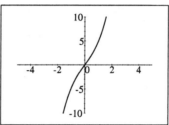

(c) For the x-intercepts of $P(x) = x^3 - ax$, we solve $x^3 - ax = 0$. Then we have $x(x^2 - a) = 0$
$\Leftrightarrow$ $x = 0$ or $x^2 = a$. If $x^2 = a$, then $x = \pm\sqrt{a}$. So P has 3 x-intercepts. Since
$P(x) = x(x^2 - a) = x(x + \sqrt{a})(x - \sqrt{a})$, by part (c) of problem 67, P has 2 local extrema.
For the x-intercepts of $Q(x) = x^3 + ax$, we solve $x^3 + ax = 0$. Then we have $x(x^2 + a) = 0$
$\Leftrightarrow$ $x = 0$ or $x^2 = -a$. The equation $x^2 = -a$ has no real solutions because $a > 0$. So Q
has 1 x-intercept. We now show that Q is always increasing and hence has no extrema. If
$x_1 < x_2$, then $ax_1 < ax_2$ (because $a > 0$) and $x_1^3 < x_2^3$. So we have $x_1^3 + ax_1 < x_2^3 + ax_2$, and
hence $Q(x_1) < Q(x_2)$. Thus Q is increasing, that is, its graph always rises, and so it has no
local extrema.

84. $P(t) = 120t - 0.4t^4 + 1000$

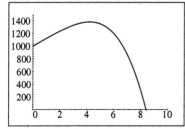

(a) A maximum population of approximately
1380 is attained after 4.22 months.

(b) The rabbit population disappears after
approximately 8.42 months.

86. (a) Let h = height of the box. Then the total length of all 12 edges is $8x + 4h = 144$ in. Thus,
$8x + 4h = 144$ $\Leftrightarrow$ $2x + h = 36$ $\Leftrightarrow$ $h = 36 - 2x$. The volume of the box is equal to

(area of base) $\times$ (height) $= (x^2) \times (36 - 2x) = -2x^3 + 36x^2$. Therefore, the volume of the box is $V(x) = -2x^3 + 36x^2 = 2x^2(18 - x)$.

(b)   Since the length of the base is x, we must have $x > 0$. Likewise, the height must be positive so $36 - 2x > 0 \quad \Leftrightarrow \quad x < 18$. Putting these together, we get that the domain of V is $0 < x < 18$.

(c)   Using the domain from part (b), we graph V in the viewing rectangle $[0, \ 18]$ by $[0, \ 2000]$. The maximum volume is $V = 1728$ in^3 when $x = 12$ in.

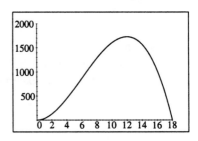

88.   Since the polynomial shown has five zeros, it has at least five factors, and so the degree of the polynomial is greater than or equal to 5.

90.   No, it is not possible. Clearly a polynomial must have a local minimum between any two local maxima.

Exercises 4.2

2. (a)

$$
\begin{array}{r}
x^2 + 5x\ -1 \\
x-1\overline{)x^3 + 4x^2 - 6x + 1} \\
\underline{x^3\ -x^2} \\
5x^2 - 6x \\
\underline{5x^2 - 5x} \\
-x+1 \\
\underline{-x+1} \\
0
\end{array}
$$

Thus $\dfrac{P(x)}{Q(x)} = x^2 + 5x - 1.$

(b) $P(x) = x^3 + 4x^2 - 6x + 1 = (x-1) \cdot (x^2 + 5x - 1)$

4. (a)

$$
\begin{array}{r}
2x^3 + 4x^2\qquad +8 \\
x^2-2\overline{)x^5 + 4x^4 - 4x^3 + 0x^2 - x\ -3} \\
\underline{x^5\qquad -4x^3} \\
4x^4\qquad +0x^2 \\
\underline{4x^4\qquad -8x^2} \\
8x^2 - x\ -3 \\
\underline{8x^2\qquad -16} \\
-x+13
\end{array}
$$

Thus $\dfrac{P(x)}{Q(x)} = 2x^3 + 4x^2 + 8 + \dfrac{-x+13}{x^2-2}.$

(b) $P(x) = 2x^5 + 4x^4 - 4x^3 - x - 3 = (x^2 - 2) \cdot (2x^3 + 4x^2 + 8) + (-x + 13)$

6.

$$
\begin{array}{r}
x^2\ +x \\
x-2\overline{)x^3\ -x^2 - 2x + 6} \\
\underline{x^3 - 2x^2} \\
x^2 - 2x \\
\underline{x^2 - 2x} \\
0+6
\end{array}
$$

Thus the quotient is $x^2 + x$, and the remainder is 6.

8.

$$
\begin{array}{r}
\frac{1}{3}x^2\ +\frac{1}{3}x\ +\frac{2}{3} \\
3x+6\overline{)x^3 + 3x^2 + 4x + 3} \\
\underline{x^3 + 2x^2} \\
x^2 + 4x \\
\underline{x^2 + 2x} \\
2x + 3 \\
\underline{2x + 4} \\
-1
\end{array}
$$

Thus the quotient is $\frac{1}{3}x^2 + \frac{1}{3}x + \frac{2}{3}$, and the remainder is -1.

10.

$$
\begin{array}{r}
3x^2 \ -8x \ \ -1 \\
x^2 + x + 3 \overline{\smash{\big)}\ 3x^4 - 5x^3 + 0x^2 - 20x - 5} \\
\underline{3x^4 + 3x^3 + 9x^2} \\
-8x^3 - 9x^2 - 20x \\
\underline{-8x^3 - 8x^2 - 24x} \\
-x^2 \ + 4x - 5 \\
\underline{-x^2 \ \ - x - 3} \\
5x - 2
\end{array}
$$

Thus the quotient is $3x^2 - 8x - 1$, and the remainder is $5x - 2$.

12.

$$
\begin{array}{r}
3 \\
3x^2 - 7x \overline{\smash{\big)}\ 9x^2 \ \ - x + 5} \\
\underline{9x^2 - 21x} \\
20x + 5
\end{array}
$$

Thus the quotient is 3, and the remainder is $20x + 5$.

14.

$$
\begin{array}{r}
\frac{1}{2}x^3 \ -x^2 \ -\frac{5}{2}x \ -\frac{7}{4} \\
4x^2 - 6x + 8 \overline{\smash{\big)}\ 2x^5 - 7x^4 + 0x^3 \ + 0x^2 \ + 0x - 13} \\
\underline{2x^5 - 3x^4 + 4x^3} \\
-4x^4 - 4x^3 \ + 0x^2 \\
\underline{-4x^4 + 6x^3 \ - 8x^2} \\
-10x^3 \ + 8x^2 \ + 0x \\
\underline{-10x^3 + 15x^2 - 20x} \\
-7x^2 + 20x - 13 \\
\underline{-7x^2 + \frac{21}{2}x - 14} \\
\frac{19}{2}x \ + 1
\end{array}
$$

Thus the quotient is $\frac{1}{2}x^3 - x^2 - \frac{5}{2}x - \frac{7}{4}$, and the remainder is $\frac{19}{2}x + 1$.

16. The synthetic division table for this problem takes the following form.

$$
\begin{array}{r|rrr}
1 & 1 & -5 & 4 \\
& & 1 & -4 \\
\hline
& 1 & -4 & 0
\end{array}
$$

Thus the quotient is $x - 4$, and the remainder is 0.

18. The synthetic division table for this problem takes the following form.

$$
\begin{array}{r|rrr}
-5 & 4 & 0 & -3 \\
& & -20 & 100 \\
\hline
& 4 & -20 & 97
\end{array}
$$

Thus the quotient is $4x - 20$, and the remainder is 97.

20. The synthetic division table for this problem takes the following form.

$$
\begin{array}{r|rrrr}
5 & 3 & -12 & -9 & 1 \\
& & 15 & 15 & 30 \\
\hline
& 3 & 3 & 6 & 31
\end{array}
$$

Thus the quotient is $3x^2 + 3x + 6$, and the remainder is 31.

22. The synthetic division table for this problem takes the following form.

$$
\begin{array}{r|rrrrr}
2 & 1 & -1 & 1 & -1 & 2 \\
  &   & 2 & 2 & 6 & 10 \\
\hline
  & 1 & 1 & 3 & 5 & 12
\end{array}
$$

Thus the quotient is $x^3 + x^2 + 3x + 5$, and the remainder is 12.

24. The synthetic division table for this problem takes the following form.

$$
\begin{array}{r|rrrr}
3 & 1 & -9 & 27 & -27 \\
  &   & 3 & -18 & 27 \\
\hline
  & 1 & -6 & 9 & 0
\end{array}
$$

Thus the quotient is $x^2 - 6x + 9$, and the remainder is 0.

26. The synthetic division table for this problem takes the following form.

$$
\begin{array}{r|rrrrr}
-\frac{2}{3} & 6 & 10 & 5 & 1 & 1 \\
  &   & -4 & -4 & -\frac{2}{3} & -\frac{2}{9} \\
\hline
  & 6 & 6 & 1 & \frac{1}{3} & \frac{7}{9}
\end{array}
$$

Thus the quotient is $6x^3 + 6x^2 + x + \frac{1}{3}$, and the remainder is $\frac{7}{9}$.

28. The synthetic division table for this problem takes the following form.

$$
\begin{array}{r|rrrrr}
-2 & 1 & 0 & 0 & 0 & -16 \\
  &   & -2 & 4 & -8 & 16 \\
\hline
  & 1 & -2 & 4 & -8 & 0
\end{array}
$$

Thus the quotient is $x^3 - 2x^2 + 4x - 8$, and the remainder is 0.

30. $P(x) = 2x^2 + 9x + 1$, $c = \frac{1}{2}$

$$
\begin{array}{r|rrr}
\frac{1}{2} & 2 & 9 & 1 \\
  &   & 1 & 5 \\
\hline
  & 2 & 10 & 6
\end{array}
$$

Therefore, by the Remainder Theorem, $P\left(\frac{1}{2}\right) = 6$.

32. $P(x) = x^3 - x^2 + x + 5$, $c = -1$

$$
\begin{array}{r|rrrr}
-1 & 1 & -1 & 1 & 5 \\
  &   & -1 & 2 & -3 \\
\hline
  & 1 & -2 & 3 & 2
\end{array}
$$

Therefore, by the Remainder Theorem, $P(-1) = 2$.

34. $P(x) = 2x^3 - 21x^2 + 9x - 200$, $c = 11$

$$
\begin{array}{r|rrrr}
11 & 2 & -21 & 9 & -200 \\
  &   & 22 & 11 & 220 \\
\hline
  & 2 & 1 & 20 & 20
\end{array}
$$

Therefore, by the Remainder Theorem, $P(11) = 20$.

36. $P(x) = 6x^5 + 10x^3 + x + 1$, $c = -2$

$$
\begin{array}{r|rrrrrr}
-2 & 6 & 0 & 10 & 0 & 1 & 1 \\
  &   & -12 & 24 & -68 & 136 & -274 \\
\hline
  & 6 & -12 & 34 & -68 & 137 & -273
\end{array}
$$

Therefore, by the Remainder Theorem, $P(-2) = -273$.

38. $P(x) = -2x^6 + 7x^5 + 40x^4 - 7x^2 + 10x + 112$, $c = -3$

$$
\begin{array}{r|rrrrrrr}
-3 & -2 & 7 & 40 & 0 & -7 & 10 & 112 \\
  &   & 6 & -39 & -3 & 9 & -6 & -12 \\
\hline
  & -2 & 13 & 1 & -3 & 2 & 4 & 100
\end{array}
$$

Therefore, by the Remainder Theorem, $P(-3) = 100$.

40. $P(x) = x^3 - x + 1, c = \frac{1}{4}$

$$\begin{array}{r|rrrr} \frac{1}{4} & 1 & 0 & -1 & 1 \\ & & \frac{1}{4} & \frac{1}{16} & -\frac{15}{64} \\ \hline & 1 & \frac{1}{4} & -\frac{15}{16} & \frac{49}{64} \end{array}$$ Therefore, by the Remainder Theorem, $P\left(\frac{1}{4}\right) = \frac{49}{64}$.

42. (a) $P(x) = 6x^7 - 40x^6 + 16x^5 - 200x^4 - 60x^3 - 69x^2 + 13x - 139, c = 7$

$$\begin{array}{r|rrrrrrrr} 7 & 6 & -40 & 16 & -200 & -60 & -69 & 13 & -139 \\ & & 42 & 14 & 210 & 70 & 70 & 7 & 140 \\ \hline & 6 & 2 & 30 & 10 & 10 & 1 & 20 & 1 \end{array}$$

Therefore, by the Remainder Theorem, $P(7) = 1$.

   (b) $P(7) = 6(7)^7 - 40(7)^6 + 16(7)^5 - 200(7)^4 - 60(7)^3 - 69(7)^2 + 13(7) - 139$
   $= 6(823543) - 40(117649) + 16(16807) - 200(2401) - 60(343)$
   $$- 69(49) + 13(7) - 139$$
   $= 1,$

   which agrees with the value obtained by synthetic division, but requires more work.

44. $P(x) = x^3 + 2x^2 - 3x - 10, c = 2$

$$\begin{array}{r|rrrr} 2 & 1 & 2 & -3 & -10 \\ & & 2 & 8 & 10 \\ \hline & 1 & 4 & 5 & 0 \end{array}$$ Since the remainder is 0, $x - 2$ is a factor.

46. $P(x) = x^4 + 3x^3 - 16x^2 - 27x + 63, c = 3, -3$

$$\begin{array}{r|rrrrr} 3 & 1 & 3 & -16 & -27 & 63 \\ & & 3 & 18 & 6 & -63 \\ \hline & 1 & 6 & 2 & -21 & 0 \end{array}$$ Since the remainder is 0, $x - 3$ is a factor.

We next show that $x + 3$ is also a factor by using synthetic division on the quotient of the above synthetic division, $x^3 + 6x^2 - 2x - 21$.

$$\begin{array}{r|rrrr} -3 & 1 & 6 & 2 & -21 \\ & & -3 & -9 & 21 \\ \hline & 1 & 3 & -7 & 0 \end{array}$$ Since the remainder is 0, $x + 3$ is a factor.

48. $P(x) = 3x^4 - x^3 - 21x^2 - 11x + 6, c = \frac{1}{3}, -2$

$$\begin{array}{r|rrrrr} \frac{1}{3} & 3 & -1 & -21 & -11 & 6 \\ & & 1 & 0 & -7 & -6 \\ \hline & 3 & 0 & -21 & -18 & 0 \end{array}$$ Since the remainder is 0, it follows that $\frac{1}{3}$ is a zero.

Thus $P(x) = \left(x - \frac{1}{3}\right)(3x^3 - 21x - 18)$. Next, we use synthetic division on the quotient.

$$\begin{array}{r|rrrr} -2 & 3 & 0 & -21 & -18 \\ & & -6 & 12 & 18 \\ \hline & 3 & -6 & -9 & 0 \end{array}$$ Since the remainder is 0, it follows that 2 is a zero.

So $P(x) = \left(x - \frac{1}{3}\right)(x + 2)(3x^2 - 6x - 9) = 3\left(x - \frac{1}{3}\right)(x + 2)(x^2 - 2x - 3)$
$= 3\left(x - \frac{1}{3}\right)(x + 2)(x - 3)(x + 1)$.
Hence, the zeros are $\frac{1}{3}, -2, -1,$ and 3.

50. Since the zeros are $x = -2, x = 0, x = 2,$ and $x = 4$, the factors are $x + 2, x, x - 2,$ and $x - 4$.
Thus $P(x) = c(x + 2)x(x - 2)(x - 4)$. If we let $c = 1$, then $P(x) = x^4 - 4x^3 - 4x^2 + 16x$.

52. Since the zeros are $x = -2$, $x = -1$, $x = 0$, $x = 1$, and $x = 2$, the factors are $x + 2$, $x + 1$, x, $x - 1$, and $x - 2$. Thus $P(x) = c(x + 2)(x + 1)x(x - 1)(x - 2)$. If we let $c = 1$, then $P(x) = x^5 - 5x^3 + 4x$.

54. Since the zeros of the polynomial are 1, -1, 2, and $\frac{1}{2}$, it follows that $P(x) = C(x - 1)(x + 1)(x - 2)\left(x - \frac{1}{2}\right) = C(x^2 - 1)(x^2 - \frac{5}{2}x + 1)$. To ensure integer coefficients, we choose C to be any nonzero multiple of 2. When $C = 2$, we have $P(x) = 2(x^2 - 1)(x^2 - \frac{5}{2}x + 1) = (x^2 - 1)(2x^2 - 5x + 2) = 2x^4 - 5x^3 + 5x - 2$. (Note: This is just one of many polynomials with the desired zeros and integer coefficients. We can choose C to be any even integer.)

56. The y-intercept is 4 and the zeros of the polynomial are -1 and 2 with 2 being degree two. It follows that $P(x) = C(x + 1)(x - 2)^2 = C(x^3 - 3x^2 + 4)$. Since $P(0) = 4$ we have $4 = C[(0)^3 - 3(0)^2 + 4] \quad \Leftrightarrow \quad 4 = 4C \quad \Leftrightarrow \quad C = 1$ and $P(x) = x^3 - 3x^2 + 4$.

58. The y-intercept is 2 and the zeros of the polynomial are -2, -1, and 1 with 1 being degree two. It follows that $P(x) = C(x + 2)(x + 1)(x - 1)^2 = C(x^4 + x^3 - 3x^2 - x + 2)$. Since $P(0) = 2$ we have $4 = C[(0)^4 + (0)^3 - 3(0)^2 - (0) + 2] \quad \Leftrightarrow \quad 2 = 2C$ so $C = 1$ and $P(x) = x^4 + x^3 - 3x^2 - x + 2$.

60. $R(x) = x^5 - 2x^4 + 3x^3 - 2x^2 + 3x + 4 = (x^4 - 2x^3 + 3x^2 - 2x + 3)x + 4$
$$= ((x^3 - 2x^2 + 3x - 2)x + 3)x + 4$$
$$= (((x^2 - 2x + 3)x - 2)x + 3)x + 4$$
$$= ((((x - 2)x + 3)x - 2)x + 3)x + 4$$

So to calculate $R(3)$, we start with 3, then:

    subtract 2; multiply by 3;

    add 3; multiply by 3;

    subtract 2; multiply by 3;

    add 3; multiply by 3; and

    add 4; to get 157.

Exercises 4.3

2. $Q(x) = x^4 - 3x^3 - 6x + 8$ has possible rational zeros $\pm 1, \pm 2, \pm 4, \pm 8$.

4. $S(x) = 6x^4 - x^2 + 2x + 12$ has possible rational zeros $\pm 1, \pm 2, \pm 3, \pm 4, \pm 6, \pm 12, \pm\frac{1}{2}, \pm\frac{3}{2}, \pm\frac{1}{3},$
$\pm\frac{2}{3}, \pm\frac{4}{3}, \pm\frac{1}{6}$.

6. $U(x) = 12x^5 + 6x^3 - 2x - 8$ has possible rational zeros $\pm 1, \pm 2, \pm 4, \pm 8, \pm\frac{1}{2}, \pm\frac{1}{3}, \pm\frac{2}{3}, \pm\frac{4}{3}, \pm\frac{8}{3},$
$\pm\frac{1}{4}, \pm\frac{1}{6}, \pm\frac{1}{12}$.

8.   (a) $P(x) = 3x^3 + 4x^2 - x - 2$ has possible rational zeros $\pm 1, \pm 2, \pm\frac{1}{3}, \pm\frac{2}{3}$.

      (b)   From the graph, the actual zeroes are -1 and $\frac{2}{3}$.

10.   (a) $P(x) = 4x^4 - x^3 - 4x + 1$ has possible rational zeros $\pm 1, \pm\frac{1}{2}, \pm\frac{1}{4}$.

      (b)   From the graph, the actual zeroes are $\frac{1}{4}$ and 1.

12. $P(x) = x^3 - 7x^2 + 14x - 8$. The possible rational zeros are $\pm 1, \pm 2, \pm 4, \pm 8$. $P(x)$ has 3 variations in sign and hence 1 or 3 positive real zeros. $P(-x) = -x^3 - 7x^2 - 14x - 8$ has 0 variations in sign and hence no negative real zeros.

$$1 \, \underline{\big|\,1 \quad -7 \quad 14 \quad -8}$$
$$1 \quad -6 \quad 8$$
$$\overline{1 \quad -6 \quad 8 \quad 0} \; \Rightarrow \quad x = 1 \text{ is a zero.}$$

So $P(x) = x^3 - 7x^2 + 14x - 8 = (x - 1)(x^2 - 6x + 8) = (x - 1)(x - 2)(x - 4)$. Therefore, the zeros are $x = 1, 2, 4$.

14. $P(x) = x^3 + 4x^2 - 3x - 18$. The possible rational zeros are $\pm 1, \pm 2, \pm 3, \pm 6, \pm 9, \pm 18$. $P(x)$ has 1 variation in sign and hence 1 positive real zero. $P(-x) = -x^3 + 4x^2 + 3x - 18$ has 2 variations in sign and hence 0 or 2 negative real zeros.

$$1 \, \underline{\big|\,1 \quad 4 \quad -3 \quad -18}\qquad\qquad 2 \, \underline{\big|\,1 \quad 4 \quad -3 \quad -18}$$
$$1 \quad 5 \quad 2 \qquad\qquad\qquad 2 \quad 12 \quad 18$$
$$\overline{1 \quad 5 \quad 2 \quad -16}\qquad\qquad \overline{1 \quad 6 \quad 9 \quad 0} \; \Rightarrow \quad x = 2 \text{ is a zero.}$$

$P(x) = x^3 + 4x^2 - 3x - 18 = (x - 2)(x^2 + 6x + 9) = (x - 2)(x + 3)^2$. Therefore, the zeros are $x = -3, 2$.

16. $P(x) = x^3 - x^2 - 8x + 12$. The possible rational zeros are $\pm 1, \pm 2, \pm 3, \pm 4, \pm 6, \pm 12$. $P(x)$ has 2 variations in sign and hence 0 or 2 positive real zeros. $P(-x) = -x^3 - x^2 + 8x + 12$ has 1 variation in sign and hence 1 negative real zero.

$$1 \, \underline{\big|\,1 \quad -1 \quad -8 \quad 12}\qquad\qquad 2 \, \underline{\big|\,1 \quad -1 \quad -8 \quad 12}$$
$$1 \quad 0 \quad -8 \qquad\qquad\qquad 2 \quad 2 \quad -12$$
$$\overline{1 \quad 0 \quad -8 \quad 4}\qquad\qquad \overline{1 \quad 1 \quad -6 \quad 0} \; \Rightarrow \quad x = 2 \text{ is a zero.}$$

$P(x) = x^3 - x^2 - 8x + 12 = (x - 2)(x^2 + x - 6) = (x - 2)(x + 3)(x - 2)$. Therefore, the zeros are $x = -3, 2$.

18. $P(x) = x^3 - 4x^2 - 7x + 10$. The possible rational zeros are $\pm 1, \pm 2, \pm 5, \pm 10$. $P(x)$ has 2 variations in sign and hence 0 or 2 positive real zeros. $P(-x) = -x^3 - 4x^2 + 7x + 10$ has 1 variation in sign and hence 1 negative real zero.

$$
\begin{array}{r|rrrr}
1 & 1 & -4 & -7 & 10 \\
  &   & 1 & -3 & -10 \\
\hline
  & 1 & -3 & -10 & 0
\end{array}
\quad \Rightarrow \quad x = 1 \text{ is a zero.}
$$

So $P(x) = x^3 - 4x^2 - 7x + 10 = (x-1)(x^2 - 3x - 10) = (x-1)(x-5)(x+2)$. Therefore, the zeros are $x = -2, 1, 5$.

20. $P(x) = x^3 - 2x^2 - 2x - 3$. The possible rational zeros are $\pm 1, \pm 3$. $P(x)$ has 1 variation in sign and hence 1 positive real zero. $P(-x) = -x^3 - 2x^2 + 2x - 3$ has 2 variations in sign and hence 0 or 2 negative real zeros.

$$
\begin{array}{r|rrrr}
1 & 1 & -2 & -2 & -3 \\
  &   & 1 & -1 & -3 \\
\hline
  & 1 & -1 & -3 & -6
\end{array}
\qquad
\begin{array}{r|rrrr}
3 & 1 & -2 & -2 & -3 \\
  &   & 3 & 3 & 3 \\
\hline
  & 1 & 1 & 1 & 0
\end{array}
\quad \Rightarrow \quad x = 3 \text{ is a zero.}
$$

So $P(x) = x^3 - 2x^2 - 2x - 3 = (x-3)(x^2 + x + 1)$. Now, $Q(x) = x^2 + x + 1$ has no real zeros, since the discriminant is $b^2 - 4ac = (1)^2 - 4(1)(1) = -3 < 0$. Thus, the only real zero is $x = 3$.

22. $P(x) = x^4 - 2x^3 - 3x^2 + 8x - 4$. Using synthetic division, we see that $(x-1)$ is a factor of $P(x)$:

$$
\begin{array}{r|rrrrr}
1 & 1 & -2 & -3 & 8 & -4 \\
  &   & 1 & -1 & -4 & 4 \\
\hline
  & 1 & -1 & -4 & 4 & 0
\end{array}
\quad \Rightarrow \quad x = 1 \text{ is a zero.}
$$

We continue by factoring the quotient, and we see that $(x-1)$ is again a factor:

$$
\begin{array}{r|rrrr}
1 & 1 & -1 & -4 & 4 \\
  &   & 1 & 0 & -4 \\
\hline
  & 1 & 0 & -4 & 0
\end{array}
\quad \Rightarrow \quad x = 1 \text{ is a zero.}
$$

$P(x) = x^4 - 2x^3 - 3x^2 + 8x - 4 = (x-1)(x-1)(x^2 - 4) = (x-1)^2(x-2)(x+2)$. Therefore, the zeros are $x = 1, \pm 2$.

24. $P(x) = x^4 - x^3 - 23x^2 - 3x + 90$. The possible rational zeros are $\pm 1, \pm 2, \pm 3, \pm 5, \pm 6, \pm 9, \pm 10, \pm 15, \pm 18, \pm 30, \pm 45, \pm 90$. Since $P(x)$ has 2 variations in sign, P has 0 or 2 positive real zeros. Since $P(-x) = x^4 + x^3 - 23x^2 + 3x + 90$ has 2 variations in sign, P has 0 or 2 negative real zeros.

$$
\begin{array}{r|rrrrr}
1 & 1 & -1 & -23 & -3 & 90 \\
  &   & 1 & 0 & -23 & -26 \\
\hline
  & 1 & 0 & -23 & -26 & 64
\end{array}
$$

$$
\begin{array}{r|rrrrr}
2 & 1 & -1 & -23 & -3 & 90 \\
  &   & 2 & 2 & -42 & -90 \\
\hline
  & 1 & 1 & -21 & -45 & 0
\end{array}
\quad \Rightarrow \quad x = 2 \text{ is a zero.}
$$

$P(x) = (x-2)(x^3 + x^2 - 21x - 45)$. Continuing with the quotient we have:

$$
\begin{array}{r|rrrr}
3 & 1 & 1 & -21 & -45 \\
  &   & 3 & 12 & -27 \\
\hline
  & 1 & 4 & -9 & -72
\end{array}
\qquad
\begin{array}{r|rrrr}
5 & 1 & 1 & -21 & -45 \\
  &   & 5 & 30 & 45 \\
\hline
  & 1 & 6 & 9 & 0
\end{array}
\quad \Rightarrow \quad x = 5 \text{ is a zero.}
$$

$P(x) = (x-2)(x-5)(x^2+6x+9) = (x-2)(x-5)(x+3)^2$. Therefore, the zeros are $x = -3, 2, 5$.

26. $P(x) = x^4 - x^3 - 5x^2 + 3x + 6$. The possible rational zeros are $\pm 1, \pm 2, \pm 3, \pm 6$. Since $P(x)$ has 2 variations in sign, P has 0 or 2 positive real zeros. Since $P(-x) = x^4 + x^3 - 5x^2 - 3x + 6$ has 2 variations in sign, P has 0 or 2 negative real zeros.

$$
\begin{array}{r|rrrrr}
1 & 1 & -1 & -5 & 3 & 6 \\
  &   & 1 & 0 & -5 & -2 \\
\hline
  & 1 & 0 & -5 & -2 & 4
\end{array}
\qquad
\begin{array}{r|rrrrr}
2 & 1 & -1 & -5 & 3 & 6 \\
  &   & 2 & 2 & -6 & -6 \\
\hline
  & 1 & 1 & -3 & -3 & 0
\end{array}
\Rightarrow \quad x = 2 \text{ is a zero.}
$$

$P(x) = (x-2)(x^3 + x^2 - 3x - 3)$. Continuing with the quotient we have:

$$
\begin{array}{r|rrrr}
3 & 1 & 1 & -3 & -3 \\
  &   & 3 & 12 & 27 \\
\hline
  & 1 & 4 & 9 & 24
\end{array}
\Rightarrow \quad x = 3 \text{ is an upper bound, thus we try negative zeros.}
$$

$$
\begin{array}{r|rrrr}
-1 & 1 & 1 & -3 & -3 \\
   &   & -1 & 0 & 3 \\
\hline
   & 1 & 0 & -3 & 0
\end{array}
\Rightarrow \quad x = -1 \text{ is a zero.}
$$

$P(x) = (x-2)(x+1)(x^2-3)$. Therefore, the rational zeros are $x = -1, 2$.

28. $P(x) = 2x^3 + 7x^2 + 4x - 4$. The possible rational zeros are $\pm 1, \pm 2, \pm 4, \pm \frac{1}{2}$. Since $P(x)$ has 1 variation in sign, P has 1 positive real zero. Since $P(-x) = -2x^3 + 7x^2 - 4x - 4$ has 2 variations in sign, P has 0 or 2 negative real zeros.

$$
\begin{array}{r|rrrr}
1 & 2 & 7 & 4 & -4 \\
  &   & 2 & 9 & 13 \\
\hline
  & 2 & 9 & 13 & 9
\end{array}
\Rightarrow x = 1 \text{ is an upper bound.}
\qquad
\begin{array}{r|rrrr}
\frac{1}{2} & 2 & 7 & 4 & -4 \\
   &   & 1 & 4 & 4 \\
\hline
   & 2 & 8 & 8 & 0
\end{array}
\Rightarrow x = \tfrac{1}{2} \text{ is a zero.}
$$

$P(x) = \left(x - \frac{1}{2}\right)(2x^2 + 8x + 8) = 2\left(x - \frac{1}{2}\right)(x^2 + 4x + 4) = 2\left(x - \frac{1}{2}\right)(x+2)^2$. Therefore, the zeros are $x = -2, \frac{1}{2}$.

30. We use factoring by grouping: $P(x) = 2x^3 - 3x^2 - 2x + 3 = 2x(x^2 - 1) - 3(x^2 - 1)$
$= (x^2 - 1)(2x - 3) = (x-1)(x+1)(2x-3)$. Therefore, the zeros are $x = \frac{3}{2}, \pm 1$.

32. $P(x) = 8x^3 + 10x^2 - x - 3$. The possible rational zeros are $\pm 1, \pm 3, \pm \frac{1}{2}, \pm \frac{3}{2}, \pm \frac{1}{4}, \pm \frac{3}{4}, \pm \frac{1}{8}, \pm \frac{3}{8}$. Since $P(x)$ has 1 variation in sign, P has 1 positive real zero. Since $P(-x) = -8x^3 + 10x^2 + x - 3$ has 2 variations in sign, P has 0 or 2 negative real zeros.

$$
\begin{array}{r|rrrr}
1 & 8 & 10 & -1 & -3 \\
  &   & 8 & 18 & 17 \\
\hline
  & 8 & 18 & 17 & 14
\end{array}
\Rightarrow \quad x = 1 \text{ is an upper bound, thus we try fractions.}
$$

$$
\begin{array}{r|rrrr}
\frac{1}{2} & 8 & 10 & -1 & -3 \\
   &   & 4 & 7 & 3 \\
\hline
   & 8 & 14 & 6 & 0
\end{array}
\Rightarrow \quad x = \tfrac{1}{2} \text{ is a zero.}
$$

$P(x) = 8x^3 + 10x^2 - x - 3 = \left(x - \frac{1}{2}\right)(8x^2 + 14x + 6) = 2\left(x - \frac{1}{2}\right)(4x^2 + 7x + 3)$
$= (2x - 1)(x+1)(4x+3) = 0$. Therefore, the zeros are $x = -1, -\frac{3}{4}, \frac{1}{2}$.

34. $P(x) = 6x^4 - 7x^3 - 12x^2 + 3x + 2$. The possible rational zeros are $\pm 1, \pm 2, \pm\frac{1}{2}, \pm\frac{1}{3}, \pm\frac{2}{3}, \pm\frac{1}{6}$.
Since $P(x)$ has 2 variations in sign, P has 0 or 2 positive real zeros. Since
$P(-x) = 6x^4 + 7x^3 - 12x^2 - 3x + 2$ has 2 variations in sign, P has 0 or 2 negative real zeros.

1	6	−7	−12	3	2
		6	−1	−13	−10
	6	−1	−13	−10	−8

2	6	−7	−12	3	2
		12	10	−4	−2
	6	5	−2	−1	0	$\Rightarrow$ $x = 2$ is a zero.

$P(x) = 6x^4 - 7x^3 - 12x^2 + 3x + 2 = (x - 2)(6x^3 + 5x^2 - 2x - 1)$. Continuing by factoring the quotient, we first note that the possible rational zeros are $-1, \pm\frac{1}{2}, \pm\frac{1}{3}, \pm\frac{1}{6}$. We have:

$\frac{1}{2}$	6	5	−2	−1
		3	4	1
	6	8	2	0	$\Rightarrow$ $x = \frac{1}{2}$ is a zero.

$P(x) = (x - 2)\left(x - \frac{1}{2}\right)(6x^2 + 8x + 2) = 2(x - 2)\left(x - \frac{1}{2}\right)(3x^2 + 4x + 1)$
$= (x - 2)\left(x - \frac{1}{2}\right)(x + 1)(3x + 1)$. Therefore, the zeros are $x = -1, -\frac{1}{3}, \frac{1}{2}, 2$.

36. $P(x) = x^5 - 4x^4 - 3x^3 + 22x^2 - 4x - 24$ has possible rational zeros $\pm 1, \pm 2, \pm 3, \pm 4, \pm 6, \pm 8,$
$\pm 12, \pm 24$. Since $P(x)$ has 3 variations in sign, there are 1 or 3 positive real zeros. Since
$P(-x) = -x^5 - 4x^4 + 3x^3 + 22x^2 + 4x - 24$ has 2 variations in sign, there are 0 or 2 negative real zeros.

1	1	−4	−3	22	−4	−24
		1	−3	−6	16	12
	1	−3	−6	16	12	−12

2	1	−4	−3	22	−4	−24
		2	−4	−14	16	24
	1	−2	−7	8	12	0	$\Rightarrow$ $x = 2$ is a zero.

$P(x) = (x - 2)(x^4 - 2x^3 - 7x^2 + 8x + 12)$

2	1	−2	−7	8	12
		2	0	−14	−12
	1	0	−7	−6	0	$\Rightarrow$ $x = 2$ is a zero again.

$P(x) = (x - 2)^2(x^3 - 7x - 6)$

2	1	0	−7	−6
		2	4	−6
	1	2	−3	−12

3	1	0	−7	−6
		3	9	6
	1	3	2	0	$\Rightarrow$ $x = 3$ is a zero.

$P(x) = (x - 2)^2(x - 3)(x^2 + 3x + 2) = (x - 2)^2(x - 3)(x + 1)(x + 2) = 0$. Therefore, the zeros
are $x = -1, \pm 2, 3$.

38. $P(x) = 2x^6 - 3x^5 - 13x^4 + 29x^3 - 27x^2 + 32x - 12$ has possible rational zeros $\pm 1, \pm 2, \pm 3, \pm 4,$
$\pm 6, \pm 12, \pm\frac{1}{2}, \pm\frac{3}{2}$. Since $P(x)$ has 5 variations in sign, there are 1 or 3 or 5 positive real zeros.
Since $P(-x) = 2x^6 + 3x^5 - 13x^4 - 29x^3 + 27x^2 - 32x - 12$ has 3 variations in sign, there are 1
or 3 negative real zeros.

1	2	−3	−13	29	−27	32	−12
		2	−1	−14	15	−12	20
	2	−1	−14	15	−12	20	8

$$\begin{array}{r|rrrrrr} 2 & 2 & -3 & -13 & 29 & -27 & 32 & -12 \\ & & 4 & 2 & -22 & 14 & -26 & 12 \\ \hline & 2 & 1 & -11 & 7 & -13 & 6 & 0 \end{array} \Rightarrow \quad x = 2 \text{ is a zero.}$$

$P(x) = (x - 2)(2x^5 + x^4 - 11x^3 + 7x^2 - 13x + 6)$. We continue with the quotient:

$$\begin{array}{r|rrrrrr} 2 & 2 & 1 & -11 & 7 & -13 & 6 \\ & & 4 & 10 & -2 & 10 & -6 \\ \hline & 2 & 5 & -1 & 5 & -3 & 0 \end{array} \Rightarrow \quad x = 2 \text{ is a zero again.}$$

$P(x) = (x - 2)^2(2x^4 + 5x^3 - x^2 + 5x - 3)$. We continue with the quotient, first noting 2 is no longer a possible rational solution:

$$\begin{array}{r|rrrrr} 3 & 2 & 5 & -1 & 5 & -3 \\ & & 6 & 22 & 42 & 94 \\ \hline & 2 & 11 & 21 & 47 & 91 \end{array} \Rightarrow x = 3 \text{ is an upper bound.}$$

We know that there is at least 1 more positive zero.

$$\begin{array}{r|rrrrr} \frac{1}{2} & 2 & 5 & -1 & 5 & -3 \\ & & 1 & 2 & 1 & 3 \\ \hline & 2 & 6 & 2 & 6 & 0 \end{array} \Rightarrow x = \frac{1}{2} \text{ is a zero.}$$

$P(x) = (x - 2)^2(x - \frac{1}{2})(2x^3 + 6x^2 + 2x + 6)$.

We can factor $2x^3 + 6x^2 + 2x + 6$ by grouping; $2x^3 + 6x^2 + 2x + 6 = (2x^3 + 6x^2) + (2x + 6)$ $= (2x + 6)(x^2 + 1)$. So $P(x) = 2(x - 2)^2(x - \frac{1}{2})(x + 3)(x^2 + 1)$. Since $x^2 + 1$ has no real zeros, the zeros of P are $x = -3, 2, \frac{1}{2}$.

40. $P(x) = x^3 - 5x^2 + 2x + 12$. The possible rational zeros are $\pm 1, \pm 2, \pm 3, \pm 4, \pm 6, \pm 12$. $P(x)$ has 2 variations in sign and hence 0 or 2 positive real zeros. $P(-x) = -x^3 - 5x^2 - 2x + 12$ has 1 variation in sign and hence 1 negative real zero.

$$\begin{array}{r|rrrr} 1 & 1 & -5 & 2 & 12 \\ & & 1 & -4 & -2 \\ \hline & 1 & -4 & -2 & 10 \end{array} \qquad\qquad \begin{array}{r|rrrr} 2 & 1 & -5 & 2 & 12 \\ & & 2 & -6 & -8 \\ \hline & 1 & -3 & -4 & 4 \end{array}$$

$$\begin{array}{r|rrrr} 3 & 1 & -5 & 2 & 12 \\ & & 3 & -6 & -12 \\ \hline & 1 & -2 & -4 & 0 \end{array} \Rightarrow x = 3 \text{ is a zero.}$$

So $P(x) = (x - 3)(x^2 - 2x - 4)$. Using the quadratic formula on the second factor, we have:
$x = \dfrac{-(-2) \pm \sqrt{(-2)^2 - 4(1)(-4)}}{2(1)} = \dfrac{2 \pm \sqrt{20}}{2} = \dfrac{2 \pm 2\sqrt{5}}{2} = 1 \pm \sqrt{5}$. Therefore, the zeros are $x = 3, 1 \pm \sqrt{5}$.

42. $P(x) = x^4 + 2x^3 - 2x^2 - 3x + 2$. The possible rational zeros are $\pm 1, \pm 2$. $P(x)$ has 2 variations in sign and hence 0 or 2 positive real zeros. $P(-x) = x^4 - 2x^3 - 2x^2 + 3x + 2$ has 2 variations in sign and hence 0 or 2 negative real zeros.

$$\begin{array}{r|rrrrr} 1 & 1 & 2 & -2 & -3 & 2 \\ & & 1 & 3 & 1 & -2 \\ \hline & 1 & 3 & 1 & -2 & 0 \end{array} \Rightarrow x = 1 \text{ is a zero.}$$

$P(x) = (x - 1)(x^3 + 3x^2 + x - 2)$. Continuing with the quotient:

$$\begin{array}{r|rrrr} 1 & 1 & 3 & 1 & -2 \\ & & 1 & 4 & 5 \\ \hline & 1 & 4 & 5 & 3 \end{array} \Rightarrow x = 1 \text{ is an upper bound.}$$

$$\begin{array}{r|rrrr} -1 & 1 & 3 & 1 & -2 \\ & & -1 & -2 & 1 \\ \hline & 1 & 2 & -1 & -1 \end{array} \qquad \begin{array}{r|rrrr} -2 & 1 & 3 & 1 & -2 \\ & & -2 & -2 & 2 \\ \hline & 1 & 1 & -1 & 0 \end{array} \Rightarrow x = -2 \text{ is a zero.}$$

So $P(x) = (x-1)(x+2)(x^2+x-1)$. Using the quadratic formula on the third factor, we have:
$x = \dfrac{-1 \pm \sqrt{1^2 - 4(1)(-1)}}{2(1)} = \dfrac{1 \pm \sqrt{5}}{2}$. Therefore, the zeros are $x = 1, -2, \dfrac{1 \pm \sqrt{5}}{2}$.

44. $P(x) = x^5 - 4x^4 - x^3 + 10x^2 + 2x - 4$. The possible rational zeros are $\pm 1, \pm 2, \pm 4$. $P(x)$ has 3 variations in sign and hence 1 or 3 positive real zeros. $P(-x) = -x^5 - 4x^4 + x^3 + 10x^2 - 2x - 4$ has 2 variations in sign and hence 0 or 2 negative real zeros.

$$\begin{array}{r|rrrrrr} 1 & 1 & -4 & -1 & 10 & 2 & -4 \\ & & 1 & -3 & -4 & 6 & 8 \\ \hline & 1 & -3 & -4 & 6 & 8 & 4 \end{array} \qquad \begin{array}{r|rrrrrr} 2 & 1 & -4 & -1 & 10 & 2 & -4 \\ & & 2 & -4 & -10 & 0 & 4 \\ \hline & 1 & -2 & -5 & 0 & 2 & 0 \end{array} \Rightarrow x = 2 \text{ is a zero.}$$

So $P(x) = (x-2)(x^4 - 2x^3 - 5x^2 + 2)$. Since the constant term of the second factor is 2, ± 4 are no longer possible zeros. Continuing by factoring the quotient, we have:

$$\begin{array}{r|rrrrr} 2 & 1 & -2 & -5 & 0 & 2 \\ & & 2 & 0 & -10 & -20 \\ \hline & 1 & 0 & -5 & -10 & 18 \end{array} \qquad \begin{array}{r|rrrrr} -1 & 1 & -2 & -5 & 0 & 2 \\ & & -1 & 3 & 2 & -2 \\ \hline & 1 & -3 & -2 & 2 & 0 \end{array} \Rightarrow x = -1 \text{ is a zero.}$$

So $P(x) = (x-2)(x+1)(x^3 - 3x^2 - 2x - 2)$. Continuing by factoring the quotient, we have:

$$\begin{array}{r|rrrr} -1 & 1 & -3 & -2 & 2 \\ & & -1 & 4 & -2 \\ \hline & 1 & -4 & 2 & 0 \end{array} \Rightarrow x = -1 \text{ is a zero again.}$$

So $P(x) = (x-2)(x+1)^2(x^2 - 4x + 2)$. Using the quadratic formula on the second factor, we have: $x = \dfrac{-(-4) \pm \sqrt{(-4)^2 - 4(1)(2)}}{2(1)} = \dfrac{4 \pm \sqrt{8}}{2} = \dfrac{4 \pm 2\sqrt{2}}{2} = 2 \pm \sqrt{2}$. Therefore, the zeros are $x = -1, 2, 2 \pm \sqrt{2}$.

46. $P(x) = 3x^3 - 5x^2 - 8x - 2$. The possible rational zeros are $\pm 1, \pm 2, \pm \frac{1}{3}, \pm \frac{2}{3}$. $P(x)$ has 1 variation in sign and hence 1 positive real zero. $P(-x) = -3x^3 - 5x^2 + 8x - 2$ has 2 variations in sign and hence 0 or 2 negative real zeros.

$$\begin{array}{r|rrrr} 1 & 3 & -5 & -8 & -2 \\ & & 3 & -2 & -10 \\ \hline & 3 & -2 & -10 & -12 \end{array} \qquad \begin{array}{r|rrrr} 2 & 3 & -5 & -8 & -2 \\ & & 6 & 2 & -12 \\ \hline & 3 & 1 & -6 & -14 \end{array}$$

$$\begin{array}{r|rrrr} \frac{1}{3} & 3 & -5 & -8 & -2 \\ & & 1 & -\frac{4}{3} & -\frac{28}{9} \\ \hline & 3 & -4 & -\frac{28}{3} & -\frac{46}{9} \end{array} \qquad \begin{array}{r|rrrr} \frac{2}{3} & 3 & -5 & -8 & -2 \\ & & 2 & -2 & -\frac{20}{3} \\ \hline & 3 & -3 & -10 & -\frac{26}{3} \end{array}$$

Thus we have tried all the positive rational zeros, so we try the negative zeros.

$$\begin{array}{r|rrrr} -1 & 3 & -5 & -8 & -2 \\ & & -3 & 8 & 0 \\ \hline & 3 & -8 & 0 & -2 \end{array} \qquad \begin{array}{r|rrrr} -2 & 3 & -5 & -8 & -2 \\ & & -6 & 22 & -28 \\ \hline & 3 & -11 & 14 & -30 \end{array}$$

$$-\tfrac{1}{3} \begin{array}{|rrrr} 3 & -5 & -8 & -2 \\ & -1 & 2 & 2 \\ \hline 3 & -6 & -6 & 0 \end{array} \Rightarrow \quad x = -\tfrac{1}{3} \text{ is a zero.}$$

So $P(x) = \left(x + \tfrac{1}{3}\right)\left(3x^2 - 6x - 6\right) = 3\left(x + \tfrac{1}{3}\right)\left(x^2 - 2x - 2\right)$. Using the quadratic formula on the second factor, we have: $x = \dfrac{-(-2) \pm \sqrt{(-2)^2 - 4(1)(-2)}}{2(1)} = \dfrac{2 \pm \sqrt{12}}{2} = \dfrac{2 \pm 2\sqrt{3}}{2} = 1 \pm \sqrt{3}$.

Therefore, the zeros are $x = -\tfrac{1}{3}, 1 \pm \sqrt{3}$.

48. $P(x) = 4x^5 - 18x^4 - 6x^3 + 91x^2 - 60x + 9$. The possible rational zeros are ± 1, ± 3, ± 9, $\pm \tfrac{1}{2}$, $\pm \tfrac{3}{2}$, $\pm \tfrac{9}{2}$, $\pm \tfrac{1}{4}$, $\pm \tfrac{3}{4}$, $\pm \tfrac{9}{4}$. $P(x)$ has 4 variations in sign and hence 0 or 2 or 4 positive real zeros. $P(-x) = -4x^5 - 18x^4 + 6x^3 + 91x^2 + 60x + 9$ has 1 variation in sign and hence 1 negative real zero.

$$1 \begin{array}{|rrrrrr} 4 & -18 & -6 & 91 & -60 & 9 \\ & 4 & -14 & -20 & 71 & 1 \\ \hline 4 & -14 & -20 & 71 & 11 & 10 \end{array}$$

$$3 \begin{array}{|rrrrrr} 4 & -18 & -6 & 91 & -60 & 9 \\ & 12 & -18 & -72 & 57 & -9 \\ \hline 4 & -6 & -24 & 19 & -3 & 0 \end{array} \Rightarrow \quad x = 3 \text{ is a zero.}$$

So $P(x) = (x - 3)(4x^4 - 6x^3 - 24x^2 + 19x - 3)$. Continuing by factoring the quotient, we have:

$$3 \begin{array}{|rrrrr} 4 & -6 & -24 & 19 & -3 \\ & 12 & 18 & -18 & 3 \\ \hline 4 & 6 & -6 & 1 & 0 \end{array} \Rightarrow \quad x = 3 \text{ is a zero again.}$$

So $P(x) = (x - 3)^2(4x^3 + 6x^2 - 6x + 1)$. Continuing by factoring the quotient, we have:

$$3 \begin{array}{|rrrr} 4 & 6 & -6 & 1 \\ & 12 & 54 & 144 \\ \hline 4 & 18 & 48 & 1445 \end{array} \Rightarrow \quad x = 3 \text{ is an upper bound.}$$

$$\tfrac{1}{2} \begin{array}{|rrrr} 4 & 6 & -6 & 1 \\ & 2 & 4 & -1 \\ \hline 4 & 8 & -2 & 0 \end{array} \Rightarrow \quad x = \tfrac{1}{2} \text{ is a zero.}$$

So $P(x) = (x - 3)^2\left(x - \tfrac{1}{2}\right)(4x^2 + 8x - 2) = 2(x - 3)^2\left(x - \tfrac{1}{2}\right)(2x^2 + 4x - 1)$. Using the quadratic formula on the second factor, we have:

$x = \dfrac{-4 \pm \sqrt{4^2 - 4(2)(-1)}}{2(2)} = \dfrac{-4 \pm \sqrt{8}}{4} = \dfrac{-4 \pm 2\sqrt{2}}{4} = \dfrac{-2 \pm \sqrt{2}}{2}$. Therefore, the zeros are $x = \tfrac{1}{2}, 3, \dfrac{-2 \pm \sqrt{2}}{2}$.

50. (a) $P(x) = -x^3 - 2x^2 + 5x + 6$ has possible rational zeros ± 1, ± 2, ± 3, ± 6.

$$1 \begin{array}{|rrrr} -1 & -2 & 5 & 6 \\ & -1 & -3 & 2 \\ \hline -1 & -3 & 2 & 8 \end{array} \qquad 2 \begin{array}{|rrrr} -1 & -2 & 5 & 6 \\ & -2 & -8 & -6 \\ \hline -1 & -4 & -3 & 0 \end{array} \Rightarrow \quad x = 2 \text{ is a zero.}$$

So $P(x) = (x - 2)(-x^2 - 4x - 3) = -(x - 2)(x^2 + 4x + 3) = -(x - 2)(x + 1)(x + 3)$. The real zeros of P are $2, -1, -3$.

(b)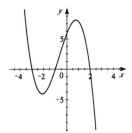

52. (a) $P(x) = 3x^3 + 17x^2 + 21x - 9$ has possible rational zeros ± 1, ± 3, ± 9, $\pm\frac{1}{3}$, $\pm\frac{2}{3}$.

$$\begin{array}{r|rrrr} 1 & 3 & 17 & 21 & -9 \\ & & 3 & 20 & 41 \\ \hline & 3 & 20 & 41 & 32 \end{array}$$ $\Rightarrow x = 1$ is an upper bound.

$$\begin{array}{r|rrrr} \frac{1}{3} & 3 & 17 & 21 & -9 \\ & & 1 & 6 & 9 \\ \hline & 3 & 18 & 27 & 0 \end{array}$$ $\Rightarrow x = \frac{1}{3}$ is a zero.

So $P(x) = \left(x - \frac{1}{3}\right)(3x^2 + 18x + 27) = 3\left(x - \frac{1}{3}\right)(x^2 + 6x + 9) = 3\left(x - \frac{1}{3}\right)(x + 3)^2$. The real zeros of P are -3, $\frac{1}{3}$.

(b)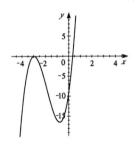

54. (a) $P(x) = -x^4 + 10x^2 + 8x - 8$ has possible rational zeros ± 1, ± 2, ± 4, ± 8.

$$\begin{array}{r|rrrrr} 1 & -1 & 0 & 10 & 8 & -8 \\ & & -1 & -1 & 9 & 17 \\ \hline & -1 & -1 & 9 & 17 & 9 \end{array} \qquad \begin{array}{r|rrrrr} 2 & -1 & 0 & 10 & 8 & -8 \\ & & -2 & -4 & 12 & 40 \\ \hline & -1 & -2 & 6 & 20 & 32 \end{array}$$

$$\begin{array}{r|rrrrr} 4 & -1 & 0 & 10 & 8 & -8 \\ & & -4 & -16 & -24 & -64 \\ \hline & -1 & -4 & -6 & -16 & -72 \end{array}$$ $\Rightarrow x = 4$ is an upper bound.

$$\begin{array}{r|rrrrr} -1 & -1 & 0 & 10 & 8 & -8 \\ & & 1 & -1 & -9 & 1 \\ \hline & -1 & 1 & 9 & -1 & -7 \end{array}$$

$$\begin{array}{r|rrrrr} -2 & -1 & 0 & 10 & 8 & -8 \\ & & 2 & -4 & -12 & 8 \\ \hline & -1 & 2 & 6 & -4 & 0 \end{array}$$ $\Rightarrow x = -2$ is a zero.

So $P(x) = (x + 2)(-x^3 + 2x^2 + 6x - 4)$. Continuing, we have:

$$\begin{array}{r|rrrr}
-2 & -1 & 2 & 6 & -4 \\
 & & 2 & -8 & 4 \\
\hline
 & -1 & 4 & -2 & 0
\end{array} \Rightarrow x = -2 \text{ is a zero again.}$$

$P(x) = (x+2)^2(-x^2 + 4x - 2)$. Using the quadratic formula on the second factor, we have

$$x = \frac{-4 \pm \sqrt{4^2 - 4(-1)(-2)}}{2(-1)} = \frac{-4 \pm \sqrt{8}}{-2} = \frac{-4 \pm 2\sqrt{2}}{-2} = 2 \pm \sqrt{2}. \text{ So the real zeros of } P$$

are $-2, 2 \pm \sqrt{2}$.

(b)

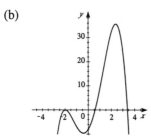

56. (a) $P(x) = x^5 - x^4 - 6x^3 + 14x^2 - 11x + 3$ has possible rational zeros $\pm 1, \pm 3$.

$$\begin{array}{r|rrrrrr}
1 & 1 & -1 & -6 & 14 & -11 & 3 \\
 & & 1 & 0 & -6 & 8 & -3 \\
\hline
 & 1 & 0 & -6 & 8 & -3 & 0
\end{array} \Rightarrow \quad x = 1 \text{ is a zero.}$$

So $P(x) = (x-1)(x^4 - 6x^2 + 8x - 3)$:

$$\begin{array}{r|rrrrr}
1 & 1 & 0 & -6 & 8 & -3 \\
 & & 1 & 1 & -5 & 3 \\
\hline
 & 1 & 1 & -5 & 3 & 0
\end{array} \Rightarrow \quad x = 1 \text{ is a zero again.}$$

So $P(x) = (x-1)^2(x^3 + x^2 - 5x + 3)$:

$$\begin{array}{r|rrrr}
1 & 1 & 1 & -5 & 3 \\
 & & 1 & 2 & -3 \\
\hline
 & 1 & 2 & -3 & 0
\end{array} \Rightarrow \quad x = 1 \text{ is a zero again.}$$

So $P(x) = (x-1)^3(x^2 + 2x - 3) = (x-1)^4(x+3)$, and the real zeros of P are 1 and -3.

(b)

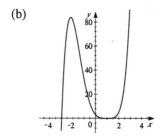

58. $P(x) = 2x^3 - x^2 + 4x - 7$. Since $P(x)$ has 3 variations in signs, P has 3 or 1 positive real zeros. Since $P(-x) = -2x^3 - x^2 - 4x - 7$ has no variations in sign, there are no negative real zeros. Thus, P has 1 or 3 real zeros.

60. $P(x) = x^4 + x^3 + x^2 + x + 12$. Since $P(x)$ has no variations in sign, P has no positive real zeros. Since $P(-x) = x^4 - x^3 + x^2 - x + 12$ has 4 variations in sign, P has 4, 2, or 0 negative real zeros. Therefore, $P(x)$ has 0, 2, or 4 real zeros.

62. $P(x) = x^8 - x^5 + x^4 - x^3 + x^2 - x + 1$. Since $P(x)$ has 6 variations in sign, the polynomial has 6, 4, 2, or 0 positive real zeros. Since $P(-x)$ has no variations in sign, the polynomial has no negative real zeros. Therefore, P has 6, 4, 2, or 0 real zeros.

64. $P(x) = x^4 - 2x^3 - 9x^2 + 2x + 8$; $a = -3, b = 5$

```
-3 | 1   -2   -9    2     8
   |      -3   15  -18    48
   ----------------------------
     1   -5    6  -16    56     Alternating signs   ⇒   lower bound.
```

```
 5 | 1   -2   -9    2     8
   |       5   15   30   160
   ----------------------------
     1    3    6   32   168     All nonnegative   ⇒   upper bound.
```

Therefore $a = -3, b = 5$ are lower and upper bounds, respectively.

66. $P(x) = 3x^4 - 17x^3 + 24x^2 - 9x + 1$; $a = 0, b = 6$

```
 0 | 3   -17   24   -9    1
   |        0    0    0    0
   ----------------------------
     3   -17   24   -9    1     Alternating signs   ⇒   lower bound.
```

```
 6 | 3   -17   24    -9      1
   |       18    6   180   1026
   --------------------------------
     3     1   30   171   1027   All nonnegative   ⇒   upper bound.
```

Therefore $a = 0, b = 6$ are lower and upper bounds, respectively. Note, since $P(x)$ alternates in sign, by Descartes' Rule of Signs, 0 is automatically a lower bound.

There are many possible solutions to Exercises 68 and 70 since we only asked to find 'an upper bound' and 'a lower bound'.

68. $P(x) = 2x^3 - 3x^2 - 8x + 12$ and using the Upper and Lower Bounds Theorem:

```
-2 | 2   -3   -8    12
   |      -4   14   -12
   ----------------------
     2   -7    6     0     Alternating signs   ⇒   x = -2 is a lower bound (and a zero).
```

```
 3 | 2   -3   -8   12
   |       6    9    3
   ---------------------
     2    3    1   15     All nonnegative   ⇒   x = 3 is an upper bound.
```

70. Set $P(x) = x^5 - x^4 + 1$.

```
 1 | 1   -1   0   0   0   1
   |       1   0   0   0   0
   ----------------------------
     1    0   0   0   0   1     All nonnegative   ⇒   x = 1 is an upper bound.
```

```
-1 | 1   -1    0    0    0    1
   |      -1    2   -2    2   -2
   --------------------------------
     1   -2    2   -2    2   -1     Alternating signs   ⇒   x = -1 is a lower bound.
```

72. $P(x) = 2x^4 + 15x^3 + 31x^2 + 20x + 4$. The possible rational zeros are $\pm 1, \pm 2, \pm 4, \pm \frac{1}{2}$. Since all of the coefficients are positive, there are no positive zeros. Since $P(-x) = 2x^4 - 15x^3 + 31x^2 - 20x + 4$ has 4 variations in sign, there are 0, 2, or 4 negative real zeros.

$$
\begin{array}{r|rrrrr}
-1 & 2 & 15 & 31 & 20 & 4 \\
& & -2 & -13 & -18 & -2 \\
\hline
& 2 & 13 & 18 & 2 & 2
\end{array}
$$

$$
\begin{array}{r|rrrrr}
-2 & 2 & 15 & 31 & 20 & 4 \\
& & -4 & -22 & -18 & -4 \\
\hline
& 2 & 11 & 9 & 2 & 0
\end{array}
$$

$\Rightarrow$ $x = -2$ is a zero, and $P(x) = (x+2)(2x^3 + 11x^2 + 9x + 2)$.

$$
\begin{array}{r|rrrr}
-2 & 2 & 11 & 9 & 2 \\
& & -4 & -14 & 10 \\
\hline
& 2 & 7 & -5 & 12
\end{array}
\qquad
\begin{array}{r|rrrr}
-4 & 2 & 11 & 9 & 2 \\
& & -8 & -12 & 12 \\
\hline
& 2 & 3 & -3 & 14
\end{array}
$$

$$
\begin{array}{r|rrrr}
-\frac{1}{2} & 2 & 11 & 9 & 2 \\
& & -1 & -5 & -2 \\
\hline
& 2 & 10 & 4 & 0
\end{array}
$$
$x = -\frac{1}{2}$ is a zero, $P(x) = (x+2)(2x+1)(x^2 + 5x + 2)$.

Now, if $x^2 + 5x + 2 = 0$, then $x = \dfrac{-5 \pm \sqrt{25 - 4(1)(2)}}{2} = \dfrac{-5 \pm \sqrt{17}}{2}$. Thus, the zeros are -2, $-\frac{1}{2}$, and $\frac{-5 \pm \sqrt{17}}{2}$.

74. $P(x) = 6x^4 - 7x^3 - 8x^2 + 5x = x(6x^3 - 7x^2 - 8x + 5)$. So $x = 0$ is a zero. Continuing with the quotient, $Q(x) = 6x^3 - 7x^2 - 8x + 5$. The possible rational zeros are $\pm 1, \pm 5, \pm \frac{1}{2}, \pm \frac{5}{2}, \pm \frac{1}{3}, \pm \frac{5}{3}, \pm \frac{1}{6}, \pm \frac{5}{6}$. Since $Q(x)$ has 2 variations in sign, there are 0 or 2 positive real zeros. Since $Q(-x) = 6x^4 + 7x^3 - 8x^2 - 5x$ has 1 variation in sign, there is 1 negative real zero.

$$
\begin{array}{r|rrrr}
1 & 6 & -7 & -8 & 5 \\
& & 6 & -1 & -9 \\
\hline
& 6 & -1 & -9 & -4
\end{array}
$$

$$
\begin{array}{r|rrrr}
5 & 6 & -7 & -8 & 5 \\
& & 30 & 115 & 535 \\
\hline
& 6 & 23 & 107 & 540
\end{array}
$$
All positive $\Rightarrow$ upper bound.

$$
\begin{array}{r|rrrr}
\frac{1}{2} & 6 & -7 & -8 & 5 \\
& & 3 & -2 & -5 \\
\hline
& 6 & -4 & -10 & 0
\end{array}
$$
$\Rightarrow$ $x = \frac{1}{2}$ is a zero.

$P(x) = x(2x - 1)(3x^2 - 2x - 5) = x(2x - 1)(3x - 5)(x + 1)$. Therefore, the zeros are $0, -1, \frac{1}{2}$ and $\frac{5}{3}$.

76. $P(x) = 8x^5 - 14x^4 - 22x^3 + 57x^2 - 35x + 6$. The possible rational zeros are $\pm 1, \pm 2, \pm 3, \pm 6, \pm \frac{1}{2}, \pm \frac{3}{2}, \pm \frac{1}{4}, \pm \frac{3}{4}, \pm \frac{1}{8}, \pm \frac{3}{8}$. Since $P(x)$ has 4 variations in sign, there are 0, 2, or 4 positive real zeros. Since $P(-x) = -8x^5 - 14x^4 + 22x^3 + 57x^2 + 35x + 6$ has 1 variation in sign, there is 1 negative real zero.

$$-1 \begin{array}{|rrrrrr} 8 & -14 & -22 & 57 & -35 & 6 \\ & -8 & 22 & 0 & -57 & 92 \\ \hline 8 & -22 & 0 & 57 & -92 & 98 \end{array}$$

$$-2 \begin{array}{|rrrrrr} 8 & -14 & -22 & 57 & -35 & 6 \\ & -16 & 60 & -76 & 38 & -6 \\ \hline 8 & -30 & 38 & -19 & 3 & 0 \end{array} \quad \Rightarrow \quad x = -2 \text{ is a zero.}$$

$P(x) = (x + 2)(8x^4 - 30x^3 + 38x^2 - 19x + 3)$. All the other real zeros are positive.

$$1 \begin{array}{|rrrrr} 8 & -30 & 38 & -19 & 3 \\ & 8 & -22 & 16 & -3 \\ \hline 8 & -22 & 16 & -3 & 0 \end{array} \quad \Rightarrow \quad x = 1 \text{ is a zero.}$$

$P(x) = (x + 2)(x - 1)(8x^3 - 22x^2 + 16x - 3)$.

$$1 \begin{array}{|rrrr} 8 & -22 & 16 & -3 \\ & 8 & -14 & 2 \\ \hline 8 & -14 & 2 & -1 \end{array} \qquad\qquad \tfrac{1}{2} \begin{array}{|rrrr} 8 & -22 & 16 & -3 \\ & 4 & -9 & \tfrac{7}{2} \\ \hline 8 & -18 & 7 & \tfrac{1}{2} \end{array}$$

Since $f\left(\tfrac{1}{2}\right) > 0 > f(1)$, there must be a zero between $\tfrac{1}{2}$ and 1. We try $\tfrac{3}{4}$:

$$\tfrac{3}{4} \begin{array}{|rrrr} 8 & -22 & 16 & -3 \\ & 6 & -12 & 3 \\ \hline 8 & -16 & 4 & 0 \end{array} \quad \Rightarrow \quad x = \tfrac{3}{4} \text{ is a zero.}$$

$P(x) = (x + 2)(x - 1)(4x - 3)(2x^2 - 4x + 1)$. Now, $2x^2 - 4x + 1 = 0$ when
$x = \frac{4 \pm \sqrt{16 - 4(2)(1)}}{2(2)} = \frac{2 \pm \sqrt{2}}{2}$. Thus, the zeros are 1, $\tfrac{3}{4}$, -2, and $\frac{2 \pm \sqrt{2}}{2}$.

78. $P(x) = 2x^4 - x^3 + x + 2$. The only possible rational zeros of $P(x)$ are ± 1, ± 2, $\pm\tfrac{1}{2}$.

$$\tfrac{1}{2} \begin{array}{|rrrrr} 2 & -1 & 0 & 1 & 2 \\ & 1 & 0 & 0 & \tfrac{1}{2} \\ \hline 2 & 0 & 0 & 1 & \tfrac{5}{2} \end{array} \quad \text{All nonnegative} \quad \Rightarrow \quad x = \tfrac{1}{2} \text{ is an upper bound.}$$

$$-1 \begin{array}{|rrrrr} 2 & -1 & 0 & 1 & 2 \\ & -2 & 3 & -3 & 2 \\ \hline 2 & -3 & 3 & -2 & 4 \end{array} \quad \text{Alternating signs} \quad \Rightarrow \quad x = -1 \text{ is a lower bound.}$$

$$-\tfrac{1}{2} \begin{array}{|rrrrr} 2 & -1 & 0 & 1 & 2 \\ & -1 & 1 & -\tfrac{1}{2} & -\tfrac{1}{4} \\ \hline 2 & -2 & 1 & \tfrac{1}{2} & \tfrac{7}{4} \end{array}$$

Therefore, there are no rational zeros.

80. $P(x) = x^{50} - 5x^{25} + x^2 - 1$. The only possible rational zeros of $P(x)$ are ± 1. Since $P(1) = (1)^{50} - 5(1)^{25} + (1)^2 - 1 = -4$ and $P(-1) = (-1)^{50} - 5(-1)^{25} + (-1)^2 - 1 = 6$, $P(x)$ does not have a rational zero.

82. $x^4 - 5x^2 + 4 = 0$, $[-4, 4]$ by $[-30, 30]$.

The possible rational solutions are ± 1, ± 2, ± 4.

By observing the graph of the equation, the solutions of the given equation are $x = \pm 1$, ± 2.

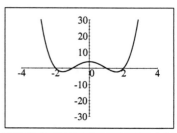

84. $3x^3 + 8x^2 + 5x + 2 = 0$; $[-3, 3]$ by $[-10, 10]$

The possible rational solutions are ± 1, ± 2, $\pm \frac{1}{3}$, $\pm \frac{2}{3}$.

By observing the graph of the equation, the only solution of the given equation is $x = -2$.

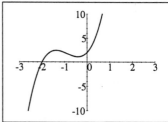

86. $2x^3 - 8x^2 + 9x - 9 = 0$. Possible rational solutions are ± 1, ± 3, ± 9, $\pm \frac{1}{2}$, $\pm \frac{3}{2}$, $\pm \frac{9}{2}$.

$$
\begin{array}{r|rrrr}
1 & 2 & -8 & 9 & -9 \\
& & 2 & -6 & 3 \\
\hline
& 2 & -6 & 3 & -6
\end{array}
\qquad
\begin{array}{r|rrrr}
3 & 2 & -8 & 9 & -9 \\
& & 6 & -6 & 9 \\
\hline
& 2 & -2 & 3 & 0
\end{array}
\Rightarrow \quad x = 3 \text{ is a zero.}
$$

$2x^3 - 8x^2 + 9x - 9 = (x - 3)(2x^2 - 2x + 3)$. Since the quotient is a quadratic expression, we can use the quadratic formula to locate the other possible solutions: $x = \frac{2 \pm \sqrt{2^2 - 4(2)(3)}}{2(2)}$, which are not real solutions. You can also use a graphing device to see that $2x^2 - 2x + 3 = 0$ has no solution. So the only solution is $x = 3$.

88. $x^5 + 2x^4 + 0.96x^3 + 5x^2 + 10x + 4.8 = 0$. Since all the coefficients are positive, there is no positive solution. So $x = 0$ is an upper bound.

$$
\begin{array}{r|rrrrrr}
-2 & 1 & 2 & 0.96 & 5 & 10 & 4.8 \\
& & -2 & 0 & -1.92 & -6.16 & -7.68 \\
\hline
& 1 & 0 & 0.96 & 3.08 & 3.84 & -2.88
\end{array}
$$

$$
\begin{array}{r|rrrrrr}
-3 & 1 & 2 & 0.96 & 5 & 10 & 4.8 \\
& & -3 & 3 & -11.88 & 20.64 & -91.92 \\
\hline
& 1 & -1 & 3.96 & -6.88 & 30.64 & -87.12
\end{array}
\Rightarrow \quad x = -3 \text{ is a lower bound.}
$$

Therefore, we graph the function in the viewing rectangle $[-3, 0]$ by $[-10, 5]$ and see that there are three possible solutions.

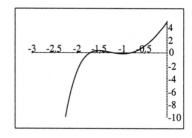

Viewing rectangle: $[-1.75, -1.7]$ by $[-0.1, 0.1]$. Solution $x \approx -1.71$.

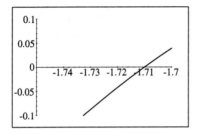

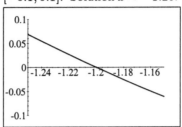

Viewing rectangle: $[-1.25, -1.15]$ by $[-0.1, 0.1]$. Solution $x \approx -1.20$.

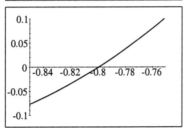

Viewing rectangle: $[-0.85, -0.75]$ by $[-0.1, 0.1]$. Solution $x \approx -0.80$.

So the solutions are $x \approx -1.71, -1.20, -0.80$.

90. $P(x) = x^5 - x^4 - x^3 - 5x^2 - 12x - 6$ has possible rational zeros $\pm 1, \pm 2, \pm 3, \pm 6$. Since $P(x)$ has 1 variation in sign, there is 1 positive real zero. Since $P(-x) = -x^5 - x^4 + x^3 - 5x^2 + 12x - 6$ has 4 variations in sign, there are 0, 2, or 4 negative real zeros.

$$
\begin{array}{r|rrrrrr}
1 & 1 & -1 & -1 & -5 & -12 & -6 \\
  &   & 1 & 0 & -1 & -6 & -18 \\
\hline
  & 1 & 0 & -1 & -6 & -18 & -24
\end{array}
\qquad
\begin{array}{r|rrrrrr}
2 & 1 & -1 & -1 & -5 & -12 & -6 \\
  &   & 2 & 2 & 2 & -6 & -36 \\
\hline
  & 1 & 1 & 1 & -3 & -18 & -42
\end{array}
$$

$$
\begin{array}{r|rrrrrr}
3 & 1 & -1 & -1 & -5 & -12 & -6 \\
  &   & 3 & 6 & 15 & 30 & 54 \\
\hline
  & 1 & 2 & 5 & 10 & 18 & 48
\end{array}
\Rightarrow \quad \text{3 is an upper bound.}
$$

$$
\begin{array}{r|rrrrrr}
-1 & 1 & -1 & -1 & -5 & -12 & -6 \\
   &   & -1 & 2 & -1 & 6   & 6 \\
\hline
   & 1 & -2 & 1 & -6 & -6 & 0
\end{array}
\Rightarrow \quad x = -1 \text{ is a zero.}
$$

$P(x) = (x + 1)(x^4 - 2x^3 + x^2 - 6x - 6)$, continuing with the quotient we have

$$
\begin{array}{r|rrrrr}
-1 & 1 & -2 & 1 & -6 & -6 \\
   &   & -1 & 3 & -4 & 10 \\
\hline
   & 1 & -3 & 4 & -10 & 4
\end{array}
\Rightarrow \quad -1 \text{ is a lower bound.}
$$

Therefore, there is 1 rational zero, namely -1. Since there are 1, 3 or 5 real zeros, and we found 1 rational zero, there must be 0, 2 or 4 irrational zeros. However, since 1 zero must be positive, there cannot be 0 irrational zeros. Therefore, there is exactly 1 rational zero and 2 or 4 irrational zeros.

92.   Given that x is the length of a side of the rectangle, we have that the length of the diagonal is $x + 10$, and the length of the other side of the rectangle is $\sqrt{(x + 10)^2 - x^2}$. Hence
$x\sqrt{(x + 10)^2 - x^2} = 5000 \quad \Rightarrow \quad x^2(20x + 100) = 25{,}000{,}000 \quad \Leftrightarrow$
$2x^3 + 10x^2 - 2{,}500{,}000 = 0 \quad \Leftrightarrow \quad x^3 + 5x^2 - 1{,}250{,}000 = 0$. The first viewing rectangle,

[0, 120] by [−1, 500], shows there is one solution. The second viewing rectangle, [106, 106.1] by [−0.1, 0.1], shows the solution is $x = 106.08$. Therefore, the dimensions of the rectangle are 47 ft by 106 ft.

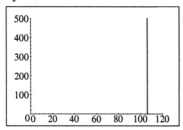   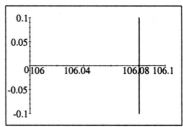

94. The volume of the box is $V = 1500 = x(20 − 2x)(40 − 2x) = 4x^3 − 120x^2 + 800x$ ⟺
$4x^3 − 120x^2 + 800x − 1500 = 4(x^3 − 30x^2 + 200x − 375) = 0$. Clearly, we must have $20 − 2x > 0$, and so $0 < x < 10$.

$$
\begin{array}{r|rrrr}
5 & 1 & -30 & 200 & -375 \\
  &   & 5   & -125 & 375 \\
\hline
  & 1 & -25 & 75 & 0
\end{array}
$$
   ⟹ $x = 5$ is a zero.

$x^3 − 30x^2 + 200x − 375 = (x − 5)(x^2 − 25x + 75) = 0$. Using the quadratic equation, we find the other zeros: $x = \frac{25 \pm \sqrt{625 − 4(1)(75)}}{2} = \frac{25 \pm \sqrt{325}}{2} = \frac{25 \pm 5\sqrt{13}}{2}$. Since $\frac{25 + 5\sqrt{13}}{2} > 10$, the two answers are:

$x = $ height $= 5$ cm, width $= 20 − 2(5) = 10$ cm, and length $= 40 − 2(5) = 30$ cm; and

$x = \frac{25 − 5\sqrt{13}}{2}$, width $= 20 − (25 − 5\sqrt{13}) = 5\sqrt{13} − 5$ cm, and length $= 40 − (25 − 5\sqrt{13})$
$= 15 + 5\sqrt{13}$ cm.

96. (a) Let x be the length, in ft, of each side of the base and let h be the height. The volume of the box is $V = 2\sqrt{2} = hx^2$, and so $hx^2 = 2\sqrt{2}$. The length of the diagonal on the base is $\sqrt{x^2 + x^2} = \sqrt{2x^2}$, and hence the length of the diagonal between opposite corners is $\sqrt{2x^2 + h^2} = x + 1$. Squaring both sides of the equation, we have $2x^2 + h^2 = x^2 + 2x + 1$
   ⟺ $h^2 = −x^2 + 2x + 1$ ⟺ $h = \sqrt{−x^2 + 2x + 1}$. Therefore,
   $2\sqrt{2} = hx^2 = \left(\sqrt{−x^2 + 2x + 1}\right)x^2$ ⟺ $(−x^2 + 2x + 1)x^4 = 8$ ⟺
   $x^6 − 2x^5 − x^4 + 8 = 0$.

   (b) We graph $y = x^6 − 2x^5 − x^4 + 8$ in the viewing rectangle [0, 5] by [−10, 10], and we see that there are two solutions. In the second viewing rectangle, [1.4, 1.5] by [−1, 1], shows the solution $x \approx 1.45$. The third viewing rectangle, [2.25, 2.35] by [−1, 1], shows the solution $x \approx 2.31$.

   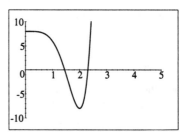

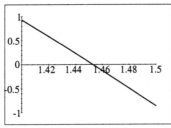 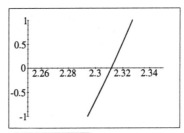

If $x = $ width $= $ length $= 1.45$ ft, then height $= \sqrt{-x^2 + 2x + 1} = 1.34$ ft, and if $x = $ width $= $ length $= 2.31$ ft, then height $= \sqrt{-x^2 + 2x + 1} = 0.53$ ft.

98. (a) An odd degree polynomial must have a real zero. The end behavior of such a polynomial requires that the graph of the polynomial heads off in opposite directions as $x \to \infty$ and $x \to -\infty$. Thus the graph must cross the x-axis.

(b) There are many possibilities one of which is $P(x) = x^4 + 1$.

(c) $P(x) = x(x - \sqrt{2})(x + \sqrt{2}) = x^3 - 2x$.

(d) $P(x) = (x - \sqrt{2})(x + \sqrt{2})(x - \sqrt{3})(x + \sqrt{3}) = x^4 - 5x^2 + 6$.

If a polynomial with integer coefficients has no real zeroes, then the polynomial must have even degree.

100. (a) Using the cubic formula:

$$x = \sqrt[3]{\frac{-2}{2} + \sqrt{\frac{2^2}{4} + \frac{(-3)^3}{27}}} + \sqrt[3]{\frac{-2}{2} - \sqrt{\frac{2^2}{4} + \frac{(-3)^3}{27}}} = \sqrt[3]{-1} + \sqrt[3]{-1} = -1 - 1 = -2.$$

$$
\begin{array}{r|rrrr}
-2 & 1 & 0 & -3 & 2 \\
   &   & -2 & 4 & -2 \\
\hline
   & 1 & -2 & 1 & 0
\end{array}
$$

So $(x + 2)(x^2 - 2x + 1) = (x + 2)(x - 1)^2 = 0 \quad \Rightarrow \quad x = -2, 1$.

Using the methods from this section, we have

$$
\begin{array}{r|rrrr}
1 & 1 & 0 & -3 & 2 \\
  &   & 1 & 1 & -2 \\
\hline
  & 1 & 1 & -2 & 0
\end{array}
$$

So $x^3 - 3x + 2 = (x - 1)(x^2 + x - 2) = (x - 1)^2(x + 2) = 0 \quad \Leftrightarrow \quad x = -2, 1$.

Since this factors easily, the factoring method was easier.

(b) Using the cubic formula:

$$x = \sqrt[3]{\frac{-(-54)}{2} + \sqrt{\frac{(-54)^2}{4} + \frac{(-27)^3}{27}}} + \sqrt[3]{\frac{-(-54)}{2} - \sqrt{\frac{(-54)^2}{4} + \frac{(-27)^3}{27}}}$$

$$= \sqrt[3]{\frac{54}{2} + \sqrt{27^2 - 27^2}} + \sqrt[3]{\frac{54}{2} - \sqrt{27^2 - 27^2}} = \sqrt[3]{27} + \sqrt[3]{27} = 3 + 3 = 6.$$

$$
\begin{array}{r|rrrr}
6 & 1 & 0 & -27 & -54 \\
  &   & 6 & 36 & 54 \\
\hline
  & 1 & 6 & 9 & 0
\end{array}
$$

$x^3 - 27x - 54 = (x - 6)(x^2 + 6x + 9) = (x - 6)(x + 3)^2 = 0 \quad \Rightarrow \quad x = -3, 6$.

Using methods from this section,

$$-1 \begin{array}{|rrrr} 1 & 0 & -27 & -54 \\ & -1 & 1 & 26 \\ \hline 1 & -1 & -26 & -28 \end{array}$$

$$-2 \begin{array}{|rrrr} 1 & 0 & -27 & -54 \\ & -2 & 4 & 46 \\ \hline 1 & -2 & -23 & -8 \end{array}$$

$$-3 \begin{array}{|rrrr} 1 & 0 & -27 & -54 \\ & -3 & 9 & 54 \\ \hline 1 & -3 & -18 & 0 \end{array}$$

So $x^3 - 27x - 54 = = (x+3)(x^2 - 3x - 18) = (x-6)(x+3)^2 = 0 \quad \Leftrightarrow \quad x = -3, 6.$

Since this factors easily, the factoring method was easier.

(c)   Using the cubic formula:

$$x = \sqrt[3]{\frac{-4}{2} + \sqrt{\frac{4^2}{4} + \frac{3^3}{27}}} + \sqrt[3]{\frac{-4}{2} - \sqrt{\frac{4^2}{4} + \frac{3^3}{27}}}$$

$$= \sqrt[3]{-2 + \sqrt{4+1}} + \sqrt[3]{-2 - \sqrt{4+1}} = \sqrt[3]{-2 + \sqrt{5}} + \sqrt[3]{-2 - \sqrt{5}}.$$

From the graphing calculator, we see that $P(x) = x^3 + 3x + 4$ has one zero.

Using methods from this section, $P(x)$ has possible rational zeros $\pm 1, \pm 2, \pm 4.$

$$1 \begin{array}{|rrrr} 1 & 0 & 3 & 4 \\ & 1 & 1 & 4 \\ \hline 1 & 1 & 4 & 4 \end{array} \quad \Rightarrow \quad \text{1 is an upper bound.}$$

$$-1 \begin{array}{|rrrr} 1 & 0 & 3 & 4 \\ & -1 & 1 & -4 \\ \hline 1 & -1 & 4 & 0 \end{array} \quad \Rightarrow x = -1 \text{ is a zero.}$$

$P(x) = x^3 + 3x + 4 = (x+1)(x^2 - x + 4).$ Using the quadratic formula we have:

$$x = \frac{-(-1) \pm \sqrt{(-1)^2 - 4(1)(4)}}{2} = \frac{1 \pm \sqrt{-15}}{2} \text{ which is not a real number.}$$

Since it is not easy to see that $\sqrt[3]{-2 + \sqrt{5}} + \sqrt[3]{-2 - \sqrt{5}} = -1$, we see that the factoring method was *much* easier.

Exercises 4.4

2. (a) $x^5 + 9x^3 = 0 \quad \Leftrightarrow \quad x^3(x^2 + 9) = 0$. So $x = 0$ or $x^2 + 9 = 0$. If $x^2 + 9 = 0$ then $x = \pm 3i$. Therefore, the zeros of P are $x = 0, \pm 3i$.

   (b) Since $-3i$ and $3i$ are the zeros from $x^2 + 9 = 0$, $x + 3i$ and $x - 3i$ are the factors of $x^2 + 9$. Thus the complete factorization is $P(x) = x^3(x^2 + 9) = x^3(x + 3i)(x - 3i)$.

4. (a) $x^3 + x^2 + x = 0 \quad x(x^2 + x + 1) = 0$. So $x = 0$ or $x^2 + x + 1 = 0$. If $x^2 + x + 1 = 0$ then
   $$x = \frac{-(1) \pm \sqrt{(1)^2 - 4(1)(1)}}{2(1)} = \frac{-1 \pm \sqrt{-3}}{2} = -\frac{1}{2} \pm i \frac{\sqrt{3}}{2}.$$ Therefore, the zeros of P are
   $x = 0, -\frac{1}{2} \pm i \frac{\sqrt{3}}{2}$.

   (b) The zeros of $x^2 + x + 1 = 0$ are $-\frac{1}{2} - \frac{\sqrt{3}}{2} i$ and $-\frac{1}{2} + \frac{\sqrt{3}}{2} i$, so factoring we get
   $$x^2 + x + 1 = \left[x - \left(-\frac{1}{2} - \frac{\sqrt{3}}{2} i\right)\right]\left[x - \left(-\frac{1}{2} + \frac{\sqrt{3}}{2} i\right)\right] = \left(x + \frac{1}{2} + \frac{\sqrt{3}}{2} i\right)\left(x + \frac{1}{2} - \frac{\sqrt{3}}{2} i\right).$$
   Thus the complete factorization is
   $$P(x) = x(x^2 + x + 1) = x\left(x + \frac{1}{2} + \frac{\sqrt{3}}{2} i\right)\left(x + \frac{1}{2} - \frac{\sqrt{3}}{2} i\right).$$

6. (a) $x^4 - x^2 - 2 = 0 \quad \Leftrightarrow \quad (x^2 - 2)(x^2 + 1) = 0$. So $x^2 - 2 = 0$ or $x^2 + 1 = 0$. If $x^2 - 2 = 0$ then $x^2 = 2 \quad \Leftrightarrow \quad x = \pm\sqrt{2}$. And if $x^2 + 1 = 0$ then $x^2 = -1 \quad \Leftrightarrow \quad x = \pm i$. Therefore, the zeros of P are $x = \pm\sqrt{2}, \pm i$.

   (b) To get the complete factorization, we factor the quadratic factors to get
   $$P(x) = (x^2 - 2)(x^2 + 1) = (x - \sqrt{2})(x + \sqrt{2})(x - i)(x + i).$$

8. (a) $x^4 + 6x^2 + 9 = 0 \quad \Leftrightarrow \quad (x^2 + 3)^2 = 0 \quad \Leftrightarrow \quad x^2 = -3$. So $x = \pm i\sqrt{3}$ are the only zeros of P (each of multiplicity 2).

   (b) To get the complete factorization, we factor the quadratic factor to get
   $$P(x) = (x^2 + 3)^2 = \left[(x - i\sqrt{3})(x + i\sqrt{3})\right]^2 = (x - i\sqrt{3})^2(x + i\sqrt{3})^2.$$

10. (a) $x^3 - 8 = 0 \quad \Leftrightarrow \quad (x - 2)(x^2 + 2x + 4) = 0$. So $x = 2$ or $x^2 + 2x + 4 = 0$. If
    $$x^2 + 2x + 4 = 0 \text{ then } x = \frac{-(2) \pm \sqrt{(2)^2 - 4(1)(4)}}{2} = \frac{-2 \pm \sqrt{-12}}{2} = \frac{-2 \pm 2i\sqrt{3}}{2}$$
    $= -1 \pm i\sqrt{3}$. Therefore, the zeros of P are $x = 2, -1 \pm i\sqrt{3}$.

    (b) Since $-1 - i\sqrt{3}$ and $-1 + i\sqrt{3}$ are the zeros from $x^2 + 2x + 4 = 0$, $x - (-1 - i\sqrt{3})$ and $x - (-1 + i\sqrt{3})$ are the factors of $x^2 - 2x + 4$. Thus the complete factorization is
    $$P(x) = (x - 2)(x^2 + 2x + 4) = (x - 2)\left[x - (-1 - i\sqrt{3})\right]\left[x - (-1 + i\sqrt{3})\right]$$
    $$= (x - 2)(x + 1 + i\sqrt{3})(x + 1 - i\sqrt{3}).$$

12. (a) $x^6 - 7x^3 - 8 = 0 \quad \Leftrightarrow \quad 0 = (x^3 - 8)(x^3 + 1) = (x - 2)(x^2 + 2x + 4)(x + 1)(x^2 - x + 1)$.
    Clearly, $x = -1$ and $x = 2$ are solutions. If $x^2 + 2x + 4 = 0$, then $x = \frac{-2 \pm \sqrt{4 - 4(1)(4)}}{2}$
    $= \frac{-2 \pm \sqrt{-12}}{2} = -\frac{2}{2} \pm \frac{2\sqrt{-3}}{2}$ so $x = -1 \pm \sqrt{3}i$. If $x^2 - x + 1 = 0$, then

$x = \frac{1 \pm \sqrt{1-4(1)(1)}}{2} = \frac{1 \pm \sqrt{-3}}{2} = \frac{1}{2} \pm \frac{\sqrt{-3}}{2} = \frac{1}{2} \pm \frac{\sqrt{3}}{2}\,i$. Therefore, the zeros of P are

$x = -1, 2, -1 \pm \sqrt{3}\,i, \frac{1}{2} \pm \frac{\sqrt{3}}{2}\,i$.

(b) From Exercise 10, $x^2 + 2x + 4 = \left(x + 1 + i\sqrt{3}\right)\left(x + 1 - i\sqrt{3}\right)$ and from Exercise 11,

$x^2 - x + 1 = \left(x - \frac{1}{2} + \frac{\sqrt{3}}{2}\,i\right)\left(x - \frac{1}{2} - \frac{\sqrt{3}}{2}\,i\right)$. Thus the complete factorization is

$P(x) = (x - 2)(x^2 + 2x + 4)(x + 1)(x^2 - x + 1)$

$= (x - 2)(x + 1)\left(x + 1 + i\sqrt{3}\right)\left(x + 1 - i\sqrt{3}\right)\left(x - \frac{1}{2} + \frac{\sqrt{3}}{2}\,i\right)\left(x - \frac{1}{2} - \frac{\sqrt{3}}{2}\,i\right)$.

14. $P(x) = 4x^2 + 9 = (2x - 3i)(2x + 3i)$. The zeros of P are $\frac{3}{2}i$ and $-\frac{3}{2}i$, both multiplicity 1.

16. $Q(x) = x^2 - 8x + 17 = (x^2 - 8x + 16) + 1 = (x - 4)^2 + 1 = [(x - 4) - i][(x - 4) + i]$
$= (x - 4 - i)(x - 4 + i)$. The zeros of Q are $4 + i$ and $4 - i$, both multiplicity 1.

18. $P(x) = x^3 + x^2 + x = x(x^2 + x + 1)$. Using the quadratic formula, we have
$x = \frac{-(1) \pm \sqrt{(1)^2 - 4(1)(1)}}{2(1)} = \frac{1 \pm \sqrt{-3}}{2} = \frac{1}{2} \pm i\frac{\sqrt{3}}{2}$. The zeros of P are 0, $\frac{1}{2} + i\frac{\sqrt{3}}{2}$, and
$\frac{1}{2} - i\frac{\sqrt{3}}{2}$, all multiplicity 1. And $P(x) = x\left(x - \frac{1}{2} - i\frac{\sqrt{3}}{2}\right)\left(x - \frac{1}{2} + i\frac{\sqrt{3}}{2}\right)$.

20. $Q(x) = x^4 - 625 = (x^2 - 25)(x^2 + 25) = (x - 5)(x + 5)(x^2 + 25)$
$= (x - 5)(x + 5)(x - 5i)(x + 5i)$. The zeros of Q are $5, -5, 5i$, and $5i$, all multiplicity 1.

22. $P(x) = x^3 - 64 = (x - 4)(x^2 + 4x + 16)$. Using the quadratic formula, we have
$x = \frac{-4 \pm \sqrt{16 - 4(1)(16)}}{2} = \frac{-4 \pm \sqrt{-48}}{2} = \frac{-4 \pm 4\sqrt{3}\,i}{2} = -2 \pm 2\sqrt{3}\,i$. The zeros of P are
$4, -2 + 2\sqrt{3}\,i$, and $-2 - 2\sqrt{3}\,i$, all multiplicity 1.
And $P(x) = (x - 4)(x + 2 - 2\sqrt{3}\,i)(x + 2 + 2\sqrt{3}\,i)$.

24. $P(x) = x^6 - 729 = (x^3 - 27)(x^3 + 27) = (x - 3)(x^2 + 3x + 9)(x + 3)(x^2 - 3x + 9)$. Using the
quadratic formula on $x^2 + 3x + 9$ we have
$x = \frac{-3 \pm \sqrt{9 - 4(1)(9)}}{2} = \frac{-3 \pm \sqrt{-27}}{2} = -\frac{3}{2} \pm \frac{\sqrt{-27}}{2}$. Using the quadratic formula on
$x^2 - 3x + 9$ we have $x = \frac{3 \pm \sqrt{9 - 4(1)(9)}}{2} = \frac{3 \pm \sqrt{-27}}{2} = \frac{3}{2} \pm \frac{\sqrt{-27}}{2} = \frac{3}{2} \pm \frac{3\sqrt{3}}{2}\,i$. The
zeros of P are $3, -3, -\frac{3}{2} + \frac{3\sqrt{3}}{2}\,i, -\frac{3}{2} - \frac{3\sqrt{3}}{2}\,i, \frac{3}{2} + \frac{3\sqrt{3}}{2}\,i$, and $\frac{3}{2} - \frac{3\sqrt{3}}{2}\,i$, all multiplicity 1. And
$P(x) = (x - 3)(x + 3)\left(x + \frac{3}{2} - \frac{3\sqrt{3}}{2}\,i\right)\left(x + \frac{3}{2} + \frac{3\sqrt{3}}{2}\,i\right)\left(x - \frac{3}{2} - \frac{3\sqrt{3}}{2}\,i\right)\left(x - \frac{3}{2} + \frac{3\sqrt{3}}{2}\,i\right)$.

26. $Q(x) = x^4 + 10x^2 + 25 = (x^2 + 5)^2 = \left[(x - i\sqrt{5})(x + i\sqrt{5})\right]^2 = (x - i\sqrt{5})^2(x + i\sqrt{5})^2$. The
zeros of Q are $i\sqrt{5}$ and $-i\sqrt{5}$, both multiplicity 2.

28. $P(x) = x^5 + 7x^3 = x^3(x^2 + 7) = x^3(x - i\sqrt{7})(x + i\sqrt{7})$. The zeros of P are 0 (multiplicity 3),
$i\sqrt{7}$ and $-i\sqrt{7}$, both multiplicity 1.

30. $P(x) = x^6 + 16x^3 + 64 = (x^3 + 8)^2 = (x + 2)^2(x^2 - 2x + 4)^2$. Using the quadratic formula, on
$x^2 - 2x + 4$ we have $x = \dfrac{-(-2) \pm \sqrt{(-2)^2 - 4(1)(4)}}{2} = \dfrac{2 \pm \sqrt{-12}}{2} = \dfrac{2 \pm 2i\sqrt{3}}{2} = 1 \pm i\sqrt{3}$.
The zeros of P are -2, $1 + i\sqrt{3}$, and $1 - i\sqrt{3}$, all multiplicity 2.

32. Since $1 + i\sqrt{2}$ and $1 - i\sqrt{2}$ are conjugates, the factorization of the polynomial must be
$P(x) = c(x - [1 + i\sqrt{2}])(x - [1 - i\sqrt{2}]) = c(x^2 - 2x + 3)$. If we let $c = 1$, we get
$P(x) = x^2 - 2x + 3$.

34. Since i is a zero, by the Conjugate Roots Theorem, $-i$ is also a zero. So the factorization of the
polynomial must be $Q(x) = b(x + 0)(x - i)(x + i) = bx(x^2 + 1) = b(x^3 + x)$. If we let $b = 1$,
we get $Q(x) = x^3 + x$.

36. Since $1 + i$ is a zero, by the Conjugate Roots Theorem, $1 - i$ is also a zero. So the factorization of
the polynomial must be $Q(x) = a(x + 3)(x - [1 + i])(x - [1 - i]) = a(x + 3)(x^2 - 2x + 2)$
$= a(x^3 + x^2 - 4x + 6)$. If we let $a = 1$, we get $Q(x) = x^3 + x^2 - 4x + 6$.

38. Since $S(x)$ has zeros $2i$ and $3i$, by the Conjugate Roots Theorem, the other zeros of $S(x)$ are $-2i$
and $-3i$. So a factorization of $S(x)$ is
$S(x) = C(x - 2i)(x + 2i)(x - 3i)(x + 3i) = C(x^2 - 4i^2)(x^2 - 9i^2) = C(x^2 + 4)(x^2 + 9)$
$= C(x^4 + 13x^2 + 36)$. If we let $C = 1$, we get $S(x) = x^4 + 13x^2 + 36$.

40. Since $U(x)$ has zeros $\frac{1}{2}$, -1 (with multiplicity two), and $-i$, by the Conjugate Roots Theorem, the
other zero is i. So a factorization of $U(x)$ is
$U(x) = c(x - \frac{1}{2})(x + 1)^2(x + i)(x - i) = \frac{1}{2}c(2x - 1)(x^2 + 2x + 1)(x^2 + 1)$
$= \frac{1}{2}c(2x^5 + 3x^4 + 2x^3 + 2x^2 - 1)$. Since the leading coefficient is 4, we have $4 = \frac{1}{2}c(2) = c$.
Thus we have $U(x) = \frac{1}{2}(4)(2x^5 + 3x^4 + 2x^3 + 2x^2 - 1) = 4x^5 + 6x^4 + 4x^3 + 4x^2 - 2$.

42. $P(x) = x^3 - 7x^2 + 17x - 15$. We start by trying the possible rational factors of the polynomial:

$$
\begin{array}{r|rrrr}
1 & 1 & -7 & 17 & -15 \\
  &   & 1 & -6 & 11 \\
\hline
  & 1 & -6 & 11 & -4
\end{array}
\qquad
\begin{array}{r|rrrr}
3 & 1 & -7 & 17 & -15 \\
  &   & 3 & -12 & 15 \\
\hline
  & 1 & -4 & 5 & 0
\end{array} \Rightarrow \quad x = 3 \text{ is a zero.}
$$

So $P(x) = (x - 3)(x^2 - 4x + 5)$. Using the quadratic formula on the second factor, we have
$x = \dfrac{4 \pm \sqrt{16 - 4(1)(5)}}{2} = \dfrac{4 \pm \sqrt{-4}}{2} = \dfrac{4 \pm 2i}{2} = 2 \pm i$. Thus the zeros are $3, 2 \pm i$.

44. $P(x) = x^3 + 7x^2 + 18x + 18$ has possible rational zeros $\pm 1, \pm 2, \pm 3, \pm 6, \pm 9, \pm 18$. Since all of
the coefficients are positive, there are no positive real zeros.

$$
\begin{array}{r|rrrr}
-1 & 1 & 7 & 18 & 18 \\
   &   & -1 & -6 & -12 \\
\hline
   & 1 & 6 & 12 & 6
\end{array}
\qquad
\begin{array}{r|rrrr}
-2 & 1 & 7 & 18 & 18 \\
   &   & -2 & -10 & -16 \\
\hline
   & 1 & 5 & 8 & 2
\end{array}
$$

$$
\begin{array}{r|rrrr}
-3 & 1 & 7 & 18 & 18 \\
   &   & -3 & -12 & -18 \\
\hline
   & 1 & 4 & 6 & 0
\end{array} \Rightarrow \quad x = -3 \text{ is a zero.}
$$

So $P(x) = (x-3)(x^2 + 4x + 6)$. Using the quadratic formula on the second factor, we have
$$x = \frac{-4 \pm \sqrt{16 - 4(1)(6)}}{2} = \frac{-4 \pm \sqrt{-8}}{2} = \frac{-4 \pm 2\sqrt{2}\,i}{2} = -2 \pm \sqrt{2}\,i. \text{ Thus the zeros are } -3,$$
$-2 \pm \sqrt{2}\,i.$

46. $P(x) = x^3 - x - 6$ has possible zeros ± 1, ± 2, ± 3.

$$
\begin{array}{r|rrrr}
1 & 1 & 0 & -1 & -6 \\
  &   & 1 & 1 & 0 \\
\hline
  & 1 & 1 & 0 & -6
\end{array}
\qquad\qquad
\begin{array}{r|rrrr}
2 & 1 & 0 & -1 & -6 \\
  &   & 2 & 4 & 6 \\
\hline
  & 1 & 2 & 3 & 0
\end{array}
\quad \Rightarrow \quad x = 2 \text{ is a zero.}
$$

$P(x) = (x-2)(x^2 + 2x + 3)$. Now $x^2 + 2x + 3$ has zeros $x = \dfrac{-2 \pm \sqrt{4 - 4(1)(3)}}{2}$
$= \frac{-2 \pm 2i\sqrt{2}}{2} = -1 \pm i\sqrt{2}$. Thus the zeros are $2, -1 \pm i\sqrt{2}$.

48. Using synthetic division, we see that $(x-3)$ is a factor of the polynomial:

$$
\begin{array}{r|rrrr}
1 & 2 & -8 & 9 & -9 \\
  &   & 2 & -6 & 3 \\
\hline
  & 2 & -6 & 3 & -6
\end{array}
\qquad\qquad
\begin{array}{r|rrrr}
3 & 2 & -8 & 9 & -9 \\
  &   & 6 & -6 & 9 \\
\hline
  & 2 & -2 & 3 & 0
\end{array}
\quad \Rightarrow \quad x = 3 \text{ is a zero.}
$$

So $P(x) = 2x^3 - 8x^2 + 9x - 9 = (x-3)(2x^2 - 2x + 3)$. Using the quadratic formula, we find
the other two solutions: $x = \dfrac{2 \pm \sqrt{4 - 4(3)(2)}}{2(2)} = \dfrac{2 \pm \sqrt{-20}}{4} = \dfrac{1}{2} \pm \dfrac{\sqrt{5}}{2}\,i$. Thus the zeros are
$3, \frac{1}{2} \pm \frac{\sqrt{5}}{2}\,i$.

50. $P(x) = x^4 - 2x^3 - 2x^2 - 2x - 3$ has possible zeros ± 1, ± 3.

$$
\begin{array}{r|rrrrr}
1 & 1 & -2 & -2 & -2 & -3 \\
  &   & 1 & -1 & -3 & -5 \\
\hline
  & 1 & -1 & -3 & -5 & -8
\end{array}
\qquad\qquad
\begin{array}{r|rrrrr}
3 & 1 & -2 & -2 & -2 & -3 \\
  &   & 3 & 3 & 3 & 3 \\
\hline
  & 1 & 1 & 1 & 1 & 0
\end{array}
\quad \Rightarrow \quad x = 3 \text{ is a zero.}
$$

$P(x) = (x-3)(x^3 + x^2 + x + 1)$. If we factor the second factor by grouping, we get
$x^3 + x^2 + x + 1 = x^2(x+1) + 1(x+1) = (x+1)(x^2+1)$. So we have
$P(x) = (x-3)(x+1)(x^2+1) = (x-3)(x+1)(x-i)(x+i)$. Thus the zeros are $3, -1, i$, and
$-i$.

52. $P(x) = x^5 + x^3 + 8x^2 + 8 = x^3(x^2+1) + 8(x^2+1) = (x^2+1)(x^3+8)$
$= (x^2+1)(x+2)(x^2 - 2x + 4)$ (factoring a sum of cubes). So $x = -2$, or $x^2 + 1 = 0$. If
$x^2 + 1 = 0$, then $x^2 = -1 \quad \Rightarrow \quad x = \pm i$. If $x^2 - 2x + 4 = 0$, then $x = \dfrac{2 \pm \sqrt{4 - 4(1)(4)}}{2}$
$= 1 \pm \dfrac{\sqrt{-12}}{2} = 1 \pm \sqrt{3}\,i$. Thus, the zeros are $-2, \pm i, 1 \pm \sqrt{3}\,i$.

54. $P(x) = x^4 - x^2 + 2x + 2$ has possible rational zeros ± 1, ± 2.

$$
\begin{array}{r|rrrrr}
1 & 1 & 0 & -1 & 2 & 2 \\
  &   & 1 & 1 & 0 & 2 \\
\hline
  & 1 & 1 & 0 & 2 & 4
\end{array}
\quad 1 \text{ is an upper bound.}
\qquad
\begin{array}{r|rrrrr}
-1 & 1 & 0 & -1 & 2 & 2 \\
   &   & -1 & 1 & 0 & -2 \\
\hline
   & 1 & -1 & 0 & 2 & 0
\end{array}
$$

$$
\begin{array}{r|rrrr}
-1 & 1 & -1 & 0 & 2 \\
   &   & -1 & 2 & -2 \\
\hline
   & 1 & -2 & 2 & 0
\end{array}
$$

$P(x) = (x+1)^2(x^2 - 2x + 2)$. Using the quadratic formula on $x^2 - 2x + 2$, we have
$x = \frac{2 \pm \sqrt{4-8}}{2} = \frac{2 \pm 2i}{2} = 1 \pm i$. Thus, the zeros of $P(x)$ are $-1, 1 \pm i$.

56. $P(x) = x^5 - 2x^4 + 2x^3 - 4x^2 + x - 2$ has possible rational zeros $\pm 1, \pm 2$.

1	1	−2	2	−4	1	−2
		1	−1	1	−3	−2
	1	−1	1	−3	−2	−4

2	1	−2	2	−4	1	−2
		2	0	4	0	2
	1	0	2	0	1	0	⟹ $x = 2$ is a zero.

$P(x) = (x - 2)(x^4 + 2x^2 + 1) = (x - 2)(x^2 + 1)^2 = (x - 2)(x - i)^2(x + i)^2$. Thus, the zeros of $P(x)$ are $2, \pm i$.

58. (a) $P(x) = x^3 - 2x - 4$

1	1	0	−2	−4
		1	1	−1
	1	1	−1	−5

2	1	0	−2	−4
		2	4	4
	1	2	2	0

$P(x) = x^3 - 2x - 4 = (x - 2)(x^2 + 2x + 2)$.

(b) $P(x) = (x - 2)(x + 1 - i)(x + 1 + i)$.

60. (a) $P(x) = x^4 + 8x^2 + 16 = (x^2 + 4)^2$.

(b) $P(x) = (x - 2i)^2(x + 2i)^2$.

62. (a) $P(x) = x^5 - 16x = x(x^4 - 16) = x(x^2 - 4)(x^2 + 4) = x(x - 2)(x + 2)(x^2 + 4)$.

(b) $P(x) = x(x - 2)(x + 2)(x - 2i)(x + 2i)$.

64. (a) $2x + 4i = 1 \quad \Leftrightarrow \quad 2x = 1 - 4i \quad \Leftrightarrow \quad x = \frac{1}{2} - 2i$.

(b) $x^2 - ix = 0 \quad \Leftrightarrow \quad x(x - i) = 0 \quad \Leftrightarrow \quad x = 0, i$.

(c) $x^2 + 2ix - 1 = 0 \quad \Leftrightarrow \quad (x + i)^2 = 0 \quad \Leftrightarrow \quad x = -i$.

(d) $ix^2 - 2x + i = 0$. Using the quadratic equation we get $x = \dfrac{2 \pm \sqrt{(-2)^2 - 4(i)(i)}}{2i}$

$= \dfrac{2 \pm \sqrt{8}}{2i} = \dfrac{2 \pm 2\sqrt{2}}{2i} = \dfrac{1 \pm \sqrt{2}}{i} = (1 \pm \sqrt{2})(-i) = (-1 \pm \sqrt{2})i$.

66. (a) Since i and $1 + i$ are zeros, $-i$ and $1 - i$ are also zeros. So
$P(x) = C(x - i)(x + i)(x - [1 + i])(x - [1 - i]) = C(x^2 + 1)(x^2 - 2x + 2)$
$= C(x^4 - 2x^3 + 2x^2 + x^2 - 2x + 2) = C(x^4 - 2x^3 + 3x^2 - 2x + 2)$. Since $C = 1$, the
polynomial is $P(x) = x^4 - 2x^3 + 3x^2 - 2x + 2$.

(b) Since i and $1 + i$ are zeros, $P(x) = C(x - i)(x - [i + 1]) = C(x^2 - xi - x - xi - 1 + i)$
$= C[x^2 - (1 + 2i)x - 1 + i]$. Since $C = 1$, the polynomial is
$P(x) = x^2 - (1 + 2i)x - 1 + i$.

68. $x^4 - 1 = 0 \quad \Leftrightarrow \quad (x^2 - 1)(x^2 + 1) = 0 \quad \Leftrightarrow \quad (x - 1)(x + 1)(x + i)(x - i) = 0 \quad \Leftrightarrow$
$x = \pm 1, \pm i$. So there are four fourth roots of 1, two that are real and two that are complex.
Consider $P(x) = x^n - 1$, where n is even. P has one change in sign so P has exactly one real
positive zero, namely $x = 1$. Since $P(-x) = P(x)$, P also has exactly one real negative zero,

namely $x = -1$. Thus P must have $n - 2$ complex roots. As a result, $x^n = 1$ has two real nth zeros and $n - 2$ complex roots.

$x^3 - 1 = 0 \quad \Leftrightarrow \quad (x - 1)(x^2 + x + 1) = 0 \quad \Leftrightarrow \quad x = 1, \frac{-1 \pm \sqrt{3} i}{2}$. So there is one real cube zero of unity and two complex roots. Now consider $Q(x) = x^k - 1$, where k is odd. Since Q has one change in sign, Q has exactly one real positive zero, namely $x = 1$. But $Q(-x) = -x^k - 1$ has no changes in sign, so there are no negative real zeros. As a result, $x^k = 1$ has one real kth zero and $k - 1$ complex roots.

Exercises 4.5

2. $s(x) = \dfrac{3x}{x-5}$. When $x = 0$, we have $s(0) = 0$, so the y-intercept is 0. The numerator is zero when $3x = 0$ or $x = 0$, so the x-intercept is 0.

4. $r(x) = \dfrac{2}{x^2 + 3x - 4}$. When $x = 0$, we have $r(0) = \frac{2}{-4} = -\frac{1}{2}$, so the y-intercept is $-\frac{1}{2}$. The numerator is never zero, so there is no x-intercept.

6. $r(x) = \dfrac{x^3 + 8}{x^2 + 4}$. When $x = 0$, we have $r(0) = \frac{8}{4} = 2$, so the y-intercept is 2. The x-intercept occurs when $x^3 + 8 = 0 \iff (x+2)(x^2 - 2x + 4) = 0 \iff x = -2$ or $x = 1 \pm i\sqrt{3}$, which has only one real solution, so the x-intercept is -2.

8. From the graph, the x-intercept is 0, the y-intercept is 0, the horizontal asymptote is $y = 0$, and the vertical asymptotes are $x = -1$ and $x = 2$.

10. From the graph, the x-intercepts are ± 2, the y-intercept is -6, the horizontal asymptote is $y = 2$, and there are no vertical asymptotes

12. $s(x) = \dfrac{2x + 3}{x - 1} = \dfrac{2 + \dfrac{3}{x}}{1 - \dfrac{1}{x}} \to 2$ as $x \to \pm\infty$. The horizontal asymptote is $y = 2$. There is a vertical asymptote when $x - 1 = 0 \iff x = 1$, so the vertical asymptote is $x = 1$.

14. $r(x) = \dfrac{2x - 4}{x^2 + 2x + 1} = \dfrac{\dfrac{2}{x} - \dfrac{4}{x^2}}{1 + \dfrac{2}{x} + \dfrac{1}{x^2}} \to 0$ as $x \to \pm\infty$. Thus, the horizontal asymptote is $y = 0$.

Also, $y = \dfrac{2(x - 2)}{(x + 1)^2}$ so there is a vertical asymptote when $x + 1 = 0 \iff x = -1$, so the vertical asymptote is $x = -1$.

16. $t(x) = \dfrac{(x - 1)(x - 2)}{(x - 3)(x - 4)} = \dfrac{x^2 - 3x + 2}{x^2 - 7x + 12} = \dfrac{1 - \dfrac{3}{x} + \dfrac{2}{x^2}}{1 - \dfrac{7}{x} + \dfrac{12}{x^2}} \to 1$ as $x \to \pm\infty$, so the horizontal asymptote is $y = 1$. Also, vertical asymptotes occur when $(x - 3)(x - 4) = 0 \Rightarrow x = 3, 4$, so the two vertical asymptotes are $x = 3$ and $x = 4$.

18. $s(x) = \dfrac{3x^2}{x^2 + 2x + 5} = \dfrac{3}{1 + \dfrac{2}{x} + \dfrac{5}{x^2}} \to 3$ as $x \to \pm\infty$, so the horizontal asymptote is $y = 3$.

Also, vertical asymptotes occur when $x^2 + 2x + 5 = 0 \Rightarrow x = \dfrac{-2 \pm \sqrt{4 - 20}}{2} = -1 \pm 2i$. Since there are no real zeros, there are no vertical asymptotes.

20. $r(x) = \dfrac{x^3 + 3x^2}{x^2 - 4} = \dfrac{x^2(x+3)}{(x-2)(x+2)}$. Because the degree of the numerator is greater than the degree

of the denominator, the function has no horizontal asymptotes. Two vertical asymptotes occur at

$x = 2$ and $x = -2$. By using long division, we see that $r(x) = x + 3 + \dfrac{4x + 12}{x^2 - 4}$ so $y = x + 3$ is a

slant asymptote. ·

In exercises 22-28, let $f(x) = \dfrac{1}{x}$.

22. $r(x) = \dfrac{1}{x+4} = f(x+4)$.

From this form we see that the graph of r is obtained from the graph
of f by shifting 4 units to the left. Thus r has vertical asymptote
$x = -4$ and horizontal asymptote $y = 0$.

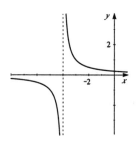

24. $s(x) = \dfrac{-2}{x-2} = -2\left(\dfrac{1}{x-2}\right) = -2f(x-2)$.

From this form we see that the graph of s is obtained from the graph
of f by shifting 2 units to the right, stretching vertically by a factor
of 2, and then reflecting about the x-axis. Thus r has vertical
asymptote $x = 2$ and horizontal asymptote $y = 0$.

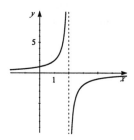

26. $t(x) = \dfrac{3x - 3}{x + 2} = 2 - \dfrac{9}{x+2} = 3 - 9\left(\dfrac{1}{x+2}\right) = -9f(x+2) + 3$.

From this form we see that the graph of t is
obtained from the graph of f by shifting 2 units
to the left, stretching vertically by a factor or 9,
reflecting about the x-axis, and then shifting 3
units vertically. Thus t has vertical asymptote
$x = -2$ and horizontal asymptote $y = 3$.

$$x + 2 \overline{\smash{)}\,3x - 3}$$
$$\underline{3x + 6}$$
$$-9$$

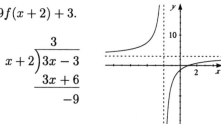

28. $t(x) = \dfrac{2x - 9}{x - 4} = 2 - \dfrac{1}{x-4} = 2 - \left(\dfrac{1}{x-4}\right) = -f(x-4) + 2$.

From this form we see that the graph of t is
obtained from the graph of f by shifting 4 units
to the right, reflecting about the x-axis, and then
shifting 2 units vertically. Thus t has vertical
asymptote $x = 4$ and horizontal asymptote $y = 2$.

$$x - 4 \overline{\smash{)}\,2x - 9}$$
$$\underline{2x - 8}$$
$$-1$$

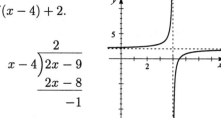

30. $r(x) = \dfrac{2x+6}{-6x+3} = \dfrac{2(x+3)}{-3(2x-1)}$. When $x = 0$, we have $y = 2$, so

the y-intercept is 2. When $y = 0$, we have $x + 3 = 0 \quad \Leftrightarrow \quad x = -3$,
so the x-intercept is -3. A vertical asymptote occurs when
$2x - 1 = 0 \quad \Leftrightarrow \quad x = \frac{1}{2}$. Because the degree of the denominator
and the numerator are the same, the horizontal asymptote is
$y = \frac{2}{-6} = -\frac{1}{3}$.

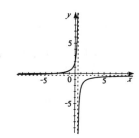

32. $s(x) = \dfrac{1-2x}{2x+3}$. When $x = 0$, $y = \frac{1}{3}$, so the y-intercept is $\frac{1}{3}$. When

$y = 0$, we have $1 - 2x = 0 \quad \Leftrightarrow \quad x = \frac{1}{2}$, so the x-intercept is $\frac{1}{2}$.
A vertical asymptote occurs when $2x + 3 = 0 \quad \Leftrightarrow \quad x = -\frac{3}{2}$,
and because the degree of the denominator and the numerator are the
same, the horizontal asymptote is $y = -1$.

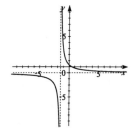

34. $r(x) = \dfrac{x-2}{(x+1)^2}$. When $x = 0$, we have $y = -2$, so the y-intercept

is -2. When $y = 0$, we have $x - 2 = 0 \Leftrightarrow x = 2$, so the
x-intercept is 2. A vertical asymptote occurs when $x = -1$, and
because the degree of the denominator is greater than the degree
of the numerator, the horizontal asymptote is $y = 0$.

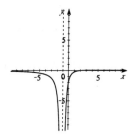

36. $s(x) = \dfrac{x+2}{(x+3)(x-1)}$. When $x = 0$, $y = \frac{2}{-3}$, so the y-intercept is $-\frac{2}{3}$

When $y = 0$, we have $x + 2 = 0 \quad \Leftrightarrow \quad x = -2$, so the x-intercept
is -2. A vertical asymptote occurs when $(x+3)(x-1) = 0$
$\Leftrightarrow \quad x = -3$ and $x = 1$. Because the degree of the denominator is
greater than the degree of the numerator, the horizontal asymptote
is $y = 0$.

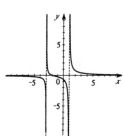

38. $s(x) = \dfrac{2x-4}{x^2+x-2} = \dfrac{2(x-2)}{(x-1)(x+2)}$. When $x = 0$, $y = 2$, so the

y-intercept is 2. When $y = 0$, we have $2x - 4 = 0 \quad \Leftrightarrow \quad x = 2$, so the
x-intercept is 2. A vertical asymptote occurs when $(x-1)(x+2) = 0$
$\Leftrightarrow \quad x = 1$ and $x = -2$. Because the degree of the denominator is
greater than the degree of the numerator, the horizontal asymptote
is $y = 0$.

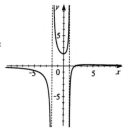

40. $t(x) = \dfrac{x-2}{x^2-4x} = \dfrac{x-2}{x(x-4)}$. Since $x=0$ is not in the domain of

$t(x)$, there is no y-intercept. When $y=0$, we have $x-2=0 \Leftrightarrow$
$x=2$, so the x-intercept is 2. A vertical asymptote occurs when
$x(x-4)=0 \Leftrightarrow x=0$ and $x=4$. Because the degree of the
denominator is greater than the degree of the numerator, the
horizontal asymptote is $y=0$.

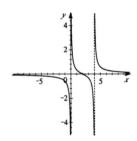

42. $r(x) = \dfrac{2x(x+2)}{(x-1)(x-4)}$. When $x=0$, we have $y=0$, so the graph

passes through the origin. Also, when $y=0$, we have $2x(x+2)=0$
$\Leftrightarrow x=0, -2$, so the x-intercepts are 0 and -2. There are two
vertical asymptotes, $x=1$ and $x=4$. Because the degree of the
denominator and numerator are the same, the horizontal asymptote
is $y = \frac{2}{1} = 2$.

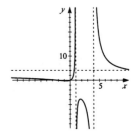

44. $r(x) = \dfrac{4x^2}{x^2-2x-3} = \dfrac{4x^2}{(x-3)(x+1)}$. When $x=0$, we have $y=0$,

so the graph passes through the origin. Vertical asymptotes occur at
$x=-1$ and $x=3$. Because the degree of the denominator and
numerator are the same, the horizontal asymptote is $y = \frac{4}{1} = 4$.

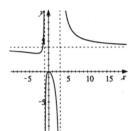

46. $r(x) = \dfrac{2x^2+2x-4}{x^2+x} = \dfrac{2(x+2)(x-1)}{x(x+1)}$. Vertical asymptotes

occur at $x=0$ and $x=-1$. Since x cannot equal zero, there is no
y-intercept. When $y=0$, we have $x=-2$ or 1, so the x-intercepts
are -2 and 1. Because the degree of the denominator and numerator
are the same, the horizontal asymptote is $y = \frac{2}{1} = 2$.

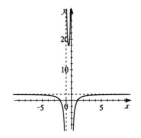

48. $r(x) = \dfrac{x^2+3x}{x^2-x-6} = \dfrac{x(x+3)}{(x-3)(x+2)}$. When $x=0$, we have $y=0$,

so the graph passes through the origin. When $y=0$, we have $x=0$
or -3, so the x-intercepts are 0 and -3. Vertical asymptotes occur
at $x=-2$ and $x=3$. Because the degree of the denominator and
numerator are the same, the horizontal asymptote is $y = \frac{1}{1} = 1$.

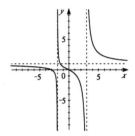

50. $r(x) = \dfrac{5x^2 + 5}{x^2 + 4x + 4} = \dfrac{5(x^2 + 1)}{(x + 2)^2}$. When $x = 0$, we have $y = \frac{5}{4}$,

so the y-intercept is $\frac{5}{4}$. Since $x^2 + 1 > 0$ for all real x, y never equals

zero, and there are no x-intercepts. The vertical asymptote is $x = -2$.

Because the degree of the denominator and numerator are the same,

the horizontal asymptote occurs at $y = \frac{5}{1} = 5$.

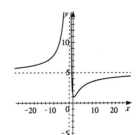

52. $r(x) = \dfrac{x^3 - x^2}{x^3 - 3x - 2} = \dfrac{x^2(x - 1)}{x^3 - 3x - 2}$. When $x = 0$, we have $y = 0$,

so the y-intercept is 0. When $y = 0$, we have $x^2(x - 1) = 0$, so the

x-intercepts are 0 and 1. Vertical asymptotes occur when

$x^3 - 3x - 2 = 0$. Since $x^3 - 3x - 2 = 0$ when $x = 2$, we can factor

$(x - 2)(x + 1)^2 = 0$, so the vertical asymptotes occur at $x = 2$ and

$x = -1$. Because the degree of the denominator and numerator are

the same, the horizontal asymptote is $y = \frac{1}{1} = 1$.

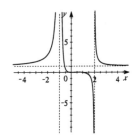

54. $r(x) = \dfrac{x^2 + 2x}{x - 1} = \dfrac{x(x + 2)}{x - 1}$. When $x = 0$, we have $y = 0$, so

the graph passes through the origin. Also, when $y = 0$, we have

$x = 0$ or -2, so the x-intercepts are -2 and 0. The vertical

asymptote is $x = 1$. There is no horizontal asymptote, and the

line $y = x + 3$ is a slant asymptote because by long division,

we have $y = x + 3 + \dfrac{2}{x - 1}$.

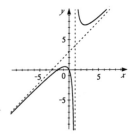

56. $r(x) = \dfrac{3x - x^2}{2x - 2} = \dfrac{x(3 - x)}{2(x - 1)}$. When $x = 0$, we have $y = 0$, so the

graph passes through the origin. Also, when $y = 0$, we have

$x = 0$ or $x = 3$, so the x-intercepts are 0 and 3. The vertical

asymptote is $x = 1$. There is no horizontal asymptote, and the line

$y = -\frac{1}{2}x + 1$ is a slant asymptote because by long division we

have $y = -\frac{1}{2}x + 1 + \dfrac{1}{x - 1}$.

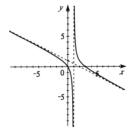

58. $r(x) = \dfrac{x^3 + 4}{2x^2 + x - 1} = \dfrac{x^3 + 4}{(2x - 1)(x + 1)}$. When $x = 0$, we have $y = \frac{0+4}{0+0-1} = -4$, so the

y-intercept is -4. Since $x^3 + 4 = 0 \Rightarrow x = -\sqrt[3]{4}$, the x-intercept

is $x = -\sqrt[3]{4}$. There are vertical asymptotes where $(2x - 1)(x + 1) = 0$

$\Rightarrow \quad x = \frac{1}{2}$ or $x = -1$. Since the degree of the numerator is greater

than the degree of the denominator, there is no horizontal asymptote.

By long division, we have $y = \frac{1}{2}x - \frac{1}{4} + \dfrac{\frac{3}{4}x + \frac{15}{4}}{2x^2 - x - 1}$,

so the line $y = \frac{1}{2}x - \frac{1}{4}$ is a slant asymptote.

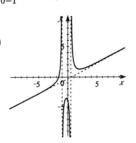

60. $r(x) = \dfrac{2x^3 + 2x}{x^2 - 1} = \dfrac{2x(x^2 + 1)}{(x - 1)(x + 1)}$. When $x = 0$, we have $y = 0$,

so the graph passes through the origin. Also, note that $x^2 + 1 > 0$, for all real x, so the only x-intercept is 0. There are two vertical asymptotes at $x = -1$ and $x = 1$. There is no horizontal asymptote, and the line $y = 2x$ is a slant asymptote because

by long division, we have $y = 2x + \dfrac{4x}{x^2 - 1}$.

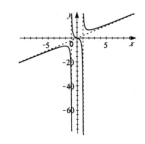

62. $f(x) = \dfrac{-x^3 + 6x^2 - 5}{x^2 - 2x}$, $g(x) = -x + 4$. The vertical asymptotes are $x = 0$ and $x = 2$.

$[-8, 8]$ by $[-20, 20]$ Graph of f. $[-20, 20]$ by $[-20, 20]$ Graph of f and g.

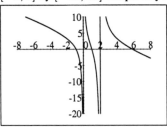 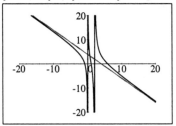

64. $f(x) = \dfrac{-x^4 + 2x^3 - 2x}{(x - 1)^2}$, $g(x) = 1 - x^2$. The vertical asymptote is $x = 1$.

$[-4, 4]$ by $[-10, 2]$ Graph of f. $[-4, 4]$ by $[-10, 2]$ Graph of f and g.

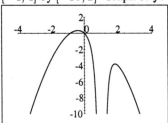 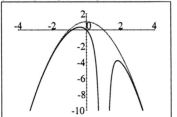

66. $f(x) = \dfrac{x^4 - 3x^3 + x^2 - 3x + 3}{x^2 - 3x}$

Vertical asymptote: $x = 0$, $x = 3$

x-intercept: 0.82

y-intercept: none

The local minima are $(-0.80, 2.63)$ and $(3.38, 14.76)$. The local maximum is $(2.56, 4.88)$.

$$
\begin{array}{r}
x^2 \qquad\qquad\qquad +1 \\
x^2 - 3x \overline{\big)\, x^4 - 3x^3 + x^2 - 3x + 3} \\
\underline{x^4 - 3x^3 \qquad\qquad\qquad} \\
0x^3 + x^2 - 3x \\
\underline{x^2 - 3x} \\
3
\end{array}
$$

By using long division, we see that $f(x) = x^2 + 1 + \dfrac{3}{x^2 - 3x}$. From the second graph, we see that the end behavior of $f(x)$ is the same as the end behavior of $g(x) = x^2 + 1$.

$[-6, 6]$ by $[-25, 25]$ Graph of f.

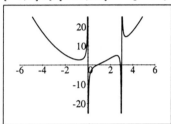

$[-8, 8]$ by $[-65, 65]$ Graph of f and g.

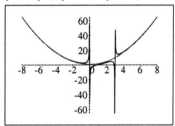

68. $f(x) = \dfrac{x^4}{x^2 - 2}$

Vertical asymptote: $x = -1.41, x = 1.41$

x-intercept: 0

y-intercept: 0

The local maximum is $(0, 0)$. The local minima are $(-2, 8)$ and $(2, 8)$.

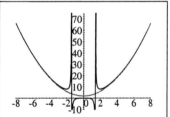

By using long division, we see that $f(x) = x^2 + 2 + \dfrac{4}{x^2 - 2}$. From the second graph, we see that the end behavior of $f(x)$ is the same as the end behavior of $g(x) = x^2 + 2$.

$[-5, 5]$ by $[-10, 20]$ Graph of $f(x)$.

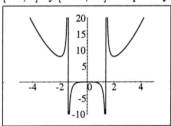

$[-8, 8]$ by $[-10, 75]$ Graph of $f(x)$ and $g(x)$.

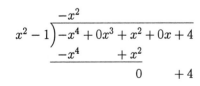

70. $r(x) = \dfrac{4 + x^2 - x^4}{x^2 - 1} = \dfrac{-x^4 + x^2 + 4}{(x-1)(x+1)}$

Vertical asymptote: $x = -1.41, x = 1.41$

x-intercept: $-1.6, 1.6$

y-intercept: -1

The local maximum is $(0, -0.4)$. No local minima

$$x^2 - 1 \overline{) -x^4 + 0x^3 + x^2 + 0x + 4} \\ \underline{-x^4 \qquad + x^2} \\ 0 \qquad + 4$$

Thus $y = -x^2 + \dfrac{6}{x^2 - 1}$. From the graphs on the next page we see that the end behavior of $f(x)$ is like the end behavior of $g(x) = -x^2$.

[−10, 10] by [−20, 20] Graph of f.

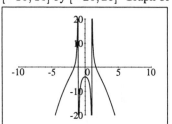

[−10, 10] by [−20, 20] Graph of f and g.

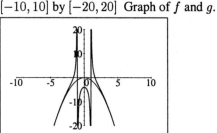

72. (a)

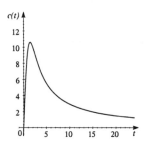

(b) $c(t) = \dfrac{30t}{t^2 + 2}$. Since the degree of the denominator is larger than the degree of the numerator, $c(t) \rightarrow 0$ as $t \rightarrow \infty$.

74. Substituting for R and g, we have $h(v) = \dfrac{(6.4 \times 10^6)\, v^2}{2(9.8)(6.4 \times 10^6) - v^2}$. The vertical asymptote is $v \approx 11{,}000$, and it represents the escape velocity from the earth's gravitational pull: $11{,}000$ m/s ≈ 1900 mi/h.

[0, 20000] by [0, 20000000]

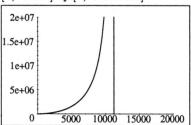

76. (a) $\dfrac{1}{x} + \dfrac{1}{y} = \dfrac{1}{F} \iff \dfrac{1}{y} = \dfrac{1}{F} - \dfrac{1}{x} \iff \dfrac{1}{y} = \dfrac{x - F}{xF}$
$\iff y = \dfrac{xF}{x - F}$. Using $F = 55$, we get $y = \dfrac{55x}{x - 55}$.
Since $y \geq 0$, we use the window $[55, 1000]$ by $[0, 250]$.

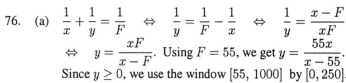

(b) y approaches 55 millimeters.

(c) y approaches ∞.

78. $r(x) = \dfrac{x^6 + 10}{x^4 + 8x^2 + 15}$ has no x-intercept since the numerator has no real roots. Likewise, $r(x)$ has no vertical asymptotes, since the denominator has no real roots. Since the degree of the numerator is two greater than the degree of the denominator, $r(x)$ has no horizontal or slant asymptotes.

80. (a) Let $f(x) = \dfrac{1}{x^2}$. Then $r(x) = \dfrac{1}{(x-2)^2} = f(x-2)$.

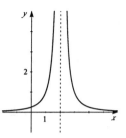

From this form we see that the graph of r is obtained
from the graph of f by shifting 2 units to the right.
Thus r has vertical asymptote $x = 2$ and horizontal
asymptote $y = 0$.

(b) $s(x) = \dfrac{2x^2 + 4x + 5}{x^2 + 2x + 1} = 2 + \dfrac{3}{(x+1)^2} = 3f(x+1) + 2.$

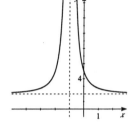

From this form we see that the
graph of s is obtained from the
graph of f by shifting 1 unit to
the left and 3 units vertically.
Thus r has vertical asymptote
$x = -1$ and horizontal asymptote $y = 2$.

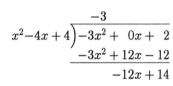

$$\begin{array}{r} 2 \\ x^2 + 2x + 1 \overline{\smash{)}\, 2x^2 + 4x + 5} \\ \underline{2x^2 + 4x + 2} \\ 3 \end{array}$$

(c) Using long division (shown to the side) we see that

$p(x) = \dfrac{2 - 3x^2}{x^2 - 4x + 4} = -3 + \dfrac{-12x + 14}{x^2 - 4x + 4}$ which

cannot be graph transforming $f(x) = \dfrac{1}{x^2}$.

$$\begin{array}{r} -3 \\ x^2 - 4x + 4 \overline{\smash{)}\, -3x^2 + 0x + 2} \\ \underline{-3x^2 + 12x - 12} \\ -12x + 14 \end{array}$$

Using long division on q we have:

$$\begin{array}{r} -3 \\ x^2 - 4x + 4 \overline{\smash{)}\, -3x^2 + 12x + 0} \\ \underline{-3x^2 + 12x - 12} \\ 12 \end{array}$$

So $q(x) = \dfrac{12x - 3x^2}{x^2 - 4x + 4} = -3 + \dfrac{12}{(x-2)^2} = 12f(x-2) - 3.$

From this form we see that the graph of q is obtained from the
graph of f by shifting 2 units to the right, stretching vertically
by a factor of 12, and then shifting 3 units vertically down.
Thus the vertical asymptote is $x = 2$ and horizontal asymptote
$y = -3$.

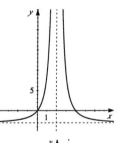

We show $y = p(x)$ just to verify that we cannot obtain $p(x)$
from $y = \dfrac{1}{x^2}$.

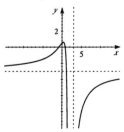

Review Exercises for Chapter 4

2. $P(x) = -2x^4 + 32 = -2(x^4 - 16)$

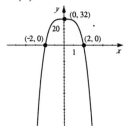

4. $P(x) = 2(x+1)^3$

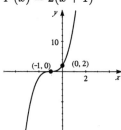

6. $P(x) = x^3 - 3x^2 - 4x = x(x-4)(x+1)$

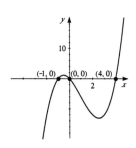

8. $P(x) = x^4 - 5x^2 + 4 = (x^2 - 1)(x^2 - 4)$
$= (x-1)(x+1)(x-2)(x+2)$

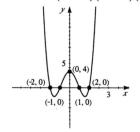

10. $P(x) = -2x^3 + 6x^2 - 2$.
x-intercepts: $-0.5, 0.7$, and 2.9.
y-intercept: -2.
Local maximum is $(2, 6)$.
Local minimum is $(0, -2)$.
$y \to \infty$ as $x \to -\infty$ and
$y \to -\infty$ as $x \to \infty$.

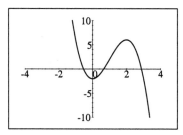

12. $P(x) = x^5 + x^4 - 7x^3 - x^2 + 6x + 3$.
x-intercepts: $-3.0, 1.3$, and 1.9.
y-intercept: -3.
Local maxima are $(-2.4, 33.2)$ and $(0.5, 5.0)$.
Local minima are $(-0.6, 0.6)$ and $(1.6, -1.6)$.
$y \to -\infty$ as $x \to -\infty$ and
$y \to \infty$ as $x \to \infty$.

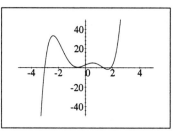

14. (a) The area of the four sides is $2x^2 + 2xy = 1200 \quad \Leftrightarrow \quad 2xy = 1200 - 2x^2 \quad \Leftrightarrow$
$y = \dfrac{600 - x^2}{x}$. Substituting we get $V = x^2 y = x^2 \left(\dfrac{600 - x^2}{x} \right) = 600x - x^3$.

(b)

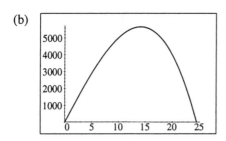

(c) V is maximized when $x \approx 14.14$, $y \approx \dfrac{600 - (14.14)^2}{14.14} = 28.28$

16. $\dfrac{x^2 + x - 12}{x - 3} = \dfrac{(x+4)(x-3)}{x-3} = x + 4.$ Factoring we have $Q(x) = x + 4$ and $R(x) = 0$. This
can be confirmed with synthetic division

$$
\begin{array}{r|rrr}
3 & 1 & 1 & -12 \\
  &   & 3 & 12 \\
\hline
  & 1 & 4 & 0
\end{array}
$$

18. $\dfrac{x^3 + 2x^2 - 10}{x + 3}$

$$
\begin{array}{r|rrrr}
-3 & 1 & 2 & 0 & -10 \\
   &   & -3 & 3 & -9 \\
\hline
   & 1 & -1 & 3 & -19
\end{array}
$$

Using synthetic division, we see that $Q(x) = x^2 - x + 3$ and $R(x) = -19$.

20. $\dfrac{2x^4 + 3x^3 - 12}{x + 4}$

$$
\begin{array}{r|rrrrr}
-4 & 2 & 3 & 0 & 0 & -12 \\
   &   & -4 & 4 & -16 & 64 \\
\hline
   & 2 & -1 & 4 & -16 & 52
\end{array}
$$

Using synthetic division, we see that $Q(x) = 2x^3 - x^2 + 4x - 16$ and $R(x) = 52$.

22. $\dfrac{x^4 - 2x^2 + 7x}{x^2 - x + 3}$

$$
\require{enclose}
\begin{array}{r}
x^2 + x - 4 \\
x^2 - x + 3 \enclose{longdiv}{x^4 + 0x^3 - 2x^2 + 7x + 0} \\
\underline{x^4 - x^3 + 3x^3} \\
x^3 - 5x^2 + 7x \\
\underline{x^3 - x^2 + 3x} \\
-4x^2 + 4x + 0 \\
\underline{-4x^2 + 4x - 12} \\
12
\end{array}
$$

Therefore, $Q(x) = x^2 + x - 4$, and $R(x) = 12$.

24. $Q(x) = x^4 + 4x^3 + 7x^2 + 10x + 15$; find $Q(-3)$

$$
\begin{array}{r|rrrrr}
-3 & 1 & 4 & 7 & 10 & 15 \\
   &   & -3 & -3 & -12 & 6 \\
\hline
   & 1 & 1 & 4 & -2 & 21
\end{array}
$$

By the Remainder Theorem, we have $Q(-3) = 21$.

26. $x + 4$ is a factor of $P(x) = x^5 + 4x^4 - 7x^3 - 23x^2 + 23x + 12$ if $P(-4) = 0$.

$$
\begin{array}{r|rrrrrr}
-4 & 1 & 4 & -7 & -23 & 23 & 12 \\
   &   & -4 & 0 & 28 & -20 & -12 \\
\hline
   & 1 & 0 & -7 & 5 & 3 & 0
\end{array}
$$

Since $P(-4) = 0$, $x + 4$ is a factor of the polynomial.

28. Let $P(x) = x^{101} - x^4 + 2$. The remainder from dividing $P(x)$ by $x + 1$ is
$P(-1) = (-1)^{101} - (-1)^4 + 2 = 0$.

30. (a) $P(x) = 6x^4 + 3x^3 + x^2 + 3x + 4$ has possible rational zeros ± 1, ± 2, ± 4, $\pm\frac{1}{2}$, $\pm\frac{1}{3}$, $\pm\frac{2}{3}$, $\pm\frac{4}{3}$, $\pm\frac{1}{6}$.

(b) Since $P(x)$ has no variations in sign, there are no positive real zeros. Since
$P(-x) = 6x^4 - 3x^3 + x^2 - 3x + 4$ has 4 variations in sign, there are 0, 2, or 4 negative real zeros.

32. Since the zeros are $\pm 3i$ and 4 (which is a double zero), a factorization is
$P(x) = C(x - 4)^2(x - 3i)(x + 3i) = C(x^2 - 8x + 16)(x^2 + 9) = $
$C(x^4 - 8x^3 + 25x^2 - 72x + 144)$. Since all of the coefficients are integers, we can choose $C = 1$,
so $P(x) = x^4 - 8x^3 + 25x^2 - 72x + 144$.

34. $P(x) = 3x^4 + 5x^2 + 2 = (3x^2 + 2)(x^2 + 1)$. Since $3x^2 + 2 = 0$ and $x^2 + 1 = 0$ have no real
zeros, it follows that $3x^4 + 5x^2 + 2$ has no real zeros.

36. $P(x) = 2x^3 + 5x^2 - 6x - 9$ has possible rational zeros ± 1, ± 3, ± 9, $\pm\frac{1}{2}$, $\pm\frac{3}{2}$, $\pm\frac{9}{2}$. Since there is
one variation in sign, there is a positive real zero.

$$
\begin{array}{r|rrrr}
1 & 2 & 5 & -6 & -9 \\
  &   & 2 & 7 & 1 \\
\hline
  & 2 & 7 & 1 & -8 \\
3 & 2 & 5 & -6 & -9 \\
  &   & 6 & 33 & 81 \\
\hline
  & 2 & 11 & 27 & 72 \\
\frac{3}{2} & 2 & 5 & -6 & -9 \\
  &   & 3 & 12 & 9 \\
\hline
  & 2 & 8 & 6 & 0
\end{array}
$$

$\Rightarrow \quad x = 3$ is an upper bound, and there is a zero between 1 and 3.

$\Rightarrow \quad x = \frac{3}{2}$ is a zero.

So $P(x) = 2x^3 + 5x^2 - 6x - 9 = (2x - 3)(x^2 + 4x + 3) = (2x - 3)(x + 3)(x + 1)$. Therefore,
the zeros are -3, -1 and $\frac{3}{2}$.

38. $P(x) = x^4 + 7x^3 + 9x^2 - 17x - 20$ has possible rational zeros ± 1, ± 2, ± 4, ± 5, ± 10, ± 20.

$$
\begin{array}{r|rrrrr}
1 & 1 & 7 & 9 & -17 & -20 \\
  &   & 1 & 8 & 17 & 0 \\
\hline
  & 1 & 8 & 17 & 0 & -20 \\
2 & 1 & 7 & 9 & -17 & -20 \\
  &   & 2 & 18 & 54 & 74 \\
\hline
  & 1 & 9 & 27 & 37 & 54
\end{array}
$$

$\Rightarrow \quad x = 2$ is an upper bound.

$$\begin{array}{r|rrrrr} -1 & 1 & 7 & 9 & -17 & -20 \\ & & -1 & -6 & -3 & 20 \\ \hline & 1 & 6 & 3 & -20 & 0 \end{array} \Rightarrow \quad x = -1 \text{ is a zero.}$$

So $P(x) = x^4 + 7x^3 + 9x^2 - 17x - 20 = (x+1)(x^3 + 6x^2 + 3x - 20)$. Continuing with the quotient, we have

$$\begin{array}{r|rrrr} -1 & 1 & 6 & 3 & -20 \\ & & -1 & -5 & 2 \\ \hline & 1 & 5 & -2 & -18 \end{array} \qquad\qquad \begin{array}{r|rrrr} -2 & 1 & 6 & 3 & -20 \\ & & -2 & -8 & 10 \\ \hline & 1 & 4 & -5 & -10 \end{array}$$

$$\begin{array}{r|rrrr} -4 & 1 & 6 & 3 & -20 \\ & & -4 & -8 & 20 \\ \hline & 1 & 2 & -5 & 0 \end{array} \Rightarrow \quad x = -4 \text{ is a zero.}$$

So $P(x) = x^4 + 7x^3 + 9x^2 - 17x - 20 = (x+1)(x+4)(x^2 + 2x - 5)$. Now using the quadratic formula on $x^2 + 2x - 5$ we have: $x = \dfrac{-2 \pm \sqrt{4 - 4(1)(-5)}}{2} = \dfrac{-2 \pm 2\sqrt{6}}{2} = -1 \pm \sqrt{6}$. Thus, the zeros are -4, -1, and $-1 \pm \sqrt{6}$.

40. $P(x) = x^4 - 81 = (x^2 - 9)(x^2 + 9) = (x - 3)(x + 3)(x^2 + 9) = (x - 3)(x + 3)(x - 3i)(x + 3i)$. Thus, the zeros are ± 3, $\pm 3i$.

42. $P(x) = 18x^3 + 3x^2 - 4x - 1$ has possible rational zeros $\pm 1, \pm\frac{1}{2}, \pm\frac{1}{3}, \pm\frac{1}{6}, \pm\frac{1}{9}, \pm\frac{1}{18}$.

$$\begin{array}{r|rrrr} 1 & 18 & 3 & -4 & -1 \\ & & 18 & 21 & 17 \\ \hline & 18 & 21 & 17 & 16 \end{array} \Rightarrow \quad x = 1 \text{ is an upper bound.}$$

$$\begin{array}{r|rrrr} \frac{1}{2} & 18 & 3 & -4 & -1 \\ & & 9 & 6 & 1 \\ \hline & 18 & 12 & 2 & 0 \end{array} \Rightarrow \quad x = \tfrac{1}{2} \text{ is a zero.}$$

So $P(x) = 18x^3 + 3x^2 - 4x - 1 = (2x - 1)(9x^2 + 6x + 1) = (2x - 1)(3x + 1)^2$. Thus the zeros are $\frac{1}{2}$ and $-\frac{1}{3}$ (multiplicity 2).

44. $P(x) = x^4 + 15x^2 + 54 = (x^2 + 9)(x^2 + 6)$. If $x^2 = -9$, then $x = \pm 3i$. If $x^2 = -6$, then $x = \pm\sqrt{6}\,i$. Therefore, the zeros are $\pm 3i$ and $\pm\sqrt{6}\,i$.

46. Let $P(x) = x^3 + x^2 - 14x - 24$. The solutions to $P(x) = 0$ are $x = -3, -2$, and 4.

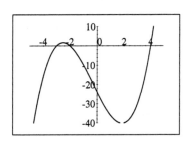

48. $x^5 = x + 3 \quad \Leftrightarrow \quad x^5 - x - 3 = 0$.
We graph $P(x) = x^5 - x - 3$. The only real
solution is 1.34. We graph $P(x) = x^5 - x - 3$.

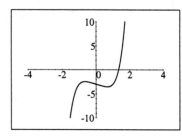

50. $r(x) = \dfrac{1}{(x+2)^2}$. When $x = 0$, we have $r(0) = \dfrac{1}{2^2} = \dfrac{1}{4}$, so the

y-intercept is $\frac{1}{4}$. Since the numerator is 1, y never equals zero and
there are no x-intercepts. A vertical asymptote occurs when $x = -2$.
The horizontal asymptote is $y = 0$ because the degree of the
denominator is greater than the degree of the numerator.

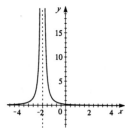

52. $r(x) = \dfrac{2x^2 - 6x - 7}{x - 4}$. When $x = 0$, we have $r(0) = \dfrac{-7}{-4} = \dfrac{7}{4}$, so the

y-intercept is $y = \frac{7}{4}$. We use the quadratic formula to find the

x-intercepts: $x = \dfrac{-(-6) \pm \sqrt{(-6)^2 - 4(2)(-7)}}{2(2)} = \dfrac{6 \pm \sqrt{92}}{4}$

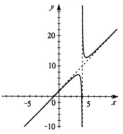

$= \frac{3 \pm \sqrt{23}}{2}$. Thus the x-intercepts are $x \approx 3.9$ and $x \approx -0.9$.
The vertical asymptote is $x = 4$. Because the degree of the
numerator is greater than the degree of the denominator, there is no horizontal asymptote. By long

division, we have $r(x) = 2x + 2 + \dfrac{1}{x - 4}$, so the slant asymptote is $r(x) = 2x + 2$.

54. $r(x) = \dfrac{x^3 + 27}{x + 4}$. When $x = 0$, we have $r(0) = \frac{27}{4}$, so the y-intercept

is $y = \frac{27}{4}$. When $y = 0$, we have $x^3 + 27 = 0 \quad \Leftrightarrow \quad x^3 = -27$
$\Rightarrow \quad x = -3$. Thus the x-intercept is $x = -3$. The vertical
asymptote is $x = -4$. Because the degree of the numerator is
greater than the degree of the denominator, there is no horizontal

asymptote. By long division, we have $r(x) = x^2 - 4x + 16 - \dfrac{37}{x + 4}$.

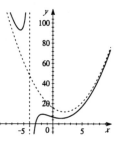

So the end behavior of y is like the end behavior of $g(x) = x^2 - 4x - 16$.

56. $r(x) = \dfrac{2x - 7}{x^2 + 9}$. From the graph we see that
x-intercept: 3.5
y-intercept: -0.78
Vertical asymptote: none
Horizontal asymptote: $y = 0$
Local minimum is $(-1.11, -0.90)$.
Local maximum is $(8.11, 0.12)$.

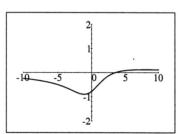

58. $r(x) = \dfrac{2x^3 - x^2}{x+1}$. From the graph we see that

     x-intercepts: $0, \frac{1}{2}$

     y-intercept: 0

     Vertical asymptote: $x = -1$

     Local minimum is $(0.425, -3.599)$.

     Local maxima are $(-1.57, 17.90)$ and $(0.32, -0.03)$.

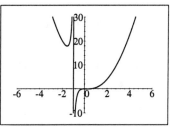

By using long division, we see that $r(x) = 2x^2 - 3x + 3 - \dfrac{3}{x+1}$. So the end behavior of r is the same as the end behavior of $g(x) = 2x^2 - 3x + 3$.

Principles of Modeling

2. (a) Using a graphing calculator, we obtain the quadratic polynomial
        $y = -0.0000002783333x^2 + 0.0184655x - 166.732$.

    (b)

    (c) Moving the cursor along the path of the polynomial, we find that yield when 37,000 plants are planted per acre is about 135 bushels/acre.

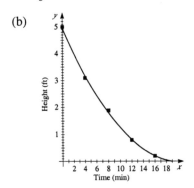

4. (a) Using a graphing calculator, we obtain the quartic polynomial
        $y = -55.908x^4 + 1414.88x^3 - 12199.1x^2 + 42577.2x - 25714.6$.

    (b)

    (c) Yes since there appears to be two local maximums and one local minimum.

6. (a) Using a graphing calculator, we obtain the quadratic polynomial
        $y = 0.0120536x^2 - 0.490357x + 4.96571$.

    (b)

    (c) Moving the cursor along the path of the polynomial, we find that the tank should drain is 19.0 minutes.

Chapter Five

Exercises 5.1

2. $g(x) = 8^x$

x	y
-2	$\frac{1}{64}$
-1	$\frac{1}{8}$
0	1
1	8
2	64

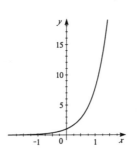

4. $h(x) = (1.1)^x$

x	y
-5	0.620921323
-1	$0.\overline{90}$
0	1
5	1.61051
10	2.59374246

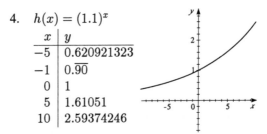

6. $g(x) = 2e^{-0.5x}$

x	$g(x) = 2e^{-0.5x}$
-4	14.7761
-3	8.96338
-2	5.43656
-1	3.29744
0	2
1	1.21306
2	0.735759
3	0.44626
4	0.270671

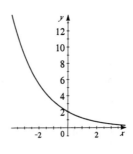

8. $y = \left(\frac{2}{3}\right)^x$ and $y = \left(\frac{4}{3}\right)^x$.

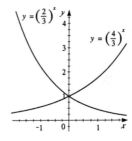

10. From the graph, $f(-1) = a^{-1} = \frac{1}{5}$, so $a = 5$. Thus $f(x) = 5^x$.

12. From the graph, $f(-3) = a^{-3} = 8$, so $a = \frac{1}{2}$. Thus $f(x) = \left(\frac{1}{2}\right)^x$.

14. V **16.** VI **18.** IV

20. $f(x) = 10^{-x}$
The graph of f is obtained by reflecting the graph of $y = 10^x$ about the y-axis.
Domain: $(-\infty, \infty)$
Range: $(0, \infty)$
Asymptote: $y = 0$

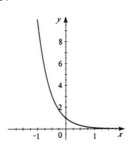

22. $g(x) = 2^{x-3}$
The graph of g is obtained by shifting the graph of $y = 2^x$
to the right 3 units.
Domain: $(-\infty, \infty)$
Range: $(0, \infty)$
Asymptote: $y = 0$

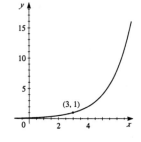

24. $h(x) = 6 - 3^x$
The graph of h is obtained by reflecting the graph of $y = 3^x$
about the x-axis and shifting upward 6 units.
Domain: $(-\infty, \infty)$
Range: $(-\infty, 6)$
Asymptote: $y = 6$

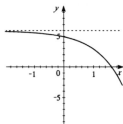

26. $f(x) = -\left(\frac{1}{5}\right)^x$
Note that $f(x) = -\left(\frac{1}{5}\right)^x = -5^{-x}$. So the graph of f is
obtained by reflecting the graph of $y = 5^x$ about the y-axis and
about the x-axis.
Domain: $(-\infty, \infty)$
Range: $(-\infty, 0)$
Asymptote: $y = 0$

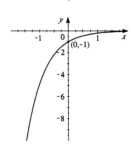

28. $f(x) = 1 - e^x$
The graph of $f(x) = 1 - e^x$ is obtained by reflecting the graph
of $y = e^x$ about the x-axis and then shifting upward 1 unit.
Domain: $(-\infty, \infty)$
Range: $(-\infty, 1)$
Asymptote: $y = 1$

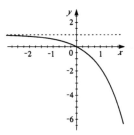

30. $f(x) = -e^{-x}$
The graph of $f(x) = -e^{-x}$ is obtained by reflecting the
graph of $y = e^x$ about the y-axis and then about the x-axis.
Domain: $(-\infty, \infty)$
Range: $(-\infty, 0)$
Asymptote: $y = 0$

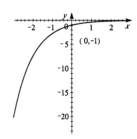

32. $f(x) = e^{x-3} + 4$

The graph of $f(x) = e^{x-3} + 4$ is obtained by shifting
the graph of $y = e^x$ to the right 3 units, and then upward
4 units.

Domain: $(-\infty, \infty)$

Range: $(4, \infty)$

Asymptote: $y = 4$

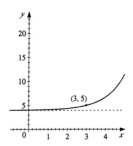

34. Using the points $(-1, 15)$ and $(0, 5)$ we have $f(0) = Ca^0 = 5 \quad \Leftrightarrow \quad C = 5$. Then
$f(-1) = 5a^{-1} = 15 \quad \Leftrightarrow \quad a^{-1} = 3 \quad \Leftrightarrow \quad a = \frac{1}{3}$. Thus $f(x) = 5\left(\frac{1}{3}\right)^x$.

36. (a)

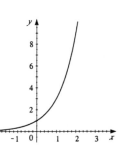

(b) $f(x) = 9^{x/2} = (3^2)^{x/2} = 3^{2 \cdot x/2} = 3^x = g(x)$.
So $f(x) = g(x)$, and the graphs are the same.

38.

x	$f(x) = x^3$	$g(x) = 3^x$
0	0	1
1	1	3
2	8	9
3	27	27
4	64	81
5	125	243
6	216	729
7	343	2187
8	512	6561
9	729	19683
10	1000	59049
15	3375	14348907
20	8000	3486784401

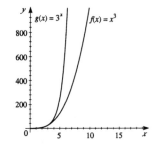

40.

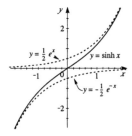

42. $\sinh(-x) = \dfrac{e^{-x} - e^{-(-x)}}{2} = \dfrac{e^{-x} - e^{x}}{2} = -\dfrac{-e^{-x} + e^{x}}{2} = -\dfrac{e^{x} - e^{-x}}{2} = -\sinh(x).$

44. $\sinh(x)\cosh(y) + \cosh(x)\sinh(y) = \dfrac{e^{x} - e^{-x}}{2} \cdot \dfrac{e^{y} + e^{-y}}{2} + \dfrac{e^{x} + e^{-x}}{2} \cdot \dfrac{e^{y} - e^{-y}}{2}$

$= \dfrac{e^{x+y} + e^{x-y} - e^{y-x} - e^{-(x+y)}}{4} + \dfrac{e^{x+y} - e^{x-y} + e^{y-x} - e^{-(x+y)}}{4}$

$= \dfrac{e^{x+y} - e^{-(x+y)}}{2} = \sinh(x + y).$

46. (a) (i) $[-4, 4]$ by $[0, 20]$ (ii) $[0, 10]$ by $[0, 5000]$

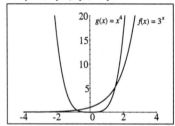   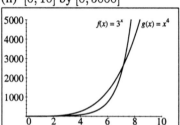

(iii) $[0, 20]$ by $[0, 10^5]$

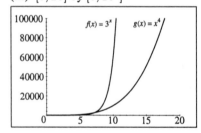

(b) From the graphs in parts (i) and (ii), we
see that the solutions of $3^x = x^4$ are
$x \approx -0.80$, $x \approx 1.52$ and $x \approx 7.17$.

48.

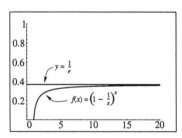

The larger the value of c, the more rapidly the
graph of $f(x) = 2^{cx}$ increases. In general,
$f(x) = 2^{cx} = (2^c)^x$; so, for example,
$f(x) = 2^{2x} = (2^2)^x = 4^x$.

50.

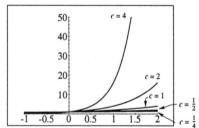

From the graph, we see that
$$f(x) = \left(1 - \frac{1}{x}\right)^{x} \quad \text{approaches } \frac{1}{e} \text{ as } x \text{ get}$$
large.

52. $y = 2^{1/x}$

Vertical Asymptote: $x = 0$

Horizontal Asymptote: $y = 1$

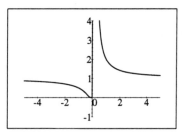

54. $g(x) = x^x$. Notice that $g(x)$ is only defined for $x \geq 0$. The graph of $g(x)$ is shown in the viewing rectangle $[0, 1.5]$ by $[0, 1.5]$. From the graphs, we see that there is a local minimum ≈ 0.69 when $x \approx 0.37$.

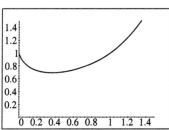

56. $y = 10^{x - x^2}$

(a) From the graph, we see that the function is increasing on $(-\infty, 0.50]$ and decreasing on $[0.50, \infty)$.

(b) From the graph, we see that the range is approximately $(0, 1.78]$.

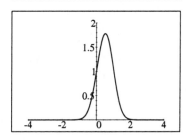

58. $D(t) = 50e^{-0.2t}$. So when $t = 3$ we have $D(3) = 50e^{-0.2(3)} \approx 27.4$ milligrams.

60. (a) $m(0) = 6\,e^{-0.087(0)} \approx 6$ grams

(b) $m(20) = 6\,e^{-0.087(20)} \approx 6(0.1755) = 1.053$. Thus approximately 1 gram of radioactive iodine remains after 20 days.

62. (a) $Q(5) = 15\bigl(1 - e^{-0.04(5)}\bigr) \approx 15(0.1813)$
$= 2.7345$. Thus approximately 2.7 lb of salt are in the barrel after 5 minutes.

(b) $Q(10) = 15\bigl(1 - e^{-0.04(10)}\bigr) \approx 15(0.3297)$
$= 4.946$. Thus approximately 4.9 lb of salt are in the barrel after 10 minutes.

(c)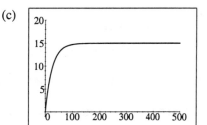

(d) 15 lb. Yes, since 50 gal $\times$ 0.3 lb/gal $= 15$ lb.

64. $n(t) = \dfrac{5600}{0.5 + 27.5\,e^{-0.044t}}$

(a) $n(0) = \frac{5600}{28} = 200$.

(b)

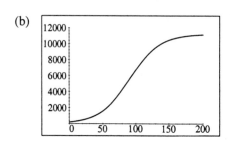

(c) From the graph, we see that $n(t)$ approaches about 11,200 as t gets large.

66. (a) Substituting $n_0 = 50$ and $t = 12$, we have $n(12) = \dfrac{300}{0.05 + \left(\frac{300}{50} - 0.05\right)e^{-0.55(12)}} \approx 5164.$

(b)

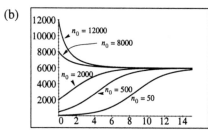

(c) The population approaches a size of 6000 rabbits.

68. $P = 4{,}000, r = 0.16,$ and $n = 4.$ So $A(t) = 4000\left(1 + \frac{0.16}{4}\right)^{4t} = 4000 \cdot 1.04^{4t}.$

   (a) $A(4) = 4000 \cdot (1.04)^{16} \approx 7491.92,$ and so the amount due is \$7,491.92.

   (b) $A(6) = 4000 \cdot (1.04)^{24} \approx 10{,}253.22,$ and so the amount due is \$10,253.22.

   (c) $A(8) = 4000 \cdot (1.04)^{32} \approx 14{,}032.23,$ and so the amount due is \$14,032.23.

70. $P = 4000, n = 4,$ and $t = 5.$ Then $A(5) = 4000\left(1 + \frac{r}{4}\right)^{4(5)},$ and so $A(5) = 4000\left(1 + \frac{r}{4}\right)^{20}.$

   (a) If $r = 0.06, A(5) = 4000\left(1 + \frac{0.06}{4}\right)^{20} = 4000 \cdot (1.015)^{20} \approx \$5,387.42.$

   (b) If $r = 0.065, A(5) = 4000\left(1 + \frac{0.065}{4}\right)^{20} = 4000 \cdot (1.01625)^{20} \approx \$5,521.68.$

   (c) If $r = 0.07, A(5) = 4000\left(1 + \frac{0.07}{4}\right)^{20} = 4000 \cdot (1.0175)^{20} \approx \$5,659.11.$

   (d) If $r = 0.08, A(5) = 4000\left(1 + \frac{0.08}{4}\right)^{20} = 4000 \cdot (1.02)^{20} \approx \$5,943.79.$

72. We find the effective rate for $P = 1$ and $t = 1.$

   (i) If $r = 0.0925$ and $n = 2,$ then $A(2) = \left(1 + \frac{0.0925}{2}\right)^2 = (1.04625)^2 \approx 1.0946.$

   (ii) If $r = 0.09$ and interest is compounded continuously, then $A(1) = e^{0.09} \approx 1.0942.$

Since the effective rate in (i) is greater than the effective rate in (ii), we can see that the account paying $9\frac{1}{4}\%$ per year compounded semiannually is the better investment.

74. (a) $A(t) = 5000\left(1 + \frac{0.09}{2}\right)^{2t} = 5000(1.045)^{2t} = 5000(1.092025)^t$.

(b)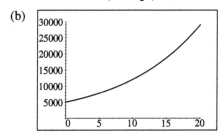

(c) $A(t) = 25000$ when $t \approx 18.28$ years.

76. Since $f(40) = 2^{40} = 1099511627776$, it would take a sheet of paper 4 inches by 1099511627776 inches. Since there are 12 inches in every foot and 5,280 feet in ever mile, 1099511627776 inches ≈ 1.74 million miles. So the dimensions of the sheet of paper required are 4 inches by about 1.74 million miles.

Exercises 5.2

2. (a) $10^{-1} = 0.1$ (b) $8^3 = 512$

4. (a) $3^4 = 81$ (b) $8^{2/3} = 4$

6. (a) $e^2 = x + 1$ (b) $e^4 = x - 1$

8. (a) $\log_{10} 1000 = 3$ (b) $\log_{81} 9 = \frac{1}{2}$

10. (a) $\log_4 0.125 = -\frac{3}{2}$ (b) $\log_7 343 = 3$

12. (a) $\ln 0.5 = x + 1$ (b) $\ln t = 0.5x$

14. (a) $\log_5 5^4 = 4$ (b) $\log_4 64 = \log_4 4^3 = 3$

    (c) $\log_9 9 = 1$

16. (a) $\log_2 32 = \log_2 2^5 = 5$ (b) $\log_8 8^{17} = 17$

    (c) $\log_6 1 = \log_6 6^0 = 0$

18. (a) $\log_5 125 = \log_5 5^3 = 3$ (b) $\log_{49} 7 = \log_{49} 49^{1/2} = \frac{1}{2}$

    (c) $\log_9 \sqrt{3} = \log_9 3^{1/2} = \log_9 \left(9^{1/2}\right)^{1/2} = \log_9 9^{1/4} = \frac{1}{4}$

20. (a) $e^{\ln \pi} = \pi$ (b) $10^{\log 5} = 5$

    (c) $10^{\log 87} = 87$

22. (a) $\log_4 \sqrt{2} = \log_4 2^{1/2} = \log_4 \left(4^{1/2}\right)^{1/2} = \log_4 4^{1/4} = \frac{1}{4}$

    (b) $\log_4 \left(\frac{1}{2}\right) = \log_4 2^{-1} = \log_4 \left(4^{1/2}\right)^{-1} = \log_4 4^{-1/2} = -\frac{1}{2}$

    (c) $\log_4 8 = = \log_4 2^3 = \log_4 \left(4^{1/2}\right)^3 = \log_4 4^{3/2} = \frac{3}{2}$

24. (a) $\log_5 x = 4 \quad \Leftrightarrow \quad x = 5^4 = 625$ (b) $x = \log_{10}(0.1) = \log_{10} 10^{-1} = -1$

26. (a) $x = \log_4 2 = \log_4 4^{1/2} = \frac{1}{2}$ (b) $\log_4 x = 2 \quad \Leftrightarrow \quad x = 4^2 = 16$

28. (a) $\log_x 1000 = 3 \quad \Leftrightarrow \quad x^3 = 1000 \quad \Leftrightarrow \quad x = 10$

    (b) $\log_x 25 = 2 \quad \Leftrightarrow \quad x^2 = 25 \quad \Leftrightarrow \quad x = 5$

30. (a) $\log_x 6 = \frac{1}{2} \quad \Leftrightarrow \quad x^{1/2} = 6 \quad \Leftrightarrow \quad x = 36$

    (b) $\log_x 3 = \frac{1}{3} \quad \Leftrightarrow \quad x^{1/3} = 3 \quad \Leftrightarrow \quad x = 27$

32. (a) $\log 50 \approx 1.6990$ (b) $\log \sqrt{2} \approx 0.1505$

    (c) $\log \left(3\sqrt{2}\right) \approx 0.6276$

34. (a) $\ln 27 \approx 3.2958$ (b) $\ln 7.39 \approx 2.0001$

      (c) $\ln 54.6 \approx 4.0000$

36. Since the point $\left(\frac{1}{2}, -1\right)$ is on the graph, we have $-1 = \log_a\left(\frac{1}{2}\right)$ $\Leftrightarrow$ $a^{-1} = \frac{1}{2}$ $\Leftrightarrow$ $a = 2$.
     Thus the function is $y = \log_2 x$.

38. Since the point $(9, 2)$ is on the graph, we have $2 = \log_a 9$ $\Leftrightarrow$ $a^2 = 9$ $\Leftrightarrow$ $a = 3$. Thus the
     function is $y = \log_3 x$.

40. V   42. IV   44. I

46. The graph of $y = \log_3 x$ is obtained from the graph of $y = 3^x$
     by reflecting it about the line $y = x$.

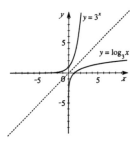

48. $f(x) = -\log_{10} x$
     The graph of f is obtained from the graph of $y = \log_{10} x$ by
     reflecting it about the x-axis.
     Domain: $(0, \infty)$
     Range: $(-\infty, \infty)$
     Vertical asymptote: $x = 0$

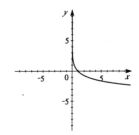

50. $g(x) = \ln(x + 2)$
     The graph of g is obtained from the graph of $y = \ln x$ by
     shifting it to the left 2 units.
     Domain: $(-2, \infty)$
     Range: $(-\infty, \infty)$
     Vertical asymptote: $x = -2$

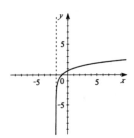

52. $y = \log_3(x - 1) - 2$
     The graph of $y = \log_3(x - 1) - 2$ is obtained from the graph
     of $y = \log_3 x$ by shifting it to the right 1 unit and then
     downward 2 units.
     Domain: $(1, \infty)$
     Range: $(-\infty, \infty)$
     Vertical asymptote: $x = 1$

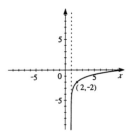

54. $y = 1 + \ln(-x)$
The graph of $y = 1 + \ln(-x)$ is obtained from the graph of
$y = \ln x$ by reflecting it about the y-axis and then shifting it
upward 1 unit.
Domain: $(-\infty, 0)$
Range: $(-\infty, \infty)$
Vertical asymptote: $x = 0$

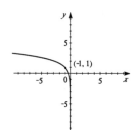

56. $y = \ln |x|$
Note that $y = \begin{cases} \ln x & x > 0 \\ \ln(-x) & x < 0 \end{cases}$.
The graph of $y = \ln |x|$ is obtained by reflecting the graph of
$y = \ln x$ about the x-axis.
Domain: $(-\infty, 0) \cup (0, \infty)$
Range: $(-\infty, \infty)$
Vertical asymptote: $x = 0$

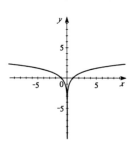

58. $f(x) = \log_5(8 - 2x)$. Then we must have $8 - 2x > 0 \quad \Leftrightarrow \quad 8 > 2x \quad \Leftrightarrow \quad 4 > x$, and so the
domain is $(-\infty, 4)$.

60. $g(x) = \ln(x - x^2)$. Then we must have $x - x^2 > 0 \quad \Leftrightarrow \quad x(1 - x) > 0$. Using the methods from
Chapter 1 with the endpoints 0 and 1, we have

Interval	$(-\infty, 0)$	$(0, 1)$	$(1, \infty)$
Sign of x	$-$	$+$	$+$
Sign of $1 - x$	$+$	$+$	$-$
Sign of $x(1 - x)$	$-$	$+$	$-$

Thus the domain is $(0, 1)$.

62. $h(x) = \sqrt{x - 2} - \log_5(10 - x)$. Then we must have $x - 2 \geq 0$ and $10 - x > 0 \quad \Leftrightarrow \quad x \geq 2$ and
$10 > x \quad \Leftrightarrow \quad 2 \leq x < 10$. So the domain is $[2, 10)$.

64. $y = \ln(x^2 - x) = \ln(x(x - 1))$
Domain: $(-\infty, 0) \cup (1, \infty)$
Vertical asymptote: $x = 0$ and $x = 1$.
No local extrema.

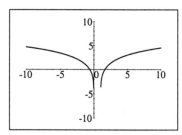

66. $y = x(\ln x)^2$
Domain: $(0, \infty)$
Vertical asymptote: none.
Local minimum $y = 0$ at $x = 1$.
Local maximum $y \approx 0.54$ at $x \approx 0.14$.

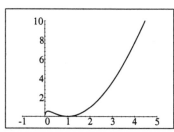

68. $y = x \log_{10}(x + 10)$
Domain: $(-10, \infty)$
Vertical asymptote: $x = -10$.
Local minimum $y \approx -3.62$ at $x \approx -5.87$.

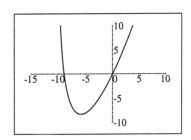

70. (a)

$g(x) = \sqrt{x}$

$f(x) = 1 + \ln(1 + x)$

(b) From the graph, we see that the solution
to the equation $\sqrt{x} = 1 + \ln(1 + x)$ is
$x \approx 13.50$.

72. (a)

$c = 4$

$c = 3$

$c = 2$

$c = 1$

(b) As c increases, the graph of
$f(x) = c \log(x)$ stretches vertically by a
factor of c.

74. (a) $f(x) = \ln(\ln(\ln x))$. We must have $\ln(\ln x) > 0 \quad \Leftrightarrow \quad \ln x > 1 \quad \Leftrightarrow \quad x > e$. So the
domain of f is (e, ∞).

(b) $y = \ln(\ln(\ln x)) \quad \Leftrightarrow \quad e^y = \ln(\ln x) \quad \Leftrightarrow \quad e^{e^y} = \ln x \quad \Leftrightarrow \quad e^{e^{e^y}} = x$. Thus the inverse
function is $f^{-1}(x) = e^{e^{e^x}}$.

76. Using $I = 0.7I_0$ we have $C = -2500 \ln\left(\dfrac{I}{I_0}\right) = -2500 \ln 0.7 = 891.69$ moles/liter.

78. Substituting $N = 1,000,000$ we get $t = 3 \dfrac{\log(N/50)}{\log 2} = 3 \dfrac{\log 20,000}{\log 2} \approx 42.86$ hours.

80. Using $k = 0.25$ and substituting $C = 0.9C_0$ we have $t = -0.25 \ln\left(1 - \dfrac{C}{C_0}\right) = -0.25 \ln(1 - 0.9)$
$= -0.25 \ln 0.1 \approx 0.58$ hours.

82. (a) Since 2 feet $= 24$ inches, the height of the graph is $2^{24} = 1677216$ inches. Now, since there
are 12 inches per foot and 5280 feet per mile, there are $12(5280) = 63360$ inches per mile. So
the height of the graph is $\frac{1677216}{63360} \approx 264.8$, or about 265 miles.

(b) Since $\log_2(2^{24}) = 24$, we must be about 2^{24} inches ≈ 265 miles to the right of the origin before
the height of the graph of $y = \log_2 x$ reaches 24 inches or 2 feet.

84. Notice that $\log_a x$ is increasing for $a > 1$. So we have $\log_4 17 > \log_4 16 = \log_4 4^2 = 2$. Also, we have $\log_5 24 < \log_5 25 = \log_5 5^2 = 2$. Thus, $\log_5 24 < 2 < \log_4 17$.

Exercises 5.3

2. $\log_2 112 - \log_2 7 = \log_2 \frac{112}{7} = \log_2 16 = \log_2 2^4 = 4$

4. $\log\sqrt{0.1} = \log 0.1^{1/2} = \frac{1}{2}\log 0.1 = \frac{1}{2}\log 10^{-1} = -\frac{1}{2}$

6. $\log_{12} 9 + \log_{12} 16 = \log_{12}(9 \cdot 16) = \log_{12} 144 = \log_{12} 12^2 = 2$

8. $\log_3 100 - \log_3 18 - \log_3 50 = \log_3\left(\frac{100}{18 \cdot 50}\right) = \log_3\left(\frac{1}{9}\right) = \log_3 3^{-2} = -2$

10. $\log_2 8^{33} = \log_2(2^3)^{33} = \log_2 2^{99} = 99$

12. $\ln\left(\ln e^{e^{200}}\right) = \ln(e^{200}\ln e) = \ln e^{200} = 200\ln e = 200$

14. $\log_3(5y) = \log_3 5 + \log_3 y$

16. $\log_5\left(\frac{x}{2}\right) = \log_5 x - \log_5 2$

18. $\ln(\sqrt{z}) = \ln\left(z^{1/2}\right) = \frac{1}{2}\ln z$

20. $\log_6 \sqrt[4]{17} = \frac{1}{4}\log_6 17$

22. $\log_2(xy)^{10} = 10\log_2(xy) = 10(\log_2 x + \log_2 y)$

24. $\log_a\left(\frac{x^2}{yz^3}\right) = \log_a x^2 - \log_a\left(yz^3\right) = 2\log_a x - (\log_a y + 3\log_a z)$

26. $\ln\sqrt[3]{3r^2 s} = \frac{1}{3}\ln(3r^2 s) = \frac{1}{3}[\ln 3 + \ln r^2 + \ln s] = \frac{1}{3}(\ln 3 + 2\ln r + \ln s)$

28. $\log\frac{a^2}{b^4\sqrt{c}} = \log a^2 - \log(b^4\sqrt{c}) = 2\log a - \left(4\log b + \frac{1}{2}\log c\right)$

30. $\log_5\sqrt{\frac{x-1}{x+1}} = \frac{1}{2}\log_5\left(\frac{x-1}{x+1}\right) = \frac{1}{2}\left[\log_5(x-1) - \log_5(x+1)\right]$

32. $\ln\frac{3x^2}{(x+1)^{10}} = \ln(3x^2) - \ln(x+1)^{10} = \ln 3 + 2\ln x - 10\ln(x+1)$

34. $\log\frac{x}{\sqrt[3]{1-x}} = \log x - \log\sqrt[3]{1-x} = \log x - \frac{1}{3}\log(1-x)$

36. $\log\sqrt{x\sqrt{y\sqrt{z}}} = \frac{1}{2}\log\left(x\sqrt{y\sqrt{z}}\right) = \frac{1}{2}\left(\log x + \log\sqrt{y\sqrt{z}}\right) = \frac{1}{2}\left[\log x + \frac{1}{2}\log\left(y\sqrt{z}\right)\right]$
$= \frac{1}{2}\left[\log x + \frac{1}{2}\left(\log y + \frac{1}{2}\log z\right)\right] = \frac{1}{2}\log x + \frac{1}{4}\log y + \frac{1}{8}\log z$

38. $\log \dfrac{10^x}{x(x^2+1)(x^4+2)} = \log 10^x - \log\left[x(x^2+1)(x^4+2)\right]$
    $= x - \left[\log x + \log(x^2+1) + \log(x^4+2)\right]$

40. $\log 12 + \frac{1}{2}\log 7 - \log 2 = \log\left(12\sqrt{7}\right) - \log 2 = \log\frac{12\sqrt{7}}{2} = \log\left(6\sqrt{7}\right)$

42. $\log_5(x^2-1) - \log_5(x-1) = \log_5\dfrac{x^2-1}{x-1} = \log_5\dfrac{(x-1)(x+1)}{x-1} = \log_5(x+1)$

44. $\ln(a+b) + \ln(a-b) - 2\ln c = \ln[(a+b)(a-b)] - \ln(c^2) = \ln\dfrac{a^2-b^2}{c^2}$

46. $2\left[\log_5 x + 2\log_5 y - 3\log_5 z\right] = 2\log_5\dfrac{xy^2}{z^3} = \log_5\left(\dfrac{xy^2}{z^3}\right)^2 = \log_5\dfrac{x^2y^4}{z^6}$

48. $\log_a b + c\log_a d - r\log_a s = \log_a(bd^c) - \log_a s^r = \log_a\dfrac{bd^c}{s^r}$

50. $\log_5 2 = \dfrac{\log 2}{\log 5} \approx 0.430677$

52. $\log_6 92 = \dfrac{\log 92}{\log 6} \approx 2.523658$

54. $\log_6 532 = \dfrac{\log 532}{\log 6} \approx 3.503061$

56. $\log_{12} 2.5 = \dfrac{\log 2.5}{\log 12} \approx 0.368743$

58.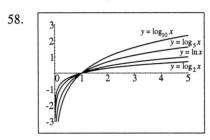

Note that $\log_c x = \left(\frac{1}{\ln c}\right)\ln x$ (by the change of base formula). So the graph of $y = \log_c x$ is obtained from the graph of $y = \ln x$ by either shrinking or stretching vertically by a factor of $\dfrac{1}{\ln c}$ depending on whether $\ln c > 1$ or $\ln c < 1$ (respectively).

60. $(\log_2 5)(\log_5 7) = \dfrac{\log 5}{\log 2}\cdot\dfrac{\log 7}{\log 5} = \dfrac{\log 7}{\log 2} = \log_2 7$

62. From Example 5, Part a, $P = \dfrac{P_0}{(t+1)^c}$. Substituting $P_0 = 80$, $t = 24$, and $c = 0.3$ we have
    $P = \dfrac{80}{(24+1)^{0.3}} \approx 30.5$. So the student should get a score of 30.

64. (a) $\log S = \log c + k\log A \quad\Leftrightarrow\quad \log S = \log c + \log A^k \quad\Leftrightarrow\quad \log S = \log(cA^k) \quad\Leftrightarrow$
    $S = cA^k$.

    (b) If $A = 2A_0$ when $k = 3$ we get $S = c(2A_0)^3 = c\cdot 2^3\cdot A_0^3 = 8\cdot cA_0^3$. Thus doubling the area increases the species eightfold.

66. (a) False; $\log\left(\dfrac{x}{y}\right) = \log x - \log y \neq \dfrac{\log x}{\log y}$

(b) False; $\log_2 x - \log_2 y = \log_2\left(\dfrac{x}{y}\right) \neq \log_2(x - y)$

(c) True; the equation is an identity: $\log_5 \dfrac{a}{b^2} = \log_5 a - \log_5 b^2 = \log_5 a - 2\log_5 b$.

(d) True; the equation is an identity: $\log 2^z = z \log 2$.

(e) False; $\log P + \log Q = \log(PQ) \neq (\log P)(\log Q)$.

(f) False; $\log a - \log b = \log\left(\dfrac{a}{b}\right) \neq \dfrac{\log a}{\log b}$.

(g) False; $x \log_2 7 = \log_2 7^x \neq (\log_2 7)^x$.

(h) True; the equation is an identity. $\log_a a^a = a \log_a a = a \cdot 1 = a$.

(i) False; $\log(x - y) \neq \dfrac{\log x}{\log y}$. For example, $0 = \log(3 - 2) \neq \dfrac{\log 3}{\log 2}$.

(j) True; the equation is an identity: $-\ln\left(\frac{1}{A}\right) = -\ln A^{-1} = -1(-\ln A) = \ln A$.

68. Let $f(x) = x^2$. Then $f(2x) = (2x)^2 = 4x^2 = 4f(x)$. Now the graph of $f(2x)$ is the same as the graph of f shrunk horizontally by a factor of $\frac{1}{2}$, whereas the graph of $4f(x)$ is the same as the graph of $f(x)$ stretched vertically by a factor of 4.
Let $g(x) = e^x$. Then $g(x + 2) = e^{x+2} = e^2 e^x = e^2 g(x)$. This shows that a horizontal shift of 2 units to the right is the same as a vertical stretch by a factor of e^2.
Let $h(x) = \ln x$. Then $h(2x) = \ln 2x = \ln 2 + \ln x = \ln 2 + h(x)$. This shows that a horizontal shrinking by a factor of $\frac{1}{2}$ is the same as a vertical shift by an amount $\ln 2$ upward.

Exercises 5.4

2. $10^{-x} = 2 \quad \Leftrightarrow \quad \log 10^{-x} = \log 2 \quad \Leftrightarrow \quad -x = \log 2 \quad \Leftrightarrow \quad x = -\log 2 \approx -0.3010$

4. $e^{-2x} = \frac{1}{10} \quad \Leftrightarrow \quad \ln e^{-2x} = \ln\left(\frac{1}{10}\right) \quad \Leftrightarrow \quad -2x = \ln\left(\frac{1}{10}\right) \quad \Leftrightarrow \quad x = \frac{\ln(1/10)}{-2} = 1.1513$

6. $3^{2x-1} = 5 \quad \Leftrightarrow \quad \log 3^{2x-1} = \log 5 \quad \Leftrightarrow \quad (2x-1)\log 3 = \log 5 \quad \Leftrightarrow \quad 2x - 1 = \frac{\log 5}{\log 3} \quad \Leftrightarrow$
    $2x = 1 + \frac{\log 5}{\log 3} \quad \Leftrightarrow \quad x = \frac{1}{2}\left(1 + \frac{\log 5}{\log 3}\right) \approx 1.2325$

8. $2\,e^{12x} = 17 \quad \Leftrightarrow \quad e^{12x} = \frac{17}{2} \quad \Leftrightarrow \quad 12x = \ln\left(\frac{17}{2}\right) \quad \Leftrightarrow \quad x = \frac{1}{12}\left[\ln\left(\frac{17}{2}\right)\right] \approx 0.1783$

10. $4(1 + 10^{5x}) = 9 \quad \Leftrightarrow \quad 1 + 10^{5x} = \frac{9}{4} \quad \Leftrightarrow \quad 10^{5x} = \frac{5}{4} \quad \Leftrightarrow \quad 5x = \log\left(\frac{5}{4}\right) \quad \Leftrightarrow$
    $x = \frac{1}{5}[\log 5 - \log 4] \approx 0.0194.$

12. $2^{3x} = 34 \quad \Leftrightarrow \quad \log 2^{3x} = \log 34 \quad \Leftrightarrow \quad 3x \log 2 = \log 34 \quad \Leftrightarrow \quad x = \frac{\log 34}{3\log 2} \approx 1.6958$

14. $3^{x/14} = 0.1 \quad \Leftrightarrow \quad \log 3^{x/14} = \log 0.1 \quad \Leftrightarrow \quad \left(\frac{x}{14}\right)\log 3 = \log 0.1 \quad \Leftrightarrow$
    $x = \frac{14 \log 0.1}{\log 3} \approx -29.3426$

16. $e^{3-5x} = 16 \quad \Leftrightarrow \quad 3 - 5x = \ln 16 \quad \Leftrightarrow \quad -5x = \ln 16 - 3 \quad \Leftrightarrow \quad x = -\frac{1}{5}(\ln 16 - 3) \approx 0.0455$

18. $\left(\frac{1}{4}\right)^x = 75 \quad \Leftrightarrow \quad 4^{-x} = 75 \quad \Leftrightarrow \quad \log 4^{-x} = \log 75 \quad \Leftrightarrow \quad (-x)(\log 4) = \log 75 \quad \Leftrightarrow$
    $-x = \frac{\log 75}{\log 4} \quad \Leftrightarrow \quad x = -\frac{\log 75}{\log 4} \approx -3.1144$

20. $10^{1-x} = 6^x \quad \Leftrightarrow \quad \log 10^{1-x} = \log 6^x \quad \Leftrightarrow \quad 1 - x = x(\log 6) \quad \Leftrightarrow \quad 1 = x(\log 6) + x \quad \Leftrightarrow$
    $1 = x(\log 6 + 1) \quad \Leftrightarrow \quad x = \frac{1}{\log 6 + 1} \approx 0.5624$

22. $7^{x/2} = 5^{1-x} \quad \Leftrightarrow \quad \log 7^{x/2} = \log 5^{1-x} \quad \Leftrightarrow \quad \left(\frac{x}{2}\right)\log 7 = (1 - x)\log 5 \quad \Leftrightarrow$

    $\left(\frac{x}{2}\right)\log 7 = \log 5 - x \log 5 \quad \Leftrightarrow \quad \left(\frac{x}{2}\right)\log 7 + x\log 5 = \log 5 \quad \Leftrightarrow \quad x(\frac{1}{2}\log 7 + \log 5) = \log 5$
    $\Leftrightarrow \quad x = \frac{\log 5}{\frac{1}{2}\log 7 + \log 5} \approx 0.6232$

24. $\dfrac{10}{1 + e^{-x}} = 2 \quad \Leftrightarrow \quad 10 = 2 + 2e^{-x} \quad \Leftrightarrow \quad 8 = 2e^{-x} \quad \Leftrightarrow \quad 4 = e^{-x} \quad \Leftrightarrow \quad \ln 4 = -x \quad \Leftrightarrow$
    $x = -\ln 4 \approx -1.3863$

26. $(1.00625)^{12t} = 2 \quad \Leftrightarrow \quad \log 1.00625^{12t} = \log 2 \quad \Leftrightarrow \quad 12t \log 1.00625 = \log 2 \quad \Leftrightarrow$
    $t = \frac{\log 2}{12 \log 1.00625} \approx 9.2708$

28. $x^2 10^x - x 10^x = 2(10^x) \quad \Leftrightarrow \quad x^2 10^x - x 10^x - 2(10^x) = 0 \quad \Leftrightarrow \quad 10^x(x^2 - x - 2) = 0 \quad \Rightarrow$
    $10^x = 0 \text{ (never) or } x^2 - x - 2 = 0. \text{ If } x^2 - x - 2 = 0, \text{ then } (x - 2)(x + 1) = 0 \quad \Rightarrow \quad x = 2, -1.$
    So the only solutions are $x = 2, -1.$

30. $x^2 e^x + x e^x - e^x = 0 \quad \Leftrightarrow \quad e^x(x^2 + x - 1) = 0 \quad \Rightarrow \quad e^x = 0 \text{ (impossible) or } x^2 + x - 1 = 0.$
    If $x^2 + x - 1 = 0$, then $x = \frac{-1 \pm \sqrt{5}}{2}$. So the solutions are $x = \frac{-1 \pm \sqrt{5}}{2}$.

32. $e^{2x} - e^x - 6 = 0$ $\Leftrightarrow$ $(e^x - 3)(e^x + 2) = 0$ $\Rightarrow$ $e^x + 2 = 0$ (impossible) or $e^x - 3 = 0$. If
$e^x - 3 = 0$, then $e^x = 3$ $\Leftrightarrow$ $x = \ln 3 \approx 1.0986$. So the only solution is $x \approx 1.0986$.

34. $e^x - 12e^{-x} - 1 = 0$ $\Leftrightarrow$ $e^x - 1 - 12e^{-x} = 0$ $\Leftrightarrow$ $e^x(e^x - 1 - 12e^{-x}) = 0 \cdot e^x$ $\Leftrightarrow$
$e^{2x} - e^x - 12 = 0$ $\Leftrightarrow$ $(e^x - 4)(e^x + 3) = 0$ $\Rightarrow$ $e^x + 3 = 0$ (impossible) or $e^x - 4 = 0$.
If $e^x - 4 = 0$, then $e^x = 4$ $\Leftrightarrow$ $x = \ln 4 \approx 1.3863$. So the only solution is $x \approx 1.3863$.

36. $\ln(2 + x) = 1$ $\Leftrightarrow$ $2 + x = e^1$ $\Leftrightarrow$ $x = e - 2 \approx 0.7183$

38. $\log(x - 4) = 3$ $\Leftrightarrow$ $x - 4 = 10^3 = 1000$ $\Leftrightarrow$ $x = 1004$

40. $\log_3(2 - x) = 3$ $\Leftrightarrow$ $2 - x = 3^3 = 27$ $\Leftrightarrow$ $-x = 25$ $\Leftrightarrow$ $x = -25$

42. $\log_2(x^2 - x - 2) = 2$ $\Leftrightarrow$ $x^2 - x - 2 = 2^2 = 4$ $\Leftrightarrow$ $x^2 - x - 6 = 0$ $\Leftrightarrow$
$(x - 3)(x + 2) = 0$ $\Leftrightarrow$ $x = 3$ or $x = -2$. Thus the solutions are $x = 3$ and $x = -2$.

44. $2 \log x = \log 2 + \log(3x - 4)$ $\Leftrightarrow$ $\log(x^2) = \log(6x - 8)$ $\Leftrightarrow$ $x^2 = 6x - 8$ $\Leftrightarrow$
$x^2 - 6x + 8 = 0$ $\Leftrightarrow$ $(x - 4)(x - 2) = 0$ $\Leftrightarrow$ $x = 4$ or $x = 2$. Thus the solutions are $x = 4$
and $x = 2$.

46. $\log_5 x + \log_5(x + 1) = \log_5 20$ $\Leftrightarrow$ $\log_5(x^2 + x) = \log_5 20$ $\Leftrightarrow$ $x^2 + x = 20$ $\Leftrightarrow$
$x^2 + x - 20 = 0$ $\Leftrightarrow$ $(x + 5)(x - 4) = 0$ $\Leftrightarrow$ $x = -5$ or $x = 4$. Since $\log_5(-5)$ is
undefined, the only solution is $x = 4$.

48. $\log x + \log(x - 3) = 1$ $\Leftrightarrow$ $\log[x(x - 3)] = 1$ $\Leftrightarrow$ $x^2 - 3x = 10$ $\Leftrightarrow$ $x^2 - 3x - 10 = 0$
$\Leftrightarrow$ $(x + 2)(x - 5) = 0$ $\Leftrightarrow$ $x = -2$ or $x = 5$. Since $\log(-2)$ is undefined, the only solution
is $x = 5$.

50. $\ln(x - 1) + \ln(x + 2) = 1$ $\Leftrightarrow$ $\ln[(x - 1)(x + 2)] = 1$ $\Leftrightarrow$ $x^2 + x - 2 = e$ $\Leftrightarrow$
$x^2 + x - (2 + e) = 0$ $\Rightarrow$ $x = \frac{-1 \pm \sqrt{1 + 4(2 + e)}}{2} = \frac{-1 \pm \sqrt{9 + 4e}}{2}$. Since $x - 1 < 0$ when
$x = \frac{-1 - \sqrt{9 + 4e}}{2}$, the only solution is $x = \frac{-1 + \sqrt{9 + 4e}}{2} \approx 1.7290$

52. $(\log x)^3 = 3 \log x$ $\Leftrightarrow$ $(\log x)^3 - 3 \log x = 0$ $\Leftrightarrow$ $(\log x)((\log x)^2 - 3)$ $\Leftrightarrow$ $(\log x) = 0$
or $(\log x)^2 - 3 = 0$. Now $\log x = 0$ $\Leftrightarrow$ $x = 1$. Also $(\log x)^2 - 3 = 0$ $\Leftrightarrow$ $(\log x)^2 = 3$
$\Leftrightarrow$ $\log x = \pm \sqrt{3}$ $\Leftrightarrow$ $x = 10^{\pm\sqrt{3}}$, so $x = 10^{\sqrt{3}} \approx 53.9574$ or $x = 10^{-\sqrt{3}} \approx 0.0185$. Thus
the solutions to the equation are $x = 1$, $x = 10^{\sqrt{3}} \approx 53.9574$ and $x = 10^{-\sqrt{3}} \approx 0.0185$.

54. $\log_2(\log_3 x) = 4$ $\Leftrightarrow$ $\log_3 x = 2^4 = 16$ $\Leftrightarrow$ $x = 3^{16} = 43{,}046{,}721$

56. $\log x = x^2 - 2$ $\Leftrightarrow$ $\log x - x^2 + 2 = 0$.
Let $f(x) = \log x - x^2 + 2$. We need to solve the
equation $f(x) = 0$. From the graph of f, we get $x \approx 0.01$
or $x \approx 1.47$.

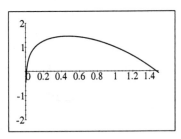

58. $x = \ln(4 - x^2)$ $\Leftrightarrow$ $x - \ln(4 - x^2) = 0$.
Let $f(x) = x - \ln(4 - x^2)$. We need to solve the
equation $f(x) = 0$. From the graph of f, we get x
$x \approx -1.96$ or $x \approx 1.06$.

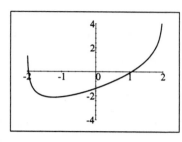

60. $2^{-x} = x - 1$ $\Leftrightarrow$ $2^{-x} - x + 1 = 0$.
Let $f(x) = 2^{-x} - x + 1$. We need to solve the
equation $f(x) = 0$. From the graph of f, we get
$x \approx 1.38$.

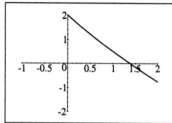

62. $e^{x^2} - 2 = x^3 - x$ $\Leftrightarrow$ $e^{x^2} - 2 - x^3 + x = 0$.
Let $f(x) = e^{x^2} - 2 - x^3 + x$. We need to solve the
equation $f(x) = 0$. From the graph of f, we get
$x \approx -0.89$ or $x \approx 0.71$.

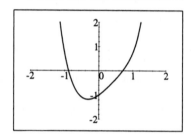

64. $3 \le \log_2 x \le 4$ $\Leftrightarrow$ $2^3 \le x \le 2^4$ $\Leftrightarrow$ $8 \le x \le 16$.

66. $x^2 e^x - 2 e^x < 0$ $\Leftrightarrow$ $e^x(x^2 - 2) < 0$ $\Leftrightarrow$ $e^x(x - \sqrt{2})(x + \sqrt{2}) < 0$. We use the methods
of Chapter 1 with the endpoints $-\sqrt{2}$ and $\sqrt{2}$ and note that $e^x > 0$ for all x. We have

Interval	$(-\infty, -\sqrt{2})$	$(-\sqrt{2}, \sqrt{2})$	$(\sqrt{2}, \infty)$
Sign of e^x	$+$	$+$	$+$
Sign of $(x - \sqrt{2})$	$-$	$-$	$+$
Sign of $(x + \sqrt{2})$	$-$	$+$	$+$
Sign of $e^x(x - \sqrt{2})(x + \sqrt{2})$	$+$	$-$	$+$

Thus $-\sqrt{2} < x < \sqrt{2}$.

68. (a) $A(2) = 6500\, e^{0.06(2)} \approx \$7328.73$

(b) $8000 = 6500\, e^{0.06t}$ $\Leftrightarrow$ $\frac{16}{13} = e^{0.06t}$ $\Leftrightarrow$ $\ln\left(\frac{16}{13}\right) = 0.06t$ $\Leftrightarrow$ $t = \frac{1}{0.06}\ln\left(\frac{16}{13}\right) \approx 3.46$.
So the investment doubles in about $3\frac{1}{2}$ years.

70. $5000 = 4000\left(1 + \frac{0.0975}{2}\right)^{2t}$ $\Leftrightarrow$ $1.25 = (1.04875)^{2t}$ $\Leftrightarrow$

$\log 1.25 = 2t \log 1.04875$ $\Leftrightarrow$ $t = \frac{\log 1.25}{2 \log 1.04875} \approx 2.344$. So it takes about $2\frac{1}{3}$ years to save \$5000.

72. $1435.77 = 1000\left(1 + \frac{r}{2}\right)^{2(4)}$ $\Leftrightarrow$ $1.43577 = \left(1 + \frac{r}{2}\right)^{8}$ $\Leftrightarrow$ $1 + \frac{r}{2} = \sqrt[8]{1.43577}$ $\Leftrightarrow$
$\frac{r}{2} = \sqrt[8]{1.43577} - 1$ $\Leftrightarrow$ $r = 2\left(\sqrt[8]{1.43577} - 1\right) \approx 0.0925$. Thus the rate was about 9.25%.

74. $r_{APY} = e^{r} - 1$. Here $r = 0.05.5$ so $r_{APY} = e^{0.055} - 1 \approx 1.0565 - 1 = 0.565$. So the annual percentage yield is about 5.65%.

76. We want to solve for t in the equation $80(e^{-0.2t} - 1) = -70$ (when motion is downwards, the velocity is negative). Then $80(e^{-0.2t} - 1) = -70$ $\Leftrightarrow$ $e^{-0.2t} - 1 = -\frac{7}{8}$ $\Leftrightarrow$ $e^{-0.2t} = \frac{1}{8}$ $\Leftrightarrow$ $-0.2t = \ln\left(\frac{1}{8}\right)$ $\Leftrightarrow$ $t = \frac{\ln\left(\frac{1}{8}\right)}{-0.2} \approx 10.4$ seconds. Thus the velocity is 70 ft/sec after about 10 seconds.

78. (a) $I = 10\,e^{-0.008(30)} = 10\,e^{-0.24} = 7.87$. So at 30 ft the intensity is 7.87 lumens.

    (b) $5 = 10\,e^{-0.008\,x}$ $\Leftrightarrow$ $e^{-0.008x} = \frac{1}{2}$ $\Leftrightarrow$ $-0.008x = \ln\left(\frac{1}{2}\right)$ $\Leftrightarrow$ $x = \frac{\ln(1/2)}{-0.008} \approx 86.6$. So the intensity will drop to 25 lumens at 86.6 ft.

80. (a) $\ln\left(\dfrac{T - 20}{200}\right) = -0.11t$ $\Leftrightarrow$ $\dfrac{T - 20}{200} = e^{-0.11t}$ $\Leftrightarrow$ $T - 20 = 200e^{-0.11t}$ $\Leftrightarrow$ $T = 20 + 200e^{-0.11t}$.

    (b) When $t = 20$ we have $T = 20 + 200e^{-0.11(20)} = 20 + 200e^{-2.2} \approx 42.2°F$.

82. (a) $P = M - Ce^{-kt}$ $\Leftrightarrow$ $Ce^{-kt} = M - P$ $\Leftrightarrow$ $e^{-kt} = \dfrac{M - P}{C}$ $\Leftrightarrow$ $-kt = \ln\left(\dfrac{M - P}{C}\right)$ $\Leftrightarrow$ $t = -\dfrac{1}{k}\ln\left(\dfrac{M - P}{C}\right)$.

    (b) $P(t) = 20 - 14e^{-0.024t}$. Substituting $M = 20$, $C = 14$, $k = 0.024$, and $P = 12$ into $t = -\frac{1}{k}\ln\left(\frac{M-P}{C}\right)$, we have $t = -\frac{1}{0.024}\ln\left(\frac{20-12}{14}\right) \approx 23.32$. So it takes about 23 months.

    (c)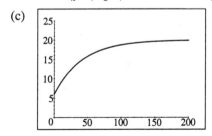

84. Notice that $\log\left(x^{1/\log x}\right) = \dfrac{1}{\log x}\log x = 1$, so $x^{1/\log x} = 10^{1}$ for all $x > 0$. So $x^{1/\log x} = 5$ has no solution, and $x^{1/\log x} = k$ has a solution only when $k = 10$. This is verified by the graph of $f(x) = x^{1/\log x}$.

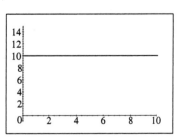

Exercises 5.5

2. (a) The relative growth rate is $0.012 = 1.2\%$.

    (b) $n(5) = 12\,e^{0.012(5)} = 12\,e^{0.06} \approx 12.74$
         million fish.

    (c) $30 = 12\,e^{0.012t} \iff 2.5 = e^{0.012t} \iff$
         $0.012t = \ln 2.5 \iff t = \frac{\ln 2.5}{0.012} \approx 76.36$.
         Thus the fish population reaches 30 million
         after about 76 years.

    (d)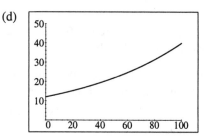

4. $n(t) = n_0\,e^{rt}$; $n_0 = 110$ million, $t = 2020 - 1995 = 25$

    (a) $r = 0.03$; $n(25) = 110{,}000{,}000\,e^{0.03(25)} = 110{,}000{,}000\,e^{0.75} \approx 232{,}870{,}000$. Thus at a 3%
         growth rate, the projected population will be approximately 233 million people by the year
         2020.

    (b) $r = 0.02$; $n(25) = 110{,}000{,}000\,e^{0.02(25)} = 110{,}000{,}000\,e^{0.50} \approx 181{,}359{,}340$. Thus at a 2%
         growth rate, the projected population will be approximately 181 million people by the year
         2020.

6. (a) $n(t) = n_0 e^{rt}$ with $n_0 = 85$ and $r = 0.18$. Thus $n(t) = 85\,e^{0.18t}$.

    (b) $n(3) = 85\,e^{0.18(3)} \approx 146$ frogs.

    (c) $600 = 85\,e^{0.18t} \iff \frac{120}{17} = e^{0.18t} \iff 0.18t = \ln\left(\frac{120}{17}\right) \iff t = \frac{1}{0.18}\ln\left(\frac{120}{17}\right) \approx 10.86$.
         So the population will reach 600 frogs in about 11 years.

8. (a) Since the population grows exponentially, the population is represented by $n(t) = n_0\,e^{rt}$, with
         $n_0 = 1500$ and $n(30) = 3000$. Solving for r, we have $3000 = 1500\,e^{30r} \iff 2 = e^{30r}$
         $\iff 30r = \ln 2 \iff r = \frac{\ln 2}{30} \approx 0.023$. Thus $n(t) = 1500\,e^{0.023t}$.

    (b) Since 2 hours is 120 minutes, the number of bacteria in 2 hours is
         $n(120) = 1500\,e^{0.023(120)} \approx 24{,}000$.

    (c) We need to solve $4000 = 1500\,e^{0.023t}$ for t. So $4000 = 1500\,e^{0.023t} \iff \frac{8}{3} = e^{0.023t} \iff$
         $0.023t = \ln\frac{8}{3} \iff t = \frac{\ln(8/3)}{0.023} \approx 42.6$. Thus the bacteria population will reach 4000 in
         about 43 minutes.

10. (a) Using $n(t) = n_0\,e^{rt}$ with $n(2) = 400$ and $n(6) = 25{,}600$, we have $n_0\,e^{2r} = 400$ and
         $n_0\,e^{6r} = 25{,}600$. Dividing the second equation by the first gives $\dfrac{n_0\,e^{6r}}{n_0\,e^{2r}} = \dfrac{25{,}600}{400} = 64 \iff$
         $e^{4r} = 64 \iff 4r = \ln 64 \iff r = \frac{1}{4}\ln 64 \approx 1.04$. Thus the relative rate of growth is
         about 104%.

    (b) Since $r = \frac{1}{4}\ln 64 = \frac{1}{2}\ln 8$, we have from part (a) $n(t) = n_0\,e^{\left(\frac{1}{2}\ln 8\right)t}$. Since $n(2) = 400$, we
         have $400 = n_0\,e^{\ln 8} \iff n_0 = \frac{400}{e^{\ln 8}} = \frac{400}{8} = 50$. So the initial size of the culture was 50.

    (c) Substituting $n_0 = 50$ and $r = 1.04$, we have $n(t) = n_0\,e^{rt} = 50\,e^{1.04t}$.

    (d) $n(4.5) = 50e^{1.04(4.5)} = 50e^{4.68} \approx 5388.5$, so the size after 4.5 hours is approximately 5400.

(e) $n(t) = 50000 = 50e^{1.04t}$ $\Leftrightarrow$ $e^{1.04t} = 1000$ $\Leftrightarrow$ $1.04t = \ln 1000$ $\Leftrightarrow$

$t = \frac{\ln 1000}{1.04} \approx 6.64$. Hence the population will reach 50,000 after roughly $6\frac{2}{3}$ hours.

12. (a) Calculating dates relative to 1950 gives $n_0 = 10{,}586{,}223$ and $n(30) = 23{,}668{,}562$. Then

$n(30) = 10{,}586{,}223e^{30r} = 23{,}668{,}562$ $\Leftrightarrow$ $e^{30r} = \frac{23668562}{10586223} \approx 2.2358$ $\Leftrightarrow$

$30r = \ln 2.2358$ $\Leftrightarrow$ $r = \frac{1}{30} \ln 2.2358 \approx 0.0268$. Thus $n(t) = 10{,}586{,}223\,e^{0.0268\,t}$.

(b) $2(10{,}586{,}223) = 10{,}586{,}223\,e^{0.0268\,t}$ $\Leftrightarrow$ $2 = e^{0.0268\,t}$ $\Leftrightarrow$ $\ln 2 = 0.0268\,t$ $\Leftrightarrow$

$t = \frac{\ln 2}{0.0268} \approx 25.86$. So the population doubles in about 26 years.

(c) $t = 2000 - 1950 = 50$; $n(50) \approx 10586223\,e^{0.0268(50)} \approx 40{,}429{,}246$ and so the population in the year 2000 will be approximately 40,429,000.

14. From the formula for radioactive decay, we have $m(t) = m_0\,e^{-rt}$, where $r = \frac{\ln 2}{h}$.

(a) We have $m_0 = 22$ and $h = 1600$, so $r = \frac{\ln 2}{1600} \approx 0.000433$ and the amount after t years is given by $m(t) = 22\,e^{-0.000433t}$.

(b) $m(4000) = 22\,e^{-0.000433(4000)} \approx 3.89$, so the amount after 4000 years is about 4 mg.

(c) We have to solve for t in the equation $18 = 22\,e^{-0.000433t}$. This gives $18 = 22\,e^{-0.000433t}$

$\Leftrightarrow$ $\frac{9}{11} = e^{-0.000433t}$ $\Leftrightarrow$ $-0.000433t = \ln\left(\frac{9}{11}\right)$ $\Leftrightarrow$ $t = \frac{\ln\left(\frac{9}{11}\right)}{-0.000433} \approx 463.4$, so it takes about 463 years.

16. (a) $m(60) = 40\,e^{-0.0277(60)} \approx 7.59$, so the mass remaining after 60 days is about 8 g.

(b) $10 = 40\,e^{-0.0277\,t}$ $\Leftrightarrow$ $0.25 = e^{-0.0277\,t}$ $\Leftrightarrow$ $\ln 0.25 = -0.0277t$ $\Leftrightarrow$

$t = -\frac{\ln 0.25}{0.0277} \approx 50.05$, so it takes about 50 days.

(c) We need to solve for t in the equation $20 = 40\,e^{-0.0277\,t}$. We have $20 = 40\,e^{-0.0277\,t}$ $\Leftrightarrow$

$e^{-0.277\,t} = \frac{1}{2}$ $\Leftrightarrow$ $-0.0277t = \ln\frac{1}{2}$ $\Leftrightarrow$ $t = \frac{\ln\frac{1}{2}}{-0.0277} \approx 25.02$. Thus the half-life of thorium-234 is about 25 days.

18. From the formula for radioactive decay, we have $m(t) = m_0\,e^{-rt}$, where $r = \frac{\ln 2}{h}$. Since $h = 30$, we have $r = \frac{\ln 2}{30} \approx 0.0231$ and $m(t) = m_0e^{-0.0231t}$. In this exercise we have to solve for t in the equation $0.05\,m_0 = m_0e^{-0.0231t}$ $\Leftrightarrow$ $e^{-0.0231t} = 0.05$ $\Leftrightarrow$ $-0.0231t = \ln 0.05$ $\Leftrightarrow$

$t = \frac{\ln 0.05}{-0.0231} \approx 129.7$. So it will take about 130 s.

20. From the formula for radioactive decay, we have $m(t) = m_0\,e^{-rt}$, where $r = \frac{\ln 2}{h}$. In other words, $m(t) = m_0\,e^{-\frac{\ln 2}{h}\,t}$.

(a) Using $m(3) = 0.58\,m_0$, we have to solve for h in the equation $0.58\,m_0 = m(3) = m_0\,e^{-\frac{\ln 2}{h}3}$.

Then $0.58\,m_0 = m_0\,e^{-\frac{3\ln 2}{h}}$ $\Leftrightarrow$ $e^{-\frac{3\ln 2}{h}} = 0.58$ $\Leftrightarrow$ $-\frac{3\ln 2}{h} = \ln 0.58$ $\Leftrightarrow$

$h = -\frac{3\ln 2}{\ln 0.58} \approx 3.82$ days. Thus the half-life of Radon-222 is about 3.82 days.

(b) Here we have to solve for t in the equation $0.2\,m_0 = m_0\,e^{-\frac{\ln 2}{3.82}\,t}$. So we have

$0.2\,m_0 = m_0\,e^{-\frac{\ln 2}{3.82}\,t}$ $\Leftrightarrow$ $0.2 = e^{-\frac{\ln 2}{3.82}\,t}$ $\Leftrightarrow$ $-\frac{\ln 2}{3.82}\,t = \ln 0.2$ $\Leftrightarrow$

$t = -\frac{3.82\ln 0.2}{\ln 2} \approx 8.87$. So it takes roughly 9 days for a sample of Radon-222 to decay to 20% of its original mass.

22. From the formula for radioactive decay, we have $m(t) = m_0 \, e^{-rt}$ where $r = \frac{\ln 2}{h}$. Since $h = 5730$, $r = \frac{\ln 2}{5730} \approx 0.000121$ and $m(t) = m_0 \, e^{-0.000121 \, t}$. We need to solve for t in the equation
$$0.59 \, m_0 = m_0 \, e^{-0.000121 \, t} \quad \Leftrightarrow \quad e^{-0.000121 \, t} = 0.59 \quad \Leftrightarrow \quad -0.000121 \, t = \ln 0.59 \quad \Leftrightarrow$$
$t = \frac{\ln 0.59}{-0.000121} \approx 4360.6$. So it will take about 4360 years.

24. (a) We use Newton's Law of Cooling: $T(t) = T_s + D_0 \, e^{-kt}$ with $k = 0.1947$, $T_s = 60$, and $D_0 = 98.6 - 60 = 38.6$. So $T(t) = 60 + 38.6 \, e^{-0.1947t}$.

    (b) Solve $T(t) = 72$. So $72 = 60 + 38.6 \, e^{-0.1947t} \quad \Leftrightarrow \quad 38.6 \, e^{-0.1947 \, t} = 12 \quad \Leftrightarrow$
    $e^{-0.1947 \, t} = \frac{12}{38.6} \quad \Leftrightarrow \quad -0.1947 \, t = \ln\left(\frac{12}{38.6}\right) \quad \Leftrightarrow \quad t = -\frac{1}{0.1947} \ln\left(\frac{12}{38.6}\right) \approx 6.00$, and the
    time of death was about 6 hours ago.

26. We use Newton's Law of Cooling: $T(t) = T_s + D_0 \, e^{-kt}$, with
$T_s = 20$ and $D_0 = 100 - 20 = 80$. So $T(t) = 20 + 80 \, e^{-kt}$.
Since $T(15) = 75$, we have $20 + 80 \, e^{-15k} = 75 \quad \Leftrightarrow$
$80 \, e^{-15k} = 55 \quad \Leftrightarrow \quad e^{-15k} = \frac{11}{16} \quad \Leftrightarrow \quad -15k = \ln\left(\frac{11}{16}\right)$
$\Leftrightarrow \quad k = -\frac{1}{15} \ln\left(\frac{11}{16}\right)$. Thus $T(25) = 20 + 80$
$e^{(25/15)\cdot \ln(11/16)} \approx 62.8$, and so the temperature after another 10
min is 63°C. The function $T(t) = 20 + 80 \, e^{(1/15)\cdot \ln(11/16)t}$ is
shown in the viewing rectangle $[0, 30]$ by $[50, 100]$.

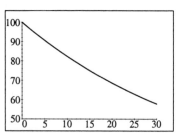

28. $\text{pH} = -\log\left[H^+\right] = -\log(3.1 \times 10^{-8}) \approx 7.5$ and the substance is basic.

30. (a) $\text{pH} = -\log\left[H^+\right] = 4.6 \quad \Leftrightarrow \quad \left[H^+\right] = 10^{-4.6} \text{ M} \approx 2.5 \times 10^{-5} \text{ M}$

    (b) $\text{pH} = -\log\left[H^+\right] = 7.3 \quad \Leftrightarrow \quad \left[H^+\right] = 10^{-7.3} \text{ M} \approx 5.0 \times 10^{-8} \text{ M}$

32. $2.8 \le \text{pH} \le 3.8 \quad \Leftrightarrow \quad -2.8 \ge -\text{pH} \ge -3.8 \quad \Leftrightarrow \quad 10^{-2.8} \ge 10^{-\text{pH}} \ge 10^{-3.8} \quad \Leftrightarrow$
$1.58 \times 10^{-3} \ge \left[H^+\right] \ge 1.58 \times 10^{-4}$. The range of $\left[H^+\right]$ is 1.58×10^{-4} to 1.58×10^{-3}.

34. Let the subscript SF represent the San Francisco earthquake and J the Japan earthquake. Then we
have $M_{SF} = \log\left(\frac{I_{SF}}{S}\right) = 8.3 \quad \Leftrightarrow \quad I_{SF} = S \cdot 10^{8.3}$ and $M_J = \log\left(\frac{I_J}{S}\right) = 4.9 \quad \Leftrightarrow \quad I_J = S \cdot 10^{4.9}$.
So $\frac{I_{SF}}{I_J} = \frac{10^{8.3}}{10^{4.9}} = 10^{3.4} \approx 2511.9$, and so the San Francisco earthquake was 2500 times more intense
than the Japan earthquake.

36. Let the subscript N represent the Northridge, California earthquake and K the Kobe, Japan
earthquake. Then $M_N = \log\left(\frac{I_N}{S}\right) = 6.8 \quad \Leftrightarrow \quad I_N = S \cdot 10^{6.8}$ and $M_K = \log\left(\frac{I_K}{S}\right) = 7.2 \quad \Leftrightarrow$
$I_K = S \cdot 10^{7.2}$. So $\frac{I_K}{I_N} = \frac{10^{7.2}}{10^{6.8}} = 10^{0.4} \approx 2.51$, and so the Kobe, Japan earthquake was 2.5 times
more intense than the Northridge, California earthquake.

38. $\beta = 10\log\left(\frac{I}{I_0}\right) = 10\log\left(\frac{2.0 \times 10^{5}}{1.0 \times 10^{-12}}\right) = 10\log(2 \times 10^7) = 10(\log 2 + \log 10^7) = 10(\log 2 + 7)$
≈ 73. Therefore the intensity level was 73 dB.

40. Let the subscripts PM represent the power mower and RC the rock concert. Then
$106 = 10\log\left(\frac{I_{PM}}{10^{-12}}\right) \quad \Leftrightarrow \quad \log(I_{PM} \cdot 10^{12}) = 10.6 \quad \Leftrightarrow \quad I_{PM} \cdot 10^{12} = 10^{10.6}$. Also
$120 = 10\log\left(\frac{I_{RC}}{10^{-12}}\right) \quad \Leftrightarrow \quad \log(I_{RC} \cdot 10^{12}) = 12.0 \quad \Leftrightarrow \quad I_{RC} \cdot 10^{12} = 10^{12.0}$. So
$\frac{I_{RC}}{I_{PM}} = \frac{10^{12}}{10^{10.6}} = 10^{1.4} \approx 25.12$, and so the ratio of intensity is roughly 25.

Review Exercises for Chapter 5

2. $g(x) = 2^{x-1}$.
   Domain: $(-\infty, \infty)$
   Range: $(0, \infty)$
   Asymptote: $y = 0$.

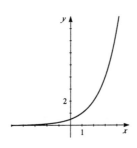

4. $y = 1 + 5^{-x}$.
   Domain: $(-\infty, \infty)$
   Range: $(1, \infty)$
   Asymptote: $y = 1$.

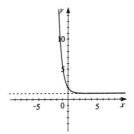

6. $g(x) = \log(-x)$.
   Domain: $(-\infty, 0)$
   Range: $(-\infty, \infty)$
   Asymptote: $x = 0$.

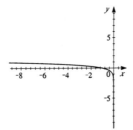

8. $y = 3 + \log_5(x + 4)$.
   Domain: $(-4, \infty)$
   Range: $(-\infty, \infty)$
   Asymptote: $x = -4$.

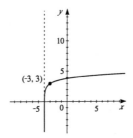

10. $G(x) = \frac{1}{2} e^{x-1}$.
    Domain: $(-\infty, \infty)$
    Range: $(0, \infty)$
    Asymptote: $y = 0$.

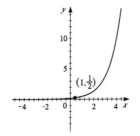

12. $y = \ln(x^2)$.
      Domain: $\{x \mid x \neq 0\} = (-\infty, 0) \cup (0, \infty)$
      Range: $(-\infty, \infty)$
      Asymptote: $x = 0$.

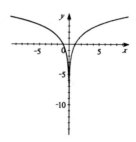

14. $g(x) = \ln(2 + x - x^2)$. We must have $2 + x - x^2 > 0$ (since $\ln y$ is defined only for $y > 0$)
      $\Leftrightarrow$ $x^2 - x - 2 < 0$ $\Leftrightarrow$ $(x - 2)(x + 1) < 0$. The endpoints of the intervals are 2 and -1.

Interval	$(-\infty, -1)$	$(-1, 2)$	$(2, \infty)$
Sign of $x - 2$	$-$	$-$	$+$
Sign of $x + 1$	$-$	$+$	$+$
Sign of $(x - 2)(x + 1)$	$+$	$-$	$+$

Thus the domain is $(-1, 2)$.

16. $k(x) = \ln|x|$. We must have $|x| > 0$. So $|x| > 0$ $\Leftrightarrow$ $-x > 0$ or $x > 0$. Since $-x > 0$ $\Leftrightarrow$
      $x < 0$, the domain is $x < 0$ or $x > 0$ which is equivalent to $x \neq 0$. In interval notation,
      $(-\infty, 0) \cup (0, \infty)$.

18. $\log_6 37 = x$ $\Leftrightarrow$ $6^x = 37$

20. $\ln c = 17$ $\Leftrightarrow$ $e^{17} = c$

22. $49^{-1/2} = \frac{1}{7}$ $\Leftrightarrow$ $\log_{49} \frac{1}{7} = -\frac{1}{2}$

24. $e^k = m$ $\Leftrightarrow$ $\ln m = k$

26. $\log_8 1 = \log_8(8^0) = 0$

28. $\log 0.000001 = \log 10^{-6} = -6$

30. $\log_4 8 = \log_4(4^{3/2}) = \frac{3}{2}$

32. $2^{\log_2 13} = 13$

34. $e^{2\ln 7} = (e^{\ln 7})^2 = 7^2 = 49$

36. $\log_3 \sqrt{243} = \log_3(3^{5/2}) = \frac{5}{2}$

38. $\log_5 250 - \log_5 2 = \log_5 \frac{250}{2} = \log_5 125 = \log_5 5^3 = 3$

40. $\log_{10}(\log_{10} 10^{100}) = \log_{10} 100 = \log_{10} 10^2 = 2$

42. $\log_2\left(x\sqrt{x^2 + 1}\right) = \log_2 x + \log_2 \sqrt{x^2 + 1} = \log_2 x + \frac{1}{2}\log_2(x^2 + 1)$

44. $\log\left(\dfrac{4x^3}{y^2(x-1)^5}\right) = \log(4x^3) - \log[y^2(x-1)^5] = \log 4 + 3\log x - [2\log y + 5\log(x-1)]$

46. $\ln\left(\dfrac{\sqrt[3]{x^4+12}}{(x+16)\sqrt{x-3}}\right) = \frac{1}{3}\ln(x^4+12) - [\ln(x+16) + \frac{1}{2}\ln(x-3)]$

48. $\log x + \log(x^2 y) + 3\log y = \log(x \cdot x^2 y \cdot y^3) = \log(x^3 y^4)$

50. $\log_5 2 + \log_5(x+1) - \frac{1}{3}\log_5(3x+7) = \log_5[2(x+1)] - \log_5(3x+7)^{1/3} = \log_5\left(\dfrac{2(x+1)}{\sqrt[3]{3x+7}}\right)$

52. $\frac{1}{2}[\ln(x-4) + 5\ln(x^2+4x)] = \frac{1}{2}\ln[(x-4)(x^2+4x)^5] = \ln\sqrt{(x-4)(x^2+4x)^5}$

54. $2^{3x-5} = 7 \quad\Leftrightarrow\quad \log 2^{3x-5} = \log 7 \quad\Leftrightarrow\quad 3x - 5 = \frac{\log 7}{\log 2} \quad\Leftrightarrow\quad x = \frac{1}{3}\left(5 + \frac{\log 7}{\log 2}\right) \approx 2.60$

56. $\ln(2x-3) = 14 \quad\Leftrightarrow\quad e^{\ln(2x-3)} = e^{14} \quad\Leftrightarrow\quad 2x - 3 = e^{14} \quad\Leftrightarrow\quad x = \frac{1}{2}(3 + e^{14}) \approx 601303.64$

58. $2^{1-x} = 3^{2x+5} \quad\Leftrightarrow\quad \log 2^{1-x} = \log 3^{2x+5} \quad\Leftrightarrow\quad (1-x)\log 2 = (2x+5)\log 3 \quad\Leftrightarrow$

$x(2\log 3 + \log 2) = \log 2 - 5\log 3 \quad\Leftrightarrow\quad x = \dfrac{\log 2 - 5\log 3}{\log 2 + 2\log 3} = \dfrac{\log \frac{2}{3^5}}{\log(2 \cdot 9)} \approx -1.66$

60. $\log_8(x+5) - \log_8(x-2) = 1 \quad\Leftrightarrow\quad \log_8\left(\dfrac{x+5}{x-2}\right) = 1 \quad\Leftrightarrow\quad \dfrac{x+5}{x-2} = 8^1 = 8 \quad\Leftrightarrow$

$x + 5 = 8x - 16 \quad\Leftrightarrow\quad 7x = 21 \quad\Leftrightarrow\quad x = 3$

62. $2^{3^x} = 5 \quad\Leftrightarrow\quad \log 2^{3^x} = \log 5 \quad\Leftrightarrow\quad 3^x \log 2 = \log 5 \quad\Leftrightarrow\quad 3^x = \dfrac{\log 5}{\log 2} \quad\Leftrightarrow$

$\log 3^x = \log\left(\dfrac{\log 5}{\log 2}\right) \quad\Leftrightarrow\quad x\log 3 = \log\left(\dfrac{\log 5}{\log 2}\right) \quad\Leftrightarrow\quad x = \dfrac{1}{\log 3} \cdot \log\left(\dfrac{\log 5}{\log 2}\right) \approx 0.77$

64. $2^{3x-5} = 7 \quad\Leftrightarrow\quad (3x-5)\log 2 = \log 7 \quad\Leftrightarrow\quad x = \frac{1}{3}\left(5 + \frac{\log 7}{\log 2}\right) \approx 2.602452$

66. $e^{-15k} = 10000 \quad\Leftrightarrow\quad -15k = \ln 10000 \quad\Leftrightarrow\quad k = -\frac{1}{15}\ln 10000 \approx -0.614023$

68. $y = 2x^2 - \ln x$.
Vertical Asymptote: $x = 0$
Horizontal Asymptote: none
Local minimum ≈ 1.19 at $x \approx 0.5$

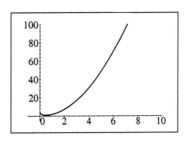

70. $y = 10^x - 5^x$.

Vertical Asymptote: none

Horizontal Asymptote: $y = 0$

Local minimum ≈ -0.13 at $x \approx -0.5$

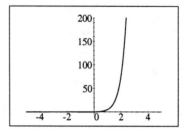

72. $4 - x^2 = e^{-2x}$.

From the graphs, we see that the solutions are
$x \approx -0.64$ and $x \approx 2$.

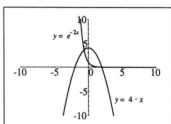

74. $e^x < 4x^2 \iff e^x - 4x^2 < 0$.

We graph the function $f(x) = e^x - 4x^2$, and we see
that the graph lies below the x-axis for
$(-\infty, -0.41) \cup (0.71, 4.31)$.

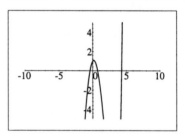

76. The line has x-intercept at $x = e^0 = 1$. When $x = e^a$, $y = \ln e^a = a$. Therefore, using the point-slope equation, we have $y - 0 = \dfrac{a - 0}{e^a - 1}(x - 1) \iff y = \dfrac{a}{e^a - 1}(x - 1)$.

78. $0.2 \le \log x < 2 \iff 10^{0.2} \le x < 10^2 \iff \sqrt[5]{10} \le x < 100$.

80. $f(x) = 2^{3^x}$. Then $y = 2^{3^x} \iff \log_2 y = 3^x \iff \log_3(\log_2 y) = x$, and so the inverse function is $f^{-1}(x) = \log_3(\log_2 x)$. Since $\log_3 y$ is defined only when $y > 0$, we have $\log_2 x > 0 \iff x > 1$. Therefore the domain is $(1, \infty)$, and the range is $(-\infty, \infty)$.

82. $P = 5000$, $r = 0.085$, and $n = 2$.

   (a) For $t = 1.5$, $A = 5000\left(1 + \frac{0.085}{2}\right)^{2(1.5)} = 5000 \cdot 1.0425^3 \approx \$5664.98$.

   (b) We want to find t such that $A = 7000$. Then $A = 5000 \cdot 1.0425^{2t} = 7000 \iff$
   $1.0425^{2t} = \frac{7000}{5000} = \frac{7}{5} \iff 2t = \dfrac{\log\left(\frac{7}{5}\right)}{\log 1.0425} \iff t = \dfrac{\log\left(\frac{7}{5}\right)}{2\log 1.0425} \approx 4.04$, and so the
   investment will amount to $7000 after approximately 4 years.

84. Using the model $n(t) = n_0\, e^{rt}$, with $n_0 = 10000$ and $n(1) = 25000$, we have
   $25000 = n(1) = 10000 e^{r \cdot 1} \iff e^r = \frac{5}{2} \iff r = \ln \frac{5}{2} \approx 0.916$. So $n(t) = 10000 e^{0.916\, t}$.

(a) Here we must solve the equation $n(t) = 20000$ for t. So $n(t) = 10000e^{0.916t} = 20000 \Leftrightarrow$ $e^{0.916t} = 2 \Leftrightarrow 0.916t = \ln 2 \Leftrightarrow t = \frac{\ln 2}{0.916} \approx 0.756$. Thus the doubling period is about 45 minutes.

(b) $n(3) = 10000e^{0.916 \cdot 3} \approx 156250$, so the population after 3 hours is about 156,250.

86. From the formula for radioactive decay, we have $m(t) = m_0\, e^{-rt}$, where $r = \frac{\ln 2}{h}$. So $m(t) = m_0\, e^{-\frac{\ln 2}{h}t}$.

(a) Using $m(8) = 0.33\, m_0$, we solve for h. We have $0.33\, m_0 = m(8) = m_0\, e^{-\frac{8\ln 2}{h}} \Leftrightarrow$ $0.33 = e^{-\frac{8\ln 2}{h}} \Leftrightarrow -\frac{8\ln 2}{h} = \ln 0.33 \Leftrightarrow h = -\frac{8\ln 2}{\ln 0.33} \approx 5.002$. So the half-life of this element is roughly 5 days.

(b) $m(12) = m_0\, e^{-\frac{\ln 2}{5} 12} \approx 0.19\, m_0$, so about 19% of the original mass remains.

88. From the formula for radioactive decay, we have $m(t) = m_0\, e^{-rt}$, where $r = \frac{\ln 2}{h}$. Since $h = 4$, we have $r = \frac{\ln 2}{4} \approx 0.173$ and $m(t) = m_0\, e^{-0.173t}$.

(a) Using $m(20) = 0.375$, we solve for m_0. We have $0.375 = m(20) = m_0\, e^{-0.173 \cdot 20} \Leftrightarrow$ $0.03125 m_0 = 0.375 \Leftrightarrow m_0 = \frac{0.375}{0.03125} \approx 12$. So the initial mass of the sample was about 12 g.

(b) $m(t) = 12\, e^{-0.173t}$.

(c) $m(3) = 12\, e^{-0.173 \cdot 3} \approx 7.135$. So there are about 7.1 g remaining after 3 days.

(d) Here we solve $m(t) = 0.15$ for t: $0.15 = 12\, e^{-0.173t} \Leftrightarrow 0.0125 = e^{-0.173t} \Leftrightarrow$ $-0.173t = \ln 0.0125 \Leftrightarrow t = \frac{\ln 0.0125}{-0.173} \approx 25.3$. So it will take about 25 days until only 15% of the substance remains.

90. We use Newton's Law of Cooling: $T(t) = T_s + D_0\, e^{-kt}$ with $k = 0.0341$, $T_s = 60$ and $D_0 = 190 - 60 = 130$. So $90 = T(t) = 60 + 130\, e^{-0.0341t} \Leftrightarrow 90 = 60 + 130\, e^{-0.0341t} \Leftrightarrow$ $130\, e^{-0.0341t} = 30 \Leftrightarrow e^{-0.0341t} = \frac{3}{13} \Leftrightarrow -0.0341\, t = \ln\left(\frac{3}{13}\right) \Leftrightarrow t = \frac{-\ln(3/13)}{0.0341} \approx 43.0$, so the engine cools to 90°F in about 43 minutes.

92. pH $= 1.9 = -\log[\text{H}^+]$. Then $[\text{H}^+] = 10^{-1.9} \approx 1.26 \times 10^{-2}$ M.

94. Let the subscript JH represent the jackhammer and W the whispering: $\beta_{\text{JH}} = 132 = 10\log\left(\frac{I_{\text{JH}}}{I_0}\right)$ $\Leftrightarrow \log\left(\frac{I_{\text{JH}}}{I_0}\right) = 13.2 \Leftrightarrow \frac{I_{\text{JH}}}{I_0} = 10^{13.2}$. Similarly $\frac{I_{\text{W}}}{I_0} = 10^{2.8}$. So $\frac{I_{\text{JH}}}{I_{\text{W}}} = \frac{10^{13.2}}{10^{2.8}} = 10^{10.4} \approx 2.51 \times 10^{10}$, and so the ratio of intensities is 2.51×10^{10}.

Focus on Modeling

2. (a)

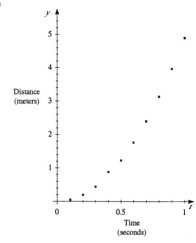

(b) We let t represent the time (in seconds) and y the distance fallen (in meters). Using a calculator, we obtain the power model: $y = 4.9622t^{2.0027}$.

(c) When $t = 3$ the model predicts that $y = 44.792$ m.

4. (a)

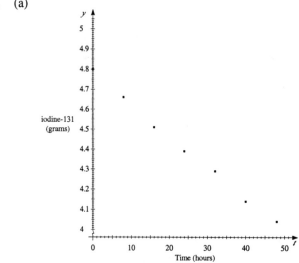

(b) Let t be the time (in hours) and y be amount of iodine-131 (in grams). Using a calculator, we obtain the exponential model $y = ab^t$, where $a = 4.79246$ and $b = 0.99642$.

(c) To find the half-life of iodine-131, we must find the time when the sample has decayed to half its original mass. Setting $y = 2.40$ g, we get $2.40 = 4.79246 \cdot (0.99642)^t \quad \Leftrightarrow$

$$\ln 2.40 = \ln 4.79246 + t \ln 0.99642 \quad \Leftrightarrow \quad t = \frac{\ln 2.40 - \ln 4.79246}{\ln 0.99642} \approx 192.8 \text{ h.}$$

6. (a) Using a graphing calculator, we find the power function model $y = 49.70030t^{-0.15437}$
         and the exponential model $y = 44.82418 \cdot (0.99317^t)$.

    (b)

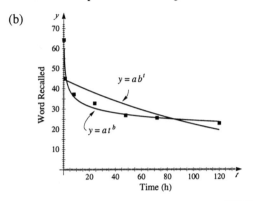

    (c) The power function, $y = 49.70030t^{-0.15437}$, seems to provide a better model.

8. Let x be the reduction in emissions (in percent), and y be the cost (in dollars). First we
    make a scatter plot of the data. A linear model does not appear appropriate, so we try an
    exponential model. Using a calculator, we get the model $y = ab^x$, where $a = 2.414$ and
    $b = 1.05452$. This model is graphed on the scatter plot.

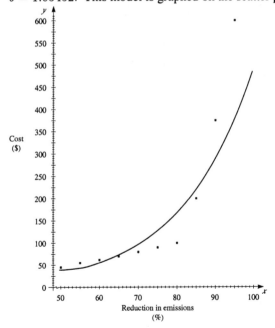

10. (a)

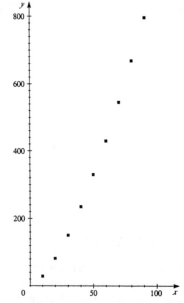

(b)

x	y	$\ln x$	$\ln y$
10	29	2.30259	3.36730
20	82	2.99573	4.40672
30	151	3.40120	5.01728
40	235	3.68888	5.45959
50	330	3.91202	5.79909

x	y	$\ln x$	$\ln y$
60	430	4.09434	6.06379
70	546	4.24850	6.30262
80	669	4.38203	6.50578
90	797	4.49981	6.68085

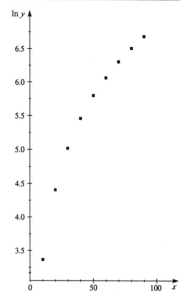

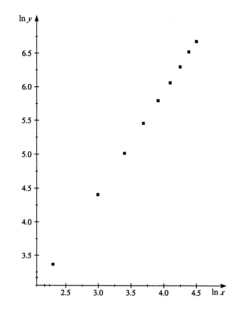

(c) The power function.

(d) $y = ax^n$ where $a = 0.893421326$ and $n = 1.50983$.

12. (a) Using the LnReg command on a TI-83 we get $y = a + b \ln x$ where $a = -7154.888128$ and $b = 1061.006551$.

(b) Using the model $C = -7154.888128 + (1061.006551)\ln 2005 \approx 912$ metric tons.

Chapter Six
Exercises 6.1

2. $\begin{cases} 2x + y = 7 \\ x + 2y = 2 \end{cases}$ Solving the first equation for y, we get $y = 7 - 2x$ and substituting this into the
second equation, gives $x + 2(7 - 2x) = 2$ $\Leftrightarrow$ $x + 14 - 4x = 2$ $\Leftrightarrow$ $-3x = -12$ $\Leftrightarrow$
$x = 4$. Substituting for x we get $y = 7 - 2x = 7 - 2(4) = -1$. Thus the solution is $(4, -1)$.

4. $\begin{cases} x^2 + y^2 = 25 \\ y = 2x \end{cases}$ Substituting for y in the first equation gives $x^2 + (2x)^2 = 25$ $\Leftrightarrow$ $5x^2 = 25$
$\Leftrightarrow$ $x^2 = 5$ $\Rightarrow$ $x = \pm\sqrt{5}$ When $x = \sqrt{5}$ then $y = 2\sqrt{5}$, and when $x = -\sqrt{5}$ then
$y = 2\left(-\sqrt{5}\right) = -2\sqrt{5}$. Thus the solutions are $\left(\sqrt{5}, 2\sqrt{5}\right)$ and $\left(-\sqrt{5}, -2\sqrt{5}\right)$.

6. $\begin{cases} x^2 + y = 9 \\ x - y + 3 = 0 \end{cases}$ Solving the first equation for y, we get $y = 9 - x^2$. Substituting this into the
second equation gives $x - (9 - x^2) + 3 = 0$ $\Leftrightarrow$ $x^2 + x - 6 = 0$ $\Leftrightarrow$ $(x + 3)(x - 2) = 0$
$\Leftrightarrow$ $x = -3$ or $x = 2$. If $x = -3$, then $y = 9 - (-3)^2 = 0$, and if $x = 2$, then $y = 9 - (2)^2 = 5$.
Thus the solutions are $(-3, 0)$ and $(2, 5)$.

8. $\begin{cases} x^2 - y = 1 \\ 2x^2 + 3y = 17 \end{cases}$ Solving the first equation for y, we get $y = x^2 - 1$. Substituting this into the
second equation gives $2x^2 + 3(x^2 - 1) = 17$ $\Leftrightarrow$ $2x^2 + 3x^2 - 3 = 17$ $\Leftrightarrow$ $5x^2 = 20$ $\Leftrightarrow$
$x^2 = 4$ $\Leftrightarrow$ $x = \pm 2$. If $x = -2$, then $y = (-2)^2 - 1 = 3$, and if $x = 2$, then $y = (2)^2 - 1 = 3$.
Thus the solutions are $(-2, 3)$ and $(2, 3)$.

10. $\begin{cases} 4x - 3y = 11 \\ 8x + 4y = 12 \end{cases}$ Multiplying the first equation by 2 gives the system $\begin{cases} 8x - 6y = 22 \\ 8x + 4y = 12 \end{cases}$.
Subtracting the equations gives $-10y = 10$ $\Leftrightarrow$ $y = -1$. Substituting this value into the second
equation gives $8x + 4(-1) = 12$ $\Leftrightarrow$ $8x = 16$ $\Leftrightarrow$ $x = 2$. Thus the solution is $(2, -1)$.

12. $\begin{cases} 3x^2 + 4y = 17 \\ 2x^2 + 5y = 2 \end{cases}$ Multiplying the first equation by 2 and the second by 3 gives the system
$\begin{cases} 6x^2 + 8y = 34 \\ -6x^2 - 15y = -6 \end{cases}$. Adding we get $-7y = 28$ $\Leftrightarrow$ $y = -4$. Substituting this value into the
second equation gives $2x^2 + 5(-4) = 2$ $\Rightarrow$ $2x^2 = 22$ $\Leftrightarrow$ $x^2 = 11$ $\Leftrightarrow$ $x = \pm\sqrt{11}$.
Thus the solutions are $\left(\sqrt{11}, -4\right)$ and $\left(-\sqrt{11}, -4\right)$.

14. $\begin{cases} 2x^2 + 4y = 13 \\ x^2 - y^2 = \frac{7}{2} \end{cases}$ Multiplying the second equation by 2 gives the system $\begin{cases} 2x^2 + 4y = 13 \\ 2x^2 - 2y^2 = 7 \end{cases}$.
Subtracting the equations gives $4y + 2y^2 = 6$ $\Leftrightarrow$ $y^2 + 2y - 3 = 0$ $\Leftrightarrow$ $(y + 3)(y - 1) = 0$
$\Leftrightarrow$ $y = -3, y = 1$. If $y = -3$, then $2x^2 + 4(-3) = 13$ $\Leftrightarrow$ $x^2 = \frac{25}{2}$ $\Leftrightarrow$ $x = \pm\frac{5\sqrt{2}}{2}$. If
$y = 1$, then $2x^2 + 4(1) = 13$ $\Leftrightarrow$ $x^2 = \frac{9}{2}$ $\Leftrightarrow$ $x = \pm\frac{3\sqrt{2}}{2}$. Hence, the solutions are
$\left(\pm\frac{5\sqrt{2}}{2}, -3\right)$ and $\left(\pm\frac{3\sqrt{2}}{2}, 1\right)$.

16. $\begin{cases} x^2 - y^2 = 1 \\ 2x^2 - y^2 = x + 3 \end{cases}$ Subtracting the first equation from the second equation gives $x^2 = x + 2$
$\Leftrightarrow \quad x^2 - x - 2 = 0 \quad \Leftrightarrow \quad (x-2)(x+1) = 0 \quad \Leftrightarrow \quad x = 2, x = -1.$ Solving the first
equation for y^2 we have $y^2 = x^2 - 1.$ When $x = -1,$ $y^2 = (-1)^2 - 1 = 0$ so $y = 0$ and when
$x = 2,$ $y^2 = (2)^2 - 1 = 3$ so $y = \pm\sqrt{3}.$ Thus, the solutions are $(-1, 0), (2, -\sqrt{3}),$ and $(2, \sqrt{3}).$

18. $\begin{cases} x + y = 2 \\ 2x + y = 5 \end{cases}$ By inspection of the graph, it appears that $(3, -1)$ is the solution to the system. We
check this in both equations to verify that it is a solution. $3 + (-1) = 2 \checkmark$ and
$2(3) + (-1) = 6 - 1 = 5 \checkmark.$ Since both equations are satisfied, the solution is $(3, -1).$

20. $\begin{cases} x - y^2 = -4 \\ x - y = 2 \end{cases}$ By inspection of the graph, it appears that $(0, -2)$ and $(5, 3)$ are solutions to the
system. We check each point in both equations to verify that it is a solution.
$(0, -2):$ $\quad (0) - (-2)^2 = -4 \checkmark$ $\qquad (0) - (-2) = 2 \checkmark$
$(5, 3):$ $\qquad 5 - 3^2 = 5 - 9 = -4 \checkmark$ $\qquad 5 - 3 = 2 \checkmark$
Thus the solutions are $(0, -2)$ and $(5, 3).$

22. $\begin{cases} x^2 + y^2 = 4x \\ x = y^2 \end{cases}$ By inspection of the graph, it appears that $(0, 0)$ is a solution, but is difficult to
get accurate values for the other points. Substituting for y^2 we have $x^2 + x = 4x \quad \Leftrightarrow$
$x^2 - 3x = 0 \quad \Leftrightarrow \quad x(x - 3) = 0.$ So $x = 0$ or $x = 3.$ If $x = 0,$ then $y^2 = 0$ so $y = 0.$ And is
$x = 3$ then $y^2 = 3$ so $y = \pm\sqrt{3}.$ Hence, the solutions are $(0, 0), (3, -\sqrt{3}),$ and $(3, \sqrt{3}).$

24. $\begin{cases} x - y^2 = 0 \\ y - x^2 = 0 \end{cases}$ Solving the first equation for x and the second equation for y gives $\begin{cases} x = y^2 \\ y = x^2 \end{cases}.$
Substituting for y in the first equation gives $x = x^4 \quad \Leftrightarrow \quad x(x^3 - 1) = 0 \quad \Leftrightarrow \quad x = 0, x = 1.$
Thus, the solutions are $(0, 0)$ and $(1, 1).$

26. $\begin{cases} y = 4 - x^2 \\ y = x^2 - 4 \end{cases}$ Setting the two equations equal, we get $4 - x^2 = x^2 - 4 \quad \Leftrightarrow \quad 2x^2 = 8 \quad \Leftrightarrow$
$x = \pm 2.$ Therefore, the solutions are $(2, 0)$ and $(-2, 0).$

28. $\begin{cases} xy = 24 \\ 2x^2 - y^2 + 4 = 0 \end{cases}$ Since $x = 0$ is not a solution, from the first equation we get $y = \dfrac{24}{x}.$
Substituting into the second equation, we get $2x^2 + 4 = \left(\dfrac{24}{x}\right)^2 \quad \Rightarrow \quad 2x^4 + 4x^2 = 576 \quad \Leftrightarrow$
$x^4 + 2x^2 - 288 = 0 \quad \Leftrightarrow \quad (x^2 + 18)(x^2 - 16) = 0.$ Since $x^2 + 18$ cannot be 0 if x is real, we
have $x^2 - 16 = 0 \quad \Leftrightarrow \quad x = \pm 4.$ When $x = 4,$ we have $y = \frac{24}{4} = 6$ and when $x = -4,$ we have
$y = \frac{24}{-4} = -6.$ Thus the solutions are $(4, 6)$ and $(-4, -6).$

30. $\begin{cases} x + \sqrt{y} = 0 \\ y^2 - 4x^2 = 12 \end{cases}$ Solving the first equation for $x,$ we get $x = -\sqrt{y}.$ Substituting for x gives
$y^2 - 4(-\sqrt{y})^2 = 12 \quad \Leftrightarrow \quad y^2 - 4y - 12 = 0 \quad \Leftrightarrow \quad (y - 6)(y + 2) = 0 \quad \Rightarrow \quad y = 6, y = -2.$
Since $x = -\sqrt{-2}$ is not a real solution, the only solution is $(-\sqrt{6}, 6).$

32. $\begin{cases} x^2 + 2y^2 = 2 \\ 2x^2 - 3y = 15 \end{cases}$ Multiplying the first equation by 2 gives the system $\begin{cases} 2x^2 + 4y^2 = 4 \\ 2x^2 - 3y = 15 \end{cases}$.

Subtracting the two equations gives $4y^2 + 3y = -11$ $\Leftrightarrow$ $4y^2 + 3y + 11 = 0$ $\Rightarrow$

$y = \frac{-3 \pm \sqrt{9 - 4(4)(11)}}{2(4)}$ which is not a real number. Therefore, there are no real solutions.

34. $\begin{cases} x^4 - y^3 = 15 \\ 3x^4 + 5y^3 = 53 \end{cases}$ Multiplying the first equation by 3 gives the system $\begin{cases} 3x^4 - 3y^3 = 45 \\ 3x^4 + 5y^3 = 53 \end{cases}$.

Subtracting the equations gives $8y^3 = 8$ $\Leftrightarrow$ $y^3 = 1$ $\Rightarrow$ $y = 1$, and then $x^4 - 1 = 15$ $\Leftrightarrow$ $x = \pm 2$. Therefore, the solutions are $(2, 1)$ and $(-2, 1)$.

36. $\begin{cases} \dfrac{4}{x^2} + \dfrac{6}{y^4} = \dfrac{7}{2} \\ \dfrac{1}{x^2} - \dfrac{2}{y^4} = 0 \end{cases}$ If we let $u = \dfrac{1}{x^2}$ and $v = \dfrac{1}{y^4}$, the system is equivalent to $\begin{cases} 4u + 6v = \frac{7}{2} \\ u - 2v = 0 \end{cases}$, and

multiplying the second equation by 3, gives $\begin{cases} 4u + 6v = \frac{7}{2} \\ 3u - 6v = 0 \end{cases}$. Adding the equations gives $7u = \frac{7}{2}$

$\Leftrightarrow$ $u = \frac{1}{2}$, and $v = \frac{1}{4}$. Therefore, $x^2 = \frac{1}{u} = 2$ $\Leftrightarrow$ $x = \pm\sqrt{2}$, and $y^4 = \frac{1}{v} = 4$ $\Leftrightarrow$ $y = \pm\sqrt{2}$. Thus, the solutions are $(\sqrt{2}, \sqrt{2})$, $(\sqrt{2}, -\sqrt{2})$, $(-\sqrt{2}, \sqrt{2})$, and $(-\sqrt{2}, -\sqrt{2})$.

38. $\begin{cases} y = -2x + 12 \\ y = x + 3 \end{cases}$

The solution is $(3, 6)$. Solution checked.

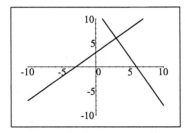

40. $\begin{cases} y = x^2 - 4x \\ 2x - y = 2 \end{cases}$ $\Leftrightarrow$ $\begin{cases} y = x^2 - 4x \\ y = 2x - 2 \end{cases}$

The solutions are approximately $(0.35, -1.30)$ and $(5.65, 9.30)$.

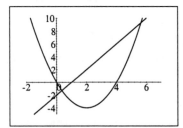

42. $\begin{cases} x^2 + y^2 = 17 \\ x^2 - 2x + y^2 = 13 \end{cases}$ $\Leftrightarrow$

$\begin{cases} y = \pm\sqrt{17 - x^2} \\ y = \pm\sqrt{13 + 2x - x^2} \end{cases}$

The solution are approximately $(2, \pm 3.61)$.

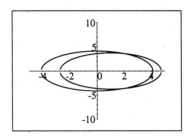

44. $\begin{cases} x^2 - y^2 = 3 \\ \qquad y = x^2 - 2x - 8 \end{cases}$ ⇔

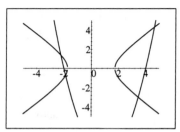

$\begin{cases} y = \pm\sqrt{x^2 - 3} \\ y = x^2 - 2x - 8 \end{cases}$

The solutions are approximately $(-2.22, 1.40)$,
$(-1.88, -0.72)$, $(3.45, -2.99)$, and $(4.65, 4.31)$.

46. $\begin{cases} y = e^x + e^{-x} \\ y = 5 - x^2 \end{cases}$

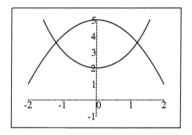

The solution are approximately $(1.19, 3.59)$
and $(-1.19, 3.59)$.

48. Let b be the length of the base of the triangle, in feet, and h be the height of the triangle, in feet.
Then, $\begin{cases} \frac{1}{2}bh = 84 \\ b^2 + h^2 = 25^2 = 625 \end{cases}$. The first equation gives $b = \dfrac{168}{h}$. By substitution,

$\left(\dfrac{168}{h}\right)^2 + h^2 = 625$ ⇔ $h^4 - 625h^2 + 168^2 = 0$ ⇔ $(h^2 - 49)(h^2 - 576) = 0$ ⇒
$h = 7$ or $h = 24$. Thus, the lengths of the other two sides are 7 ft and 24 ft.

50. Let w be the width and l be the length of the rectangle, in inches. From the figure, the diagonals of
the rectangle are simply diameters of the circle. Then, $\begin{cases} wl = 160 \\ w^2 + l^2 = 20^2 = 400 \end{cases}$ ⇔ $w = \dfrac{160}{l}$.

By substitution, $\dfrac{160^2}{l^2} + l^2 = 400$ ⇔ $l^4 - 400l^2 + 160^2 = 0$ ⇔ $(l^2 - 80)(l^2 - 320) = 0$
⇒ $l = \sqrt{80} = 4\sqrt{5}$ or $l = \sqrt{320} = 8\sqrt{5}$. Therefore, the dimensions of the rectangle are $4\sqrt{5}$
in. by $8\sqrt{5}$ in..

52. Let x be the circumference and y be length of the stove pipe. Using the circumference we can
determine the radius, $2\pi r = x$ ⇔ $r = \dfrac{x}{2\pi}$. Thus the volume is $\pi\left(\dfrac{x}{2\pi}\right)^2 y = \dfrac{1}{4\pi}x^2 y$. So the
system is given by $\begin{cases} xy = 1200 \\ \frac{1}{4\pi}x^2 y = 600 \end{cases}$. Substituting for xy in the second equation gives

$\frac{1}{4\pi}x^2 y = \frac{1}{4\pi}x(xy) = \frac{1}{4\pi}x(1200) = 600$ ⇔ $x = 2\pi$. So $y = \dfrac{1200}{x} = \dfrac{1200}{2\pi} = \dfrac{600}{\pi}$. Thus the
dimensions of the sheet metal are 2π by $\frac{600}{\pi}$.

54. The graphs of $y = x^2$ and $y = x + k$ for various values of k are
    shown. If we solve the system $\begin{cases} y = x^2 \\ y = x + k \end{cases}$, we get $x^2 - x - k = 0$.

    Using the quadratic formula, we have $x = \frac{-1 \pm \sqrt{1+4k}}{2}$. So there will

    be no solution when $\sqrt{1 + 4k}$ is undefined, that is, when $1 + 4k < 0$

    $\Leftrightarrow$ $k < -\frac{1}{4}$. There will be exactly one solution when $1 + 4k = 0$

    $\Leftrightarrow$ $k < -\frac{1}{4}$, and there will be two solutions when $1 + 4k > 0$

    $\Leftrightarrow$ $k > -\frac{1}{4}$.

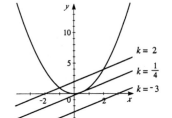

Exercises 6.2

2. $\begin{cases} 2x + y = 11 \\ x - 2y = 4 \end{cases}$

The solution is $x = 5.2$, $y = 0.6$.

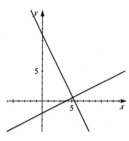

4. $\begin{cases} 2x + 6y = 0 \\ -3x - 9y = 18 \end{cases}$

No solution. The lines are parallel, so there is no intersection.

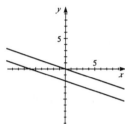

6. $\begin{cases} 12x + 15y = -18 \\ 2x + \frac{5}{2}y = -3 \end{cases}$

There are infinitely many solutions. The lines are the same.

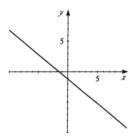

8. $\begin{cases} x - y = 3 \\ x + 3y = 7 \end{cases}$ Subtracting the first equation from the second equation gives

$\begin{aligned} -x + y &= -3 \\ x + 3y &= 7 \\ \hline 4y &= 4 \end{aligned}$ $\Leftrightarrow$ $y = 1..$

Substituting, we have $x - 1 = 3$ $\Leftrightarrow$ $x = 4$. Hence, the solution is $(4, 1)$.

10. $\begin{cases} 3x + 2y = 0 \\ -x - 2y = 8 \end{cases}$ Adding the two equations gives $2x = 8$ $\Leftrightarrow$ $x = 4$. Substituting for x in the

second equation gives $3(4) + 2y = 0$ $\Leftrightarrow$ $12 + 2y = 0$ $\Leftrightarrow$ $y = -6$. Hence, the solution is $(4, -6)$.

12. $\begin{cases} x + y = 7 \\ 2x - 3y = -1 \end{cases}$ Adding 3 times the first equation to the second equation gives

$\begin{aligned} 3x + 3y &= 21 \\ 2x - 3y &= -1 \\ \hline 5x &= 20 \end{aligned}$ $\Leftrightarrow$ $x = 4$.

So $4 + y = 7$ $\Leftrightarrow$ $y = 3$, and the solution is $(4, 3)$.

14. $9x - y = -6$ $\Leftrightarrow$ $y = 9x + 6$. Substituting for y into $4x - 3y = 28$ gives $4x - 3(9x + 6) = 28$
     $\Leftrightarrow$ $-23x = 46$ $\Leftrightarrow$ $x = -2$, and so $y = 9(-2) + 6 = -12$. Thus, the solution is $(-2, -12)$.

16. $-4x + 12y = 0$ $\Leftrightarrow$ $x = 3y$. Substituting for x into $12x + 4y = 160$ gives $12(3y) + 4y = 160$
     $\Leftrightarrow$ $40y = 160$ $\Leftrightarrow$ $y = 4$, and so $x = 3(4) = 12$. Therefore, the solution is $(12, 4)$.

18. $0.2x - 0.2y = -1.8$ $\Leftrightarrow$ $x = y - 9$. Substituting for x into $-0.3x + 0.5y = 3.3$ gives
     $-0.3(y - 9) + 0.5y = 3.3$ $\Leftrightarrow$ $0.2y = 0.6$ $\Leftrightarrow$ $y = 3$, and so $x = (3) - 9 = -6$. Hence,
     the solution is $(-6, 3)$.

20. $\begin{cases} 4x + 2y = 16 \\ x - 5y = 70 \end{cases}$ Adding the first equation to -4 times the second equation gives

     $4x + 2y = 16$
     $\underline{-4x + 20y = -280}$
     $22y = -264$ $\Leftrightarrow$ $y = -12$.
     So $4x + 2(-12) = 16$ $\Leftrightarrow$ $x = 10$, and the solution is $(10, -12)$.

22. $\begin{cases} -3x + 5y = 2 \\ 9x - 15y = 6 \end{cases}$ Adding 3 times the first equation to the second equation gives

     $-9x + 15y = 6$
     $\underline{9x - 15y = 6}$
     $ 0 = 12$, which is false. Therefore, there is no solution to this system.

24. $\begin{cases} 2x - 3y = -8 \\ 14x - 21y = 3 \end{cases}$ Adding 7 times the first equation to -1 times the second equation gives

     $14x - 21y = -56$
     $\underline{-14x + 21y = -3}$
     $ 0 = -59$, which is false. Therefore, there is no solution to this system.

26. $\begin{cases} 25x - 75y = 100 \\ -10x + 30y = -40 \end{cases}$ Adding $\frac{1}{25}$ times the first equation to $\frac{1}{10}$ times the second equation gives

     $x - 3y = 4$
     $\underline{-x + 3y = -4}$
     $ 0 = 0$, which is always true.
     We now put the equation in slope-intercept form. We have $x - 3y = 4$ $\Leftrightarrow$ $-3y = -x + 4$
     $\Leftrightarrow$ $y = \frac{1}{3}x - \frac{4}{3}$, so a solution is any pair of the form $\left(x, \frac{1}{3}x - \frac{4}{3}\right)$, where x is any real number.

28. $\begin{cases} u - 30v = -5 \\ -3u + 80v = 5 \end{cases}$ Adding 3 times the first equation to the second equation gives

     $3u - 90v = -15$
     $\underline{-3u + 80v = 5}$
     $-10v = -10$ $\Leftrightarrow$ $v = 1$.
     So $u - 30(1) = -5$ $\Leftrightarrow$ $u = 25$. Thus, the solution is $(u, v) = (25, 1)$.

30. $\begin{cases} \frac{3}{2}x - \frac{1}{3}y = \frac{1}{2} \\ 2x - \frac{1}{2}y = -\frac{1}{2} \end{cases}$ Adding -6 times the first equation to 4 times the second equation gives

$$\begin{array}{r} -9x + 2y = -3 \\ 8x - 2y = -2 \\ \hline -x \qquad = -5 \end{array}$$ $\Leftrightarrow$ $\quad x = 5$. So $9(5) - 2y = 3$ $\quad \Leftrightarrow \quad y = 21$. Thus the solution is $(5, 21)$.

32. $\begin{cases} 26x - 10y = -4 \\ -0.6x + 1.2y = 3 \end{cases}$ Adding 3 times the first equation to 25 times the second equation gives

$$\begin{array}{r} 78x - 30y = -12 \\ -15x + 30y = 75 \\ \hline 63x \qquad = 63 \end{array}$$ $\Leftrightarrow$ $\quad x = 1$. So $26(1) - 10y = -4$ $\quad \Leftrightarrow \quad -10y = -30$ $\quad \Leftrightarrow \quad y = 3$.
Thus the solution is $(1, 3)$.

34. $\begin{cases} -\frac{1}{10}x + \frac{1}{2}y = 4 \\ 2x - 10y = -80 \end{cases}$ Adding 20 times the first equation to the second equation gives

$$\begin{array}{r} -2x + 10y = 80 \\ 2x - 10y = 80 \\ \hline 0 = 0 \end{array}$$, which is always true.

We now put the equation in slope-intercept form. We have $2x - 10y = -80$ $\quad \Leftrightarrow$
$-10y = -2x - 80$ $\quad \Leftrightarrow \quad y = \frac{1}{5}x + 8$, so a solution is any pair of the form $\left(x, \frac{1}{5}x + 8\right)$, where x is
any real number.

36. $\begin{cases} 18.72x - 14.91y = 12.33 & l_1 \\ 6.21x - 12.92y = 17.82 & l_2 \end{cases}$
The solution is approximately $(-0.71, -1.72)$.

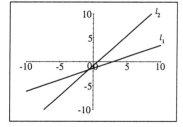

38. $\begin{cases} -435x + 912y = 0 & l_1 \\ 132x + 455y = 994 & l_2 \end{cases}$
The solution is approximately $(2.85, 1.36)$.

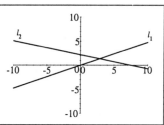

40. $ax + by = 0 \qquad \times -1 \qquad\qquad -ax - by = 0$
$ x + y = 1 \qquad \times a \quad \Rightarrow \qquad \underline{ax + ay = a}$
$$(a - b)y = a \qquad \Leftrightarrow \quad y = \frac{a}{a - b}, a \neq b.$$

So $x + \left(\dfrac{a}{a - b}\right) = 1$ $\quad \Leftrightarrow \quad x = \dfrac{b}{b - a}$. Hence, the solution is $\left(\dfrac{b}{b - a}, \dfrac{a}{a - b}\right)$.

42. $ax + by = 0$ $\times -a$ $-a^2x - aby = 0$

      $a^2x + b^2y = 1$ $\times 1$ $\Rightarrow$ $\underline{a^2x + b^2y = 1}$

$$b(b-a)y = 1 \quad \Leftrightarrow \quad y = \frac{1}{b(b-a)}.$$

So $ax + \dfrac{b}{b(b-a)} = 0 \quad \Leftrightarrow \quad x = -\dfrac{1}{a(b-a)}$. Hence, the solution is

$$\left(-\frac{1}{a(b-a)}, \frac{1}{b(b-a)}\right) = \left(\frac{1}{a^2-ab}, \frac{1}{b^2-ab}\right).$$

44.   Let x be the larger number and y be the other number. This gives

      $x + y = 2(x-y)$ $-x + 3y = 0$

         $x = 6 + 2y$ $\Rightarrow$ $\underline{x - 2y = 6}$

                                $y = 6.$

So $x = 6 + 2(6) = 18$. Therefore, the two numbers are 18 and 6.

46.   Let c be the number of children and a be the number of adults. This gives

           $c + a = 2200$ $\times -3$ $-3c - 3a = -6600$

    $1.50c + 4.00a = 5050$ $\times 2$ $\Rightarrow$ $\underline{3c + 8a = 10100}$

                                 $5a = 3500$ $\Leftrightarrow$ $a = 700.$

So $c + 700 = 2200 \quad \Leftrightarrow \quad c = 1500$. Therefore, the number of children admitted was 1500 and the number of adults was 700.

48.   Let x be speed of the boat in still water and y be speed of the river flow.

    Down river: $x + y = 20$ $\times 5$ $5x + 5y = 100$

      Up river: $2.5x - 2.5y = 20$ $\times 2$ $\Rightarrow$ $\underline{5x - 5y = 40}$

                               $10x \quad\quad = 140$ $\Leftrightarrow$ $x = 14$

So $14 + y = 20 \quad \Leftrightarrow \quad y = 6$. Therefore, the boat's speed is 14 mph and the current in the river flows at 6 mph.

50.   Let x and y be the number of milliliters of the two brine solutions.

      Quantity: $x + y = 1000$ $\times -1$ $-x - y = -1000$

   Concentrations: $0.05x + 0.20y = 0.14$ $\times 20$ $\Rightarrow$ $\underline{x + 4y = 2800}$

                                       $3y = 1800$ $\Leftrightarrow$ $y = 600$

So $x + 600 = 1000 \quad \Leftrightarrow \quad x = 400$. Therefore, 400 milliliters of the 5% solution and 600 milliliters of the 20% solution should be mixed.

52.   Let x be the number of pounds of Kenyan coffee and y be the number of pounds of Sri Lankan coffee. This gives

    $3.50x + 5.60y = 11.55$ $\times -10$ $35x + 56y = 115.5$

             $x + y = 3$ $\times -35$ $\Rightarrow$ $\underline{-35x - 35y = 105}$

                                 $21y = 10.5$ $\Leftrightarrow$ $y = 0.5$

So $x + (2.5) = 3 \quad \Leftrightarrow \quad x = 2.5$. Thus, 2.5 pounds of Kenyan coffee and 0.5 pounds of Sri Lankan coffee should be mixed.

54. Let x be the amount she invests at 5% and let y be the amount she invests at 8%.

Total invested: $x + y = 20,000$ $\times -5$ $-5x - 5y = -100,000$

Interest earned: $0.05x + 0.08y = 1180$ $\times 100$ $\Rightarrow$ $\underline{5x + 8y = 118,000}$

$3y = 18,000$ $\Leftrightarrow$ $y = 6,000$

So $x + 6,000 = 20,000$ $\Leftrightarrow$ $x = 14,000$. She invests \$14,000 at 5% and \$6,000 at 8%.

56. Let x be the length of time John drives and y be the length of time Mary drives. Then $y = x + 0.25$, so $-x + y = 0.25$, and multiplying by 40, we get $-40x + 40y = 10$. Comparing the distances, we get $60x = 40y + 35$, or $60x - 40y = 35$. This gives the system $\begin{cases} -40x + 40y = 10 \\ 60x - 40y = 35 \end{cases}$. Subtracting, we get $-40x + 40y = 10$

$\underline{60x - 40y = 35}$

$20x\qquad = 45$ $\Leftrightarrow$ $x = 2.25$.

So $y = 2.25 + 0.25 = 2.5$. Thus, John travels for $2\frac{1}{4}$ hours and Mary travels for $2\frac{1}{2}$ hours.

58. First, let us find the intersection point of the two lines. The y-coordinate of the intersection point is the height of the triangle. We have

$y = 2x - 4$ $2y = 4x - 8$

$y = -4x + 20$ Adding 2 times the first equation to the second equation gives $\underline{y = -4x + 20}$

$3y = 12$

So the triangle has height 4. Furthermore, $y = 2x - 4$ intersects the x-axis at $x = 2$, and $y = -4x + 20$ intersects the x-axis at $x = 5$. Thus the base has length $5 - 2 = 3$. Therefore, the area of the triangle is $A = \frac{1}{2} \cdot base \cdot height = \frac{1}{2} \cdot 3 \cdot 4 = 6$ square units.

Exercises 6.3

2. The equation $x^2 + y^2 + z^2 = 4$ is not linear, since it contains squares of variables.

4. The system $\begin{cases} x - 2y + 3z = 10 \\ 2x + 5y \quad\;\; = 2 \\ \quad\;\; y + 2z = 4 \end{cases}$ is linear.

6. $\begin{cases} x + y - 3z = 8 \\ \quad\; y - 3z = 5 \\ \quad\quad\quad z = -1 \end{cases}$ Substituting $z = -1$ into the second equation gives $y - 3(-1) = 5$ $\Leftrightarrow$

    $y = 2$. Substituting $z = -1$ and $y = 2$ into the first equation gives $x + 2 - 3(-1) = 8$ $\Leftrightarrow$
    $x = 3$. Thus the solution is $(3, 2, -1)$.

8. $\begin{cases} x - 2y + 3z = 10 \\ \quad\; 2y - z = 2 \\ \quad\quad\; 3z = 12 \end{cases}$ Solving we get $3z = 12$ $\Leftrightarrow$ $z = 4$. Substituting $z = 4$ into the second

    equation gives $2y - 4 = 2$ $\Leftrightarrow$ $y = 3$. Substituting $z = 4$ and $y = 3$ into the first equation gives
    $x - 2(3) + 3(4) = 10$ $\Leftrightarrow$ $x = 4$. Thus the solution is $(4, 3, 4)$.

10. $\begin{cases} 4x \quad\; + 3z = 10 \\ 2y - z = -6 \\ \quad\; \frac{1}{2}z = 2 \end{cases}$ Solving we get $\frac{1}{2}z = 4$ $\Leftrightarrow$ $z = 8$. Substituting $z = 8$ into the second

    equation gives $2y - 8 = -6$ $\Leftrightarrow$ $2y = 2$ $\Leftrightarrow$ $y = 1$. Substituting $z = 8$ into the first
    equation gives $4x + 3(8) = 10$ $\Leftrightarrow$ $4x = -14$ $\Leftrightarrow$ $x = -\frac{7}{2}$. Thus the solution is $\left(-\frac{7}{2}, 1, 8\right)$.

12. $\begin{cases} x + y - 3z = 3 \\ -2x + 3y + z = 2 \\ x - y + 2z = 0 \end{cases}$ Using the first equation we $\begin{matrix} 2x + 2y - 6z = 6 \\ -2x + 3y + z = 2 \\ \hline 5y - 5z = 8 \end{matrix}$ This gives the

    add 2 times the first equation
    to the second equation.

    system $\begin{cases} x + y - 3z = 3 \\ \quad\; 5y - 5z = 8. \\ x - y + 2z = 0 \end{cases}$

    $\begin{cases} x + y - 3z = 3 \\ -2x + 3y + z = 2 \\ x - y + 2z = 0 \end{cases}$ Or using the third equation, we $\begin{matrix} 2x - 2y + 4z = 0 \\ -2x + 3y + z = 2 \\ \hline y + 5z = 2 \end{matrix}$ This gives the

    add 2 times the third equation
    to the second equation.

    system $\begin{cases} x + y - 3z = 3 \\ \quad\; y + 5z = 2. \\ x - y + 2z = 0 \end{cases}$

14. $\begin{cases} x - 4y + z = 3 \\ \quad\; y - 3z = 10 \\ \quad\; 3y - 8z = 24 \end{cases}$ Using the first equation we $\begin{matrix} \frac{3}{4}x - 3y + \frac{3}{4}z = \frac{9}{4} \\ 3y - 8z = 24 \\ \hline \frac{3}{4}x \quad\; - \frac{29}{4}z = \frac{105}{4} \end{matrix}$ This gives the

    add $\frac{3}{4}$ times the first equation
    to the third equation.

    system $\begin{cases} x - 4y + z = 3 \\ \quad\; y - 3z = 10 \\ \frac{3}{4}x - \frac{29}{4}z = \frac{105}{4} \end{cases}$.

$$\begin{cases} x - 4y + z = 3 \\ \quad\ y - 3z = 10 \\ \quad\ 3y - 8z = 24 \end{cases}$$ Using the second equation, we subtract 3 times the second equation
from the second equation.

$$\begin{array}{r} 3y - 8z = 24 \\ -3y + 9z = -30 \\ \hline z = -6 \end{array}$$ This gives the

system $\begin{cases} x - 4y + z = 3 \\ \quad\ y - 3z = 10 \ . \\ \qquad\quad z = -6 \end{cases}$.

16. $\begin{cases} \quad x + y + \ z = 0 \\ -x + 2y + 5z = 3 \\ \quad 3x - y \qquad = 6 \end{cases}$ Using the first equation we
eliminate the x term from the
second and the third equation.

$$\begin{array}{l} x + y + z = 0 \\ -x + 2y + 5z = 3 \quad \text{and} \\ \hline \qquad\quad 3y + 6z = 3 \end{array}$$

$$\begin{array}{l} -3x - 3y - 3z = 0 \\ \ \ \ 3x - y \qquad = 6 \\ \hline -4y - 3z = 6 \end{array}$$ So we get
the system $\begin{cases} x + y + z = 0 \\ 3y + 6z = 3 \ , \\ -4y + -3z = 6 \end{cases}$ which we can
write as $\begin{cases} x + y + z = 0 \\ \ \ y + 2z = 1 \ . \\ -4y + -3z = 6 \end{cases}$

Eliminate the y
term from the
third equation. $\begin{array}{l} 4y + 8z = 4 \\ -4y + -3z = 6 \\ \hline 5z = 10 \end{array}$ To get
the system $\begin{cases} x + y + z = 0 \\ \ \ y + 2z = 1 \\ \quad\ 5z = 10 \end{cases}$

So $z = 2$ and $y + 2(2) = 1 \ \Leftrightarrow \ y = -3$. Then $x + (-3) + 2 = 0 \ \Leftrightarrow \ x = 1$. So the
solution is $(1, -3, 2)$.

18. $\begin{cases} \ x - y + 2z = 2 \\ 3x + y + 5z = 8 \\ 2x - y - 2z = -7 \end{cases}$ Using the first equation we
eliminate the x term from the
second and the third equation.

$$\begin{array}{l} -3x + 3y - 6z = -6 \\ \ \ \ 3x + y + 5z = 8 \quad \text{and} \\ \hline \qquad\quad 4y + -z = 2 \end{array}$$

$$\begin{array}{l} -2x + 2y - 4z = -4 \\ \ \ \ 2x - y - 2z = -7 \\ \hline y - 6z = -11 \end{array}$$ So we get
the system $\begin{cases} x - y + 2z = 2 \\ 4y + -z = 2 \ , \\ \ \ y - 6z = -11 \end{cases}$ which we can
write as $\begin{cases} x - y + 2z = 2 \\ \ \ y - 6z = -11 \ . \\ 4y + -z = 2 \end{cases}$

Eliminate the y
term from the
third equation. $\begin{array}{l} -4y + 24z = 44 \\ \ \ 4y + -z = 2 \\ \hline 23z = 46 \end{array}$ To get
the system $\begin{cases} x - y + 2z = 2 \\ \ \ y - 6z = -11 \\ \quad 23z = 46 \end{cases}$

So $z = 2$ and $y - 6(2) = -11 \ \Leftrightarrow \ y = 1$. Then $x + 1 + 2 = 2 \ \Leftrightarrow \ x = -1$. So the
solution is $(-1, 1, 2)$.

20. $\begin{cases} \ \ 2x + y \ - z = -8 \\ -x + y \ + z = 3 \\ -2x \qquad + 4z = 18 \end{cases}$ First
rewrite
as $\begin{cases} -x + y \ + z = 3 \\ \ \ 2x + y \ - z = -8 \ . \\ -2x \qquad + 4z = 18 \end{cases}$ Using the first equation we
eliminate the x term from the
second and the third equation.

$$\begin{array}{l} -2x + 2y + 2z = 6 \\ \ \ 2x + y - z = -8 \\ \hline 3y + z = -2 \end{array}$$ and $\begin{array}{l} 2x - 2y - 2z = -6 \\ -2x \qquad + 4z = 18 \\ \hline -2y + 2z = 12 \end{array}$ $\Leftrightarrow \ y - z = -6$ So we get
the system

$\begin{cases} -x + y + z = 3 \\ \ \ 3y + z = -2 \ , \\ \ \ y - z = -6 \end{cases}$ which we can
write as $\begin{cases} -x + y + z = 3 \\ \ \ y - z = -6 \ . \\ 3y + z = -2 \end{cases}$ Eliminate the y
term from the
third equation. $\begin{array}{l} -3y + 3z = 18 \\ \ \ 3y + z = -2 \\ \hline 4z = 16 \end{array}$

To get
the system $\begin{cases} -x + y + z = 3 \\ \ \ y - z = -6 \ . \\ \quad\ 4z = 16 \end{cases}$

So $z = 4$ and $y - 4 = -6 \ \Leftrightarrow \ y = -2$. Then $-x + (-2) + 4 = 3 \ \Leftrightarrow \ x = -1$. So the
solution is $(-1, -2, 4)$.

22. $\begin{cases} 2y + z = 3 \\ 5x + 4y + 3z = -1 \\ x - 3y = -2 \end{cases}$ First rewrite as $\begin{cases} x - 3y = -2 \\ 2y + z = 3 \\ 5x + 4y + 3z = -1 \end{cases}$ Using the first equation we eliminate the x term from the third equation.

$\begin{array}{l} -5x + 15y = 10 \\ \underline{5x + 4y + 3z = -1} \\ 19y + 3z = 9 \end{array}$ to get the system $\begin{cases} x - 3y = -2 \\ 2y + z = 3 \\ 19y + 3z = 9 \end{cases}$.

Eliminate the y term from the third equation by adding -19 times the second equation to 2 times the third equation

$\begin{array}{l} -38y - 19z = -57 \\ \underline{38y + 6z = 18} \\ -13z = -39 \end{array}$ To get the system

$\begin{cases} x - 3y = -2 \\ 2y + z = 3 \\ -13z = -39 \end{cases}$. So $z = 3$ and $2y + 3 = 3$ $\Leftrightarrow$ $y = 0$. Then $x - 3(0) = -2$ $\Leftrightarrow$ $x = -2$. So the solution is $(-2, 0, 3)$.

24. $\begin{cases} -x + 2y + 5z = 4 \\ x - 2z = 0 \\ 4x - 2y - 11z = 2 \end{cases}$ First rewrite as $\begin{cases} x - 2z = 0 \\ -x + 2y + 5z = 4 \\ 4x - 2y - 11z = 2 \end{cases}$. Using the first equation we eliminate the x term from the second and the third equation.

$\begin{array}{l} x - 2z = 0 \\ \underline{-x + 2y + 5z = 4} \\ 2y + 3z = 4 \end{array}$ and $\begin{array}{l} -4x + 8z = 0 \\ \underline{4x - 2y - 11z = 2} \\ -2y - 3z = 2 \end{array}$. So we get the system $\begin{cases} x - 2z = 0 \\ 2y + 3z = 4 \\ -2y - 3z = 2 \end{cases}$.

Eliminate the y term from the third equation. $\begin{array}{l} 2y + 3z = 4 \\ \underline{-2y - 3z = 2} \\ 0 = 6 \end{array}$ Since $0 = 6$ is false, this system is inconsistent.

26. $\begin{cases} x - 2y - 3z = 5 \\ 2x + y - z = 5 \\ 4x - 3y - 7z = 5 \end{cases}$ Using the first equation we eliminate the x term from the second and the third equation. $\begin{array}{l} -2x + 4y + 6z = -10 \\ \underline{2x + y - z = 5} \\ 5y + 5z = -5 \end{array}$ and

$\begin{array}{l} -4x + 8y + 12z = -20 \\ \underline{4x - 3y - 7z = 5} \\ 5y + 5z = -15 \end{array}$ So we get the system $\begin{cases} x - 2y - 3z = 5 \\ 5y + 5z = -5 \\ 5y + 5z = -15 \end{cases}$. Eliminate the y term from the third equation.

$\begin{array}{l} -5y - 5z = 5 \\ \underline{5y + 5z = -15} \\ 0 = -10 \end{array}$ Since $0 = -10$ is false, this system is inconsistent.

28. $\begin{cases} x - 2y + z = 3 \\ 2x - 5y + 6z = 7 \\ 2x - 3y - 2z = 5 \end{cases}$ Using the first equation we eliminate the x term from the second and the third equation. $\begin{array}{l} -2x + 4y - 2z = -6 \\ \underline{2x - 5y + 6z = 7} \\ -y + 4z = 1 \end{array}$ and

$\begin{array}{l} -2x + 4y - 2z = -6 \\ \underline{2x - 3y - 2z = 5} \\ y - 4z = -1 \end{array}$ So we get the system $\begin{cases} x - 2y + z = 3 \\ -y + 4z = 1 \\ y - 4z = -1 \end{cases}$. Eliminate the y term from the third equation. $\begin{array}{l} -y + 4z = 1 \\ \underline{y - 4z = -1} \\ 0 = 0 \end{array}$

Then we have $-y + 4t = 1$ $\Leftrightarrow$ $y = 4t - 1$. Substituting into the first equation, we have $x - 2(4t - 1) + t = 3$ $\Leftrightarrow$ $x = 7t + 1$. So the solutions are $(7t + 1, 4t - 1, t)$, where t is any real number.

30. $\begin{cases} 2x + 4y - z = 3 \\ x + 2y + 4z = 6 \\ x + 2y - 2z = 0 \end{cases}$ First rewrite as $\begin{cases} x + 2y - 2z = 0 \\ 2x + 4y - z = 3 \\ x + 2y + 4z = 6 \end{cases}$ Using the first equation we eliminate the x term from the second and the third equation.

$$-2x - 4y + 4z = 0$$
$$\underline{2x + 4y - z = 3}$$ and
$$3z = 3$$

$$-x - 2y + 2z = 0$$
$$\underline{x + 2y + 4z = 6}$$
$$6z = 6$$

So we get the system

$$\begin{cases} x + 2y - 2z = 0 \\ 3z = 3 \\ 6z = 6 \end{cases}.$$

Eliminate the z term from the third equation.

$$-6z = -6$$
$$\underline{6z = 6}$$
$$0 = 0$$

To get the system

$$\begin{cases} x + 2y - 2z = 0 \\ 3z = 3 \\ 0 = 0 \end{cases}$$

So $z = 1$, and substituting into the first equation we have $x + 2t - 2(1) = 0 \quad \Leftrightarrow \quad x = -2t + 2$. Thus, the solutions are $(-2t + 2, t, 1)$, where t is any real number.

32.

$$\begin{cases} x + y + z + w = 0 \\ x + y + 2z + 2w = 0 \\ 2x + 2y + 3z + 4w = 1 \\ 2x + 3y + 4z + 5w = 2 \end{cases}$$

Using the first equation we eliminate the x term from the other three equations.

$$-x - y - z - w = 0$$
$$\underline{x + y + 2z + 2w = 0}$$ and
$$z + w = 0$$

$$-2x - 2y - 2z - 2w = 0$$
$$\underline{2x + 2y + 3z + 4w = 1}$$
$$z + 2w = 1$$ and

$$-2x - 2y - 2z - 2w = 0$$
$$\underline{2x + 3y + 4z + 5w = 2}$$
$$y + 2z + 3w = 2$$

So we get the system

$$\begin{cases} x + y + z + w = 0 \\ z + w = 0 \\ z + 2w = 1 \\ y + 2z + 3w = 2 \end{cases},$$

which we can write as

$$\begin{cases} x + y + z + w = 0 \\ y + 2z + 3w = 2 \\ z + w = 0 \\ z + 2w = 1 \end{cases}.$$

Eliminate the z term from the fourth equation.

$$-z - w = 0$$
$$\underline{z + 2w = 1}$$
$$w = 1$$

To get the system

$$\begin{cases} x + y + z + w = 0 \\ y + 2z + 3w = 2 \\ z + w = 0 \\ w = 1 \end{cases}.$$

So $w = 1$ and $z + 1 = 0 \quad \Leftrightarrow \quad z = -1$. Then $y + 2(-1) + 3(1) = 2 \quad \Leftrightarrow \quad y = 1$ and $x + 1 + (-1) + 1 = 0 \quad \Leftrightarrow \quad x = -1$. So the solution is $(-1, 1, -1, 1)$.

34. Let x be the amount she invests at 4%, y be the amount she invests at 6%, and let z be the amount she invests at 8%. Modeling these we get the equations:

Total money $x + y + z = 100{,}000$
Annual income $0.04x + 0.06y + 0.08z = 6{,}700$ $\Leftrightarrow$ $\begin{cases} x + y + z = 100{,}000 \\ 4x + 6y + 8z = 670{,}000 \\ y - z = 0 \end{cases}.$
Equal amounts $y = z$

Using the first equation we eliminate the x term from the second equation.

$$-4x - 4y - 4z = -400{,}000$$
$$\underline{4x + 6y + 8z = \quad 670{,}000}$$
$$2y + 4z = \quad 270{,}000$$

So we get the system

$$\begin{cases} x + y + z = 100{,}000 \\ 2y + 4z = 270{,}000 \\ y - z = 0 \end{cases}$$

Eliminate the y term from the third equation.

$$-2y - 4z = -270{,}000$$
$$\underline{2y - 2z = 0}$$
$$-6z = -270{,}000$$

Finally, we get the system

$$\begin{cases} x + y + z = 100{,}000 \\ 2y + 4z = 270{,}000 \\ -6z = -270{,}000 \end{cases}.$$

So $z = 45{,}000$ and $y = z = 45{,}000$. Since $x + 45{,}000 + 45{,}000 = 100{,}000 \quad \Leftrightarrow \quad x = 10{,}000$. She must invest $10,000 in short-term bonds, $45,000 in intermediate-term bonds, and $45,000 in long-term bonds.

36. $\begin{cases} I_1 + I_2 - I_3 = 0 \\ 16I_1 - 8I_2 \quad\quad = 4 \\ \quad\quad 8I_2 + 4I_3 = 5 \end{cases}$ Using the first equation we eliminate the I_1 term from the second equation. $\quad \begin{array}{l} -16I_1 - 16I_2 + 16I_3 = 0 \\ \underline{16I_1 - 8I_2 \quad\quad\quad = 4} \\ \quad\quad -24I_2 + 16I_3 = 4 \end{array}$ So we get the system

$\begin{cases} I_1 + I_2 - I_3 = 0 \\ -24I_2 + 16I_3 = 4 \\ \quad\quad 8I_2 + 4I_3 = 5 \end{cases}$ Eliminate the I_2 term from the third equation. $\quad \begin{array}{l} -24I_2 + 16I_3 = 4 \\ \underline{24I_2 + 12I_3 = 15} \\ \quad\quad 28I_3 = 19 \end{array}$ To get the system

$\begin{cases} I_1 + I_2 - I_3 = 0 \\ \quad\quad 8I_2 + 4I_3 = 5 \\ \quad\quad\quad 28I_3 = 19 \end{cases}$

So $I_3 = \frac{19}{28} \approx 0.68$ and $8I_2 + 4(\frac{19}{28}) = 5 \iff I_2 = \frac{2}{7} \approx 0.29$. Then $I_1 + \frac{2}{7} - \frac{19}{28} = 0 \iff I_1 = \frac{11}{28} \approx 0.39$.

38. Let a, b, and c be the number of shares of Stock A, Stock B, and Stock C in the investor's portfolio. Since the total value remains unchanged we get the following system

$\begin{cases} 10a + 25b + 29c = 74{,}000 \\ 12a + 20b + 32c = 74{,}000 \\ 16a + 15b + 32c = 74{,}000 \end{cases}$ Using the first equation we eliminate the a term from the second and the third equation. $\quad \begin{array}{l} 60a + 150b + 174c = \quad 444{,}000 \\ \underline{-60a - 100b - 160c = -370{,}000} \\ \quad\quad 50b + 14c = \quad 74{,}000 \end{array}$

and $\begin{array}{l} 80a + 200b + 232c = \quad 592{,}000 \\ \underline{-80a - 75b - 160c = -370{,}000} \\ \quad\quad 125b + 72c = \quad 222{,}000 \end{array}$ So we get the system $\begin{cases} 10a + 25b + 29c = \quad 74{,}000 \\ \quad\quad 50b + 14c = \quad 74{,}000 \\ \quad\quad 125b + 72c = 222{,}000 \end{cases}$

Eliminate the b term from the third equation. $\quad \begin{array}{l} 250b + 70c = \quad 370{,}000 \\ \underline{-250b - 144c = -444{,}000} \\ \quad\quad -74c = \quad -74{,}000 \end{array}$ So $-74c = -74{,}000 \iff c = 1{,}000$. Back substituting we have $50b + 14(1000) = 74{,}000 \iff 50b = 60{,}000 \iff b = 1{,}200$. And finally $10a + 25(1200) + 29(1000) = 74{,}000 \quad 10a + 30{,}000 + 29{,}000 = 74{,}000 \iff 10a = 15{,}000 \iff a = 1{,}500$. Thus the portfolio consists of 1,500 shares of Stock A, 1,200 shares of Stock B, and 1,000 shares of Stock C.

Exercises 6.4

2. $y \geq -2$

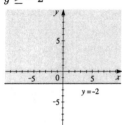

4. $y < x + 2$

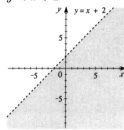

6. $y < -x + 5$

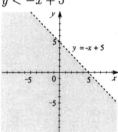

8. $3x + 4y + 12 > 0$

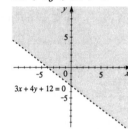

10. $-x^2 + y \geq 10$

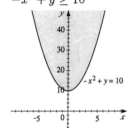

12. $x^2 + y^2 \geq 9$

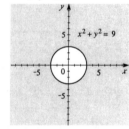

14. $x^2 + (y-1)^2 \leq 1$

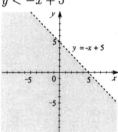

16. The boundary is a solid curve, so we have the inequality $y \leq x^2 + 2$. We take the test point $(0,0)$ and verify that it satisfies the inequality: $0 \leq 0^2 + 2$.

18. The boundary is a solid curve, so we have the inequality $y \geq x^3 - 4x$. We take the test point $(1,1)$ and verify that it satisfies the inequality: $1 \geq 1^3 - 4(1)$.

20. $\begin{cases} 2x + 3y > 12 \\ 3x - y < 21 \end{cases}$ The vertices occur where $\begin{cases} 2x + 3y = 12 \\ 3x - y = 21 \end{cases}$ $\Leftrightarrow$

$\begin{cases} 2x + 3y = 12 \\ 9x - 3y = 63 \end{cases}$, and adding the two equations gives $11x = 75$

$\Leftrightarrow \quad x = \frac{75}{11}$. Then $2\left(\frac{75}{11}\right) + 3y = 12 \quad \Leftrightarrow$

$3y = \frac{132 - 150}{11} = \frac{-18}{11} \quad \Leftrightarrow \quad y = -\frac{6}{11}$. Therefore, the vertex is

$\left(\frac{75}{11}, -\frac{6}{11}\right)$, and the solution set is not bounded.

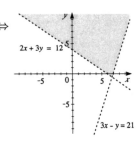

22. $\begin{cases} x - y > 0 \\ 4 + y \le 2x \end{cases}$

The vertices occur where $\begin{cases} x - y = 0 \quad \Leftrightarrow \quad y = x \\ 4 + y = 2x \end{cases}$.

Substituting for y gives $4 + x = 2x \quad \Leftrightarrow \quad x = 4$, so $y = 4$.

Hence, the vertex is $(4, 4)$, and the solution set is not bounded.

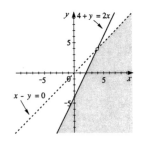

24. $\begin{cases} x > 2 \\ y < 12 \\ 2x - 4y > 8 \end{cases}$ From the graph, the vertices occur at $(2, -1)$

and $(28, 12)$. The solution set is not bounded.

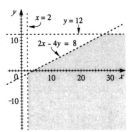

26. $\begin{cases} y \ge x^2 \\ x + y \ge 6 \end{cases}$ The vertices occur where $\begin{cases} y = x^2 \\ x + y = 6 \end{cases}$.

Substituting for y gives $x^2 + x = 6 \quad \Leftrightarrow \quad x^2 + x - 6 = 0 \quad \Leftrightarrow$

$(x + 3)(x - 2) = 0 \quad \Rightarrow \quad x = -3, x = 2$. Since $y = x^2$, the

vertices are $(-3, 9)$ and $(2, 4)$, and the solution set is not bounded.

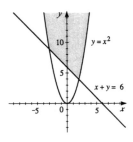

28. $\begin{cases} x > 0 \\ y > 0 \\ x + y < 10 \\ x^2 + y^2 > 9 \end{cases}$ From the graph, the vertices are $(0, 3)$, $(0, 10)$,

$(3, 0)$, and $(10, 0)$. The solution set is bounded.

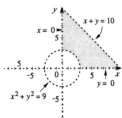

30. $\begin{cases} x^2 + y^2 < 9 \\ 2x + y^2 \geq 1 \end{cases}$ The vertices occur where $\begin{cases} x^2 + y^2 = 9 \\ 2x + y^2 = 1 \end{cases}$.

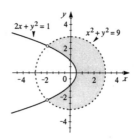

Subtracting the equations gives $x^2 - 2x = 8$ ⇔
$x^2 - 2x - 8 = 0$ ⇔ $(x-4)(x+2) = 0$ ⇔ $x = -2$,
$x = 4$. Therefore, the vertices are $(-2, -\sqrt{5})$ and $(-2, \sqrt{5})$,
since $x = 4$ does not give a real solution for y. The solution set
is bounded.

32. $\begin{cases} y < x + 6 \\ 3x + 2y \geq 12 \\ x - 2y \leq 2 \end{cases}$

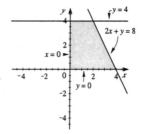

To find where the line $y = x + 6$ intersects the lines $3x + 2y = 12$
and $x - 2y = 2$ we substitute for y: $3x + 2(x+6) = 12$ ⇔
$x = 0$, and $y = 6$; $x - 2(x+6) = 2$ ⇔ $x = -14$, and
$y = -8$. Next, adding the equations $3x + 2y = 12$ and $x - 2y = 2$
yields $4x = 14$ ⇔ $x = \frac{7}{2}$. So these lines intersect at the point
$(\frac{7}{2}, \frac{3}{4})$. Since the vertex $(-14, -8)$ is not part of the solution set, the vertices are $(0, 6)$ and $(\frac{7}{2}, \frac{3}{4})$,
and the solution set is not bounded.

34. $\begin{cases} x \geq 0 \\ y \geq 0 \\ y \leq 4 \\ 2x + y \leq 8 \end{cases}$

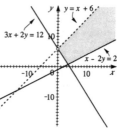

The points of intersection are $(0, 8)$, $(0, 4)$, $(4, 0)$, $(2, 4)$, and $(3, 2)$.
However, the point $(0, 8)$ is not in the solution set. Therefore, the
vertices are $(0, 4)$, $(2, 4)$, $(4, 0)$, and $(0, 0)$. The solution set is
bounded.

36. $\begin{cases} x + y > 12 \\ y < \frac{1}{2}x - 6 \\ 3x + y < 6 \end{cases}$

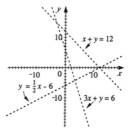

Graphing these inequalities, we see that there are no points that
satisfy all three, and hence the solution set is empty.

38. $\begin{cases} x^2 - y \geq 0 \\ x + y < 6 \\ x - y < 6 \end{cases}$

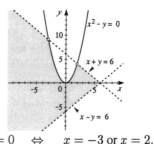

Adding the equations $x + y = 6$ and $x - y = 6$ yields $2x = 12$
⇔ $x = 6$. So these equations intersect at the point $(6, 0)$. To
find where $x^2 - y = 0$ and $x + y = 6$ intersect, we solve the first
for y, giving $y = x^2$, and then substitute into the second equation
to get $x + x^2 = 6$ ⇔ $x^2 + x - 6 = 0$ ⇔ $(x+3)(x-2) = 0$ ⇔ $x = -3$ or $x = 2$.

When $x = -3$, we have $y = 9$, and when $x = 2$, we have $y = 4$, so the points of intersection are $(-3, 9)$ and $(2, 4)$. Substituting for $y = x^2$ into the equation $x - y = 6$ gives $x - x^2 = 6$, which has no solution. Thus the vertices are $(6, 0)$, $(-3, 9)$, and $(2, 4)$ (however, they are not in the solution set). The solution set is not bounded.

40. $\begin{cases} y \geq x^3 \ \cdot \\ y \leq 2x + 4 \\ x + y \geq 0 \end{cases}$

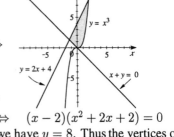

The curves $y = x^3$ and $x + y = 0$ intersect when $x^3 + x = 0$ $\Leftrightarrow$ $x = 0$ $\Rightarrow$ $y = 0$. The lines $x + y = 0$ and $y = 2x + 4$ intersect when $-x = 2x + 4$ $\Leftrightarrow$ $3x = -4$ $\Leftrightarrow$ $x = -\frac{4}{3}$ $\Rightarrow$ $y = \frac{4}{3}$. To find where $y = x^3$ and $y = 2x + 4$ intersect, we substitute for y and get $x^3 = 2x + 4$ $\Leftrightarrow$ $x^3 - 2x - 4 = 0$ $\Leftrightarrow$ $(x - 2)(x^2 + 2x + 2) = 0$ $\Rightarrow$ $x = 2$ (the other factor has no real solution). When $x = 2$ we have $y = 8$. Thus the vertices of the region are $(0, 0)$, $(-\frac{4}{3}, \frac{4}{3})$, and $(2, 8)$. The solution set is bounded.

42. $\begin{cases} x + y \geq 12 \\ 2x + y \leq 24 \\ x - y \geq -6 \end{cases}$

Using the graphing calculator we find the region as shown. The vertices are $(3, 9)$, $(6, 12)$, and $(12, 0)$.

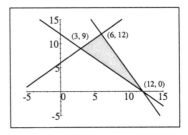

44. $\begin{cases} y \geq x^3 \\ 2x + y \geq 0 \\ y \leq 2x + 6 \end{cases}$

Using the graphing calculator we find the region as shown. The vertices are $(0, 0)$, $(2.2, 10.3)$ and $(-1.5, 3)$.

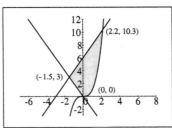

46. Let x = number of chairs made and y = number of tables made. Then, the following system of inequalities holds:

$\begin{cases} 2x + 3y \leq 12 \\ 2 + yx \leq 8 \\ x \geq 0 \\ y \geq 0 \end{cases}$. The intersection points are $(0, 4)$, $(0, 8)$, $(6, 0)$, $(4, 0)$, $(0, 0)$, and $(3, 2)$. Since the points $(0, 8)$ and $(6, 0)$ are not in the solution set, the vertices are $(0, 4)$, $(0, 0)$, $(4, 0)$, and $(3, 2)$.

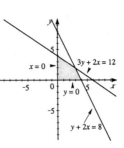

48. Let $x =$ ounces of fish used in each can and $y =$ ounces of beef used in each can. Then, the following system of inequalities holds:

$$\begin{cases} 12x + 6y \geq 60 \\ 3x + 9y \geq 45 \\ x \geq 0 \\ y \geq 0 \end{cases}$$

From the graph, we see that the vertices are $(15, 0)$, $(3, 4)$, and $(0, 10)$.

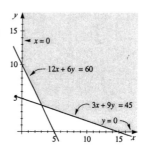

Exercises 6.5

2. $\dfrac{x}{x^2+3x-4} = \dfrac{x}{(x-1)(x+4)} = \dfrac{A}{(x-1)} + \dfrac{B}{(x+4)}$

4. $\dfrac{1}{x^4-x^3} = \dfrac{1}{x^3(x-1)} = \dfrac{A}{x} + \dfrac{B}{x^2} + \dfrac{C}{x^3} + \dfrac{D}{x-1}$

6. $\dfrac{1}{x^4-1} = \dfrac{1}{(x^2-1)(x^2+1)} = \dfrac{1}{(x-1)(x+1)(x^2+1)} = \dfrac{A}{x-1} + \dfrac{B}{x+1} + \dfrac{Cx+D}{x^2+1}$

8. $\dfrac{x^4+x^2+1}{x^2(x^2+4)^2} = \dfrac{A}{x} + \dfrac{B}{x^2} + \dfrac{Cx+D}{x^2+4} + \dfrac{Ex+F}{(x^2+4)^2}$

10. Since $(x^3-1)(x^2-1) = (x-1)(x^2+x+1)(x-1)(x+1) = (x-1)^2(x+1)(x^2+x+1)$, we have

$$\frac{1}{(x^3-1)(x^2-1)} = \frac{1}{(x-1)^2(x+1)(x^2+x+1)} = \frac{A}{x-1} + \frac{B}{(x-1)^2} + \frac{C}{x+1} + \frac{Dx+E}{x^2+x+1}.$$

12. $\dfrac{2x}{(x-1)(x+1)} = \dfrac{A}{x-1} + \dfrac{B}{x+1}$. Multiplying by $(x-1)(x+1)$, we get

$2x = A(x+1) + B(x-1)$ $\Leftrightarrow$ $2x = Ax + A + Bx - B$. Thus

$\begin{cases} A+B = 2 \\ A-B = 0 \end{cases}$. Adding we $2A = 2$ $\Leftrightarrow$ $A = 1$. Since $A - B = 0$ $\Leftrightarrow$ $B = A, B = 1$. The

required partial fraction decomposition is $\dfrac{2x}{(x-1)(x+1)} = \dfrac{1}{x-1} + \dfrac{1}{x+1}$.

14. $\dfrac{x+6}{x(x+3)} = \dfrac{A}{x} + \dfrac{B}{x+3}$. Multiplying by $x(x+3)$ we get

$x+6 = A(x+3) + xB$ $\Leftrightarrow$ $x+6 = Ax + 3A + Bx = (A+B)x + 3A$. Thus

$\begin{cases} A+B = 1 \\ \quad 3A = 6 \end{cases}$. Now $3A = 6$ $\Leftrightarrow$ $A = 2$, and $2 + B = 1$ $\Leftrightarrow$ $B = -1$. The required

partial fraction decomposition is $\dfrac{x+6}{x(x+3)} = \dfrac{2}{x} - \dfrac{1}{x+3}$.

16. $\dfrac{x-12}{x^2-4x} = \dfrac{x-12}{x(x-4)} = \dfrac{A}{x} + \dfrac{B}{x-4}$. Multiplying by $x(x-4)$, we get $x - 12 = A(x-4) + Bx$

$= Ax - 4A + Bx = (A+B)x - 4A$. Thus we must solve the system $\begin{cases} A+B = 1 \\ -4A = -12 \end{cases}$. This

gives $-4A = -12$ $\Leftrightarrow$ $A = 3$, and $3 + B = 1$ $\Leftrightarrow$ $B = -2$. Thus $\dfrac{x-12}{x^2-4x} = \dfrac{3}{x} - \dfrac{2}{x-4}$.

18. $\dfrac{2x+1}{x^2+x-2} = \dfrac{2x+1}{(x+2)(x-1)} = \dfrac{A}{x+2} + \dfrac{B}{x-1}$. Thus, $2x+1 = A(x-1) + B(x+2)$

$= (A+B)x + (-A+2B)$, and so $\begin{cases} A+B=2 \\ -A+2B=1 \end{cases}$. Adding the two equations, we get $3B = 3$

$\Leftrightarrow \quad B=1$. Thus $A+1=2 \quad \Leftrightarrow \quad A=1$. Therefore, $\dfrac{2x+1}{x^2+x-2} = \dfrac{1}{x+2} + \dfrac{1}{x-1}$.

20. $\dfrac{8x-3}{2x^2-x} = \dfrac{8x-3}{x(2x-1)} = \dfrac{A}{x} + \dfrac{B}{2x-1}$. Hence, $8x-3 = A(2x-1) + Bx = (2A+B)x + (-A)$.

$\begin{cases} 2A+B=8 \\ -A=-3 \end{cases}$. So $-A = -3 \quad \Leftrightarrow \quad A=3$, and $2(3)+B=8 \quad \Leftrightarrow \quad B=2$. Therefore,

$\dfrac{8x-3}{2x^2-x} = \dfrac{3}{x} + \dfrac{2}{2x-1}$.

22. $\dfrac{7x-3}{x^3+2x^2-3x} = \dfrac{7x-3}{x(x+3)(x-1)} = \dfrac{A}{x} + \dfrac{B}{x+3} + \dfrac{C}{x-1}$. Hence,

$7x-3 = A(x+3)(x-1) + Bx(x-1) + Cx(x+3)$

$= A(x^2+2x-3) + B(x^2-x) + C(x^2+3x) = (A+B+C)x^2 + (2A-B+3C)x - 3A$.

Thus $\begin{cases} A+B+C=0 & \text{Eq. 1: Coefficients of } x^2 \\ 2A-B+3C=7 & \text{Eq. 2: Coefficients of } x \\ -3A=-3 & \text{Eq. 3: Constant terms} \end{cases}$. So $-3A = -3 \quad \Leftrightarrow \quad A=1$.

Substituting, the system reduces to

$\begin{cases} 1+B+C=0 \\ 2-B+3C=7 \end{cases}$ Adding these two equations, we get $3+4C = 7 \quad \Leftrightarrow \quad C=1$. Thus

$1+B+1=0 \Leftrightarrow B=-2$. Therefore, $\dfrac{7x-3}{x^3+2x^2-3x} = \dfrac{1}{x} - \dfrac{2}{x+3} + \dfrac{1}{x-1}$.

24. $\dfrac{-3x^2-3x+27}{(x+2)(2x^2+3x-9)} = \dfrac{-3x^2-3x+27}{(x+2)(2x-3)(x+3)} = \dfrac{A}{x+2} + \dfrac{B}{2x-3} + \dfrac{C}{x+3}$. Thus,

$-3x^2-3x+27 = A(2x-3)(x+3) + B(x+2)(x+3) + C(x+2)(2x-3)$

$= A(2x^2+3x-9) + B(x^2+5x+6) + C(2x^2+x-6)$

$= (2A+B+2C)x^2 + (3A+5B+C)x + (-9A+6B-6C)$.

This leads to the following system:

$\begin{cases} 2A+B+2C=-3 & \text{Eq. 1: Coefficients of } x^2 \\ 3A+5B+C=-3 & \text{Eq. 2: Coefficients of } x \\ -9A+6B-6C=27 & \text{Eq. 3: Constant terms} \end{cases}$ We start by dividing the third equation by 3 to get the system

$\begin{cases} \text{Eq. 1} & 2A+B+2C=-3 \\ \text{Eq. 2:} & 3A+5B+C=-3 \\ \text{Eq. 3:} & -3A+2B-2C=9 \end{cases}$ Then $\begin{array}{r} -5 \times \text{Eq. 1} \quad -10A-5B-10C=15 \\ + \text{Eq. 2} \quad 3A+5B+C=-3 \\ \hline \text{Eq. 4} \quad -7A \quad -9C=12 \end{array}$

$\begin{array}{r} 2 \times \text{Eq. 1} \quad 4A+2B+4C=-6 \\ - \text{Eq. 2} \quad 3A-2B+2C=-9 \\ \hline \text{Eq. 5} \quad -7A \quad +6C=-15 \end{array}$ Thus we get the system $\begin{cases} \text{Eq. 1} & 2A+B+2C=-3 \\ \text{Eq. 4:} & -7A \quad -9C=12 \\ \text{Eq. 5:} & 7A \quad +6C=-15 \end{cases}$

Eq. 4 $\quad -7A - 9C = 12$

Now $\quad$ +Eq. 5 $\quad 7A + 6C = -15$ $\quad$ Hence, $-3C = -3$ $\quad \Leftrightarrow \quad C = 1$; then $-7A - 6(1) = 15$

$\quad$ Eq. 6 $\quad\quad\quad -3C = -3$

$\Leftrightarrow \quad -7A = 21 \quad \Leftrightarrow \quad A = -3$; and $2(-3) + B + 2(1) = -3 \quad \Leftrightarrow \quad B = 1$. Therefore,

$$\frac{-3x^2 - 3x + 27}{(x+2)(2x^2 + 3x - 9)} = \frac{-3}{x+2} + \frac{1}{2x-3} + \frac{1}{x+3}.$$

26. $\quad \dfrac{3x^2 + 5x - 13}{(3x + 2)(x^2 - 4x + 4)} = \dfrac{3x^2 + 5x - 13}{(3x + 2)(x - 2)^2} = \dfrac{A}{3x + 2} + \dfrac{B}{x - 2} + \dfrac{C}{(x - 2)^2}$. Thus,

$3x^2 + 5x - 13 = A(x - 2)^2 + B(3x + 2)(x - 2) + C(3x + 2)$

$\quad = A(x^2 - 4x + 4) + B(3x^2 - 4x - 4) + C(3x + 2)$

$\quad = (A + 3B)x^2 + (-4A - 4B + 3C)x + (4A - 4B + 2C).$

This leads to the following system:

$\begin{cases} A + 3B = 3 & \text{Eq. 1: Coefficients of } x^2 \\ -4A - 4B + 3C = 5 & \text{Eq. 2: Coefficients of } x \quad\quad \text{Then} \\ 4A - 4B + 2C = -13 & \text{Eq. 3: Constant terms} \end{cases}$

$4 \times$ Eq. 1 $\quad 4A + 12B = 12$

$\quad$ + Eq. 2 $\quad -4A - 4B + 3C = 5$

$\quad$ Eq. 4 $\quad\quad\quad 8B + 3C = 17$

and $\quad \begin{array}{l} 4 \times \text{Eq. 1} \quad 4A + 12B = 12 \\ - \text{Eq. 3} \quad -4A + 4B + 2C = 13 \\ \hline \text{Eq. 5} \quad\quad\quad 16B + 2C = 25 \end{array}$ $\quad \begin{array}{l} \text{From which we} \\ \text{get the system} \end{array}$ $\quad \begin{cases} \text{Eq. 1} \quad A + 3B = -3 \\ \text{Eq. 4:} \quad\quad 8B + 3C = 17 \\ \text{Eq. 5:} \quad\quad 16B - 2C = 25 \end{cases}$

Then $\quad \begin{array}{l} 2 \times \text{Eq. 4} \quad 16B + 6C = 34 \\ - \text{Eq. 5} \quad -16B + 2C = -25 \\ \hline \text{Eq. 6} \quad\quad\quad 8C = 9 \end{array}$ $\quad$ Hence, $8C = 9 \Leftrightarrow C = \frac{9}{8}$;

$8B + 3(\frac{9}{8}) = 8B + \frac{27}{8} = 17 \quad \Leftrightarrow \quad B = \frac{109}{64}$; and $A + 3(\frac{109}{64}) = A + \frac{327}{64} = 3 \quad \Leftrightarrow \quad A = -\frac{135}{64}.$

Therefore, $\dfrac{3x^2 + 5x - 13}{(3x + 2)(x - 2)^2} = \dfrac{-\frac{135}{64}}{3x + 2} + \dfrac{\frac{109}{64}}{x - 2} + \dfrac{\frac{9}{8}}{(x - 2)^2}.$

28. $\quad \dfrac{x - 4}{(2x - 5)^2} = \dfrac{A}{2x - 5} + \dfrac{B}{(2x - 5)^2}$. Hence, $x - 4 = A(2x - 5) + B = 2Ax + (-5A + B)$, and so

$\begin{cases} 2A = 1 \\ -5A + B = -4 \end{cases}$. $A = \frac{1}{2}$, and $-5(\frac{1}{2}) + B = -4 \Leftrightarrow B = -\frac{3}{2}$. Therefore,

$$\frac{x - 4}{(2x - 5)^2} = \frac{1/2}{2x - 5} - \frac{3/2}{(2x - 5)^2}.$$

30. $\quad \dfrac{x^3 - 2x^2 - 4x + 3}{x^4} = \dfrac{A}{x} + \dfrac{B}{x^2} + \dfrac{C}{x^3} + \dfrac{D}{x^4}$. Hence, $x^3 - 2x^2 - 4x + 3 = Ax^3 + Bx^2 + Cx + D.$

Thus $A = 1$; $B = -2$; $C = -4$; and $D = 3$. Therefore,

$$\frac{x^3 - 2x^2 - 4x + 3}{x^4} = \frac{1}{x} - \frac{2}{x^2} - \frac{4}{x^3} + \frac{3}{x^4}.$$

32. $\dfrac{-2x^2 + 5x - 1}{x^4 - 2x^3 + 2x - 1} = \dfrac{-2x^2 + 5x - 1}{(x - 1)(x^3 - x^2 - x + 1)} = \dfrac{-2x^2 + 5x - 1}{(x - 1)^3(x + 1)}$

$= \dfrac{A}{x + 1} + \dfrac{B}{x - 1} + \dfrac{C}{(x - 1)^2} + \dfrac{D}{(x - 1)^3}$. Thus,

$-2x^2 + 5x - 1 = A(x - 1)^3 + B(x + 1)(x - 1)^2 + C(x + 1)(x - 1) + D(x + 1)$

$= A(x^3 - 3x^2 + 3x - 1) + B(x + 1)(x^2 - 2x + 1) + C(x^2 - 1) + D(x + 1)$

$= A(x^3 - 3x^2 + 3x - 1) + B(x^3 - x^2 - x + 1) + C(x^2 - 1) + D(x + 1)$

$= (A + B)x^3 + (-3A - B + C)x^2 + (3A - B + D)x + (-A + B - C + D)$, which leads to the

system:

$$\begin{cases} A + B &= 0 \quad \text{Eq. 1: Coefficients of } x^3 \\ -3A - B + C &= -2 \quad \text{Eq. 2: Coefficients of } x^2 \\ 3A - B \quad + D = 5 \quad \text{Eq. 3: Coefficients of } x \\ -A + B - C + D = -1 \quad \text{Eq. 4: Constant terms} \end{cases}$$ Then

$\begin{array}{ll} 3 \times \text{Eq. 1} & 3A + 3B \quad\quad = 0 \\ + \text{Eq. 2} & -3A - \;\; B + C = -2 \, , \\ \hline \text{Eq. 5} & 2B + C = -2 \end{array}$

$\begin{array}{ll} -3 \times \text{Eq. 1} & -3A - 3B \quad\quad = 0 \\ + \;\; \text{Eq. 3} & 3A - \;\; B + D = 5 \\ \hline \text{Eq. 6} & -4B + D = 5 \end{array}$ and $\begin{array}{ll} \text{Eq. 1} & A + B \quad\quad\quad = 0 \\ + \text{Eq. 4} & -A + B - C + D = -1 \\ \hline \text{Eq. 7} & 2B - C + D = -1 \end{array}$

Thus we get the system $\begin{cases} \text{Eq. 1} \quad A + B \quad\quad\quad = 0 \\ \text{Eq. 5:} \quad\quad 2B + C \quad\quad = -2 \\ \text{Eq. 6:} \quad\quad -4B \quad + D = 5 \\ \text{Eq. 7:} \quad\quad 2B - C + D = -1 \end{cases}$ Then $\begin{array}{ll} 2 \times \text{Eq. 5} & 4B + 2C \quad\quad = -4 \\ + \text{Eq. 6} & -4B \quad\quad + D = 5 \\ \hline \text{Eq. 8} & 2C + D = 1 \end{array}$ and

$\begin{array}{ll} \text{Eq. 5} & 2B + C \quad\quad = -2 \\ - \text{Eq. 7} & -2B + C - D = 1 \\ \hline \text{Eq. 9} & 2C - D = -1 \end{array}$ So we get the system $\begin{cases} \text{Eq. 1} \quad A + B \quad\quad\quad = 0 \\ \text{Eq. 5:} \quad\quad 2B + C \quad\quad = -2 \\ \text{Eq. 8:} \quad\quad\quad 2C + D = 1 \\ \text{Eq. 9:} \quad\quad\quad 2C - D = -1 \end{cases}$ Finally,

$\begin{array}{ll} \text{Eq. 8} & 2C + D = 1 \\ -\text{Eq. 9} & -2C + D = 1 \\ \hline \text{Eq. 10} & 2D = 2 \end{array}$ Hence, $2D = 2 \quad \Leftrightarrow \quad D = 1; \;\; 2C + 1 = 1 \quad \Leftrightarrow \quad C = 0;$

$2B + 0 = -2 \quad \Leftrightarrow \quad B = -1;$ and $A - 1 = 0 \quad \Leftrightarrow \quad A = 1.$ Therefore,

$\dfrac{-2x^2 + 5x - 1}{x^4 - 2x^3 + 2x - 1} = \dfrac{1}{x + 1} - \dfrac{1}{x - 1} + \dfrac{1}{(x - 1)^3}.$

34. $\dfrac{3x^2 + 12x - 20}{x^4 - 8x^2 + 16} = \dfrac{3x^2 + 12x - 20}{(x^2 - 4)^2} = \dfrac{3x^2 + 12x - 20}{(x + 2)^2(x - 2)^2}$

$= \dfrac{A}{x + 2} + \dfrac{B}{(x + 2)^2} + \dfrac{C}{x - 2} + \dfrac{D}{(x - 2)^2}$. Thus,

$3x^2 + 12x - 20 = A(x + 2)(x - 2)^2 + B(x - 2)^2 + C(x + 2)^2(x - 2) + D(x + 2)^2$

$= A(x^3 - 2x^2 - 4x + 8) + B(x^2 - 4x + 4) + C(x^3 + 2x^2 - 4x - 8) + D(x^2 + 4x + 4)$

$= (A + C)x^3 + (-2A + B + 2C + D)x^2 + (-4A - 4B - 4C + 4D)x$

$+ (8A + 4B - 8C + 4D)$, which leads to the system:

$$\begin{cases} A & + C & = 0 \\ -2A + B + 2C + D = 3 \\ -4A - 4B - 4C + 4D = 12 \\ 8A + 4B - 8C + 4D = -20 \end{cases} \begin{matrix} \text{Eq. 1: Coefficients of } x^3 \\ \text{Eq. 2: Coefficients of } x^2 \\ \text{Eq. 3: Coefficients of } x \\ \text{Eq. 4: Constant terms} \end{matrix} \quad \text{Then}$$

$$\begin{matrix} 2 \times \text{Eq. 1} \\ + \text{Eq. 2} \\ \hline \text{Eq. 5} \end{matrix} \quad \begin{matrix} A & + 2C & = 0 \\ -2A + B + 2C + D = 3 \\ \hline B + 4C + D = 3 \end{matrix}, \quad \begin{matrix} 4 \times \text{Eq. 1} \\ + \text{Eq. 3} \\ \hline \text{Eq. 6} \end{matrix} \quad \begin{matrix} 4A & + 4C & = 0 \\ -4A - 4B - 4C + 4D = 12 \\ \hline -4B & + 4D = 12 \end{matrix} \quad \text{and}$$

$$\begin{matrix} 8\text{Eq. 1} \\ - \text{Eq. 4} \\ \hline \text{Eq. 7} \end{matrix} \quad \begin{matrix} 8A & + 8C & = 0 \\ -8A - 4B + 8C - 4D = 20 \\ \hline -4B + 16C - 4D = 20 \end{matrix} \quad \begin{matrix} \text{Thus we get} \\ \text{the system} \end{matrix} \quad \begin{cases} \text{Eq. 1} & A & + C & = 0 \\ \text{Eq. 5:} & B + 4C + D = 3 \\ \text{Eq. 6:} & -4B & + 4D = 12 \\ \text{Eq. 7:} & -4B + 16C - 4D = 20 \end{cases}$$

$$\text{Then} \quad \begin{matrix} 4 \times \text{Eq. 5} \\ + \text{Eq. 6} \\ \hline \text{Eq. 8} \end{matrix} \quad \begin{matrix} 4B + 16C + 4D = 12 \\ -4B & + 4D = 12 \\ \hline 16C + 8D = 24 \end{matrix} \quad \text{and} \quad \begin{matrix} 4 \times \text{Eq. 5} \\ + \text{Eq. 7} \\ \hline \text{Eq. 9} \end{matrix} \quad \begin{matrix} 4B + 16C + 4D = 12 \\ -4B + 16C - 4D = 20 \\ \hline 32C & = 32 \end{matrix} .$$

Hence, $32C = 32 \iff C = 1$; $16(1) + 8D = 16 + 8D = 24 \iff D = 1$; $B + 4 + 1 = 3 \iff B = -2$; and $A + 1 = 0 \iff A = -1$. Therefore,

$$\frac{3x^2 + 12x - 20}{x^4 - 8x^2 + 16} = -\frac{1}{x + 2} - \frac{2}{(x + 2)^2} + \frac{1}{x - 2} + \frac{1}{(x - 2)^2}.$$

36. $$\frac{3x^2 - 2x + 8}{x^3 - x^2 + 2x - 2} = \frac{3x^2 - 2x + 8}{(x^2 + 2)(x - 1)} = \frac{Ax + B}{x^2 + 2} + \frac{C}{x - 1}. \text{ Thus,}$$

$$3x^2 - 2x + 8 = (Ax + B)(x - 1) + C(x^2 + 2) = (A + C)x^2 + (-A + B)x + (-B + 2C),$$

which leads to the system:

$$\begin{cases} A & + C = 3 \\ -A + B & = -2 \\ -B + 2C = 8 \end{cases} \begin{matrix} \text{Eq. 1: Coefficients of } x^2 \\ \text{Eq. 2: Coefficients of } x \\ \text{Eq. 3: Constant terms} \end{matrix} \quad \begin{matrix} \text{Eq. 1} \\ + \text{Eq. 2} \\ \hline \text{Eq. 4} \end{matrix} \quad \begin{matrix} A & + C = 3 \\ -A + B & = -2 \\ \hline B + C = 1 \end{matrix}$$

$$\begin{matrix} \text{Thus we get} \\ \text{the system} \end{matrix} \quad \begin{cases} \text{Eq. 1 } A & + C = 3 \\ \text{Eq. 4:} & B + C = 1 \\ \text{Eq. 3:} & -B + 2C = 8 \end{cases} \quad \text{Then} \quad \begin{matrix} \text{Eq. 4} \\ + \text{Eq. 3} \\ \hline \text{Eq. 5} \end{matrix} \quad \begin{matrix} B + C = 1 \\ -B + 2C = 8 \\ \hline 3C = 9 \end{matrix}$$

Hence, $3C = 9 \iff C = 3$; $B + 3 = 1 \iff B = -2$; and $A + 3 = 3 \iff A = 0$.

Therefore, $$\frac{3x^2 - 2x + 8}{x^3 - x^2 + 2x - 2} = -\frac{2}{x^2 + 2} + \frac{3}{x - 1}.$$

38. $$\frac{x^2 + x + 1}{2x^4 + 3x^2 + 1} = \frac{x^2 + x + 1}{(2x^2 + 1)(x^2 + 1)} = \frac{Ax + B}{2x^2 + 1} + \frac{Cx + D}{x^2 + 1}. \text{ Thus,}$$

$$x^2 + x + 1 = (Ax + B)(x^2 + 1) + (Cx + D)(2x^2 + 1)$$
$$= Ax^3 + Ax + Bx^2 + B + 2Cx^3 + 2Dx^2 + Cx + D$$
$$= (A + 2C)x^3 + (B + 2D)x^2 + (A + C)x + (B + D), \text{ which leads to the system:}$$

$$\begin{cases} A & + 2C & = 0 \\ B & + 2D = 1 \\ A & + C & = 1 \\ B & + D = 1 \end{cases} \begin{matrix} \text{Eq. 1: Coefficients of } x^3 \\ \text{Eq. 2: Coefficients of } x^2 \\ \text{Eq. 3: Coefficients of } x \\ \text{Eq. 4: Constant terms} \end{matrix} \quad \text{Then} \quad \begin{matrix} \text{Eq. 1} \\ - \text{Eq. 3} \\ \hline \text{Eq. 5} \end{matrix} \quad \begin{matrix} A + 2C = 0 \\ -A - C = -1 \\ \hline C = -1 \end{matrix}. \text{ Also}$$

Eq. 2 $\quad B + 2D = 1$

$-$ Eq. 4 $\quad -B - D = -1$ $\quad$ Hence, $D = 0$ and $C = -1$, then $B + 0 = 1$ $\quad \Leftrightarrow \quad B = 1$; and

Eq. 6 $\quad\qquad\qquad D = 0$

$A - 2 = 0$ $\quad \Leftrightarrow \quad A = 2$. Therefore, $\dfrac{x^2 + x + 1}{2x^4 + 3x^2 + 1} = \dfrac{2x + 1}{2x^2 + 1} - \dfrac{x}{x^2 + 1}$.

40. $\dfrac{2x^2 - x + 8}{(x^2 + 4)^2} = \dfrac{Ax + B}{x^2 + 4} + \dfrac{Cx + D}{(x^2 + 4)^2}$. Thus, $2x^2 - x + 8 = (Ax + B)(x^2 + 4) + Cx + D$

$= Ax^3 + 4Ax + Bx^2 + 4B + Cx + D = Ax^3 + Bx^2 + (4A + C)x + (4B + D)$, and so $A = 0$;

$B = 2$; $0 + C = -1 \Leftrightarrow C = -1$; and $8 + D = 8 \Leftrightarrow D = 0$. Therefore,

$\dfrac{2x^2 - x + 8}{(x^2 + 4)^2} = \dfrac{2}{x^2 + 4} - \dfrac{1}{(x^2 + 4)^2}$.

42. $\dfrac{x^5 - 3x^4 + 3x^3 - 4x^2 + 4x + 12}{(x - 2)^2(x^2 + 2)} = \dfrac{x^5 - 3x^4 + 3x^3 - 4x^2 + 4x + 12}{x^4 - 4x^3 + 6x^2 - 8x + 8}$. Next use long division to

get a proper rational function.

$$
\begin{array}{r}
x + 1 \\
x^4 - 4x^3 + 6x^2 - 8x + 8 \overline{\smash{\big)}\, x^5 - 3x^4 + 3x^3 - 4x^2 + 4x + 12}\\
\underline{x^5 - 4x^4 + 6x^3 - 8x^2 + 8x}\\
x^4 - 3x^3 + 4x^2 - 4x + 12\\
\underline{x^4 - 4x^3 + 6x^2 - 8x + 8}\\
x^3 - 2x^2 + 4x + 4
\end{array}
$$

Thus $\dfrac{x^5 - 3x^4 + 3x^3 - 4x^2 + 4x + 12}{(x - 2)^2(x^2 + 2)} = x + 1 + \dfrac{x^3 - 2x^2 + 4x + 4}{(x - 2)^2(x^2 + 2)}$. Now,

$\dfrac{x^3 - 2x^2 + 4x + 4}{(x - 2)^2(x^2 + 2)} = \dfrac{A}{x - 2} + \dfrac{B}{(x - 2)^2} + \dfrac{Cx + D}{x^2 + 2}$, and so

$x^3 - 2x^2 + 4x + 4 = A(x - 2)(x^2 + 2) + B(x^2 + 2) + (Cx + D)(x - 2)^2$

$= A(x^3 - 2x^2 + 2x - 4) + B(x^2 + 2) + (Cx + D)(x^2 - 4x + 4)$

$= Ax^3 - 2Ax^2 + 2Ax - 4A + Bx^2 + 2B + Cx^3 - 4Cx^2 + 4Cx + Dx^2 - 4Dx + 4D$

$= (A + C)x^3 + (-2A + B - 4C + D)x^2 + (2A + 4C - 4D)x + (-4A + 2B + 4D)$.

which leads to the system:

$$
\begin{cases}
A + C = 1 & \text{Eq. 1: Coefficients of } x^3\\
-2A + B - 4C + D = -2 & \text{Eq. 2: Coefficients of } x^2\\
2A + 4C - 4D = 4 & \text{Eq. 3: Coefficients of } x\\
-4A + 2B + 4D = 4 & \text{Eq. 4: Constant terms}
\end{cases}
$$

Then

$$
\begin{array}{l}
2 \times \text{Eq. 1} \quad 2A + 2C = 2\\
\underline{+ \text{Eq. 2} \quad -2A + B - 4C + D = -2}\\
\text{Eq. 5} \quad\qquad B - 2C + D = 0
\end{array}
$$,

$$
\begin{array}{l}
2 \times \text{Eq. 1} \quad 2A + 2C = 2\\
\underline{- \text{Eq. 3} \quad -2A - 4C + 4D = -4}\\
\text{Eq. 6} \quad\qquad -2C + 4D = -2
\end{array}
$$ and

$$
\begin{array}{l}
4 \times \text{Eq. 1} \quad 4A + + 4C = 4\\
\underline{+ \text{Eq. 4} \quad -4A + 2B + 4D = 4}\\
\text{Eq. 7} \quad\qquad 2B + 4C + 4D = 8
\end{array}
$$

Thus we get the system

$$
\begin{cases}
\text{Eq. 1} \quad A + C = 1\\
\text{Eq. 5:} \quad\qquad B - 2C + D = 0\\
\text{Eq. 6:} \quad\qquad\qquad -2C + 4D = -2\\
\text{Eq. 7:} \quad\qquad 2B + 4C + 4D = 8
\end{cases}
$$

Then

$$2 \times \text{Eq. 5} \quad 2B - 4C + 2D = 0$$
$$\underline{-\quad \text{Eq. 7} \quad -2B - 4C - 4D = -8} \quad \begin{array}{l}\text{which leads}\\ \text{to the system}\end{array}$$
$$\text{Eq. 8} \qquad -8C - 2D = -8$$

$$\begin{cases} \text{Eq. 1} & A \;+\; C \qquad\qquad = 1 \\ \text{Eq. 5:} & \quad B - 2C + \;\; D = 0 \\ \text{Eq. 6:} & \qquad\quad -2C + 4D = -2 \\ \text{Eq. 8:} & \qquad\quad -8C - 2D = -8 \end{cases}$$

$$4 \times \text{Eq. 6} \quad -8C + 16D = -8$$
$$\underline{-\quad \text{Eq. 8} \qquad 8C + \;\; 2D = 8} \qquad \text{Then, } D = 0; \;\; 8C + 2(0) = 8 \quad \Leftrightarrow \quad C = 1; \;\; B - 2 = 0 \quad \Leftrightarrow$$
$$\text{Eq. 9} \qquad\qquad\quad 18D = 0$$

$B = 2$; and $A + 1 = 1 \quad \Leftrightarrow \quad A = 0$. Therefore,

$$\frac{x^5 - 3x^4 + 3x^3 - 4x^2 + 4x + 12}{(x-2)^2(x^2+2)} = x + 1 + \frac{2}{(x-2)^2} + \frac{x}{x^2+2}.$$

44. $\dfrac{ax^3 + bx^2}{(x^2+1)^2} = \dfrac{Ax+B}{x^2+1} + \dfrac{Cx+D}{(x^2+1)^2}$. Hence, $ax^3 + bx^2 = (Ax+B)(x^2+1) + Cx + D$

   $= Ax^3 + Ax + Bx^2 + B + Cx + D = Ax^3 + Bx^2 + (A+C)x + (B+D)$, and so $A = a$;

   $B = b$; $a + C = 0 \quad \Leftrightarrow \quad C = -a$; and $b + D = 0 \quad \Leftrightarrow \quad D = -b$. Therefore, $A = a$, $B = b$,

   $C = -a$, and $D = -b$.

46. Combining the terms, we have

   $$\frac{2}{x-1} + \frac{1}{(x-1)^2} + \frac{1}{x+1} = \frac{2(x^2-1)}{(x-1)^2(x+1)} + \frac{1(x+1)}{(x-1)^2(x+1)} + \frac{1(x^2-2x+1)}{(x-1)^2(x+1)}$$
   $$= \frac{2x^2 - 2 + x + 1 + x^2 - 2x + 1}{(x-1)^2(x+1)} = \frac{3x^2 - x}{(x-1)^2(x+1)}.$$

   Now to find the partial fraction decomposition, we have

   $$\frac{3x^2 - x}{(x-1)^2(x+1)} = \frac{A}{x-1} + \frac{B}{(x-1)^2} + \frac{C}{x+1}, \text{ and so}$$

   $3x^2 - x = A(x-1)(x+1) + B(x+1) + C(x-1)^2 = A(x^2-1) + B(x+1) + C(x^2-2x+1)$

   $= Ax^2 - A + Bx + B + Cx^2 - 2Cx + C = (A+C)x^2 + (B-2C)x + (-A+B+C)$

   which result in the following system:

   $$\begin{cases} A \;+\; C = 3 \\ \quad B - 2C = -1 \\ -A + B + \; C = 0 \end{cases}$$
   $\begin{array}{l}\text{Eq. 1: Coefficients of } x^2 \\ \text{Eq. 2: Coefficients of } x \\ \text{Eq. 3: Constant terms}\end{array}$
   Then
   $$\begin{array}{ll}\text{Eq. 1} & A \;+\; C = 3 \\ \underline{+\text{ Eq. 3}} & \underline{-A + B + \; C = 0} \\ \text{Eq. 4} & \quad B + 2C = 3\end{array}$$

   $\begin{array}{l}\text{so we get}\\ \text{the system}\end{array}$
   $\begin{cases} \text{Eq. 1} & A \;+\; C = 3 \\ \text{Eq. 2:} & \quad B - 2C = -1. \\ \text{Eq. 4:} & \quad B + 2C = 3 \end{cases}$
   Now
   $\begin{array}{ll}\text{Eq. 2} & B - 2C = -1 \\ \underline{-\text{ Eq. 4}} & \underline{-B - 2C = -3} \\ \text{Eq. 5} & \quad -4C = -4\end{array}$
   so, $-4C = -4$

   $\Leftrightarrow \quad C = 1$. Then $B - 2(1) = -1 \quad \Leftrightarrow \quad B = 1$; and $A + 1 = 3 \quad \Leftrightarrow \quad A = 2$. Therefore we

   get back the same expression, $\dfrac{3x^2 - x}{(x-1)^2(x+1)} = \dfrac{2}{x-1} + \dfrac{1}{(x-1)^2} + \dfrac{1}{x+1}.$

Review Exercises for Chapter 6

2. $\begin{cases} 3x + y = 8 \\ y = x^2 - 5x \end{cases}$ By inspection of the graph, it appears that $(-2, 14)$ and $(4, -4)$ are solutions to the system. We check each possible solutions in both equations to verify that it is the solution.
$3(-2) + (14) = -6 + 14 = 8 \checkmark$ and $(-2)^2 - 5(-2) = 4 + 10 = 14 \checkmark$.
$3(4) + (-4) = 12 - 4 = 8 \checkmark$ and $(4)^2 - 5(4) = 16 - 20 = -4 \checkmark$.
Since both points satisfy both equations, the solutions are $(-2, 14)$ and $(4, -4)$.

4. $\begin{cases} x - y = -2 \\ x^2 + y^2 - 4y = 4 \end{cases}$ By inspection of the graph, it appears that $(-2, 0)$ and $(2, 4)$ are solutions to the system. We check each possible solutions in both equations to verify that it is the solution.
$(-2) - (0) = -2 \checkmark$ and $(-2)^2 + (0)^2 - 4(0) = 4 \checkmark$.
$(2) - (4) = -2 \checkmark$ and $(2)^2 + (4)^2 - 4(4) = 4 + 16 - 16 = 4 \checkmark$.
Since both points satisfy both equations, the solutions are $(-2, 0)$ and $(2, 4)$.

6. $\begin{cases} y = 2x + 6 \\ y = -x + 3 \end{cases}$ Subtracting the second equation from the first, we
get $0 = 3x + 3 \quad \Leftrightarrow \quad x = -1$. So $y = -(-1) + 3 = 4$. Thus the solution is $(-1, 4)$.

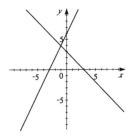

8. $\begin{cases} 6x - 8y = 15 \\ -\frac{3}{2}x + 2y = -4 \end{cases} \Leftrightarrow \begin{cases} 6x - 8y = 15 \\ -6x + 8y = -16 \end{cases}$ Adding gives
$0 = -1$ which is false. Hence, there is no solution. The lines are parallel.

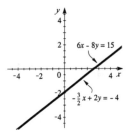

10. $\begin{cases} 2x + 5y = 9 \\ -x + 3y = 1 \\ 7x - 2y = 14 \end{cases}$ Adding the first equation to twice the second equation gives $11y = 11 \quad \Leftrightarrow \quad y = 1$. Substituting back into the second equation, we get $-x + 3(1) = 1 \quad \Leftrightarrow \quad x = 4$. Checking point $(4, 1)$ in the third equation gives $7(4) - 2(1) = 26 \neq 14$. Thus there is no solution, and the lines do not intersect at one point.

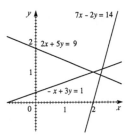

12. $\begin{cases} x^2 + y^2 = 8 \\ y = x + 2 \end{cases}$ Substituting for y in the first equation gives $x^2 + (x + 2)^2 = 8 \quad \Leftrightarrow$
$2x^2 + 4x - 4 = 0 \quad \Leftrightarrow \quad 2(x^2 + 2x - 2) = 0$. Using the quadratic formula, we have
$x = \frac{-2 \pm \sqrt{4+8}}{2} = \frac{-2 \pm 2\sqrt{3}}{2} = -1 \pm \sqrt{3}$. If $x = -1 - \sqrt{3}$. then $y = (-1 - \sqrt{3}) + 2 = 1 - \sqrt{3}$,

and if $x = -1 + \sqrt{3}$, then $y = (-1 + \sqrt{3}) + 2 = 1 + \sqrt{3}$. Thus, the solutions are $(-1 - \sqrt{3}, 1 - \sqrt{3})$ and $(-1 + \sqrt{3}, 1 + \sqrt{3})$

14. $\begin{cases} x^2 + y^2 = 10 \\ x^2 + 2y^2 - 7y = 0 \end{cases}$ Subtracting the first equation from the second equation gives

$y^2 - 7y = -10 \;\Leftrightarrow\; y^2 - 7y + 10 = 0 \;\Leftrightarrow\; (y-2)(y-5) = 0 \;\Rightarrow\; y = 2, y = 5$. If $y = 2$, then $x^2 + 4 = 10 \;\Leftrightarrow\; x^2 = 6 \;\Rightarrow\; x = \pm\sqrt{6}$, and if $y = 5$, then $x^2 + 25 = 10 \;\Leftrightarrow\; x^2 = -15$, which leads to no real solution. Thus the solutions are $(\sqrt{6},\, 2)$ and $(-\sqrt{6},\, 2)$.

16. $\begin{cases} \sqrt{12}\, x - 3\sqrt{2}\, y = 660 \\ 7137x + 3931y = 20{,}000 \end{cases} \;\Leftrightarrow\;$

$\begin{cases} y = \frac{\sqrt{12}}{3\sqrt{2}} x - \frac{660}{3\sqrt{2}} \\ y = -\frac{7137}{3931} x + \frac{20{,}000}{3931} \end{cases}$

The solution is approximately $(61.04, -105.73)$

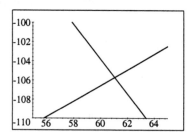

18. $\begin{cases} y = 5^x + x \\ y = x^5 + 5 \end{cases}$

The solutions are approximately $(-1.45, -1.35), (1, 6)$, and $(1.51, 12.93)$.

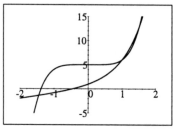

20. $\begin{cases} \text{Eq. 1:} & x - 2y + 3z = 1 \\ \text{Eq. 2:} & x - 3y - z = 0 \\ \text{Eq. 3:} & 2x - 6z = 6 \end{cases}$ Then $\begin{array}{ll} \text{Eq. 1} & x - 2y + 3z = 1 \\ - \text{ Eq. 2} & -x + 3y + z = 0 \\ \hline \text{Eq. 4} & y + 4z = 1 \end{array}$ and

$\begin{array}{ll} 2 \times \text{Eq. 1} & 2x - 4y + 6z = 2 \\ - \text{ Eq. 3} & -2x + 6z = -6 \\ \hline \text{Eq. 5} & -4y + 12z = -4 \end{array}$ This gives us the system $\begin{cases} \text{Eq. 1:} & x - 2y + 3z = 1 \\ \text{Eq. 4:} & y + 4z = 1 \\ \text{Eq. 5:} & -4y + 12z = -4 \end{cases}$ Finally,

$\begin{array}{ll} 4 \times \text{Eq. 4} & 4y + 16z = 4 \\ + \text{ Eq. 5} & -4y + 12z = -4 \\ \hline \text{Eq. 8} & 28z = 0 \end{array}$ Thus, $z = 0$; $y = 1$; and $x - 2(1) + 3(0) = 1 \;\Leftrightarrow\; x = 3$, and

so the solution is $(3, 1, 0)$.

22. $\begin{cases} \text{Eq. 1:} & x + y + z + w = 2 \\ \text{Eq. 2:} & 2x - 3z = 5 \\ \text{Eq. 3:} & x - 2y + 4w = 9 \\ \text{Eq. 4:} & x + y + 2z + 3w = 5 \end{cases}$ Then $\begin{array}{ll} 2 \times \text{Eq. 1} & 2x + 2y + 2z + 2w = 4 \\ - \text{ Eq. 2} & -2x + 3z = -5 \\ \hline \text{Eq. 5} & 2y + 5z + 2w = -1 \end{array}$,

$\begin{array}{ll} \text{Eq. 1} & x + y + z + w = 2 \\ - \text{ Eq. 3} & -x + 2y - 4w = -9 \\ \hline \text{Eq. 6} & 3y + z - 3w = -7 \end{array}$ and $\begin{array}{ll} \text{Eq. 1} & x + y + z + w = 2 \\ - \text{ Eq. 4} & -x - y - 2z - 3w = -5 \\ \hline \text{Eq. 7} & -z - 2w = -3 \end{array}$

This gives
the system
$$\begin{cases} \text{Eq. 1:} & x + y + z + w = 2 \\ \text{Eq. 5:} & 2y + 5z + 2w = -1 \\ \text{Eq. 6:} & 3y + z - 3w = -7 \\ \text{Eq. 7:} & -z - 2w = -3 \end{cases}$$

Now,
$$\begin{array}{ll} 3 \times \text{Eq. 5} & 6y + 15z + 6w = -3 \\ -2 \times \text{Eq. 6} & -6y - 2z + 6w = 14 \\ \hline \text{Eq. 8} & 13z + 12w = 11 \end{array}$$

so we get
the system
$$\begin{cases} \text{Eq. 1:} & x + y + z + w = 2 \\ \text{Eq. 5:} & 2y + 5z + 2w = -1 \\ \text{Eq. 8:} & 13z + 12w = 11 \\ \text{Eq. 7:} & -z - 2w = -3 \end{cases}$$

Then
$$\begin{array}{ll} \text{Eq. 8} & 13z + 12w = 11 \\ +13 \times \text{Eq. 3} & -13z - 26w = -39 \\ \hline \text{Eq. 6} & -14w = -28 \end{array}$$

Therefore, $-14w = -28 \iff w = 2$; $-z - 2(2) = -3 \iff z = -1$;

$2y + 5(-1) + 2(2) = -1 \iff 2y - 1 = -1 \iff y = 0$; and $x - 1 + 2 = 2 \iff x = 1$.

So, the solution is $(1, 0, -1, 2)$.

24. $\begin{cases} \text{Eq. 1:} & 2x - 3y + 4z = 3 \\ \text{Eq. 2:} & 4x - 5y + 9z = 13 \\ \text{Eq. 3:} & 2x + 7z = 0 \end{cases}$ Then
$$\begin{array}{ll} 2 \times \text{Eq. 1} & 4x - 6y + 8z = 6 \\ -\text{Eq. 2} & -4x + 5y - 9z = -13 \\ \hline \text{Eq. 4} & -y - z = -7 \end{array}$$ and

$$\begin{array}{ll} \text{Eq. 1} & 2x - 3y + 4z = 3 \\ -\text{Eq. 3} & -2x - 7z = 0 \\ \hline \text{Eq. 5} & -3y - 3z = 3 \end{array}$$
This gives us
the system
$\begin{cases} \text{Eq. 1:} & 2x - 3y + 4z = 3 \\ \text{Eq. 4:} & -y - z = -7 \\ \text{Eq. 5:} & -3y - 3z = 3 \end{cases}$ Then,

$$\begin{array}{ll} 3 \times \text{Eq. 4} & -3y - 3z = -21 \\ -\text{Eq. 5} & 3y + 3z = -3 \\ \hline \text{Eq. 6} & 0 = -24 \end{array}$$ Since this last equation is impossible, the system is inconsistent and

has no solution.

26. $\begin{cases} \text{Eq. 1:} & x - z + w = 2 \\ \text{Eq. 2:} & 2x + y - 2w = 12 \\ \text{Eq. 3:} & 3y + z + w = 4 \\ \text{Eq. 4:} & x + y - z = 10 \end{cases}$ Then
$$\begin{array}{ll} 2 \times \text{Eq. 1} & 2x - 2z + 2w = 4 \\ -\text{Eq. 2} & -2x - y + 2w = -12 \\ \hline \text{Eq. 5} & -y - 2z + 4w = -8 \end{array}$$ and

$$\begin{array}{ll} \text{Eq. 1} & x - z + w = 2 \\ -\text{Eq. 4} & -x - y + z = -10 \\ \hline \text{Eq. 6} & -y + w = -8 \end{array}$$
which gives
the system
$\begin{cases} \text{Eq. 1:} & x - z + w = 2 \\ \text{Eq. 5:} & -y - 2z + 4w = -8 \\ \text{Eq. 3:} & 3y + z + w = 4 \\ \text{Eq. 6:} & -y + w = -8 \end{cases}$ Now,

$$\begin{array}{ll} 3 \times \text{Eq. 5} & -3y - 6z + 12w = -24 \\ +\text{Eq. 3} & 3y + z + w = 4 \\ \hline \text{Eq. 7} & -5z + 13w = -20 \end{array}$$ and
$$\begin{array}{ll} \text{Eq. 5} & -y - 2z + 4w = -8 \\ -\text{Eq. 6} & y - w = 8 \\ \hline \text{Eq. 8} & -2z + 3w = 0 \end{array}$$

so we get
the system
$\begin{cases} \text{Eq. 1:} & x - z + w = 2 \\ \text{Eq. 5:} & -y - 2z + 4w = -8 \\ \text{Eq. 7:} & -5z + 13w = -20 \\ \text{Eq. 8:} & -2z + 3w = 0 \end{cases}$ Then
$$\begin{array}{ll} 2 \times \text{Eq. 7} & -10z + 26w = -40 \\ -5 \times \text{Eq. 8} & 10z - 15w = 0 \\ \hline \text{Eq. 9} & 11w = -40 \end{array}$$

Therefore, $11w = -40 \iff w = -\frac{40}{11}$; $-5z + 13\left(-\frac{40}{11}\right) = -20 \iff z = -\frac{60}{11}$;

$y + 2\left(-\frac{60}{11}\right) - 4\left(-\frac{40}{11}\right) = 8 \iff y = \frac{48}{11}$; and $x - \left(-\frac{60}{11}\right) + \left(-\frac{40}{11}\right) = 2 \iff x = \frac{2}{11}$. Hence,

the solution is $\left(\frac{2}{11}, \frac{48}{11}, -\frac{60}{11}, -\frac{40}{11}\right)$.

28. Let n = the number nickels, d = the number of dimes, and q = the number of quarter in the piggy bank. We get the following system:

$$\begin{cases} n + d + q = 50 \\ 5n + 10d + 25z = 560. \\ 10d = 5(5n) \end{cases}$$ Since $10d = 25n$ we have $d = \frac{5}{2}n$, so substituting into the first

equation we get $n + \frac{5}{2}n + q = 50 \quad \Leftrightarrow \quad \frac{7}{2}n + q = 50 \quad \Leftrightarrow \quad q = 50 - \frac{7}{2}n$. Now substituting this into the second equation we have $5n + 10(\frac{5}{2}n) + 25(50 - \frac{7}{2}n) = 560 \quad \Leftrightarrow$

$5n + 25n + 1250 - \frac{175}{2}n = 560 \quad \Leftrightarrow \quad 1250 - \frac{115}{2}n = 560 \quad \Leftrightarrow \quad \frac{115}{2}n = 650 \quad \Leftrightarrow \quad n = 12$.
Then $d = \frac{5}{2}(12) = 30$ and $q = 50 - n - d = 50 - 12 - 30 = 8$. Thus the piggy bank has 12 nickels, 30 dimes, and 8 quarters.

30. The boundary is a solid curve, so we have the inequality $x^2 + y^2 \geq 8$. We take the test point $(0, 3)$ and verify that it satisfies the inequality: $0^2 + 3^2 \geq 8$ ✓.

32.

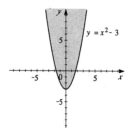

34.

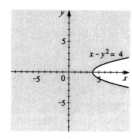

36.

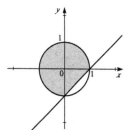

38.

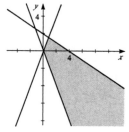

40. $\begin{cases} y - x^2 \geq 4 \\ y < 20 \end{cases}$ The vertices occur where $y = x^2 + 4$ and $y = 2$.

By substitution, $x^2 + 4 = 20 \quad \Leftrightarrow \quad x^2 = 16 \quad \Leftrightarrow \quad x = \pm 4$,
and $y = 20$. Thus, the vertices are $(\pm 4, 20)$, and the solution set is bounded.

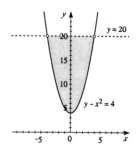

42. $\begin{cases} x \geq 4 \\ x + y \geq 24 \\ x \leq 2y + 12 \end{cases}$

The lines $x + y = 24$ and $x = 2y + 12$ intersect at the point $(20, 4)$.
The lines $x = 4$ and $x = 2y + 12$ intersect at the point $(-4, 4)$,
however, this vertex does not satisfy the other inequality. The lines
$x + y = 24$ and $x = 4$ intersect at the point $(4, 20)$. Hence, the
vertices are $(4, 20)$ and $(20, 4)$. The solution set is not bounded.

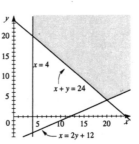

44. $\begin{cases} ax + by + cz = a - b + c \\ bx + by + cz = c \\ cx + cy + cz = c \end{cases}$ $\quad a \neq b, b \neq c, c \neq 0 \quad \Leftrightarrow \quad \begin{array}{l} ax + by + cz = a - b + c \\ bx + by + cz = c \\ x + y + z = 1 \end{array}$

Subtracting the second equation from the first gives $(a - b)x = a - b \quad \Leftrightarrow \quad x = 1$. Subtracting
the third equation from the second, $(b - c)x + (b - c)y = 0 \quad \Leftrightarrow \quad y = -x = -1$. So,
$1 - 1 + z = 1 \quad \Leftrightarrow \quad z = 1$, and the solution is $(1, -1, 1)$.

46. The system will have infinitely many solutions when the system has solutions other than $(0, 0, 0)$. So
we solve the system with $x \neq 0$, $y \neq 0$, and $z \neq 0$.

$\begin{cases} \text{Eq. 1} \quad kx + y + z = 0 \\ \text{Eq. 2:} \quad x + 2y + kz = 0 \\ \text{Eq. 3:} \quad -x + 2y + 3z = 0 \end{cases}$ $\quad$ Then eliminating the terms with x we get

$$\begin{array}{ll} \begin{array}{r} \text{Eq. 1} \quad kx + \quad y + \quad z = 0 \\ +k \times \text{Eq. 3} \quad -kx + \quad 2ky + \quad 3kz = 0 \\ \hline \text{Eq. 4} \quad (2k + 1)y + (3k + 1)z = 0 \end{array} & \text{and} \quad \begin{array}{r} \text{Eq. 2} \quad x + 2y + \quad kz = 0 \\ +\text{Eq. 3} \quad -x + 2y + \quad 3z = 0 \\ \hline \text{Eq. 5} \quad 4y + (k + 3)z = 0 \end{array} \end{array}$$

Solving Equation 5 for y in terms of z we get: $4y + (k + 3)z = 0 \quad \Leftrightarrow \quad 4y = -(k + 3)z \quad \Leftrightarrow$
$y = -\frac{1}{4}(k + 3)z$. Substituting for y in Equation 5 give us $(2k + 1)\left[-\frac{1}{4}(k + 3)z\right] + (3k + 1)z = 0$
$\Leftrightarrow \quad -(2k + 1)(k + 3)z + 4(3k + 1)z = 0 \quad \Leftrightarrow \quad (-2k^2 - 7k - 3)z + (12k + 4)z = 0 \quad \Leftrightarrow$
$(-2k^2 + 5k + 1)z = 0$. Since we assume $z \neq 0$ we must have $-2k^2 + 5k + 1 = 0$ so
$k = \dfrac{-5 \pm \sqrt{5^2 - 4(-2)(1)}}{2(-2)} = \dfrac{-5 \pm \sqrt{33}}{-4} = \dfrac{5 \pm \sqrt{33}}{4}$.

48. $\dfrac{8}{x^3 - 4x} = \dfrac{8}{x(x^2 - 4)} = \dfrac{8}{x(x + 2)(x - 2)} = \dfrac{A}{x} + \dfrac{B}{x - 2} + \dfrac{C}{x + 2}$. Then,

$8 = A(x^2 - 4) + Bx(x + 2) + Cx(x - 2) = x^2(A + B + C) + x(2B - 2C) - 4A$. Thus,
$-4A = 8 \quad \Leftrightarrow \quad A = -2; \quad 2B - 2C = 0 \quad \Leftrightarrow \quad B = C; \quad \text{and} -2 + 2B = 0 \quad \Leftrightarrow \quad B = 1$ so
$C = 1$. Therefore, $\dfrac{8}{x^3 - 4x} = -\dfrac{2}{x} + \dfrac{1}{x - 2} + \dfrac{1}{x + 2}$.

50. $\dfrac{x + 6}{x^3 - 2x^2 + 4x - 8} = \dfrac{x + 6}{(x^2 + 4)(x - 2)} = \dfrac{Ax + B}{x^2 + 4} + \dfrac{C}{x - 2}$. Thus,

$x + 6 = (Ax + B)(x - 2) + C(x^2 + 4) = Ax^2 - 2Ax + Bx - 2B + Cx^2 + 4C$
$= x^2(A + C) + x(-2A + B) + (-2B + 4C)$, and so we get the system

$$\begin{cases} A \quad + C = 0 \\ -2A + B \quad\quad = 1 \\ \quad -2B + 4C = 6 \end{cases} \text{Then} \quad \begin{array}{l} 2 \times \text{Eq. 1} \\ + \text{Eq. 2} \\ \hline \text{Eq. 4} \end{array} \quad \begin{array}{l} 2A \quad\quad + 2C = 0 \\ -2A + B \quad\quad = 1 \\ \hline B + 2C = 1 \end{array} \quad \text{to give the system}$$

$$\begin{cases} A \quad + C = 0 \\ \quad B + 2C = 1 \\ -2B + 4C = 6 \end{cases} \text{Then} \quad \begin{array}{l} 2 \times \text{Eq. 2} \\ + \text{Eq. 2} \\ \hline \text{Eq. 4} \end{array} \quad \begin{array}{l} 2B + 4C = 2 \\ -2B + 4C = 6 \\ \hline 8C = 8 \end{array} \quad \text{Thus, } 8C = 8 \quad \Leftrightarrow \quad C = 1;$$

$B + 2 = 1 \quad \Leftrightarrow \quad B = -1; \text{ and } A + 1 = 0 \quad \Leftrightarrow \quad A = -1. \text{ So,}$

$$\frac{x+6}{x^3 - 2x^2 + 4x - 8} = -\frac{x+1}{x^2 + 4} + \frac{1}{x-2}.$$

52. Since $x^4 + x^2 - 2 = (x^2 - 1)(x^2 + 2) = (x - 1)(x + 1)(x^2 + 2)$ we have

$\dfrac{5x^2 - 3x + 10}{x^4 + x^2 - 2} = \dfrac{A}{x-1} + \dfrac{B}{x+1} + \dfrac{Cx + D}{x^2 + 2}$. Thus

$5x^2 - 3x + 10 = (x + 1)(x^2 + 2)A + (x - 1)(x^2 + 2)B + (x^2 - 1)(Cx + D)$

$\quad = Ax^3 + Ax^2 + 2Ax + 2A + Bx^3 - Bx^2 + 2Bx - 2B + Cx^3 + Dx^2 - Cx - D$

$\quad = (A + B + C)x^3 = (A - B + D)x^2 + (2A + 2B - C)x + (2A - 2B - D).$ This leads to the

following system:

$$\begin{cases} A + B + C \quad\quad = 0 \\ A - B \quad + D = 5 \\ 2A + 2B - C \quad\quad = -3 \\ 2A - 2B \quad - D = 10 \end{cases} \begin{array}{l} \text{Eq. 1: Coefficients of } x^3 \\ \text{Eq. 2: Coefficients of } x^2 \\ \text{Eq. 3: Coefficients of } x \\ \text{Eq. 4: \quad Constant terms} \end{array}$$

$$\text{Then} \quad \begin{array}{l} 2 \times \text{Eq. 1} \\ - \quad \text{Eq. 3} \\ \hline \text{Eq. 5} \end{array} \quad \begin{array}{l} 2A + 2B + 2C = 0 \\ -2A - 2B + \ C = 3 \\ \hline 3C = 3 \end{array}$$

so $C = 1$. Now $\quad \begin{array}{l} 2 \times \text{Eq. 2} \\ - \quad \text{Eq. 4} \\ \hline \text{Eq. 6} \end{array} \quad \begin{array}{l} 2A - 2B + 2D = 10 \\ -2A + 2B + \ D = -10 \\ \hline 3D = 0 \end{array}$ so $D = 0$. Substituting of C and D in

the first two equations gives the system $\begin{cases} A + B = -1 \\ A - B = 5 \end{cases}$ Adding these two equations give us

$\begin{array}{l} A + B = -1 \\ A - B = 5 \\ \hline 2A \quad\quad = 4 \end{array}$. Thus $A = 2$. Finally, $2 + B = -1 \quad \Leftrightarrow \quad B = -3.$ Thus

$\dfrac{5x^2 - 3x + 10}{x^4 + x^2 - 2} = \dfrac{2}{x - 1} - \dfrac{3}{x + 1} + \dfrac{x}{x^2 + 2}.$

Focus on Modeling

2.

Vertex	$N = \frac{1}{2}x + \frac{1}{4}y + 40$
$(1,0)$	$\frac{1}{2}(1) + \frac{1}{4}(0) + 40 = 40.5$
$\left(\frac{1}{2}, \frac{1}{2}\right)$	$\frac{1}{2}\left(\frac{1}{2}\right) + \frac{1}{4}\left(\frac{1}{2}\right) + 40 = 40.375$ ← minimum value
$(2,2)$	$\frac{1}{2}(2) + \frac{1}{4}(2) + 40 = 41.5$
$(4,0)$	$\frac{1}{2}(4) + \frac{1}{4}(0) + 40 = 42$ ← maximum value

Thus the maximum value is 42, and the minimum value is 40.375.

4.
$$\begin{cases} x \geq 0, \, y \geq 0 \\ x \leq 10, \, y \leq 20 \\ x + y \geq 5 \\ x + 2y \leq 18 \end{cases}$$

The objective function is $Q = 70x + 82y$.

From the graph, the vertices are at $(0,9)$, $(0,5)$, $(5,0)$, $(10,0)$, and $(10,4)$.

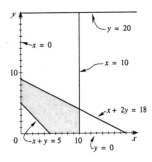

Vertex	$Q = 70x + 82y$
$(0,9)$	$70(0) + 82(9) = 738$
$(0,5)$	$70(0) + 82(5) = 410$
$(5,0)$	$70(5) + 82(0) = 350$ ← minimum value
$(10,0)$	$70(10) + 82(0) = 700$
$(10,4)$	$70(10) + 82(4) = 1028$ ← maximum value

Thus, the maximum value of Q is 1028, and the minimum value is 350.

6. Let c be the number of colonial homes built and r be the number of ranch homes built. Since there are 100 lots available, $c + r \leq 100$. From the capital restriction, we get $30{,}000c + 40{,}000r \leq 3{,}600{,}000$, or $3c + 4r \leq 360$. Thus, we wish to maximize the profit, $P = 4000c + 8000r$, subject to the constraints:
$$\begin{cases} c \geq 0, \, r \geq 0 \\ c + r \leq 100 \\ 3c + 4r \leq 360. \end{cases}$$

From the graph, the vertices occur at $(0,0)$, $(100,0)$, $(40,60)$, and $(0,90)$.

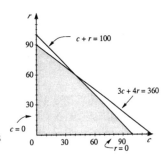

Vertex	$P = 4000c + 8000r$
$(0,0)$	$4000(0) + 8000(0) = 0$
$(100,0)$	$4000(100) + 8000(0) = 400{,}000$
$(40,60)$	$4000(40) + 8000(60) = 640{,}000$
$(0,90)$	$4000(0) + 8000(90) = 720{,}000$ ← maximum value

Therefore, he should build 0 colonial style and 90 ranch style houses for a maximum profit of \$720,000. Ten of the lots will be left vacant.

8. Let x be the daily production of standard calculators and y be the daily production of scientific calculators. Then, the following inequalities describe the constraints:

$$\begin{cases} x \geq 100, \, y \geq 80 \\ x \leq 200, \, y \leq 170 \\ x + y \geq 200. \end{cases}$$

From the graph, the vertices at $(100, 100)$, $(100, 170)$, $(200, 170)$, $(200, 80)$, and $(120, 80)$.

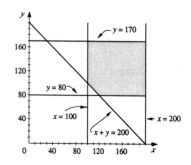

(a) Minimize the objective function (cost): $C = 5x + 7y$

Vertex	$C = 5x + 7y$	
$(100, 100)$	$5(100) + 7(100) = 1200$	
$(100, 170)$	$5(100) + 7(170) = 1690$	
$(200, 170)$	$5(200) + 7(170) = 2190$	
$(200, 80)$	$5(200) + 7(80) = 1560$	
$(120, 80)$	$5(120) + 7(80) = 1160$	← minimum cost

So, to minimize costs, they should produce 120 standard and 80 scientific calculators.

(b) Maximize the objective function (profit): $P = -2x + 5y$

Vertex	$P = -2x + 5y$	
$(100, 100)$	$-2(100) + 5(100) = 300$	
$(100, 170)$	$-2(100) + 5(170) = 650$	← maximum profit
$(200, 170)$	$-2(200) + 5(170) = 450$	
$(200, 80)$	$-2(200) + 5(80) = 0$	
$(120, 80)$	$-2(120) + 5(80) = 160$	

So, to maximize profit, they should produce 100 standard and 170 scientific calculators.

10. Let x be the number of sheets shipped from the east-side store to customer A and y be the number of sheets shipped from the east-side store to customer B. Then, $50 - x$ sheets must be shipped to customer A from the west-side store, and $70 - y$ sheets must be shipped to customer B from the west-side store. Thus we obtain the constraints:

$$\begin{cases} x \geq 0, \, y \geq 0 \\ x \leq 50, \, y \leq 70 \\ x + y \leq 80 \\ 50 - x + 70 - y \leq 45 \end{cases} \Leftrightarrow \begin{cases} x \geq 0, \, y \geq 0 \\ x \leq 50, \, y \leq 70 \\ x + y \leq 80 \\ x + y \geq 75. \end{cases}$$

The objective function is cost,
$C = 0.5x + 0.6y + 0.4(50 - x) + 0.55(70 - y)$
$= 0.1x + 0.05y + 58.5$, which we wish to minimize. From the graph, the vertices occur at $(5, 70)$, $(10, 70)$, $(50, 30)$, and $(50, 25)$.

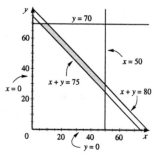

Vertex	$C = 0.1x + 0.05y + 58.5$	
$(5, 70)$	$0.1(5) + 0.05(70) + 58.5 = 62.5$	← minimum cost
$(10, 70)$	$0.1(10) + 0.05(70) + 58.5 = 63$	
$(50, 30)$	$0.1(50) + 0.05(30) + 58.5 = 65$	
$(50, 25)$	$0.1(50) + 0.05(25) + 58.5 = 64.75$	

Therefore, the minimum cost is $62.50 and occurs when $x = 5$ and $y = 70$. So 5 sheets should be shipped from the east-side to customer A, 70 sheets from the east-side store to customer B, 45 sheets from the west-side store to customer A, and 0 sheets from the west-side store to customer B.

12. Let x be the quantity, in ounces, of type I food and y be the quantity of type II food. Then the data can be summarized by the following table:

	type I	type II	required
fat	8 g	12 g	24 g
carbohydrate	12 g	12 g	36 g
protein	2 g	1 g	4 g
cost	$0.20	$0.30	

Also, the total amount of food must be no more than 5 oz. Thus the constraints are:

$$\begin{cases} x \geq 0, y \geq 0 \\ x + y \leq 5 \\ 8x + 12y \geq 24 \\ 12x + 12y \geq 36 \\ 2x + y \geq 4. \end{cases}$$

The objective function is cost, $C = 0.2x + 0.3y$, which we wish to minimize. From the graph, the vertices occur at $(1, 2)$, $(0, 4)$, $(0, 5)$, $(5, 0)$, and $(3, 0)$.

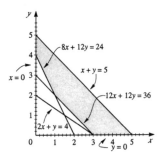

Vertex	$C = 0.2x + 0.3y$
$(1, 2)$	$0.2(1) + 0.3(2) = 0.8$
$(0, 4)$	$0.2(0) + 0.3(4) = 1.2$
$(0, 5)$	$0.2(0) + 0.3(5) = 1.5$
$(5, 0)$	$0.2(5) + 0.3(0) = 1.0$
$(3, 0)$	$0.2(3) + 0.3(0) = 0.6$ ← minimum cost

Hence, the rabbits should be fed 3 ounces of type I food and no type II food, for a minimum cost of $0.60.

14. The only change that needs to be made to the constraints in Exercise 13 is that the 2000 in the last inequality becomes 3000. Then we have

$$\begin{cases} x \geq 0, y \geq 0 \\ 12{,}000 - x - y \geq 0 \\ x \geq 3y \\ 12000 - x - y \leq 3000 \end{cases} \Leftrightarrow \begin{cases} x \geq 0, y \geq 0 \\ x + y \leq 12{,}000 \\ x \geq 3y \\ x + y \geq 9000. \end{cases}$$

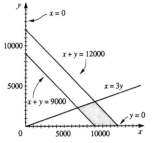

From the graph, the vertices occur at $(6750, 2250)$, $(9000, 0)$, $(12000, 0)$, and $(9000, 3000)$. The objective function is
$$Y = 0.07x + 0.08y + 0.12(12000 - x - y)$$
$$= 1440 - 0.05x - 0.04y, \text{ which we wish to maximize.}$$

Vertex	$Y = 1440 - 0.05x - 0.04y$
$(6750, 2250)$	$1440 - 0.05(6750) - 0.04(2250) = 1012.5$ ← maximum value
$(9000, 0)$	$1440 - 0.05(9000) - 0.04(0) = 990$
$(12000, 0)$	$1440 - 0.05(12000) - 0.04(0) = 840$
$(9000, 3000)$	$1440 - 0.05(9000) - 0.04(3000) = 870$

Hence, she should invest $6750 in municipal bonds, $2250 in bank certificates, and $3000 in high-risk bonds for a maximum yield of $1012.50 which is an increase of $47.50 over her yield in Exercise 13.

16.
$$\begin{cases} x \geq 0, \, x \geq y \\ x + 2y \leq 12 \\ x + y \leq 10 \end{cases}$$

(a)

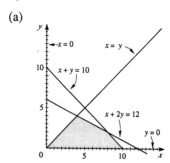

(b)

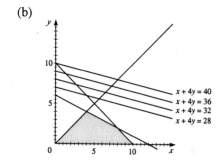

(c)   These lines will first touch the feasible region at the top vertex, $(4, 4)$.

(d)

Vertex	$P = x + 4y$
$(0, 0)$	$(0) + 4(0) = 0$
$(10, 0)$	$(10) + 4(0) = 10$
$(4, 4)$	$(4) + 4(4) = 20$ ← maximum value
$(8, 2)$	$(8) + 4(2) = 16$

So, the maximum value of the objective function occurs at the vertex $(4, 4)$.

Chapter Seven

Exercises 7.1

2. 2×4

4. 3×1

6. 2×2

8. (a) Yes, this matrix is in echelon form.

   (b) No, this matrix not in reduced echelon form. There must be a 0 above the leading 1 in the second row.

   (c) $x + 3y = -3$
   $\qquad y = 5$

10. (a) Yes, the matrix is in echelon form.

    (b) Yes, the matrix is in reduced echelon form.

    (c) $x - 7z = 0$
    $\qquad y + 3z = 0$
    $\qquad\qquad 0 = 1$

12. (a) Yes, the matrix is in echelon form.

    (b) Yes, the matrix is in reduced echelon form.

    (c) $x = 1$
    $\qquad y = 2$
    $\qquad z = 3$

14. (a) No, this matrix is not in echelon form, since the fourth column has the leading 1 of two rows.

    (b) No, this matrix is not in reduced echelon form.

    (c) $x + 3y + w \qquad\ = 0$
    $\qquad\ y + 4w \qquad\ = 0$
    $\qquad\qquad\ w + u = 2$
    $\qquad\qquad\ w \qquad\ = 0$

16. $\begin{bmatrix} 1 & 1 & 6 & 3 \\ 1 & 1 & 3 & 3 \\ 1 & 2 & 4 & 7 \end{bmatrix} \xrightarrow{R_2 - R_1 \to R_2} \begin{bmatrix} 1 & 1 & 6 & 3 \\ 0 & 0 & -3 & 0 \\ 1 & 2 & 4 & 7 \end{bmatrix} \xrightarrow[R_3 - R_1 \to R_3]{-\frac{1}{3}R_2} \begin{bmatrix} 1 & 1 & 6 & 3 \\ 0 & 0 & 1 & 0 \\ 0 & 1 & -2 & 4 \end{bmatrix}$

$\xrightarrow{R_3 \leftrightarrow R_2} \begin{bmatrix} 1 & 1 & 6 & 3 \\ 0 & 1 & -2 & 4 \\ 0 & 0 & 1 & 0 \end{bmatrix}$

Thus, $z = 0$; $y - 2z = y - 0 = y = 4$; and $x + y + 6z = 3 \quad \Leftrightarrow \quad x + 4 - 0 = 3 \quad \Leftrightarrow \quad x = -1$. Therefore, the solution is $(-1, 4, 0)$.

18.
$$
\begin{bmatrix}
1 & 1 & 1 & 4 \\
-1 & 2 & 3 & 17 \\
2 & -1 & 0 & -7
\end{bmatrix}
\quad
\begin{array}{c}
R_2+R_1 \to R_2 \\
R_3-2R_1 \to R_3
\end{array}
\quad
\begin{bmatrix}
1 & 1 & 1 & 4 \\
0 & 3 & 4 & 21 \\
0 & -3 & -2 & -15
\end{bmatrix}
\quad
R_3+R_2 \to R_3
$$

$$
\begin{bmatrix}
1 & 1 & 1 & 4 \\
0 & 3 & 4 & 21 \\
0 & 0 & 2 & 6
\end{bmatrix}
$$

Thus, $2z = 6 \iff z = 3; 3y + 4(3) = 21 \iff y = 3;$ and $x + (3) + (3) = 4 \iff$
$x = -2$. The solution is $(-2, 3, 3)$.

20.
$$
\begin{bmatrix}
0 & 2 & 1 & 4 \\
1 & 1 & 0 & 4 \\
3 & 3 & -1 & 10
\end{bmatrix}
\quad
R_1 \leftrightarrow R_2
\quad
\begin{bmatrix}
1 & 1 & 0 & 4 \\
0 & 2 & 1 & 4 \\
3 & 3 & -1 & 10
\end{bmatrix}
\quad
R_3-3R_1 \to R_3
\quad
\begin{bmatrix}
1 & 1 & 0 & 4 \\
0 & 2 & 1 & 4 \\
0 & 0 & -1 & -2
\end{bmatrix}
$$

Thus, $-z = -2 \iff z = 2; 2y + (2) = 4 \iff y = 1;$ and $x + (1) = 4 \iff x = 3.$
The solution is $(3, 1, 2)$

22.
$$
\begin{bmatrix}
2 & 1 & 0 & 7 \\
2 & -1 & 1 & 6 \\
3 & -2 & 4 & 11
\end{bmatrix}
\quad
\begin{array}{c}
R_2-R_1 \to R_2 \\
2R_3-3R_1 \to R_3
\end{array}
\quad
\begin{bmatrix}
2 & 1 & 0 & 7 \\
0 & -2 & 1 & -1 \\
0 & -7 & 8 & 1
\end{bmatrix}
\quad
2R_3-7R_2 \to R_3
$$

$$
\begin{bmatrix}
2 & 1 & 0 & 7 \\
0 & -2 & 1 & -1 \\
0 & 0 & 9 & 9
\end{bmatrix}
$$

Then, $9x_3 = 9 \iff x_3 = 1; -2x_2 + 1 = -1 \iff x_2 = 1;$ and $2x_1 + 1 = 7 \iff x_1 = 3.$
Therefore, the solution is $(x_1, x_2, x_3) = (3, 1, 1).$

24.
$$
\begin{bmatrix}
10 & 10 & -20 & 60 \\
15 & 20 & 30 & -25 \\
-5 & 30 & -10 & 45
\end{bmatrix}
\quad
\begin{array}{c}
2R_2-3R_1 \to R_2 \\
2R_3+R_1 \to R_3
\end{array}
\quad
\begin{bmatrix}
10 & 10 & -20 & 60 \\
0 & 10 & 120 & -230 \\
0 & 70 & -40 & 150
\end{bmatrix}
\quad
R_3-7R_2 \to R_3
$$

$$
\begin{bmatrix}
10 & 10 & -20 & 60 \\
0 & 10 & 120 & -230 \\
0 & 0 & -880 & 1760
\end{bmatrix}
$$

Thus, $-880z = 1760 \iff z = -2; 10y + 120(-2) = -230 \iff y = 1;$ and
$10x + 10 + 40 = 60 \iff x = 1.$ Therefore, the solution is $(1, 1, -2).$

26.
$$
\begin{bmatrix}
1 & 0 & 3 & 3 \\
2 & 1 & -2 & 5 \\
0 & -1 & 8 & 1
\end{bmatrix}
\quad
R_2-2R_1 \to R_2
\quad
\begin{bmatrix}
1 & 0 & 3 & 3 \\
0 & 1 & -8 & -1 \\
0 & -1 & 8 & 1
\end{bmatrix}
\quad
R_3+R_2 \to R_3
\quad
\begin{bmatrix}
1 & 0 & 3 & 3 \\
0 & 1 & -8 & -1 \\
0 & 0 & 0 & 0
\end{bmatrix}
$$

The system is dependent; there are infinitely many solutions, given by $x + 3t = 3 \iff$
$x = 3 - 3t;$ and $y - 8t = -1 \iff y = 8t - 1.$ The solutions are $(3 - 3t, 8t - 1, t),$ where t is
any real number.

28.
$$
\begin{bmatrix}
1 & -2 & 5 & 3 \\
-2 & 6 & -11 & 1 \\
3 & -16 & 20 & -26
\end{bmatrix}
\quad
\begin{array}{c}
R_2+2R_1 \to R_2 \\
R_3-3R_1 \to R_3
\end{array}
\quad
\begin{bmatrix}
1 & -2 & 5 & 3 \\
0 & 2 & -1 & 7 \\
0 & -10 & 5 & -35
\end{bmatrix}
\quad
R_3+5R_2 \to R_3
$$

$$
\begin{bmatrix}
1 & -2 & 5 & 3 \\
0 & 2 & -1 & 7 \\
0 & 0 & 0 & 0
\end{bmatrix}
$$

The system is dependent; there are infinitely many solutions, given by $2y - t = 7$ $\Leftrightarrow$ $2y = t + 7$ $\Leftrightarrow$ $y = \frac{1}{2}t + \frac{7}{2}$; and $x - 2\left(\frac{1}{2}t + \frac{7}{2}\right) + 5t = 3$ $\Leftrightarrow$ $x - t - 7 + 5t = 3$ $\Leftrightarrow$ $x = -4t + 10$. The solutions are $\left(-4t + 10, \frac{1}{2}t + \frac{7}{2}, t\right)$, where t is any real number.

30.
$$\begin{bmatrix} -2 & 6 & -2 & -12 \\ 1 & -3 & 2 & 10 \\ -1 & 3 & 2 & 6 \end{bmatrix} \xrightarrow{-\frac{1}{2}R_1} \begin{bmatrix} 1 & -3 & 1 & 6 \\ 1 & -3 & 2 & 10 \\ -1 & 3 & 2 & 6 \end{bmatrix} \xrightarrow[R_3+R_1 \to R_3]{R_2-R_1 \to R_2} \begin{bmatrix} 1 & -3 & 1 & 6 \\ 0 & 0 & 1 & 4 \\ 0 & 0 & 3 & 12 \end{bmatrix}$$

$$\xrightarrow{R_3-3R_2 \to R_3} \begin{bmatrix} 1 & -3 & 1 & 6 \\ 0 & 0 & 1 & 4 \\ 0 & 0 & 0 & 0 \end{bmatrix}$$

The system is dependent; there are infinitely many solutions, given by $z = 4$, and $x - 3y + z = 6$, so $x - 3t + 4 = 6$ $\Leftrightarrow$ $x = 3t + 2$. The solutions are $(3t + 2, t, 4)$, where t is any real number.

32.
$$\begin{bmatrix} 3 & 2 & -3 & 10 \\ 1 & -1 & -1 & -5 \\ 1 & 4 & -1 & 20 \end{bmatrix} \xrightarrow{R_1 \leftrightarrow R_2} \begin{bmatrix} 1 & -1 & -1 & -5 \\ 3 & 2 & -3 & 10 \\ 1 & 4 & -1 & 20 \end{bmatrix} \xrightarrow[R_3-R_1 \to R_3]{R_2-3R_1 \to R_2}$$

$$\begin{bmatrix} 1 & -1 & -1 & -5 \\ 0 & 5 & 0 & 25 \\ 0 & 5 & 0 & 25 \end{bmatrix} \xrightarrow{R_3-R_2 \to R_3} \begin{bmatrix} 1 & -1 & -1 & -5 \\ 0 & 5 & 0 & 25 \\ 0 & 0 & 0 & 0 \end{bmatrix}$$

The system is dependent; there are infinitely many solutions, given by $5s = 25$ $\Leftrightarrow$ $s = 5$, and $r - 5 - t = -5$ $\Leftrightarrow$ $r = t$. Hence, the solutions are $(t, 5, t)$, where t is any real number.

34.
$$\begin{bmatrix} 0 & 1 & -5 & 7 \\ 3 & 2 & 0 & 12 \\ 3 & 0 & 10 & 80 \end{bmatrix} \xrightarrow{R_1 \leftrightarrow R_2} \begin{bmatrix} 3 & 2 & 0 & 12 \\ 0 & 1 & -5 & 7 \\ 3 & 0 & 10 & 80 \end{bmatrix} \xrightarrow{R_3-R_1 \to R_3} \begin{bmatrix} 3 & 2 & 0 & 12 \\ 0 & 1 & -5 & 7 \\ 0 & -2 & 10 & 68 \end{bmatrix}$$

$$\xrightarrow{R_3+2R_2 \to R_3} \begin{bmatrix} 3 & 2 & 0 & 12 \\ 0 & 1 & -5 & 7 \\ 0 & 0 & 0 & 82 \end{bmatrix} \quad \text{Therefore, the system is inconsistent and has no solution.}$$

36.
$$\begin{bmatrix} 2 & -3 & 5 & 14 \\ 4 & -1 & -2 & -17 \\ -1 & -1 & 1 & 3 \end{bmatrix} \xrightarrow[-R_1]{R_1 \leftrightarrow R_3} \begin{bmatrix} 1 & 1 & -1 & -3 \\ 4 & -1 & -2 & -17 \\ 2 & -3 & 5 & 14 \end{bmatrix} \xrightarrow[R_3-2R_1 \to R_3]{R_2-4R_1 \to R_2}$$

$$\begin{bmatrix} 1 & 1 & -1 & -3 \\ 0 & -5 & 2 & -5 \\ 0 & -5 & 7 & 20 \end{bmatrix} \xrightarrow{R_3-R_2 \to R_3} \begin{bmatrix} 1 & 1 & -1 & -3 \\ 0 & -5 & 2 & -5 \\ 0 & 0 & 5 & 25 \end{bmatrix}$$

Thus $5z = 25$ $\Leftrightarrow$ $z = 5$; $-5y + 2(5) = -5$ $\Leftrightarrow$ $-5y = -15$ $\Leftrightarrow$ $y = 3$; and $x + (3) - (5) = -3$ $\Leftrightarrow$ $x = -1$. Hence, the solution is $(-1, 3, 5)$.

38.
$$\begin{bmatrix} 3 & -1 & 2 & -1 \\ 4 & -2 & 1 & -7 \\ -1 & 3 & -2 & -1 \end{bmatrix} \xrightarrow[-R_1]{R_1 \leftrightarrow R_3} \begin{bmatrix} 1 & -3 & 2 & 1 \\ 4 & -2 & 1 & -7 \\ 3 & -1 & 2 & -1 \end{bmatrix} \xrightarrow[R_3-3R_1 \to R_3]{R_2-4R_1 \to R_2}$$

$$\begin{bmatrix} 1 & -3 & 2 & 1 \\ 0 & 10 & -7 & -11 \\ 0 & 8 & -4 & -4 \end{bmatrix} \xrightarrow{R_2 \leftrightarrow \frac{1}{4}R_3} \begin{bmatrix} 1 & -3 & 2 & 1 \\ 0 & 2 & -1 & -1 \\ 0 & 10 & -7 & -11 \end{bmatrix} \xrightarrow{R_3-5R_2 \to R_3}$$

$$\begin{bmatrix} 1 & -3 & 2 & 1 \\ 0 & 2 & -1 & -1 \\ 0 & 0 & -2 & -6 \end{bmatrix}$$

Thus $-2z = -6 \iff z = 3; 2y - (3) = -1 \iff 2y = 2 \iff y = 1;$ and
$x - 3(1) + 2(3) = 1 \iff x = -2.$ Hence, the solution is $(-2, 1, 3).$

40.
$$\begin{bmatrix} 1 & 1 & -1 & -1 & 6 \\ 2 & 0 & 1 & -3 & 8 \\ 1 & -1 & 0 & 4 & -10 \\ 3 & 5 & -1 & -1 & 20 \end{bmatrix} \xrightarrow[R_4-3R_1 \to R_4]{R_2-2R_1 \to R_2,\ R_3-R_1 \to R_3} \begin{bmatrix} 1 & 1 & -1 & -1 & 6 \\ 0 & -2 & 3 & -1 & -4 \\ 0 & -2 & 1 & 5 & -16 \\ 0 & 2 & 2 & 2 & 2 \end{bmatrix}$$

$$\xrightarrow[R_3+R_4 \to R_3,\ \frac{1}{2}R_4]{R_2+R_4 \to R_2} \begin{bmatrix} 1 & 1 & -1 & -1 & 6 \\ 0 & 0 & 5 & 1 & -2 \\ 0 & 0 & 3 & 7 & -14 \\ 0 & 1 & 1 & 1 & 1 \end{bmatrix} \xrightarrow[R_3 \leftrightarrow R_4]{R_4 \leftrightarrow R_2} \begin{bmatrix} 1 & 1 & -1 & -1 & 6 \\ 0 & 1 & 1 & 1 & 1 \\ 0 & 0 & 5 & 1 & -2 \\ 0 & 0 & 3 & 7 & -14 \end{bmatrix}$$

$$\xrightarrow{5R_4-3R_3 \to R_4} \begin{bmatrix} 1 & 1 & -1 & -1 & 6 \\ 0 & 1 & 1 & 1 & 1 \\ 0 & 0 & 5 & 1 & -2 \\ 0 & 0 & 0 & 32 & -64 \end{bmatrix}$$

Thus $32w = -64 \iff w = -2; 5z + (-2) = -2 \iff 5z = 0 \iff z = 0;$
$y + 0 + (-2) = 1 \iff y = 3;$ and $x + 3 - 0 - (-2) = 6 \iff x = 1.$ Hence the solution is
$(1, 3, 0, -2).$

42.
$$\begin{bmatrix} 1 & -3 & 2 & 1 & -2 \\ 1 & -2 & 0 & -2 & -10 \\ 0 & 0 & 1 & 5 & 15 \\ 3 & 0 & 2 & 1 & -3 \end{bmatrix} \xrightarrow[R_4-3R_1 \to R_4]{R_2-R_1 \to R_2} \begin{bmatrix} 1 & -3 & 2 & 1 & -2 \\ 0 & 1 & -2 & -3 & -8 \\ 0 & 0 & 1 & 5 & 15 \\ 0 & 9 & -4 & -2 & 3 \end{bmatrix} \xrightarrow{R_4-9R_2 \to R_4}$$

$$\begin{bmatrix} 1 & -3 & 2 & 1 & -2 \\ 0 & 1 & -2 & -3 & -8 \\ 0 & 0 & 1 & 5 & 15 \\ 0 & 0 & 14 & 25 & 75 \end{bmatrix} \xrightarrow{R_4-14R_3 \to R_4} \begin{bmatrix} 1 & -3 & 2 & 1 & -2 \\ 0 & 1 & -2 & -3 & -8 \\ 0 & 0 & 1 & 5 & 15 \\ 0 & 0 & 0 & -45 & -135 \end{bmatrix}$$

Thus $-45w = -135 \iff w = 3; z + 5(3) = 15 \iff z = 0; y - 2(0) - 3(3) = -8 \iff$
$y = 1;$ and $x - 3(1) + 2(0) + (3) = -2 \iff x = -2.$ Hence, the solution is $(-2, 1, 0, 3).$

44.
$$\begin{bmatrix} 0 & 1 & -1 & 2 & 0 \\ 3 & 2 & 0 & 1 & 0 \\ 2 & 0 & 0 & 4 & 12 \\ -2 & 0 & -2 & 5 & 6 \end{bmatrix} \xrightarrow[R_4+R_3 \to R_4,\ \frac{1}{2}R_3]{R_2-R_3 \to R_2} \begin{bmatrix} 0 & 1 & -1 & 2 & 0 \\ 1 & 2 & 0 & -3 & -12 \\ 1 & 0 & 0 & 2 & 6 \\ 0 & 0 & -2 & 9 & 18 \end{bmatrix} \xrightarrow[R_2-R_1 \to R_2]{R_1 \leftrightarrow R_3}$$

$$\begin{bmatrix} 1 & 0 & 0 & 2 & 6 \\ 0 & 2 & 0 & -5 & -18 \\ 0 & 1 & -1 & 2 & 0 \\ 0 & 0 & -2 & 9 & 18 \end{bmatrix} \xrightarrow[R_2 \leftrightarrow R_3]{R_2-2R_3 \to R_2} \begin{bmatrix} 1 & 0 & 0 & 2 & 6 \\ 0 & 1 & -1 & 2 & 0 \\ 0 & 0 & 2 & -9 & -18 \\ 0 & 0 & -2 & 9 & 18 \end{bmatrix} \xrightarrow{R_4+R_3 \to R_4}$$

$$\begin{bmatrix} 1 & 0 & 0 & 2 & 6 \\ 0 & 1 & -1 & 2 & 0 \\ 0 & 0 & 2 & -9 & -18 \\ 0 & 0 & 0 & 0 & 0 \end{bmatrix}$$

Thus, the system has infinitely many solutions, given by $-2z + 9t = 18$ $\Leftrightarrow$ $z = \frac{9}{2}t - 9$; $y - \left(\frac{9}{2}t - 9\right) + 2t = 0$ $\Leftrightarrow$ $y = \frac{5}{2}t - 9$; and $x + 2t = 6$ $\Leftrightarrow$ $x = -2t + 6$. So, the solutions are $\left(-2t + 6, \frac{5}{2}t - 9, \frac{9}{2}t - 9, t\right)$, where t is any real number.

46. $\begin{bmatrix} 2 & -1 & 2 & 1 & 5 \\ -1 & 1 & 4 & -1 & 3 \\ 3 & -2 & -1 & 0 & 0 \end{bmatrix}$ $\xrightarrow[{-R_1}]{R_1 \leftrightarrow R_2}$ $\begin{bmatrix} 1 & -1 & -4 & 1 & -3 \\ 2 & -1 & 2 & 1 & 5 \\ 3 & -2 & -1 & 0 & 0 \end{bmatrix}$ $\xrightarrow[{R_3 - 3R_1 \to R_3}]{R_2 - 2R_1 \to R_2}$

$\begin{bmatrix} 1 & -1 & -4 & 1 & -3 \\ 0 & 1 & 10 & -1 & 11 \\ 0 & 1 & 11 & -3 & 9 \end{bmatrix}$ $\xrightarrow{R_3 - R_2 \to R_3}$ $\begin{bmatrix} 1 & -1 & -4 & 1 & -3 \\ 0 & 1 & 10 & -1 & 11 \\ 0 & 0 & 1 & -2 & -2 \end{bmatrix}$

Thus, the system has infinitely many solutions, given by $z - 2t = -2$ $\Leftrightarrow$ $z = -2 + 2t$; $y + 10(-2 + 2t) - t = 11$ $\Leftrightarrow$ $y = 31 - 19t$; and $x - (31 - 19t) - 4(-2 + 2t) + t = -3$ $\Leftrightarrow$ $x = 20 - 12t$. Hence, the solutions are $(20 - 12t, 31 - 19t, -2 + 2t, t)$, where t is any real number.

48. Let x be the quantity, in mL, of 10% acid, y be the quantity of 20% acid, and z be the quantity of 40% acid. Then,

$$\begin{cases} 0.1x + 0.2y + 0.4z = 18 \\ x + y + z = 100 \\ x - 4z = 0 \end{cases}$$. Writing this equation in matrix form, we get

$\begin{bmatrix} 1 & 1 & 1 & 100 \\ 0.1 & 0.2 & 0.4 & 18 \\ 1 & 0 & -4 & 0 \end{bmatrix}$ $\xrightarrow[{R_3 - R_1 \to R_3}]{10R_2, R_2 - R_1 \to R_2}$ $\begin{bmatrix} 1 & 1 & 1 & 100 \\ 0 & 1 & 3 & 80 \\ 0 & -1 & -5 & -100 \end{bmatrix}$ $\xrightarrow[{-R_3}]{R_3 + R_2 \to R_3}$

$\begin{bmatrix} 1 & 1 & 1 & 100 \\ 0 & 1 & 3 & 80 \\ 0 & 0 & 2 & 20 \end{bmatrix}$.

Thus $2z = 20$ $\Leftrightarrow$ $z = 10$; $y + 30 = 80$ $\Leftrightarrow$ $y = 50$; and $x + 50 + 10 = 100$ $\Leftrightarrow$ $x = 40$. So he should mix together 40 mL of 10% acid, 50 mL of 20% acid, and 10 mL of 40% acid.

50. Let a, b, and c be the number of students in classrooms A, B, and C, respectively, where $a, b, c \geq 0$. Then, the system of equations is

$$\begin{cases} a + b + c = 100 \\ \frac{1}{2}a = \frac{1}{5}b \\ \frac{1}{5}b = \frac{1}{3}c \end{cases} \Leftrightarrow \begin{cases} a + b + c = 100 \\ b = \frac{5}{2}a \\ c = \frac{3}{5}b = \frac{3}{2}a \end{cases}$$

By substitution, $a + \frac{5}{2}a + \frac{3}{2}a = 100$ $\Leftrightarrow$ $a = 20$; $b = \frac{5}{2}(20) = 50$; and $c = \frac{3}{2}(20) = 30$. So there are 20 students in classroom A, 50 students in classroom B, and 30 students in classroom C.

52. The number of cars entering each intersection must equal the number of cars leaving that intersection. This leads to the following equations:

$$\begin{cases} 200 + 180 = x + z \\ x + 70 = 20 + w \\ w + 200 = y + 30 \\ y + z = 400 + 200 \end{cases}$$ Simplifying and writing this in matrix form, we get:

$$\begin{bmatrix} 1 & 0 & 1 & 0 & 380 \\ 1 & 0 & 0 & -1 & -50 \\ 0 & 1 & 0 & -1 & 170 \\ 0 & 1 & 1 & 0 & 600 \end{bmatrix} \xrightarrow[R_4-R_3 \to R_4]{R_2-R_1 \to R_2} \begin{bmatrix} 1 & 0 & 1 & 0 & 380 \\ 0 & 0 & -1 & -1 & -430 \\ 0 & 1 & 0 & -1 & 170 \\ 0 & 0 & 1 & 1 & 430 \end{bmatrix} \xrightarrow{R_2 \leftrightarrow R_3}$$

$$\begin{bmatrix} 1 & 0 & 1 & 0 & 380 \\ 0 & 1 & 0 & -1 & 170 \\ 0 & 0 & -1 & -1 & -430 \\ 0 & 0 & 1 & 1 & 430 \end{bmatrix}$$

Therefore, $z + t = 430 \quad \Leftrightarrow \quad z = 430 - t$; $y - t = 170 \quad \Leftrightarrow \quad y = 170 + t$; and $x + 430 - t = 380 \quad \Leftrightarrow \quad x = t - 50$. Since $x, y, z, w \geq 0$, it follows that $50 \leq t \leq 430$, and so the solutions are $(t - 50, 170 + t, 430 - t, t)$, where $50 \leq t \leq 430$.

Exercises 7.2

2. Since $\frac{1}{4} = 0.25$, $\ln 1 = 0$, $2 = \sqrt{4}$, and $3 = \frac{6}{2}$, the corresponding entries are equal, so the matrices are equal.

4. $\begin{bmatrix} 0 & 1 & 1 \\ 1 & 1 & 0 \end{bmatrix} - \begin{bmatrix} 2 & 1 & -1 \\ 1 & 3 & -2 \end{bmatrix} = \begin{bmatrix} -2 & 0 & 2 \\ 0 & -2 & 2 \end{bmatrix}$

6. $2\begin{bmatrix} 1 & 1 & 0 \\ 1 & 0 & 1 \\ 0 & 1 & 1 \end{bmatrix} + \begin{bmatrix} 1 & 1 \\ 2 & 1 \\ 3 & 1 \end{bmatrix}$ is undefined because these matrices have incompatible dimensions

8. $\begin{bmatrix} 2 & 1 & 2 \\ 6 & 3 & 4 \end{bmatrix} \begin{bmatrix} 1 & -2 \\ 3 & 6 \\ -2 & 0 \end{bmatrix} = \begin{bmatrix} 1 & 2 \\ 7 & 64 \end{bmatrix}$

10. $\begin{bmatrix} 2 & -3 \\ 0 & 1 \\ 1 & 2 \end{bmatrix} \begin{bmatrix} 5 \\ 1 \end{bmatrix} = \begin{bmatrix} 7 \\ 1 \\ 7 \end{bmatrix}$

12. $3X - B = C$ $\Leftrightarrow$ $3X = C + B$. but C is 3×2 matrix and B is a 2×2 matrix so $C + B$ is impossible, thus no solutions exists.

14. $5\,(X - C) = D$ $\Leftrightarrow$ $X - C = \frac{1}{5}D$ $\Leftrightarrow$ $X = \frac{1}{5}D + C = \frac{1}{5}\begin{bmatrix} 10 & 20 \\ 30 & 20 \\ 10 & 0 \end{bmatrix} + \begin{bmatrix} 2 & 3 \\ 1 & 0 \\ 0 & 2 \end{bmatrix}$

$= \begin{bmatrix} 2 & 4 \\ 6 & 4 \\ 2 & 0 \end{bmatrix} + \begin{bmatrix} 2 & 3 \\ 1 & 0 \\ 0 & 2 \end{bmatrix} = \begin{bmatrix} 4 & 7 \\ 7 & 4 \\ 2 & 2 \end{bmatrix}$.

16. $2A = B - 3X$ $\Leftrightarrow$ $2A - B = -3X$ $\Leftrightarrow$ $X = -\frac{1}{3}\,(2A - B) = -\frac{1}{3}\left(2\begin{bmatrix} 4 & 6 \\ 1 & 3 \end{bmatrix} - \begin{bmatrix} 2 & 5 \\ 3 & 7 \end{bmatrix}\right)$

$= -\frac{1}{3}\left(\begin{bmatrix} 8 & 12 \\ 2 & 6 \end{bmatrix} - \begin{bmatrix} 2 & 5 \\ 3 & 7 \end{bmatrix}\right) = -\frac{1}{3}\begin{bmatrix} 6 & 7 \\ -1 & -1 \end{bmatrix} = \begin{bmatrix} -2 & -\frac{7}{3} \\ \frac{1}{3} & \frac{1}{3} \end{bmatrix}$.

In Exercises 18 – 38, the matrices A, B, C, D, E, F, and G are defined as follows:

$A = \begin{bmatrix} 2 & -5 \\ 0 & 7 \end{bmatrix}$ $B = \begin{bmatrix} 3 & \frac{1}{2} & 5 \\ 1 & -1 & 3 \end{bmatrix}$ $C = \begin{bmatrix} 2 & -\frac{5}{2} & 0 \\ 0 & 2 & -3 \end{bmatrix}$ $D = [7 \ \ 3]$

$E = \begin{bmatrix} 1 \\ 2 \\ 0 \end{bmatrix}$ $F = \begin{bmatrix} 1 & 0 & 0 \\ 0 & 1 & 0 \\ 0 & 0 & 1 \end{bmatrix}$ $G = \begin{bmatrix} 5 & -3 & 10 \\ 6 & 1 & 0 \\ -5 & 2 & 2 \end{bmatrix}$

18. $B + F$ is undefined because B (2×3) and F (3×3) have incompatible dimensions.

20. $5A = 5\begin{bmatrix} 2 & -5 \\ 0 & 7 \end{bmatrix} = \begin{bmatrix} 10 & -25 \\ 0 & 35 \end{bmatrix}$

22. $C - 5A$ is undefined because C (2×3) and A (2×2) have incompatible dimensions.

24. $DA = [7 \ \ 3] \begin{bmatrix} 2 & -5 \\ 0 & 7 \end{bmatrix} = [14 \ \ -14]$

26. BC is undefined because B (2×3) and C (2×3) have incompatible dimensions.

28. $GF = \begin{bmatrix} 5 & -3 & 10 \\ 6 & 1 & 0 \\ -5 & 2 & 2 \end{bmatrix} \begin{bmatrix} 1 & 0 & 0 \\ 0 & 1 & 0 \\ 0 & 0 & 1 \end{bmatrix} = \begin{bmatrix} 5 & -3 & 10 \\ 6 & 1 & 0 \\ -5 & 2 & 2 \end{bmatrix}$

30. $D(AB) = [7 \ \ 3] \begin{bmatrix} 2 & -5 \\ 0 & 7 \end{bmatrix} \begin{bmatrix} 3 & \frac{1}{2} & 5 \\ 1 & -1 & 3 \end{bmatrix} = [7 \ \ 3] \begin{bmatrix} 1 & 6 & -5 \\ 7 & -7 & 21 \end{bmatrix} = [28 \ \ 21 \ \ 28]$

32. $A^2 = \begin{bmatrix} 2 & -5 \\ 0 & 7 \end{bmatrix} \begin{bmatrix} 2 & -5 \\ 0 & 7 \end{bmatrix} = \begin{bmatrix} 4 & -45 \\ 0 & 49 \end{bmatrix}$

34. $DB + DC = D(B + C) = [7 \ \ 3] \left(\begin{bmatrix} 3 & \frac{1}{2} & 5 \\ 1 & -1 & 3 \end{bmatrix} + \begin{bmatrix} 2 & -\frac{5}{2} & 0 \\ 0 & 2 & -3 \end{bmatrix} \right)$

   $= [7 \ \ 3] \begin{bmatrix} 5 & -2 & 5 \\ 1 & 1 & 0 \end{bmatrix} = [38 \ \ -11 \ \ 35]$

36. $F^2 = \begin{bmatrix} 1 & 0 & 0 \\ 0 & 1 & 0 \\ 0 & 0 & 1 \end{bmatrix} \begin{bmatrix} 1 & 0 & 0 \\ 0 & 1 & 0 \\ 0 & 0 & 1 \end{bmatrix} = \begin{bmatrix} 1 & 0 & 0 \\ 0 & 1 & 0 \\ 0 & 0 & 1 \end{bmatrix}$

38. $ABE = \begin{bmatrix} 2 & -5 \\ 0 & 7 \end{bmatrix} \begin{bmatrix} 3 & \frac{1}{2} & 5 \\ 1 & -1 & 3 \end{bmatrix} \begin{bmatrix} 1 \\ 2 \\ 0 \end{bmatrix} = \begin{bmatrix} 1 & 6 & -5 \\ 7 & -7 & 21 \end{bmatrix} \begin{bmatrix} 1 \\ 2 \\ 0 \end{bmatrix} = \begin{bmatrix} 13 \\ -7 \end{bmatrix}$

40. $3 \cdot \begin{bmatrix} x & y \\ y & x \end{bmatrix} = \begin{bmatrix} 6 & -9 \\ -9 & 6 \end{bmatrix}$. Since $3 \cdot \begin{bmatrix} x & y \\ y & x \end{bmatrix} = \begin{bmatrix} 3x & 3y \\ 3y & 3x \end{bmatrix}$. $=$ Thus we must solve the system

   $\begin{cases} 3x = 6 \\ 3y = -9 \\ 3x = 6 \\ 3y = -9 \end{cases}$ So $3x = 6 \quad \Leftrightarrow \quad x = 2$ and $3y = -9 \quad \Leftrightarrow \quad y = -3$.

   So the solution is $x = 2$, $y = -3$.

42. $\begin{bmatrix} x & y \\ -y & x \end{bmatrix} - \begin{bmatrix} y & x \\ x & -y \end{bmatrix} = \begin{bmatrix} 4 & -4 \\ -6 & 6 \end{bmatrix} \quad \Leftrightarrow \quad \begin{bmatrix} x-y & y-x \\ -x-y & x+y \end{bmatrix} = \begin{bmatrix} 4 & -4 \\ -6 & 6 \end{bmatrix}$. Thus we must

   solve the system $\begin{cases} x - y = 4 \\ y - x = -4 \\ -x - y = -6 \\ x + y = 6 \end{cases}$ Adding the first equation to the last equation gives $2x = 10$

   $\Leftrightarrow \quad x = 5$ and $5 + y = 6 \quad \Leftrightarrow \quad y = 1$. So the solution is $x = 5$, $y = 1$.

44. $\begin{cases} 6x - y + z = 12 \\ 2x + z = 7 \\ y - 2z = 4 \end{cases}$ written as a matrix equation is $\begin{bmatrix} 6 & -1 & 1 \\ 2 & 0 & 1 \\ 0 & 1 & -2 \end{bmatrix} \begin{bmatrix} x \\ y \\ z \end{bmatrix} = \begin{bmatrix} 12 \\ 7 \\ 4 \end{bmatrix}$.

46. $\begin{cases} x - y + z = 2 \\ 4x - 2y - z = 2 \\ x + y + 5z = 2 \\ -x - y - z = 2 \end{cases}$ written as a matrix equation is $\begin{bmatrix} 1 & -1 & 1 \\ 4 & -2 & -1 \\ 1 & 1 & 5 \\ -1 & -1 & -1 \end{bmatrix} \begin{bmatrix} x \\ y \\ z \end{bmatrix} = \begin{bmatrix} 2 \\ 2 \\ 2 \\ 2 \end{bmatrix}$.

48. (a) Let $A = \begin{bmatrix} a & b \\ c & d \end{bmatrix}$ and $B = \begin{bmatrix} e & f \\ g & h \end{bmatrix}$. Then, $A + B = \begin{bmatrix} a+e & b+f \\ c+g & d+h \end{bmatrix}$, and

$$(A+B)^2 = \begin{bmatrix} a+e & b+f \\ c+g & d+h \end{bmatrix} \begin{bmatrix} a+e & b+f \\ c+g & d+h \end{bmatrix}$$
$$= \begin{bmatrix} (a+e)^2 + (b+f)(c+g) & (a+e)(b+f) + (b+f)(d+h) \\ (c+g)(a+e) + (d+h)(c+g) & (c+g)(b+f) + (d+h)^2 \end{bmatrix};$$

$$A^2 = \begin{bmatrix} a & b \\ c & d \end{bmatrix} \begin{bmatrix} a & b \\ c & d \end{bmatrix} = \begin{bmatrix} a^2 + bc & ab + bd \\ ac + cd & bc + d^2 \end{bmatrix};$$

$$B^2 = \begin{bmatrix} e & f \\ g & h \end{bmatrix} \begin{bmatrix} e & f \\ g & h \end{bmatrix} = \begin{bmatrix} e^2 + fg & ef + fh \\ eg + gh & fg + h^2 \end{bmatrix};$$

$$AB = \begin{bmatrix} a & b \\ c & d \end{bmatrix} \begin{bmatrix} e & f \\ g & h \end{bmatrix} = \begin{bmatrix} ae + bg & af + bh \\ ce + dg & cf + dh \end{bmatrix};$$

$$BA = \begin{bmatrix} e & f \\ g & h \end{bmatrix} \begin{bmatrix} a & b \\ c & d \end{bmatrix} = \begin{bmatrix} ae + cf & be + df \\ ag + ch & bg + dh \end{bmatrix}.$$

Then, $A^2 + AB + BA + B^2$

$$= \begin{bmatrix} a^2 + bc + ae + bg + ae + cf + e^2 + fg & ab + bd + ef + fh + af + bh + be + df \\ ac + cd + eg + gh + ce + dg + ag + ch & bc + d^2 + fg + h^2 + cf + dh + bg + dh \end{bmatrix}$$

$$= \begin{bmatrix} a^2 + 2ae + e^2 + b(c+g) + f(c+g) & a(b+f) + e(b+f) + b(d+h) + f(d+h) \\ c(a+e) + g(a+e) + d(c+g) + h(c+g) & c(b+f) + g(b+f) + d^2 + 2dh + h^2 \end{bmatrix}$$

$$= \begin{bmatrix} (a+e)^2 + (b+f)(c+g) & (a+e)(b+f) + (b+f)(d+h) \\ (c+g)(a+e) + (d+h)(c+g) & (c+g)(b+f) + (d+h)^2 \end{bmatrix} = (A+B)^2.$$

(b) No, from Exercise 46, $(A+B)^2 = (A+B)(A+B) = A^2 + AB + BA + B^2 \neq A^2 + 2AB + B^2$ unless $AB = BA$ which is, in general, not true, as we saw in Example 3.

50. (a) $AB = \begin{bmatrix} 12 & 10 & 0 \\ 4 & 4 & 20 \\ 8 & 9 & 12 \end{bmatrix} \begin{bmatrix} \$1000 & \$500 \\ \$2000 & \$1200 \\ \$1500 & \$1000 \end{bmatrix} = \begin{bmatrix} \$32,000 & \$18,000 \\ \$42,000 & \$26,800 \\ \$44,000 & \$26,800 \end{bmatrix}$

(b) The daily profit in January from the Biloxi plant is the (2nd row, 1st column) matrix entry, and hence is \$42,000.

(c) The total daily profit (from all three plants) in February was $\$18,000 + \$26,800 + \$26,800 = \$71,600.$

52. (a) $AC = \begin{bmatrix} 120 & 50 & 60 \\ 40 & 25 & 30 \\ 60 & 30 & 20 \end{bmatrix} \begin{bmatrix} 0.10 \\ 0.50 \\ 1.00 \end{bmatrix} = \begin{bmatrix} 97.00 \\ 46.50 \\ 41.00 \end{bmatrix}$ Amy's stand sold \$97 of produce on Saturday,

Beth's stand sold \$46.50 and Chad's stand sold \$41.

(b) $BC = \begin{bmatrix} 100 & 60 & 30 \\ 35 & 20 & 20 \\ 60 & 25 & 30 \end{bmatrix} \begin{bmatrix} 0.10 \\ 0.50 \\ 1.00 \end{bmatrix} = \begin{bmatrix} 70.00 \\ 33.50 \\ 48.50 \end{bmatrix}$ Amy's stand sold \$70 of produce on Sunday,

Beth's stand sold \$33.50 and Chad's stand sold \$48.50.

(c) $A + B = \begin{bmatrix} 120 & 50 & 60 \\ 40 & 25 & 30 \\ 60 & 30 & 20 \end{bmatrix} + \begin{bmatrix} 100 & 60 & 30 \\ 35 & 20 & 20 \\ 60 & 25 & 30 \end{bmatrix} = \begin{bmatrix} 220 & 110 & 90 \\ 75 & 45 & 50 \\ 120 & 55 & 50 \end{bmatrix}$ This represents the

melons, squash, and tomatoes they sold during the weekend

(d) $(A + B)C = \left(\begin{bmatrix} 120 & 50 & 60 \\ 40 & 25 & 30 \\ 60 & 30 & 20 \end{bmatrix} + \begin{bmatrix} 100 & 60 & 30 \\ 35 & 20 & 20 \\ 60 & 25 & 30 \end{bmatrix} \right) \begin{bmatrix} 0.10 \\ 0.50 \\ 1.00 \end{bmatrix}$

$= \begin{bmatrix} 220 & 110 & 90 \\ 75 & 45 & 50 \\ 120 & 55 & 50 \end{bmatrix} \begin{bmatrix} 0.10 \\ 0.50 \\ 1.00 \end{bmatrix} = \begin{bmatrix} 167.00 \\ 80.00 \\ 89.50 \end{bmatrix}$

During the weekend Amy's stand sold \$167, Beth's stand sold \$80 and Chad's stand sold \$89.50 of produce.

Notice that $(A + B)C = AC + BC = \begin{bmatrix} 97.00 \\ 46.50 \\ 41.00 \end{bmatrix} + \begin{bmatrix} 70.00 \\ 33.50 \\ 48.50 \end{bmatrix} = \begin{bmatrix} 167.00 \\ 80.00 \\ 89.50 \end{bmatrix}.$

54. Suppose A is $n \times m$ and B is $i \times j$. If the product AB is defined, then $m = i$. If the product BA is defined, then $j = n$. Thus if both products are defined and if A is $n \times m$, then B must be $m \times n$.

56. $A = \begin{bmatrix} 1 & 1 \\ 1 & 1 \end{bmatrix};$

$A^2 = \begin{bmatrix} 1 & 1 \\ 1 & 1 \end{bmatrix} \begin{bmatrix} 1 & 1 \\ 1 & 1 \end{bmatrix} = \begin{bmatrix} 2 & 2 \\ 2 & 2 \end{bmatrix};$

$A^3 = A \cdot A^2 = \begin{bmatrix} 1 & 1 \\ 1 & 1 \end{bmatrix} \begin{bmatrix} 2 & 2 \\ 2 & 2 \end{bmatrix} = \begin{bmatrix} 4 & 4 \\ 4 & 4 \end{bmatrix};$

$A^4 = A \cdot A^3 = \begin{bmatrix} 1 & 1 \\ 1 & 1 \end{bmatrix} \begin{bmatrix} 4 & 4 \\ 4 & 4 \end{bmatrix} = \begin{bmatrix} 8 & 8 \\ 8 & 8 \end{bmatrix}.$

From this pattern, we see that

$A^n = \begin{bmatrix} 2^{n-1} & 2^{n-1} \\ 2^{n-1} & 2^{n-1} \end{bmatrix}.$

Exercises 7.3

2. $A = \begin{bmatrix} 2 & -3 \\ 4 & -7 \end{bmatrix}$; $B = \begin{bmatrix} \frac{7}{2} & -\frac{3}{2} \\ 2 & -1 \end{bmatrix}$. $AB = \begin{bmatrix} 2 & -3 \\ 4 & -7 \end{bmatrix}\begin{bmatrix} \frac{7}{2} & -\frac{3}{2} \\ 2 & -1 \end{bmatrix} = \begin{bmatrix} 1 & 0 \\ 0 & 1 \end{bmatrix}$ and

$BA = \begin{bmatrix} \frac{7}{2} & -\frac{3}{2} \\ 2 & -1 \end{bmatrix}\begin{bmatrix} 2 & -3 \\ 4 & -7 \end{bmatrix} = \begin{bmatrix} 1 & 0 \\ 0 & 1 \end{bmatrix}$.

4. $A = \begin{bmatrix} 3 & 2 & 4 \\ 1 & 1 & -6 \\ 2 & 1 & 12 \end{bmatrix}$; $B = \begin{bmatrix} 9 & -10 & -8 \\ -12 & 14 & 11 \\ -\frac{1}{2} & \frac{1}{2} & \frac{1}{2} \end{bmatrix}$.

$AB = \begin{bmatrix} 3 & 2 & 4 \\ 1 & 1 & -6 \\ 2 & 1 & 12 \end{bmatrix}\begin{bmatrix} 9 & -10 & -8 \\ -12 & 14 & 11 \\ -\frac{1}{2} & \frac{1}{2} & \frac{1}{2} \end{bmatrix} = \begin{bmatrix} 1 & 0 & 0 \\ 0 & 1 & 0 \\ 0 & 0 & 1 \end{bmatrix}$ and

$BA = \begin{bmatrix} 9 & -10 & -8 \\ -12 & 14 & 11 \\ -\frac{1}{2} & \frac{1}{2} & \frac{1}{2} \end{bmatrix}\begin{bmatrix} 3 & 2 & 4 \\ 1 & 1 & -6 \\ 2 & 1 & 12 \end{bmatrix} = \begin{bmatrix} 1 & 0 & 0 \\ 0 & 1 & 0 \\ 0 & 0 & 1 \end{bmatrix}$.

6. $B = \begin{bmatrix} 1 & 3 & 2 \\ 0 & 2 & 2 \\ -2 & -1 & 0 \end{bmatrix}$. We begin with a 3×6 matrix whose left half is B and whose right half is I_3.

$\begin{bmatrix} 1 & 3 & 2 & 1 & 0 & 0 \\ 0 & 2 & 2 & 0 & 1 & 0 \\ -2 & -1 & 0 & 0 & 0 & 1 \end{bmatrix} \xrightarrow{R_3+2R_1 \to R_3} \begin{bmatrix} 1 & 3 & 2 & 1 & 0 & 0 \\ 0 & 2 & 2 & 0 & 1 & 0 \\ 0 & 5 & 4 & 2 & 0 & 1 \end{bmatrix} \xrightarrow{2R_3-5R_2 \to R_3}$

$\begin{bmatrix} 1 & 3 & 2 & 1 & 0 & 0 \\ 0 & 2 & 2 & 0 & 1 & 0 \\ 0 & 0 & -2 & 4 & -5 & 2 \end{bmatrix} \xrightarrow[\frac{1}{2}R_3]{\frac{1}{2}R_2} \begin{bmatrix} 1 & 3 & 2 & 1 & 0 & 0 \\ 0 & 1 & 1 & 0 & \frac{1}{2} & 0 \\ 0 & 0 & 1 & -2 & \frac{5}{2} & -1 \end{bmatrix} \xrightarrow[R_2-R_3 \to R_2]{R_1-3R_2 \to R_1}$

$\begin{bmatrix} 1 & 0 & -1 & 1 & -\frac{3}{2} & 0 \\ 0 & 1 & 0 & 2 & -2 & 1 \\ 0 & 0 & 1 & -2 & \frac{5}{2} & -1 \end{bmatrix} \xrightarrow{R_1+R_3 \to R_1} \begin{bmatrix} 1 & 0 & 0 & -1 & 1 & -1 \\ 0 & 1 & 0 & 2 & -2 & 1 \\ 0 & 0 & 1 & -2 & \frac{5}{2} & -1 \end{bmatrix}$

Then $B^{-1} = \begin{bmatrix} -1 & 1 & -1 \\ 2 & -2 & 1 \\ -2 & \frac{5}{2} & -1 \end{bmatrix}$;

$B^{-1}B = \begin{bmatrix} -1 & 1 & -1 \\ 2 & -2 & 1 \\ -2 & \frac{5}{2} & -1 \end{bmatrix}\begin{bmatrix} 1 & 3 & 2 \\ 0 & 2 & 2 \\ -2 & -1 & 0 \end{bmatrix} = \begin{bmatrix} 1 & 0 & 0 \\ 0 & 1 & 0 \\ 0 & 0 & 1 \end{bmatrix}$; and

$BB^{-1} = \begin{bmatrix} 1 & 3 & 2 \\ 0 & 2 & 2 \\ -2 & -1 & 0 \end{bmatrix}\begin{bmatrix} -1 & 1 & -1 \\ 2 & -2 & 1 \\ -2 & \frac{5}{2} & -1 \end{bmatrix} = \begin{bmatrix} 1 & 0 & 0 \\ 0 & 1 & 0 \\ 0 & 0 & 1 \end{bmatrix}$.

8. $\begin{bmatrix} 3 & 4 \\ 7 & 9 \end{bmatrix}^{-1} = \frac{1}{27-28}\begin{bmatrix} 9 & -4 \\ -7 & 3 \end{bmatrix} = -\begin{bmatrix} 9 & -4 \\ -7 & 3 \end{bmatrix} = \begin{bmatrix} -9 & 4 \\ 7 & -3 \end{bmatrix}$

10. $\begin{bmatrix} -7 & 4 \\ 8 & -5 \end{bmatrix}^{-1} = \frac{1}{35-32}\begin{bmatrix} -5 & -4 \\ -8 & -7 \end{bmatrix} = \begin{bmatrix} -\frac{5}{3} & -\frac{4}{3} \\ -\frac{8}{3} & -\frac{7}{3} \end{bmatrix}$

12. $\begin{bmatrix} \frac{1}{2} & \frac{1}{3} \\ 5 & 4 \end{bmatrix}^{-1} = \frac{1}{2-\frac{5}{3}}\begin{bmatrix} 4 & -\frac{1}{3} \\ -5 & \frac{1}{2} \end{bmatrix} = \begin{bmatrix} 12 & -1 \\ -15 & \frac{3}{2} \end{bmatrix}$

14. $\begin{bmatrix} 4 & 2 & 3 & 1 & 0 & 0 \\ 3 & 3 & 2 & 0 & 1 & 0 \\ 1 & 0 & 1 & 0 & 0 & 1 \end{bmatrix} \xrightarrow[4R_3-R_1 \to R_3]{4R_2-3R_1 \to R_2} \begin{bmatrix} 4 & 2 & 3 & 1 & 0 & 0 \\ 0 & 6 & -1 & -3 & 4 & 0 \\ 0 & -2 & 1 & -1 & 0 & 4 \end{bmatrix} \xrightarrow[3R_3+R_2 \to R_3]{R_1+R_3 \to R_1}$

$\begin{bmatrix} 4 & 0 & 4 & 0 & 0 & 4 \\ 0 & 6 & -1 & -3 & 4 & 0 \\ 0 & 0 & 2 & -6 & 4 & 12 \end{bmatrix} \xrightarrow[\frac{1}{2}R_3]{\frac{1}{4}R_1} \begin{bmatrix} 1 & 0 & 1 & 0 & 0 & 1 \\ 0 & 6 & -1 & -3 & 4 & 0 \\ 0 & 0 & 1 & -3 & 2 & 6 \end{bmatrix} \xrightarrow[R_2+R_3 \to R_2]{R_1-R_3 \to R_1}$

$\begin{bmatrix} 1 & 0 & 0 & 3 & -2 & -5 \\ 0 & 6 & 0 & -6 & 6 & 6 \\ 0 & 0 & 1 & -3 & 2 & 6 \end{bmatrix} \xrightarrow{\frac{1}{6}R_2} \begin{bmatrix} 1 & 0 & 0 & 3 & -2 & -5 \\ 0 & 1 & 0 & -1 & 1 & 1 \\ 0 & 0 & 1 & -3 & 2 & 6 \end{bmatrix}$

Therefore, the inverse matrix is $\begin{bmatrix} 3 & -2 & -5 \\ -1 & 1 & 1 \\ -3 & 2 & 6 \end{bmatrix}$.

16. $\begin{bmatrix} 5 & 7 & 4 & 1 & 0 & 0 \\ 3 & -1 & 3 & 0 & 1 & 0 \\ 6 & 7 & 5 & 0 & 0 & 1 \end{bmatrix} \xrightarrow[R_1 \leftrightarrow R_3]{R_3-R_1 \to R_3} \begin{bmatrix} 1 & 0 & 1 & -1 & 0 & 1 \\ 3 & -1 & 3 & 0 & 1 & 0 \\ 5 & 7 & 4 & 1 & 0 & 0 \end{bmatrix} \xrightarrow[R_3-5R_1 \to R_3]{R_2-3R_1 \to R_2}$

$\begin{bmatrix} 1 & 0 & 1 & -1 & 0 & 1 \\ 0 & -1 & 0 & 3 & 1 & -3 \\ 0 & 7 & -1 & 6 & 0 & -5 \end{bmatrix} \xrightarrow{R_3+7R_2 \to R_3} \begin{bmatrix} 1 & 0 & 1 & -1 & 0 & 1 \\ 0 & -1 & 0 & 3 & 1 & -3 \\ 0 & 0 & -1 & 27 & 7 & -26 \end{bmatrix}$

$\xrightarrow[-R_2, -R_3]{R_1+R_3 \to R_1} \begin{bmatrix} 1 & 0 & 0 & 26 & 7 & -25 \\ 0 & 1 & 0 & -3 & -1 & 3 \\ 0 & 0 & 1 & -27 & -7 & 26 \end{bmatrix}$

Therefore, the inverse matrix is $\begin{bmatrix} 26 & 7 & -25 \\ -3 & -1 & 3 \\ -27 & -7 & 26 \end{bmatrix}$.

18. $\begin{bmatrix} 2 & 1 & 0 & 1 & 0 & 0 \\ 1 & 1 & 4 & 0 & 1 & 0 \\ 2 & 1 & 2 & 0 & 0 & 1 \end{bmatrix} \xrightarrow{R_1 \leftrightarrow R_2} \begin{bmatrix} 1 & 1 & 4 & 0 & 1 & 0 \\ 2 & 1 & 0 & 1 & 0 & 0 \\ 2 & 1 & 2 & 0 & 0 & 1 \end{bmatrix} \xrightarrow[R_3-2R_1 \to R_3]{R_2-2R_1 \to R_2}$

$\begin{bmatrix} 1 & 1 & 4 & 0 & 1 & 0 \\ 0 & -1 & -8 & 1 & -2 & 0 \\ 0 & -1 & -6 & 0 & -2 & 1 \end{bmatrix} \xrightarrow[R_3-R_2 \to R_3]{R_1+R_2 \to R_1} \begin{bmatrix} 1 & 0 & -4 & 1 & -1 & 0 \\ 0 & -1 & -8 & 1 & -2 & 0 \\ 0 & 0 & 2 & -1 & 0 & 1 \end{bmatrix}$

$\xrightarrow[R_2+4R_3 \to R_2]{R_1+2R_3 \to R_1} \begin{bmatrix} 1 & 0 & 0 & -1 & -1 & 2 \\ 0 & -1 & 0 & -3 & -2 & 4 \\ 0 & 0 & 2 & -1 & 0 & 1 \end{bmatrix} \xrightarrow[\frac{1}{2}R_3]{-R_2} \begin{bmatrix} 1 & 0 & 0 & -1 & -1 & 2 \\ 0 & 1 & 0 & 3 & 2 & -4 \\ 0 & 0 & 1 & -\frac{1}{2} & 0 & \frac{1}{2} \end{bmatrix}$

Therefore, the inverse matrix is $\begin{bmatrix} -1 & -1 & 2 \\ 3 & 2 & -4 \\ -\frac{1}{2} & 0 & \frac{1}{2} \end{bmatrix}$.

20. $\begin{bmatrix} 3 & -2 & 0 & 1 & 0 & 0 \\ 5 & 1 & 1 & 0 & 1 & 0 \\ 2 & -2 & 0 & 0 & 0 & 1 \end{bmatrix} \xrightarrow{R_1-R_3 \to R_1} \begin{bmatrix} 1 & 0 & 0 & 1 & 0 & -1 \\ 5 & 1 & 1 & 0 & 1 & 0 \\ 2 & -2 & 0 & 0 & 0 & 1 \end{bmatrix} \xrightarrow[R_3-2R_1 \to R_3]{R_2-5R_1 \to R_2}$

$\begin{bmatrix} 1 & 0 & 0 & 1 & 0 & -1 \\ 0 & 1 & 1 & -5 & 1 & 5 \\ 0 & -2 & 0 & -2 & 0 & 3 \end{bmatrix} \xrightarrow{R_3+2R_2 \to R_3} \begin{bmatrix} 1 & 0 & 0 & 1 & 0 & -1 \\ 0 & 1 & 1 & -5 & 1 & 5 \\ 0 & 0 & 2 & -12 & 2 & 13 \end{bmatrix}$

$\xrightarrow[R_2-R_3 \to R_2]{\frac{1}{2}R_3} \begin{bmatrix} 1 & 0 & 0 & 1 & 0 & -1 \\ 0 & 1 & 0 & 1 & 0 & -\frac{3}{2} \\ 0 & 0 & 1 & -6 & 1 & \frac{13}{2} \end{bmatrix}$

Therefore, the inverse matrix is $\begin{bmatrix} 1 & 0 & -1 \\ 1 & 0 & -\frac{3}{2} \\ -6 & 1 & \frac{13}{2} \end{bmatrix}$.

22. $\begin{bmatrix} 1 & 0 & 1 & 0 & 1 & 0 & 0 & 0 \\ 0 & 1 & 0 & 1 & 0 & 1 & 0 & 0 \\ 1 & 1 & 1 & 0 & 0 & 0 & 1 & 0 \\ 1 & 1 & 1 & 1 & 0 & 0 & 0 & 1 \end{bmatrix} \xrightarrow[R_3-R_1 \to R_3]{R_4-R_1 \to R_4} \begin{bmatrix} 1 & 0 & 1 & 0 & 1 & 0 & 0 & 0 \\ 0 & 1 & 0 & 1 & 0 & 1 & 0 & 0 \\ 0 & 1 & 0 & 0 & -1 & 0 & 1 & 0 \\ 0 & 1 & 0 & 1 & -1 & 0 & 0 & 1 \end{bmatrix}$

$\xrightarrow[R_3-R_2 \to R_3]{R_4-R_2 \to R_4} \begin{bmatrix} 1 & 0 & 1 & 0 & 1 & 0 & 0 & 0 \\ 0 & 1 & 0 & 1 & 0 & 1 & 0 & 0 \\ 0 & 0 & 0 & -1 & -1 & -1 & 1 & 0 \\ 0 & 0 & 0 & 0 & -1 & -1 & 0 & 1 \end{bmatrix}$

Therefore, there is no inverse matrix.

24. $\begin{cases} 3x + 4y = 10 \\ 7x + 9y = 20 \end{cases}$ is equivalent to the matrix equation $\begin{bmatrix} 3 & 4 \\ 7 & 9 \end{bmatrix}\begin{bmatrix} x \\ y \end{bmatrix} = \begin{bmatrix} 10 \\ 20 \end{bmatrix}$ $\Leftrightarrow$

$\begin{bmatrix} x \\ y \end{bmatrix} = \begin{bmatrix} -9 & 4 \\ 7 & -3 \end{bmatrix}\begin{bmatrix} 10 \\ 20 \end{bmatrix} = \begin{bmatrix} -10 \\ 10 \end{bmatrix}$. Therefore, $x = -10$ and $y = 10$.

26. $\begin{cases} -7x + 4y = 0 \\ 8x - 5y = 100 \end{cases}$ is equivalent to the matrix equation $\begin{bmatrix} -7 & 4 \\ 8 & -5 \end{bmatrix}\begin{bmatrix} x \\ y \end{bmatrix} = \begin{bmatrix} 0 \\ 100 \end{bmatrix}$ $\Leftrightarrow$

$\begin{bmatrix} x \\ y \end{bmatrix} = \begin{bmatrix} -\frac{5}{3} & -\frac{4}{3} \\ -\frac{8}{3} & -\frac{7}{3} \end{bmatrix}\begin{bmatrix} 0 \\ 100 \end{bmatrix} = \begin{bmatrix} -\frac{400}{3} \\ -\frac{700}{3} \end{bmatrix}$. Therefore, $x = -\frac{400}{3}$ and $y = -\frac{700}{3}$.

28. $\begin{cases} 5x + 7y + 4z = 1 \\ 3x - y + 3z = 1 \\ 6x + 7y + 5z = 1 \end{cases}$ is equivalent to the matrix equation $\begin{bmatrix} 5 & 7 & 4 \\ 3 & -1 & 3 \\ 6 & 7 & 5 \end{bmatrix}\begin{bmatrix} x \\ y \\ z \end{bmatrix} = \begin{bmatrix} 1 \\ 1 \\ 1 \end{bmatrix}$ $\Leftrightarrow$

$\begin{bmatrix} x \\ y \\ z \end{bmatrix} = \begin{bmatrix} 26 & 7 & -25 \\ -3 & -1 & 3 \\ -27 & -7 & 26 \end{bmatrix}\begin{bmatrix} 1 \\ 1 \\ 1 \end{bmatrix} = \begin{bmatrix} 8 \\ -1 \\ -8 \end{bmatrix}$. Therefore, $x = 8$, $y = -1$, and $z = -8$.

30.
$$\begin{cases} x + 2y + 3w = 0 \\ y + z + w = 1 \\ y + w = 2 \\ x + 2y + 2w = 3 \end{cases}$$ is equivalent to the matrix equation $\begin{bmatrix} 1 & 2 & 0 & 3 \\ 0 & 1 & 1 & 1 \\ 0 & 1 & 0 & 1 \\ 1 & 2 & 0 & 2 \end{bmatrix} \begin{bmatrix} x \\ y \\ z \\ w \end{bmatrix} = \begin{bmatrix} 0 \\ 1 \\ 2 \\ 3 \end{bmatrix}$ $\Leftrightarrow$

$\begin{bmatrix} x \\ y \\ z \\ w \end{bmatrix} = \begin{bmatrix} 0 & 0 & -2 & 1 \\ -1 & 0 & 1 & 1 \\ 0 & 1 & -1 & 0 \\ 1 & 0 & 0 & -1 \end{bmatrix} \begin{bmatrix} 0 \\ 1 \\ 2 \\ 3 \end{bmatrix} = \begin{bmatrix} -1 \\ 5 \\ -1 \\ -3 \end{bmatrix}$. Therefore, $x = -1$, $y = 5$, $z = -1$, and $w = -3$.

32. Using the calculator you should get the result $(-1, 2, 3)$.

34. Using the calculator you should get the result $(6, 12, 24)$.

36. Using the calculator you should get the result $(8, 4, 2, 1)$.

38. Using the inverse matrix from Exercise 19, we see that
$$\begin{bmatrix} -\frac{9}{2} & -1 & 4 \\ 3 & 1 & -3 \\ \frac{7}{2} & 1 & -3 \end{bmatrix} \begin{bmatrix} 3 & 6 \\ 6 & 12 \\ 0 & 0 \end{bmatrix} = \begin{bmatrix} -\frac{39}{2} & -39 \\ 15 & 30 \\ \frac{33}{2} & 33 \end{bmatrix}. \text{ Hence}, \begin{bmatrix} x & u \\ y & v \\ z & w \end{bmatrix} = \begin{bmatrix} -\frac{39}{2} & -39 \\ 15 & 30 \\ \frac{33}{2} & 33 \end{bmatrix}.$$

40.
$$\begin{bmatrix} a & 0 & 0 & 0 & 1 & 0 & 0 & 0 \\ 0 & b & 0 & 0 & 0 & 1 & 0 & 0 \\ 0 & 0 & c & 0 & 0 & 0 & 1 & 0 \\ 0 & 0 & 0 & d & 0 & 0 & 0 & 1 \end{bmatrix} \xrightarrow[\frac{1}{c}R_3, \frac{1}{d}R_4]{\frac{1}{a}R_1, \frac{1}{b}R_2} \begin{bmatrix} 1 & 0 & 0 & 0 & \frac{1}{a} & 0 & 0 & 0 \\ 0 & 1 & 0 & 0 & 0 & \frac{1}{b} & 0 & 0 \\ 0 & 0 & 1 & 0 & 0 & 0 & \frac{1}{c} & 0 \\ 0 & 0 & 0 & 1 & 0 & 0 & 0 & \frac{1}{d} \end{bmatrix}$$

Thus the matrix $\begin{bmatrix} a & 0 & 0 & 0 \\ 0 & b & 0 & 0 \\ 0 & 0 & c & 0 \\ 0 & 0 & 0 & d \end{bmatrix}$ has inverse $\begin{bmatrix} \frac{1}{a} & 0 & 0 & 0 \\ 0 & \frac{1}{b} & 0 & 0 \\ 0 & 0 & \frac{1}{c} & 0 \\ 0 & 0 & 0 & \frac{1}{d} \end{bmatrix}$.

42. $\begin{bmatrix} e^x & -e^{2x} \\ e^{2x} & e^{3x} \end{bmatrix}^{-1} = \frac{1}{e^{4x} + e^{4x}} \begin{bmatrix} e^{3x} & e^{2x} \\ -e^{2x} & e^x \end{bmatrix} = \frac{1}{2} \begin{bmatrix} e^{-x} & e^{-2x} \\ -e^{-2x} & e^{-3x} \end{bmatrix}$. Inverse exists for all x.

44. $\begin{bmatrix} x & 1 \\ -x & \frac{1}{x-1} \end{bmatrix}^{-1} = \frac{1}{\frac{x}{x-1} + x} \begin{bmatrix} \frac{1}{x-1} & -1 \\ x & x \end{bmatrix} = \frac{x-1}{x^2} \begin{bmatrix} \frac{1}{x-1} & -1 \\ x & x \end{bmatrix} = \begin{bmatrix} \frac{1}{x^2} & -\frac{x-1}{x^2} \\ \frac{x-1}{x} & \frac{x-1}{x} \end{bmatrix}$.

Inverse exists for $x \neq 0, 1$.

46.
$$\begin{bmatrix} 3 & 1 & 4 & 1 & 0 & 0 \\ 4 & 2 & 6 & 0 & 1 & 0 \\ 3 & 2 & 5 & 0 & 0 & 1 \end{bmatrix} \xrightarrow[R_3 - R_1 \to R_3]{R_2 - R_1 \to R_2} \begin{bmatrix} 3 & 1 & 4 & 1 & 0 & 0 \\ 1 & 1 & 2 & -1 & 1 & 0 \\ 0 & 1 & 1 & -1 & 0 & 1 \end{bmatrix} \xrightarrow{R_1 \leftrightarrow R_2}$$

$$\begin{bmatrix} 1 & 1 & 2 & -1 & 1 & 0 \\ 3 & 1 & 4 & 1 & 0 & 0 \\ 0 & 1 & 1 & -1 & 0 & 1 \end{bmatrix} \xrightarrow{R_2 - 3R_1 \to R_2} \begin{bmatrix} 1 & 1 & 2 & -1 & 1 & 0 \\ 0 & -2 & -2 & 4 & -3 & 0 \\ 0 & 1 & 1 & -1 & 0 & 1 \end{bmatrix} \xrightarrow{R_3 + \frac{1}{2}R_2 \to R_3}$$

$$\begin{bmatrix} 1 & 1 & 2 & -1 & 1 & 0 \\ 0 & -2 & -2 & 4 & -3 & 0 \\ 0 & 0 & 0 & 1 & -\frac{3}{2} & 1 \end{bmatrix}$$

Since the inverse matrix does not exist, it would not be possible to use matrix inversion in the solution of parts (b), (c), and (d).

48. No, consider the following counterexample: $A = \begin{bmatrix} 0 & 1 \\ 0 & 0 \end{bmatrix}$ and $B = \begin{bmatrix} 0 & 2 \\ 0 & 0 \end{bmatrix}$. Then, $AB = O$, but neither $A = O$ nor $B = O$.

There are infinitely many matrices for which $A^2 = O$. One example is $A = \begin{bmatrix} 0 & 1 \\ 0 & 0 \end{bmatrix}$. Then, $A^2 = O$, but $A \neq O$.

Exercises 7.4

2. The matrix $\begin{bmatrix} 0 & -1 \\ 2 & 0 \end{bmatrix}$ has determinant $|D| = (0)(0) - (-1)(2) = 2$.

4. The matrix $\begin{bmatrix} -2 & 1 \\ 3 & -2 \end{bmatrix}$ has determinant $|D| = (-2)(-2) - (1)(3) = 1$.

6. The matrix $\begin{bmatrix} 3 \\ 0 \end{bmatrix}$ does not have a determinant because the matrix is not square.

8. The matrix $\begin{bmatrix} 2.2 & -1.4 \\ 0.5 & 1.0 \end{bmatrix}$ has determinant $|D| = (2.2)(1.0) - (0.5)(-1.4) = 2.2 + 0.7 = 2.9$.

In Exercises 10–14, the matrix is $A = \begin{bmatrix} 1 & 0 & \frac{1}{2} \\ -3 & 5 & 2 \\ 0 & 0 & 4 \end{bmatrix}$.

10. $M_{33} = 1 \cdot 5 + 3 \cdot 0 = 5$, $A_{33} = (-1)^6 M_{33} = 5$

12. $M_{13} = -3 \cdot 0 - 0 \cdot 5 = 0$, $A_{13} = (-1)^4 M_{13} = 0$

14. $M_{32} = 1 \cdot 2 + 3 \cdot \frac{1}{2} = \frac{7}{2}$, $A_{32} = (-1)^5 M_{32} = -\frac{7}{2}$

16. $M = \begin{bmatrix} 0 & -1 & 0 \\ 2 & 6 & 4 \\ 1 & 0 & 3 \end{bmatrix}$. Therefore, expanding by the first row, $|M| = -(-1)\begin{vmatrix} 2 & 4 \\ 1 & 3 \end{vmatrix} = 6 - 4 = 2$.
Since $|M| \neq 0$, the matrix has an inverse.

18. $M = \begin{bmatrix} -2 & -\frac{3}{2} & \frac{1}{2} \\ 2 & 4 & 0 \\ \frac{1}{2} & 2 & 1 \end{bmatrix}$. Therefore, $|M| = \frac{1}{2}\begin{vmatrix} 2 & 4 \\ \frac{1}{2} & 2 \end{vmatrix} + 1\begin{vmatrix} -2 & -\frac{3}{2} \\ 2 & 4 \end{vmatrix} = \frac{1}{2}(4 - 2) + (-8 + 3)$
$= 1 - 5 = -4$, and the matrix has an inverse.

20. $M = \begin{bmatrix} 1 & 2 & 5 \\ -2 & -3 & 2 \\ 3 & 5 & 3 \end{bmatrix}$. Therefore, $|M| = 1\begin{vmatrix} -3 & 2 \\ 5 & 3 \end{vmatrix} - 2\begin{vmatrix} -2 & 2 \\ 3 & 3 \end{vmatrix} + 5\begin{vmatrix} -2 & -3 \\ 3 & 5 \end{vmatrix}$
$= (-9 - 10) - 2(-6 - 6) + 5(-10 + 9) = -19 + 24 - 5 = 0$, and so the matrix does not have an inverse.

22. $M = \begin{bmatrix} 1 & 2 & 0 & 2 \\ 3 & -4 & 0 & 4 \\ 0 & 1 & 6 & 0 \\ 1 & 0 & 2 & 0 \end{bmatrix}$. Therefore, $|M| = -1\begin{vmatrix} 2 & 0 & 2 \\ -4 & 0 & 4 \\ 1 & 6 & 0 \end{vmatrix} - 2\begin{vmatrix} 1 & 2 & 2 \\ 3 & -4 & 4 \\ 0 & 1 & 0 \end{vmatrix}$
$= 6\begin{vmatrix} 2 & 2 \\ -4 & 4 \end{vmatrix} + 2\begin{vmatrix} 1 & 2 \\ 3 & 4 \end{vmatrix} = 6 \cdot 16 - 2 \cdot 2 = 92$, and so M^{-1} exists.

24. $M = \begin{bmatrix} -2 & 3 & -1 & 7 \\ 4 & 6 & -2 & 3 \\ 7 & 7 & 0 & 5 \\ 3 & -12 & 4 & 0 \end{bmatrix}$. Then, $|M| = \begin{vmatrix} -2 & 3 & -1 & 7 \\ 4 & 6 & -2 & 3 \\ 7 & 7 & 0 & 5 \\ 3 & -12 & 4 & 0 \end{vmatrix} = \begin{vmatrix} -2 & 0 & -1 & 7 \\ 4 & 0 & -2 & 3 \\ 7 & 7 & 0 & 5 \\ 3 & 0 & 4 & 0 \end{vmatrix}$, by

replacing C_2 with $C_2 + 3C_3$. So expanding about the second column, $|M| = -7 \begin{vmatrix} -2 & -1 & 7 \\ 4 & -2 & 3 \\ 3 & 4 & 0 \end{vmatrix}$

$= -7 \left(7 \begin{vmatrix} 4 & -2 \\ 3 & 4 \end{vmatrix} - 3 \begin{vmatrix} -2 & -1 \\ 3 & 4 \end{vmatrix} \right) = -7 (7 \cdot 22 + 3 \cdot 5) = -1183.$

26. $M = \begin{bmatrix} 2 & -1 & 6 & 4 \\ 7 & 2 & -2 & 5 \\ 4 & -2 & 10 & 8 \\ 6 & 1 & 1 & 4 \end{bmatrix}$. Then, $|M| = \begin{vmatrix} 2 & -1 & 6 & 4 \\ 7 & 2 & -2 & 5 \\ 4 & -2 & 10 & 8 \\ 6 & 1 & 1 & 4 \end{vmatrix} = \begin{vmatrix} 0 & -1 & 6 & 4 \\ 11 & 2 & -2 & 5 \\ 0 & -2 & 10 & 8 \\ 8 & 1 & 1 & 4 \end{vmatrix}$, by replacing

C_1 with $C_1 + 2C_2$. So $|M| = -11 \begin{vmatrix} -1 & 6 & 4 \\ -2 & 10 & 8 \\ 1 & 1 & 4 \end{vmatrix} - 8 \begin{vmatrix} -1 & 6 & 4 \\ 2 & -2 & 5 \\ -2 & 10 & 8 \end{vmatrix}$

$= -11 \begin{vmatrix} -1 & 6 & 0 \\ -2 & 10 & 0 \\ 1 & 1 & 8 \end{vmatrix} - 8 \begin{vmatrix} -1 & 6 & 0 \\ 2 & -2 & 13 \\ -2 & 10 & 0 \end{vmatrix} = -88 \begin{vmatrix} -1 & 6 \\ -2 & 10 \end{vmatrix} + 104 \begin{vmatrix} -1 & 6 \\ -2 & 10 \end{vmatrix}$

$= -88 \cdot 2 + 104 \cdot 2 = 32.$

28. $\begin{cases} x + 2y + 6z = 5 \\ -3x - 6y + 5z = 8 \\ 2x + 6y + 9z = 7 \end{cases}$

(a) If $x = -1, y = 0$, and $z = 1$, then $x + 2y + 6z = (-1) + 2(0) + 6(1) = 5$,
$-3x - 6y + 5z = -3(-1) - 6(0) + 5(1) = 8$, and $2x + 6y + 9z = 2(-1) + 6(0) + 9(1)$
$= 7$. Therefore, $x = -1, y = 0, z = 1$ is a solution of the system.

(b) $M = \begin{bmatrix} 1 & 2 & 6 \\ -3 & -6 & 5 \\ 2 & 6 & 9 \end{bmatrix}$. Then, $|M| = \begin{vmatrix} 1 & 2 & 6 \\ -3 & -6 & 5 \\ 2 & 6 & 9 \end{vmatrix} = \begin{vmatrix} 1 & 0 & 6 \\ -3 & 0 & 5 \\ 2 & 2 & 9 \end{vmatrix}$ (replacing C_2 with

$C_2 - 2C_1$), so $|M| = -2 \begin{vmatrix} 1 & 6 \\ -3 & 5 \end{vmatrix} = -2 (5 + 18) = -46.$

(c) We can write the system as a matrix equation:
$\begin{bmatrix} 1 & 2 & 6 \\ -3 & -6 & 5 \\ 2 & 6 & 9 \end{bmatrix} \begin{bmatrix} x \\ y \\ z \end{bmatrix} = \begin{bmatrix} 5 \\ 8 \\ 7 \end{bmatrix}$, or $MX = B$. Since $|M| \neq 0$, M has an inverse. If we

multiply both sides of the matrix equation by M^{-1}, then we get a unique solution for X, given
by $X = M^{-1}B$ (see Section 8.5). Thus the equation has no other solutions.

(d) Yes, since $|M| \neq 0$.

30. $\begin{cases} 6x + 12y = 33 \\ 4x + 7y = 20 \end{cases}$ Then, $|D| = \begin{vmatrix} 6 & 12 \\ 4 & 7 \end{vmatrix} = -6, |D_x| = \begin{vmatrix} 33 & 12 \\ 20 & 7 \end{vmatrix} = -9,$ and

$|D_y| = \begin{vmatrix} 6 & 33 \\ 4 & 20 \end{vmatrix} = -12.$ Hence, $x = \frac{|D_x|}{|D|} = \frac{-9}{-6} = \frac{3}{2}$ and $y = \frac{|D_y|}{|D|} = \frac{-12}{-6} = 2,$ and so the

solution is $\left(\frac{3}{2}, 2\right).$

32. $\begin{cases} \frac{1}{2}x + \frac{1}{3}y = 1 \\ \frac{1}{4}x - \frac{1}{6}y = -\frac{3}{2} \end{cases}$ Then, $|D| = \begin{vmatrix} \frac{1}{2} & \frac{1}{3} \\ \frac{1}{4} & -\frac{1}{6} \end{vmatrix} = -\frac{1}{6}, |D_x| = \begin{vmatrix} 1 & \frac{1}{3} \\ -\frac{3}{2} & -\frac{1}{6} \end{vmatrix} = \frac{1}{3},$

and $|D_y| = \begin{vmatrix} \frac{1}{2} & 1 \\ \frac{1}{4} & -\frac{3}{2} \end{vmatrix} = -1.$ Hence, $x = \frac{|D_x|}{|D|} = \frac{\frac{1}{3}}{-\frac{1}{6}} = -2, y = \frac{|D_y|}{|D|} = \frac{-1}{-\frac{1}{6}} = 6,$ and so

the solution is $(-2, 6).$

34. $\begin{cases} 10x - 17y = 21 \\ 20x - 31y = 39 \end{cases}$ Then, $|D| = \begin{vmatrix} 10 & -17 \\ 20 & -31 \end{vmatrix} = 30, |D_x| = \begin{vmatrix} 21 & -17 \\ 39 & -31 \end{vmatrix} = 12,$ and

$|D_y| = \begin{vmatrix} 10 & 21 \\ 20 & 39 \end{vmatrix} = -30.$ Hence, $x = \frac{|D_x|}{|D|} = \frac{12}{30} = \frac{2}{5}, y = \frac{|D_y|}{|D|} = \frac{-30}{30} = -1,$ and so the solution is

$\left(\frac{2}{5}, -1\right).$

36. $\begin{cases} 5x - 3y + z = 6 \\ \quad 4y - 6z = 22 \\ 7x + 10y = -13 \end{cases}$. Then,

$|D| = \begin{vmatrix} 5 & -3 & 1 \\ 0 & 4 & -6 \\ 7 & 10 & 0 \end{vmatrix} = 1 \begin{vmatrix} 0 & 4 \\ 7 & 10 \end{vmatrix} + 6 \begin{vmatrix} 5 & -3 \\ 7 & 10 \end{vmatrix} = -28 + 426 = 398,$

$|D_x| = \begin{vmatrix} 6 & -3 & 1 \\ 22 & 4 & -6 \\ -13 & 10 & 0 \end{vmatrix} = 1 \begin{vmatrix} 22 & 4 \\ -13 & 10 \end{vmatrix} + 6 \begin{vmatrix} 6 & -3 \\ -13 & 10 \end{vmatrix} = 272 + 126 = 398,$

$|D_y| = \begin{vmatrix} 5 & 6 & 1 \\ 0 & 22 & -6 \\ 7 & -13 & 0 \end{vmatrix} = 1 \begin{vmatrix} 0 & 22 \\ 7 & -13 \end{vmatrix} + 6 \begin{vmatrix} 5 & 6 \\ 7 & -13 \end{vmatrix} = -154 - 642 = -796,$ and

$|D_z| = \begin{vmatrix} 5 & -3 & 6 \\ 0 & 4 & 22 \\ 7 & 10 & -13 \end{vmatrix} = 4 \begin{vmatrix} 5 & 6 \\ 7 & -13 \end{vmatrix} - 22 \begin{vmatrix} 5 & -3 \\ 7 & 10 \end{vmatrix} = -428 - 1562 = -1990.$ Therefore, the

solution is $x = \frac{398}{398} = 1, y = \frac{-796}{398} = -2,$ and $z = \frac{-1990}{398} = -5.$

38. $\begin{cases} -2a + c = 2 \\ a + 2b - c = 9 \\ 3a + 5b + 2c = 22 \end{cases}$. Then,

$|D| = \begin{vmatrix} -2 & 0 & 1 \\ 1 & 2 & -1 \\ 3 & 5 & 2 \end{vmatrix} = -2 \begin{vmatrix} 2 & -1 \\ 5 & 2 \end{vmatrix} + 1 \begin{vmatrix} 1 & 2 \\ 3 & 5 \end{vmatrix} = -18 - 1 = -19,$

$$|D_a| = \begin{vmatrix} 2 & 0 & 1 \\ 9 & 2 & -1 \\ 22 & 5 & 2 \end{vmatrix} = 2\begin{vmatrix} 2 & -1 \\ 5 & 2 \end{vmatrix} + 1\begin{vmatrix} 9 & 2 \\ 22 & 5 \end{vmatrix} = 18 + 1 = 19,$$

$$|D_b| = \begin{vmatrix} -2 & 2 & 1 \\ 1 & 9 & -1 \\ 3 & 22 & 2 \end{vmatrix} = -2\begin{vmatrix} 9 & -1 \\ 22 & 2 \end{vmatrix} - 2\begin{vmatrix} 1 & -1 \\ 3 & 2 \end{vmatrix} + 1\begin{vmatrix} 1 & 9 \\ 3 & 22 \end{vmatrix} = -80 - 10 - 5 = -95, \text{ and}$$

$$|D_c| = \begin{vmatrix} -2 & 0 & 2 \\ 1 & 2 & 9 \\ 3 & 5 & 22 \end{vmatrix} = -2\begin{vmatrix} 2 & 9 \\ 5 & 22 \end{vmatrix} + 2\begin{vmatrix} 1 & 2 \\ 3 & 5 \end{vmatrix} = 2 - 2 = 0.$$

Hence, the solution is $a = -1$, $b = 5$, and $c = 0$.

40. $\begin{cases} 2x - y = 5 \\ 5x + 3z = 19 \\ 4y + 7z = 17 \end{cases}$ Then, $|D| = \begin{vmatrix} 2 & -1 & 0 \\ 5 & 0 & 3 \\ 0 & 4 & 7 \end{vmatrix} = 2\begin{vmatrix} 0 & 3 \\ 4 & 7 \end{vmatrix} + 1\begin{vmatrix} 5 & 3 \\ 0 & 7 \end{vmatrix} = -24 + 35 = 11,$

$$|D_x| = \begin{vmatrix} 5 & -1 & 0 \\ 19 & 0 & 3 \\ 17 & 4 & 7 \end{vmatrix} = 5\begin{vmatrix} 0 & 3 \\ 4 & 7 \end{vmatrix} + 1\begin{vmatrix} 19 & 3 \\ 17 & 7 \end{vmatrix} = -60 + 82 = 22,$$

$$|D_y| = \begin{vmatrix} 2 & 5 & 0 \\ 5 & 19 & 3 \\ 0 & 17 & 7 \end{vmatrix} = 2\begin{vmatrix} 19 & 3 \\ 17 & 7 \end{vmatrix} - 5\begin{vmatrix} 5 & 3 \\ 0 & 7 \end{vmatrix} = 164 - 175 = -11, \text{ and}$$

$$|D_z| = \begin{vmatrix} 2 & -1 & 5 \\ 5 & 0 & 19 \\ 0 & 4 & 17 \end{vmatrix} = -4\begin{vmatrix} 2 & 5 \\ 5 & 19 \end{vmatrix} + 17\begin{vmatrix} 2 & -1 \\ 5 & 0 \end{vmatrix} = -52 + 85 = 33. \text{ Thus, the solution is}$$

$x = 2$, $y = -1$, and $z = 3$.

42. $\begin{cases} 2x - 5y = 4 \\ x + y - z = 8. \text{ Then,} \\ 3x + 5z = 0 \end{cases}$

$$|D| = \begin{vmatrix} 2 & -5 & 0 \\ 1 & 1 & -1 \\ 3 & 0 & 5 \end{vmatrix} = 2\begin{vmatrix} 1 & -1 \\ 0 & 5 \end{vmatrix} + 5\begin{vmatrix} 1 & -1 \\ 3 & 5 \end{vmatrix} = 10 + 40 = 50,$$

$$|D_x| = \begin{vmatrix} 4 & -5 & 0 \\ 8 & 1 & -1 \\ 0 & 0 & 5 \end{vmatrix} = 5\begin{vmatrix} 4 & -5 \\ 8 & 1 \end{vmatrix} = 220,$$

$$|D_y| = \begin{vmatrix} 2 & 4 & 0 \\ 1 & 8 & -1 \\ 3 & 0 & 5 \end{vmatrix} = 2\begin{vmatrix} 8 & -1 \\ 0 & 5 \end{vmatrix} - 4\begin{vmatrix} 1 & -1 \\ 3 & 5 \end{vmatrix} = 80 - 32 = 48, \text{ and}$$

$$|D_z| = \begin{vmatrix} 2 & -5 & 4 \\ 1 & 1 & 8 \\ 3 & 0 & 0 \end{vmatrix} = 3\begin{vmatrix} -5 & 4 \\ 1 & 8 \end{vmatrix} = -132. \text{ Thus, } x = \frac{22}{5}, y = \frac{24}{25}, \text{ and } z = -\frac{66}{25}.$$

44. $\begin{cases} x+y=1 \\ y+z=2 \\ z+w=3 \\ w-x=4 \end{cases}$. Then $|D| = \begin{vmatrix} 1 & 1 & 0 & 0 \\ 0 & 1 & 1 & 0 \\ 0 & 0 & 1 & 1 \\ -1 & 0 & 0 & 1 \end{vmatrix} = 1\begin{vmatrix} 1 & 1 & 0 \\ 0 & 1 & 1 \\ 0 & 0 & 1 \end{vmatrix} - 1\begin{vmatrix} 0 & 1 & 0 \\ 0 & 1 & 1 \\ -1 & 0 & 1 \end{vmatrix}$

$= 1\begin{vmatrix} 1 & 1 \\ 0 & 1 \end{vmatrix} - (-1)\begin{vmatrix} 1 & 0 \\ 1 & 1 \end{vmatrix} = 1 + 1 = 2,$

$|D_x| = \begin{vmatrix} 1 & 1 & 0 & 0 \\ 2 & 1 & 1 & 0 \\ 3 & 0 & 1 & 1 \\ 4 & 0 & 0 & 1 \end{vmatrix} = 1\begin{vmatrix} 1 & 1 & 0 \\ 0 & 1 & 1 \\ 0 & 0 & 1 \end{vmatrix} - 1\begin{vmatrix} 2 & 1 & 0 \\ 3 & 1 & 1 \\ 4 & 0 & 1 \end{vmatrix} = \begin{vmatrix} 1 & 1 \\ 0 & 1 \end{vmatrix} - \left(2\begin{vmatrix} 1 & 1 \\ 0 & 1 \end{vmatrix} - 1\begin{vmatrix} 3 & 1 \\ 4 & 1 \end{vmatrix}\right)$

$= 1 - 3 = -2,$

$|D_y| = \begin{vmatrix} 1 & 1 & 0 & 0 \\ 0 & 2 & 1 & 0 \\ 0 & 3 & 1 & 1 \\ -1 & 4 & 0 & 1 \end{vmatrix} = 1\begin{vmatrix} 2 & 1 & 0 \\ 3 & 1 & 1 \\ 4 & 0 & 1 \end{vmatrix} - 1\begin{vmatrix} 0 & 1 & 0 \\ 0 & 1 & 1 \\ -1 & 0 & 1 \end{vmatrix}$

$= \left(2\begin{vmatrix} 1 & 1 \\ 0 & 1 \end{vmatrix} - 1\begin{vmatrix} 3 & 1 \\ 4 & 1 \end{vmatrix}\right) - (-1)\begin{vmatrix} 1 & 0 \\ 1 & 1 \end{vmatrix} = 3 + 1 = 4,$

$|D_z| = \begin{vmatrix} 1 & 1 & 1 & 0 \\ 0 & 1 & 2 & 0 \\ 0 & 0 & 3 & 1 \\ -1 & 0 & 4 & 1 \end{vmatrix} = 1\begin{vmatrix} 1 & 2 & 0 \\ 0 & 3 & 1 \\ 0 & 4 & 1 \end{vmatrix} + 1\begin{vmatrix} 1 & 1 & 0 \\ 1 & 2 & 0 \\ 0 & 3 & 1 \end{vmatrix} = 1\begin{vmatrix} 3 & 1 \\ 4 & 1 \end{vmatrix} + 1\begin{vmatrix} 1 & 1 \\ 1 & 2 \end{vmatrix} = -1 + 1 = 0,$

$|D_w| = \begin{vmatrix} 1 & 1 & 0 & 1 \\ 0 & 1 & 1 & 2 \\ 0 & 0 & 1 & 3 \\ -1 & 0 & 0 & 4 \end{vmatrix} = 1\begin{vmatrix} 1 & 1 & 2 \\ 0 & 1 & 3 \\ 0 & 0 & 4 \end{vmatrix} + 1\begin{vmatrix} 1 & 0 & 1 \\ 1 & 1 & 2 \\ 0 & 1 & 3 \end{vmatrix} = \begin{vmatrix} 1 & 3 \\ 0 & 4 \end{vmatrix} + \left(1\begin{vmatrix} 1 & 2 \\ 1 & 3 \end{vmatrix} - 1\begin{vmatrix} 0 & 1 \\ 1 & 3 \end{vmatrix}\right)$

$= 4 + 2 = 6.$ Hence, the solution is $x = \dfrac{|D_x|}{|D|} = \dfrac{-2}{2} = -1$, $y = \dfrac{|D_y|}{|D|} = \dfrac{4}{2} = 2$,

$z = \dfrac{|D_z|}{|D|} = \dfrac{0}{2} = 0$, and $w = \dfrac{|D_w|}{|D|} = \dfrac{6}{2} = 3$.

46. $\begin{vmatrix} a & a & a & a & a \\ 0 & a & a & a & a \\ 0 & 0 & a & a & a \\ 0 & 0 & 0 & a & a \\ 0 & 0 & 0 & 0 & a \end{vmatrix} = a\begin{vmatrix} a & a & a & a \\ 0 & a & a & a \\ 0 & 0 & a & a \\ 0 & 0 & 0 & a \end{vmatrix} = a^2\begin{vmatrix} a & a & a \\ 0 & a & a \\ 0 & 0 & a \end{vmatrix} = a^3\begin{vmatrix} a & a \\ 0 & a \end{vmatrix} = a^5.$

48. $\begin{vmatrix} x & 1 & 1 \\ 1 & 1 & x \\ x & 1 & x \end{vmatrix} = x\begin{vmatrix} 1 & x \\ 1 & x \end{vmatrix} - \begin{vmatrix} 1 & x \\ x & x \end{vmatrix} + \begin{vmatrix} 1 & 1 \\ x & 1 \end{vmatrix} = x(0) - (x - x^2) + 1 - x = x^2 - 2x + 1 = 0$

$\Leftrightarrow \quad (x-1)^2 = 0 \quad \Leftrightarrow \quad x = 1.$

50. $\begin{vmatrix} a & b & x-a \\ x & x+b & x \\ 0 & 1 & 1 \end{vmatrix} = -1 \begin{vmatrix} a & x-a \\ x & x \end{vmatrix} + \begin{vmatrix} a & b \\ x & x+b \end{vmatrix} = -1[ax - x(x-a)] + a(x+b) - bx$

$= -ax + x^2 - ax + ax + ab - bx = x^2 - ax - bx + ab = (x-a)(x-b) = 0 \quad \Leftrightarrow \quad x = a$ or $x = b$.

52. Area $= \pm\frac{1}{2} \begin{vmatrix} 1 & 0 & 1 \\ 3 & 5 & 1 \\ -2 & 2 & 1 \end{vmatrix} = \pm\frac{1}{2}\left(1\begin{vmatrix} 5 & 1 \\ 2 & 1 \end{vmatrix} + 1\begin{vmatrix} 3 & 5 \\ -2 & 2 \end{vmatrix}\right)$

$= \pm\frac{1}{2}[(5-2) + (6+10)]$

$= \pm\frac{1}{2}[3 + 16] = \frac{1}{2}(19) = \frac{19}{2}.$

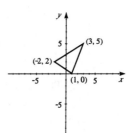

54. Area $= \pm\frac{1}{2} \begin{vmatrix} -2 & 5 & 1 \\ 7 & 2 & 1 \\ 3 & -4 & 1 \end{vmatrix}$

$= \pm\frac{1}{2}\left(-2\begin{vmatrix} 2 & 1 \\ -4 & 1 \end{vmatrix} - 5\begin{vmatrix} 7 & 1 \\ 3 & 1 \end{vmatrix} + 1\begin{vmatrix} 7 & 2 \\ 3 & -4 \end{vmatrix}\right)$

$= \pm\frac{1}{2}[-2(2+4) - 5(7-3) + (-28-6)]$

$= \pm\frac{1}{2}[-2(6) - 5(4) + (-34)]$

$= \pm\frac{1}{2}[-12 - 20 - 34] = \frac{1}{2}(66) = 33$

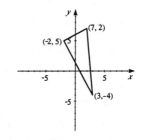

56. (a) Let $x =$ the pounds of apples, $y =$ the pounds of peaches, and $z =$ the pounds of pears. We get the following model:

$$\begin{cases} x + y + z = 18 \\ 0.75x + 0.90y + 0.60z = 13.80 \\ -0.75x + 0.90y + 0.60z = 1.80 \end{cases}$$

(b) $|D| = \begin{vmatrix} 1 & 1 & 1 \\ 0.75 & 0.90 & 0.60 \\ -0.75 & 0.90 & 0.60 \end{vmatrix} = 1 \cdot \begin{vmatrix} 0.90 & 0.60 \\ 0.90 & 0.60 \end{vmatrix} - 1 \cdot \begin{vmatrix} 0.75 & 0.60 \\ -0.75 & 0.60 \end{vmatrix} + 1 \cdot \begin{vmatrix} 0.75 & 0.90 \\ -0.75 & 0.90 \end{vmatrix}$

$= 0 - 0.90 + 1.35 = 0.45$

$|D_x| = \begin{vmatrix} 18 & 1 & 1 \\ 13.80 & 0.90 & 0.60 \\ 1.80 & 0.90 & 0.60 \end{vmatrix} = 18 \cdot \begin{vmatrix} 0.90 & 0.60 \\ 0.90 & 0.60 \end{vmatrix} - 1 \cdot \begin{vmatrix} 13.80 & 0.60 \\ 1.80 & 0.60 \end{vmatrix} + 1 \cdot \begin{vmatrix} 13.80 & 0.90 \\ 1.80 & 0.90 \end{vmatrix}$

$= 0 - 7.2 + 10.8 = 3.6$

$$|D_y| = \begin{vmatrix} 1 & 18 & 1 \\ 0.75 & 13.80 & 0.60 \\ -0.75 & 1.80 & 0.60 \end{vmatrix}$$

$$= 1 \cdot \begin{vmatrix} 13.80 & 0.60 \\ 1.80 & 0.60 \end{vmatrix} - 18 \cdot \begin{vmatrix} 0.75 & 0.60 \\ -0.75 & 0.60 \end{vmatrix} + 1 \cdot \begin{vmatrix} 0.75 & 13.80 \\ -0.75 & 1.80 \end{vmatrix}$$

$$= 7.2 - 16.2 + 11.7 = 2.7$$

$$|D_z| = \begin{vmatrix} 1 & 1 & 18 \\ 0.75 & 0.90 & 13.80 \\ -0.75 & 0.90 & 1.80 \end{vmatrix}$$

$$= 1 \cdot \begin{vmatrix} 0.90 & 13.80 \\ 0.90 & 1.80 \end{vmatrix} - 1 \cdot \begin{vmatrix} 0.75 & 13.80 \\ -0.75 & 1.80 \end{vmatrix} + 18 \cdot \begin{vmatrix} 0.75 & 0.90 \\ -0.75 & 0.90 \end{vmatrix}$$

$$= -10.8 - 11.7 + 24.3 = 1.8$$

So $x = \dfrac{|D_x|}{|D|} = \dfrac{3.6}{0.45} = 8$; $y = \dfrac{|D_y|}{|D|} = \dfrac{27}{0.45} = 6$; and $z = \dfrac{|D_z|}{|D|} = \dfrac{1.8}{0.45} = 4$.

Thus Muriel buys 8 pounds of apples, 6 pounds of peaches, and 4 pounds of pears.

58. Using the determinant formula for the area of a triangle (Exercise 59) we have

$$\text{Area} = \pm\tfrac{1}{2} \begin{vmatrix} 1000 & 2000 & 1 \\ 5000 & 4000 & 1 \\ 2000 & 6000 & 1 \end{vmatrix}$$

$$= \pm\tfrac{1}{2} \left(1 \cdot \begin{vmatrix} 5000 & 4000 \\ 2000 & 6000 \end{vmatrix} - 1 \cdot \begin{vmatrix} 1000 & 2000 \\ 2000 & 6000 \end{vmatrix} + 1 \cdot \begin{vmatrix} 1000 & 2000 \\ 5000 & 4000 \end{vmatrix} \right)$$

$$= \pm\tfrac{1}{2} \cdot (22,000,000 - 2,000,000 - 6,000,000) = \pm\tfrac{1}{2} \cdot 14,000,000 = 7,000,000.$$

Thus the area is $7,000,000 \text{ ft}^2$.

60. (a) If three points lie on a line then the area of the "triangle" they determine is 0, that is,

$$\text{Area} = \pm\tfrac{1}{2} \begin{vmatrix} a_1 & b_1 & 1 \\ a_2 & b_2 & 1 \\ a_3 & b_3 & 1 \end{vmatrix} = 0 \quad \Leftrightarrow \quad \begin{vmatrix} a_1 & b_1 & 1 \\ a_2 & b_2 & 1 \\ a_3 & b_3 & 1 \end{vmatrix} = 0. \text{ If the points are not collinear, then}$$

the point form a triangle, and the area of the triangle determined by these points $\neq 0$. If

$\begin{vmatrix} a_1 & b_1 & 1 \\ a_2 & b_2 & 1 \\ a_3 & b_3 & 1 \end{vmatrix} = 0$ then since Area $= \pm\tfrac{1}{2} \begin{vmatrix} a_1 & b_1 & 1 \\ a_2 & b_2 & 1 \\ a_3 & b_3 & 1 \end{vmatrix} = \pm\tfrac{1}{2}(0) = 0.$ So the "triangle" has no

area, and the points are collinear.

(b) (i) $\begin{vmatrix} -6 & 4 & 1 \\ 2 & 10 & 1 \\ 6 & 13 & 1 \end{vmatrix} = \begin{vmatrix} 2 & 10 \\ 6 & 13 \end{vmatrix} - \begin{vmatrix} -6 & 4 \\ 6 & 13 \end{vmatrix} + \begin{vmatrix} -6 & 4 \\ 2 & 10 \end{vmatrix}$

$$= (26 - 60) - (-78 - 24) + (-60 - 8)$$

$$= -34 + 104 - 68 = 0.$$

Thus these points are collinear.

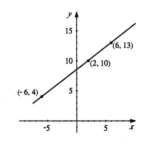

(c) (i) $\begin{vmatrix} -5 & 10 & 1 \\ 2 & 6 & 1 \\ 15 & -2 & 1 \end{vmatrix} = \begin{matrix} 2 & 6 \\ 15 & -2 \end{matrix} - \begin{matrix} -5 & 10 \\ 15 & -2 \end{matrix} + \begin{matrix} -5 & 10 \\ 2 & 6 \end{matrix}$

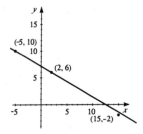

$$= (-4 - 90) - (10 - 150) + (-30 - 20)$$
$$= -94 + 140 - 50 = -4$$

These points are not collinear.

62. (a) If A is a matrix with a row or column consisting entirely of zeros, then if we expand the determinant by this row or column, we get $|A| = 0 \cdot |A_{1j}| - 0 \cdot |A_{2j}| + \cdots - 0 \cdot |A_{nj}| = 0$.

(b) Use the principle that if matrix B is a square matrix obtained from A by adding a multiple of one row to another, or a multiple of one column to another, then $|A| = |B|$. If we let B be the matrix obtained by subtracting the two rows (or columns) that are the same, then matrix B will have a row or column that consists entirely of zeros. So $|B| = 0 \quad \Rightarrow \quad |A| = 0$.

(c) Again use the principle that if matrix B is a square matrix obtained from A by adding a multiple of one row to another, or a multiple of one column to another, then $|A| = |B|$. If we let B be the matrix obtained by subtracting the proper multiple of the row (or column) from the other similar row (or column), then matrix B will have a row or column that consists entirely of zeros. So $|B| = 0 \quad \Rightarrow \quad |A| = 0$.

Review Exercises for Chapter 7

2. (a) 2×3

    (b) Yes, this matrix is in row-echelon form.

    (c) Yes, this matrix is in reduced row-echelon form.

    (d) $\begin{cases} x \quad\ = 6 \\ \quad\ y = 0 \end{cases}$

4. (a) 3×4

    (b) No, this matrix is not in row-echelon form, since the leading "one" in the second row is not to the left of the one above it.

    (c) Since this matrix is not in row-echelon form, it is not in reduced row-echelon form

    (d) $\begin{cases} x + 3y + 6z = 2 \\ 2x + \ y \quad\ = 5 \\ \quad\ z = 0 \end{cases}$

6. (a) 4×4

    (b) No, this matrix is not in echelon form, since the leading "one" in the fourth row is not to the left of the one above it.

    (c) Since this matrix is not in echelon form, it is not in reduced echelon form

    (d) $\begin{cases} x + 8y + 6z = -4 \\ \quad\ y - 3z = 5 \\ \quad\quad\ 2z = -7 \\ x + \ y + \ z = 0 \end{cases}$

8. $\begin{bmatrix} 1 & -1 & 1 & 2 \\ 1 & 1 & 3 & 6 \\ 0 & 2 & 3 & 5 \end{bmatrix} \xrightarrow{R_2 - R_1 \to R_2} \begin{bmatrix} 1 & -1 & 1 & 2 \\ 0 & 2 & 2 & 4 \\ 0 & 2 & 3 & 5 \end{bmatrix} \xrightarrow{R_3 - R_2 \to R_3} \begin{bmatrix} 1 & -1 & 1 & 2 \\ 0 & 2 & 2 & 4 \\ 0 & 0 & 1 & 1 \end{bmatrix}$

    Thus, $z = 1$; $2y + 2(1) = 4 \quad \Leftrightarrow \quad y = 1$; and $x - (1) + (1) = 2 \quad \Leftrightarrow \quad x = 2$, and so the solution is $(2, 1, 1)$.

10. $\begin{bmatrix} 1 & -1 & 1 & 2 \\ 1 & 1 & 3 & 6 \\ 3 & -1 & 5 & 10 \end{bmatrix} \xrightarrow[R_3 - 3R_1 \to R_3]{R_2 - R_1 \to R_2} \begin{bmatrix} 1 & -1 & 1 & 2 \\ 0 & 2 & 2 & 4 \\ 0 & 2 & 2 & 4 \end{bmatrix} \xrightarrow{R_3 - R_2 \to R_3} \begin{bmatrix} 1 & -1 & 1 & 2 \\ 0 & 2 & 2 & 4 \\ 0 & 0 & 0 & 0 \end{bmatrix}$

    $\xrightarrow{\frac{1}{2}R_2} \begin{bmatrix} 1 & -1 & 1 & 2 \\ 0 & 1 & 1 & 2 \\ 0 & 0 & 0 & 0 \end{bmatrix} \xrightarrow{R_1 + R_2 \to R_1} \begin{bmatrix} 1 & 0 & 2 & 4 \\ 0 & 1 & 1 & 2 \\ 0 & 0 & 0 & 0 \end{bmatrix}.$

    Let $z = t$, then $y + t = 2 \quad \Leftrightarrow \quad y = 2 - t$ and $x + 2t = 4 \quad \Leftrightarrow \quad x = 4 - 2t$, and so the solution is $(4 - 2t, 2 - t, t)$.

12.
$$\begin{bmatrix} 1 & 0 & 3 & 0 & -1 \\ 0 & 1 & 0 & -4 & 5 \\ 0 & 2 & 1 & 1 & 0 \\ 2 & 1 & 5 & -4 & 4 \end{bmatrix} \xrightarrow{R_4-2R_1 \to R_4} \begin{bmatrix} 1 & 0 & 3 & 0 & -1 \\ 0 & 1 & 0 & -4 & 5 \\ 0 & 2 & 1 & 1 & 0 \\ 0 & 1 & -1 & -4 & 6 \end{bmatrix} \xrightarrow[R_4-R_2 \to R_4]{R_3-2R_2 \to R_3}$$

$$\begin{bmatrix} 1 & 0 & 3 & 0 & -1 \\ 0 & 1 & 0 & -4 & 5 \\ 0 & 0 & 1 & 9 & -10 \\ 0 & 0 & -1 & 0 & 1 \end{bmatrix} \xrightarrow{R_3 \leftrightarrow R_4} \begin{bmatrix} 1 & 0 & 3 & 0 & -1 \\ 0 & 1 & 0 & -4 & 5 \\ 0 & 0 & -1 & 0 & 1 \\ 0 & 0 & 1 & 9 & -10 \end{bmatrix} \xrightarrow{R_4+3R_3 \to R_4}$$

$$\begin{bmatrix} 1 & 0 & 3 & 0 & -1 \\ 0 & 1 & 0 & -4 & 5 \\ 0 & 0 & -1 & 0 & 1 \\ 0 & 0 & 0 & 9 & -9 \end{bmatrix} \xrightarrow[\frac{1}{9}R_4]{-R_3} \begin{bmatrix} 1 & 0 & 3 & 0 & -1 \\ 0 & 1 & 0 & -4 & 5 \\ 0 & 0 & 1 & 0 & -1 \\ 0 & 0 & 0 & 1 & -1 \end{bmatrix} \xrightarrow[R_2+4R_4 \to R_2]{R_1-3R_3 \to R_1}$$

$$\begin{bmatrix} 1 & 0 & 0 & 0 & 2 \\ 0 & 1 & 0 & 0 & 1 \\ 0 & 0 & 1 & 0 & -1 \\ 0 & 0 & 0 & 1 & -1 \end{bmatrix}.$$ Therefore the solution is $(2, 1, -1, -1)$.

14.
$$\begin{bmatrix} 1 & -1 & 0 & 1 \\ 1 & 1 & 2 & 3 \\ 1 & -3 & -2 & -1 \end{bmatrix} \xrightarrow[R_3-R_1 \to R_3]{R_2-R_1 \to R_2} \begin{bmatrix} 1 & -1 & 0 & 1 \\ 0 & 2 & 2 & 2 \\ 0 & -2 & -2 & -2 \end{bmatrix} \xrightarrow{R_3+R_2 \to R_3}$$

$$\begin{bmatrix} 1 & -1 & 0 & 1 \\ 0 & 2 & 2 & 2 \\ 0 & 0 & 0 & 0 \end{bmatrix} \xrightarrow{\frac{1}{2}R_4} \begin{bmatrix} 1 & -1 & 0 & 1 \\ 0 & 1 & 1 & 1 \\ 0 & 0 & 0 & 0 \end{bmatrix} \xrightarrow{R_1+R_2 \to R_1} \begin{bmatrix} 1 & 0 & 1 & 2 \\ 0 & 1 & 1 & 1 \\ 0 & 0 & 0 & 0 \end{bmatrix}$$

Since the system is dependent, $z = t$; $y + t = 1$ $\Leftrightarrow$ $y = -t + 1$; $x + t = 2$ $\Leftrightarrow$ $x = -t + 2$. So, the solution is $(-t + 2, -t + 1, t)$ where t is any real number.

16.
$$\begin{bmatrix} 1 & -1 & 3 \\ 2 & 1 & 6 \\ 1 & -2 & 9 \end{bmatrix} \xrightarrow[R_3-R_1 \to R_3]{R_2-2R_1 \to R_2} \begin{bmatrix} 1 & -1 & 3 \\ 0 & 3 & 0 \\ 0 & -1 & 6 \end{bmatrix} \xrightarrow{R_2+3R_3 \to R_2} \begin{bmatrix} 1 & -1 & 3 \\ 0 & 0 & 18 \\ 0 & -1 & 6 \end{bmatrix}$$

Since the second row corresponds to the equation $0 = 18$, which is always false, this system has no solution.

18.
$$\begin{bmatrix} 1 & 2 & 3 & 2 \\ 2 & -1 & -5 & 1 \\ 4 & 3 & 1 & 6 \end{bmatrix} \xrightarrow[R_3-4R_1 \to R_3]{R_2-2R_1 \to R_2} \begin{bmatrix} 1 & 2 & 3 & 2 \\ 0 & -5 & -11 & -3 \\ 0 & -5 & -11 & -2 \end{bmatrix} \xrightarrow{R_3-R_2 \to R_3}$$

$$\begin{bmatrix} 1 & 2 & 3 & 2 \\ 0 & -5 & -11 & -3 \\ 0 & 0 & 0 & 1 \end{bmatrix}$$

Since the third row corresponds to the equation $0 = 1$, which is always false, this system has no solution.

20.
$$\begin{bmatrix} 1 & -1 & -2 & 3 & 0 \\ 0 & 1 & -1 & 1 & 1 \\ 3 & -2 & -7 & 10 & 2 \end{bmatrix} \xrightarrow{R_3-3R_1 \to R_3} \begin{bmatrix} 1 & -1 & -2 & 3 & 0 \\ 0 & 1 & -1 & 1 & 1 \\ 0 & 1 & -1 & 1 & 2 \end{bmatrix} \xrightarrow{R_3-R_2 \to R_3}$$

$$\begin{bmatrix} 1 & -1 & -2 & 3 & 0 \\ 0 & 1 & -1 & 1 & 1 \\ 0 & 0 & 0 & 0 & 1 \end{bmatrix}$$

Since the third row corresponds to the equation $0 = 1$, which is always false, this system has no solution.

In Exercises 22–32,

$A = \begin{bmatrix} 2 & 0 & -1 \end{bmatrix}$

$B = \begin{bmatrix} 1 & 2 & 4 \\ -2 & 1 & 0 \end{bmatrix}$ $\qquad D = \begin{bmatrix} 1 & 4 \\ 0 & -1 \\ 2 & 0 \end{bmatrix}$ $\qquad F = \begin{bmatrix} 4 & 0 & 2 \\ -1 & 1 & 0 \\ 7 & 5 & 0 \end{bmatrix}$

$C = \begin{bmatrix} \frac{1}{2} & 3 \\ 2 & \frac{3}{2} \\ -2 & 1 \end{bmatrix}$ $\qquad E = \begin{bmatrix} 2 & -1 \\ -\frac{1}{2} & 1 \end{bmatrix}$ $\qquad G = \begin{bmatrix} 5 \end{bmatrix}$

22. $C - D = \begin{bmatrix} \frac{1}{2} & 3 \\ 2 & \frac{3}{2} \\ -2 & 1 \end{bmatrix} - \begin{bmatrix} 1 & 4 \\ 0 & -1 \\ 2 & 0 \end{bmatrix} = \begin{bmatrix} -\frac{1}{2} & -1 \\ 2 & \frac{5}{2} \\ -4 & 1 \end{bmatrix}$

24. $5B - 2C$ cannot be performed because the matrix dimensions $(2 \times 3 \text{ and } 3 \times 2)$ are not compatible.

26. AG cannot be performed because the matrix dimensions $(1 \times 3 \text{ and } 1 \times 1)$ are not compatible.

28. $CB = \begin{bmatrix} \frac{1}{2} & 3 \\ 2 & \frac{3}{2} \\ -2 & 1 \end{bmatrix} \begin{bmatrix} 1 & 2 & 4 \\ -2 & 1 & 0 \end{bmatrix} = \begin{bmatrix} -\frac{11}{2} & 4 & 2 \\ -1 & \frac{11}{2} & 8 \\ -4 & -3 & -8 \end{bmatrix}$

30. $FC = \begin{bmatrix} 4 & 0 & 2 \\ -1 & 1 & 0 \\ 7 & 5 & 0 \end{bmatrix} \begin{bmatrix} \frac{1}{2} & 3 \\ 2 & \frac{3}{2} \\ -2 & 1 \end{bmatrix} = \begin{bmatrix} -2 & 14 \\ \frac{3}{2} & -\frac{3}{2} \\ \frac{27}{2} & \frac{57}{2} \end{bmatrix}$

32. $F(2C - D) = \begin{bmatrix} 4 & 0 & 2 \\ -1 & 1 & 0 \\ 7 & 5 & 0 \end{bmatrix} \left(\begin{bmatrix} 1 & 6 \\ 4 & 3 \\ -4 & 2 \end{bmatrix} - \begin{bmatrix} 1 & 4 \\ 0 & -1 \\ 2 & 0 \end{bmatrix} \right) = \begin{bmatrix} 4 & 0 & 2 \\ -1 & 1 & 0 \\ 7 & 5 & 0 \end{bmatrix} \begin{bmatrix} 0 & 2 \\ 4 & 4 \\ -6 & 2 \end{bmatrix}$

$= \begin{bmatrix} -12 & 12 \\ 4 & 2 \\ 20 & 34 \end{bmatrix}$

34. $AB = \begin{bmatrix} 2 & -1 & 3 \\ 2 & -2 & 1 \\ 0 & 1 & 1 \end{bmatrix} \begin{bmatrix} -\frac{3}{2} & 2 & \frac{5}{2} \\ -1 & 1 & 2 \\ 1 & -1 & -1 \end{bmatrix} = \begin{bmatrix} 1 & 0 & 0 \\ 0 & 1 & 0 \\ 0 & 0 & 1 \end{bmatrix}$

$BA = \begin{bmatrix} -\frac{3}{2} & 2 & \frac{5}{2} \\ -1 & 1 & 2 \\ 1 & -1 & -1 \end{bmatrix} \begin{bmatrix} 2 & -1 & 3 \\ 2 & -2 & 1 \\ 0 & 1 & 1 \end{bmatrix} = \begin{bmatrix} 1 & 0 & 0 \\ 0 & 1 & 0 \\ 0 & 0 & 1 \end{bmatrix}$

In exercises 36-40, $A = \begin{bmatrix} 2 & 1 \\ 3 & 2 \end{bmatrix}$, $B = \begin{bmatrix} 1 & -2 \\ -2 & 4 \end{bmatrix}$, and $C = \begin{bmatrix} 0 & 1 & 3 \\ -2 & 4 & 0 \end{bmatrix}$.

36. $\frac{1}{2}(X - 2B) = A \quad \Leftrightarrow \quad X - 2B = 2A \quad \Leftrightarrow \quad X = 2A + 2B = 2(A + B)$. Thus

$$X = 2\left(\begin{bmatrix} 2 & 1 \\ 3 & 2 \end{bmatrix} + \begin{bmatrix} 1 & -2 \\ -2 & 4 \end{bmatrix} \right) = 2 \begin{bmatrix} 3 & -1 \\ 1 & 6 \end{bmatrix} = \begin{bmatrix} 6 & -2 \\ 2 & 12 \end{bmatrix}$$

38. $2X + C = 5A \quad \Leftrightarrow \quad 2X = 5A - C$, but the difference $5A - C$ is not possible since the dimensions of A and C are not compatible.

40. $AX = B \quad \Leftrightarrow \quad A^{-1}AX = X = A^{-1}B$. Since $A = \begin{bmatrix} 2 & 1 \\ 3 & 2 \end{bmatrix}$ we have

$$A^{-1} = \frac{1}{4-3} \begin{bmatrix} 2 & -1 \\ -3 & 2 \end{bmatrix} = \begin{bmatrix} 2 & -1 \\ -3 & 2 \end{bmatrix}.$$

Thus $X = A^{-1}B = \begin{bmatrix} 2 & -1 \\ -3 & 2 \end{bmatrix} \begin{bmatrix} 1 & -2 \\ -2 & 4 \end{bmatrix} = \begin{bmatrix} 4 & -8 \\ -7 & 14 \end{bmatrix}$

42. $D = \begin{bmatrix} 2 & 2 \\ 1 & -3 \end{bmatrix}$. Then, $|D| = 2(-3) - 1(2) = -8$, and so $D^{-1} = -\frac{1}{8} \begin{bmatrix} -3 & -2 \\ -1 & 2 \end{bmatrix} = \begin{bmatrix} \frac{3}{8} & \frac{1}{4} \\ \frac{1}{8} & -\frac{1}{4} \end{bmatrix}$.

44. $D = \begin{bmatrix} 2 & 4 & 0 \\ -1 & 1 & 2 \\ 0 & 3 & 2 \end{bmatrix}$. Then, $|D| = 2 \begin{vmatrix} 1 & 2 \\ 3 & 2 \end{vmatrix} - 4 \begin{vmatrix} -1 & 2 \\ 0 & 2 \end{vmatrix} = 2(2 - 6) - 4(-2) = 0$, and so D has

no inverse.

46. $D = \begin{bmatrix} 1 & 2 & 3 \\ 2 & 4 & 5 \\ 2 & 5 & 6 \end{bmatrix}$. Then, $|D| = 1 \begin{vmatrix} 4 & 5 \\ 5 & 6 \end{vmatrix} - 2 \begin{vmatrix} 2 & 5 \\ 2 & 6 \end{vmatrix} + 3 \begin{vmatrix} 2 & 4 \\ 2 & 5 \end{vmatrix} = -1 - 4 + 6 = 1$. So D^{-1}

exists .

$$\begin{bmatrix} 1 & 2 & 3 & 1 & 0 & 0 \\ 2 & 4 & 5 & 0 & 1 & 0 \\ 2 & 5 & 6 & 0 & 0 & 1 \end{bmatrix} \xrightarrow[R_3 - 2R_1 \to R_3]{R_2 - 2R_1 \to R_2} \begin{bmatrix} 1 & 2 & 3 & 1 & 0 & 0 \\ 0 & 0 & -1 & -2 & 1 & 0 \\ 0 & 1 & 0 & -2 & 0 & 1 \end{bmatrix} \xrightarrow{-R_2 \leftrightarrow R_3}$$

$$\begin{bmatrix} 1 & 2 & 3 & 1 & 0 & 0 \\ 0 & 1 & 0 & -2 & 0 & 1 \\ 0 & 0 & 1 & 2 & -1 & 0 \end{bmatrix} \xrightarrow{R_1 - 3R_3 \to R_1} \begin{bmatrix} 1 & 2 & 0 & -5 & 3 & 0 \\ 0 & 1 & 0 & -2 & 0 & 1 \\ 0 & 0 & 1 & 2 & -1 & 0 \end{bmatrix} \xrightarrow{R_1 - 2R_2 \to R_1}$$

$$\begin{bmatrix} 1 & 0 & 0 & -1 & 3 & -2 \\ 0 & 1 & 0 & -2 & 0 & 1 \\ 0 & 0 & 1 & 2 & -1 & 0 \end{bmatrix}.$$

Thus, $D^{-1} = \begin{bmatrix} -1 & 3 & -2 \\ -2 & 0 & 1 \\ 2 & -1 & 0 \end{bmatrix}$.

48. $D = \begin{bmatrix} 1 & 0 & 1 & 0 \\ 0 & 1 & 0 & 1 \\ 1 & 1 & 1 & 2 \\ 1 & 2 & 1 & 2 \end{bmatrix}$. Thus,

$$|D| = \begin{vmatrix} 1 & 0 & 1 \\ 1 & 1 & 2 \\ 2 & 1 & 2 \end{vmatrix} + \begin{vmatrix} 0 & 1 & 1 \\ 1 & 1 & 2 \\ 1 & 2 & 2 \end{vmatrix} = \left(\begin{vmatrix} 1 & 2 \\ 1 & 2 \end{vmatrix} + \begin{vmatrix} 1 & 1 \\ 2 & 1 \end{vmatrix} \right) + \left(-\begin{vmatrix} 1 & 2 \\ 1 & 2 \end{vmatrix} + \begin{vmatrix} 1 & 1 \\ 1 & 2 \end{vmatrix} \right)$$
$$= (0 - 1) + (0 + 1) = 0.$$

Hence D^{-1} does not exists.

50. $\begin{bmatrix} 6 & -5 \\ 8 & -7 \end{bmatrix} \begin{bmatrix} x \\ y \end{bmatrix} = \begin{bmatrix} 1 \\ -1 \end{bmatrix}$. If we let $A = \begin{bmatrix} 6 & -5 \\ 8 & -7 \end{bmatrix}$, then $A^{-1} = \frac{1}{-42+40} \begin{bmatrix} -7 & 5 \\ -8 & 6 \end{bmatrix} = -\frac{1}{2} \begin{bmatrix} -7 & 5 \\ -8 & 6 \end{bmatrix}$

$= \begin{bmatrix} \frac{7}{2} & -\frac{5}{2} \\ 4 & -3 \end{bmatrix}$, and so $\begin{bmatrix} x \\ y \end{bmatrix} = \begin{bmatrix} \frac{7}{2} & -\frac{5}{2} \\ 4 & -3 \end{bmatrix} \begin{bmatrix} 1 \\ -1 \end{bmatrix} = \begin{bmatrix} 6 \\ 7 \end{bmatrix}$. Therefore, the solution is $(6, 7)$.

52. $\begin{bmatrix} 2 & 0 & 3 \\ 1 & 1 & 6 \\ 3 & -1 & 1 \end{bmatrix} \begin{bmatrix} x \\ y \\ z \end{bmatrix} = \begin{bmatrix} 5 \\ 0 \\ 5 \end{bmatrix}$. Let $A = \begin{bmatrix} 2 & 0 & 3 \\ 1 & 1 & 6 \\ 3 & -1 & 1 \end{bmatrix}$. Then,

$\begin{bmatrix} 2 & 0 & 3 & 1 & 0 & 0 \\ 1 & 1 & 6 & 0 & 1 & 0 \\ 3 & -1 & 1 & 0 & 0 & 1 \end{bmatrix} \xrightarrow{R_1 \leftrightarrow R_2} \begin{bmatrix} 1 & 1 & 6 & 0 & 1 & 0 \\ 2 & 0 & 3 & 1 & 0 & 0 \\ 3 & -1 & 1 & 0 & 0 & 1 \end{bmatrix} \begin{array}{c} R_2 - 2R_1 \to R_2 \\ \xrightarrow{\hspace{1cm}} \\ R_3 - 3R_1 \to R_3 \end{array}$

$\begin{bmatrix} 1 & 1 & 6 & 0 & 1 & 0 \\ 0 & -2 & -9 & 1 & -2 & 0 \\ 0 & -4 & -17 & 0 & -3 & 1 \end{bmatrix} \xrightarrow{R_2 - 2R_3 \to R_2} \begin{bmatrix} 1 & 1 & 6 & 0 & 1 & 0 \\ 0 & -2 & -9 & 1 & -2 & 0 \\ 0 & 0 & 1 & -2 & 1 & 1 \end{bmatrix} \begin{array}{c} R_1 + 9R_3 \to R_1 \\ \xrightarrow{\hspace{1cm}} \\ R_2 - 6R_3 \to 2R_2 \end{array}$

$\begin{bmatrix} 1 & 1 & 0 & 12 & -5 & -6 \\ 0 & -2 & 0 & -17 & 7 & 9 \\ 0 & 0 & 1 & -2 & 1 & 1 \end{bmatrix} \xrightarrow{-\frac{1}{2}R_2} \begin{bmatrix} 1 & 1 & 0 & 12 & -5 & -6 \\ 0 & 1 & 0 & \frac{17}{2} & -\frac{7}{2} & -\frac{9}{2} \\ 0 & 0 & 1 & -2 & 1 & 1 \end{bmatrix} \xrightarrow{R_1 - R_2 \to R_1}$

$\begin{bmatrix} 1 & 0 & 0 & \frac{7}{2} & -\frac{3}{2} & -\frac{3}{2} \\ 0 & 1 & 0 & \frac{17}{2} & -\frac{7}{2} & -\frac{9}{2} \\ 0 & 0 & 1 & -2 & 1 & 1 \end{bmatrix}$.

Hence, $A^{-1} = \begin{bmatrix} \frac{7}{2} & -\frac{3}{2} & -\frac{3}{2} \\ \frac{17}{2} & -\frac{7}{2} & -\frac{9}{2} \\ -2 & 1 & 1 \end{bmatrix}$ and $\begin{bmatrix} x \\ y \\ z \end{bmatrix} = \begin{bmatrix} \frac{7}{2} & -\frac{3}{2} & -\frac{3}{2} \\ \frac{17}{2} & -\frac{7}{2} & -\frac{9}{2} \\ -2 & 1 & 1 \end{bmatrix} \begin{bmatrix} 5 \\ 0 \\ 5 \end{bmatrix} = \begin{bmatrix} 10 \\ 20 \\ -5 \end{bmatrix}$, and so the

solution is $(10, 20, -5)$.

54. $|D| = \begin{vmatrix} 12 & -11 \\ 7 & 9 \end{vmatrix} = 108 + 77 = 185; \ |D_x| = \begin{vmatrix} 140 & -11 \\ 20 & 9 \end{vmatrix} = 1260 + 220 = 1480; \ $ and

$|D_y| = \begin{vmatrix} 12 & 140 \\ 7 & 20 \end{vmatrix} = 240 - 980 = -740.$ Therefore, $x = \frac{1480}{185} = 8$ and $y = \frac{-740}{185} = -4$, and so

the solution is $(8, -4)$.

56. $|D| = \begin{vmatrix} 3 & 4 & -1 \\ 1 & 0 & -4 \\ 2 & 1 & 5 \end{vmatrix} = -4 \begin{vmatrix} 1 & -4 \\ 2 & 5 \end{vmatrix} - 1 \begin{vmatrix} 3 & -1 \\ 1 & -4 \end{vmatrix} = -52 + 11 = -41;$

$$|D_x| = \begin{vmatrix} 10 & 4 & -1 \\ 20 & 0 & -4 \\ 30 & 1 & 5 \end{vmatrix} = -4 \begin{vmatrix} 20 & -4 \\ 30 & 5 \end{vmatrix} - 1 \begin{vmatrix} 10 & -1 \\ 20 & -4 \end{vmatrix} = -880 + 20 = -860;$$

$$|D_y| = \begin{vmatrix} 3 & 10 & -1 \\ 1 & 20 & -4 \\ 2 & 30 & 5 \end{vmatrix} = 3 \begin{vmatrix} 20 & -4 \\ 30 & 5 \end{vmatrix} - 1 \begin{vmatrix} 10 & -1 \\ 30 & 5 \end{vmatrix} + 2 \begin{vmatrix} 10 & -1 \\ 20 & -4 \end{vmatrix} = 660 - 80 - 40 = 540; \text{ and}$$

$$|D_z| = \begin{vmatrix} 3 & 4 & 10 \\ 1 & 0 & 20 \\ 2 & 1 & 30 \end{vmatrix} = -4 \begin{vmatrix} 1 & 20 \\ 2 & 30 \end{vmatrix} - 1 \begin{vmatrix} 3 & 10 \\ 1 & 20 \end{vmatrix} = 40 - 50 = -10.$$

Therefore, $x = \frac{-860}{-41} = \frac{860}{41}$, $y = -\frac{540}{41}$, and $z = \frac{10}{41}$, and so the solution is $\left(\frac{860}{41}, -\frac{540}{41}, \frac{10}{41} \right)$.

58. Area $= \pm \frac{1}{2} \begin{vmatrix} 5 & -2 & 1 \\ 1 & 5 & 1 \\ -4 & 1 & 1 \end{vmatrix} = \pm \frac{1}{2} \left(\begin{vmatrix} 1 & 5 \\ -4 & 1 \end{vmatrix} - \begin{vmatrix} 5 & -2 \\ -4 & 1 \end{vmatrix} + \begin{vmatrix} 5 & -2 \\ 1 & 5 \end{vmatrix} \right) = \pm \frac{1}{2} (21 + 3 + 27)$

$= \frac{51}{2}.$

60. Let $x =$ pounds of haddock, $y =$ pounds of sea bass, and $z =$ pounds of red snapper. We get the following system:

$$\begin{cases} x + y + z = 560 \\ 1.25x + 0.75y + 2.00z = 575 \\ 1.25x \qquad\quad + 2.00z = 320 \end{cases}$$

$$\begin{bmatrix} 1 & 1 & 1 & 560 \\ 1.25 & 0.75 & 2.00 & 575 \\ 1.25 & 0 & 2.00 & 320 \end{bmatrix} \xrightarrow{R_2 - R_3 \to R_2} \begin{bmatrix} 1 & 1 & 1 & 560 \\ 0 & 0.75 & 0 & 255 \\ 1.25 & 0 & 2.00 & 320 \end{bmatrix} \xrightarrow{R_3 - 1.25R_1 \to R_3}$$

$$\begin{bmatrix} 1 & 1 & 1 & 560 \\ 0 & 0.75 & 0 & 255 \\ 0 & -1.25 & 0.75 & -380 \end{bmatrix} \xrightarrow{\frac{4}{3}R_2} \begin{bmatrix} 1 & 1 & 1 & 560 \\ 0 & 1 & 0 & 340 \\ 0 & -1.25 & 0.75 & -380 \end{bmatrix} \xrightarrow{R_3 + 1.25R_2 \to R_3}$$

$$\begin{bmatrix} 1 & 1 & 1 & 560 \\ 0 & 1 & 0 & 340 \\ 0 & 0 & 0.75 & 45 \end{bmatrix} \xrightarrow{\frac{4}{3}R_3} \begin{bmatrix} 1 & 1 & 1 & 560 \\ 0 & 1 & 0 & 340 \\ 0 & 0 & 1 & 60 \end{bmatrix} \xrightarrow{R_1 - R_3 \to R_1} \begin{bmatrix} 1 & 1 & 0 & 500 \\ 0 & 1 & 0 & 340 \\ 0 & 0 & 1 & 60 \end{bmatrix}$$

$$\xrightarrow{R_1 - R_2 \to R_1} \begin{bmatrix} 1 & 0 & 0 & 160 \\ 0 & 1 & 0 & 340 \\ 0 & 0 & 1 & 60 \end{bmatrix}.$$

Thus he caught 160 of haddock, 340 of sea bass, and 60 pounds of red snapper.

Focus on Modeling

2.   The Data matrix $D = \begin{bmatrix} 0 & 1 & 1 & 0 \\ 0 & 0 & 1 & 1 \end{bmatrix}$ represents the square

Reflection: $T_1 = \begin{bmatrix} -1 & 0 \\ 0 & 1 \end{bmatrix}$

$T_1 D = \begin{bmatrix} 1 & 0 \\ 0 & -1 \end{bmatrix} \begin{bmatrix} 0 & 1 & 1 & 0 \\ 0 & 0 & 1 & 1 \end{bmatrix}$

$= \begin{bmatrix} 0 & -1 & -1 & 0 \\ 0 & 0 & 1 & 1 \end{bmatrix}$

Reflection about the y-axis.

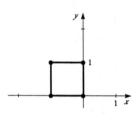

Expansion with $c = 3$: $T_2 = \begin{bmatrix} 1 & 0 \\ 0 & 3 \end{bmatrix}$

$T_2 D = \begin{bmatrix} 1 & 0 \\ 0 & 3 \end{bmatrix} \begin{bmatrix} 0 & 1 & 1 & 0 \\ 0 & 0 & 1 & 1 \end{bmatrix}$

$= \begin{bmatrix} 0 & 1 & 1 & 0 \\ 0 & 0 & 3 & 3 \end{bmatrix}$

Expansion in y-direction.

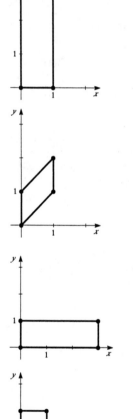

Shear with $c = 1$: $T_3 = \begin{bmatrix} 1 & 0 \\ 1 & 1 \end{bmatrix}$

$T_3 D = \begin{bmatrix} 1 & 0 \\ 1 & 1 \end{bmatrix} \begin{bmatrix} 0 & 1 & 1 & 0 \\ 0 & 0 & 1 & 1 \end{bmatrix}$

$= \begin{bmatrix} 0 & 1 & 1 & 0 \\ 0 & 1 & 2 & 1 \end{bmatrix}$

Shear in y-direction.

4.   (a) T is an expansion (by a factor of 3)
            in the x-direction.

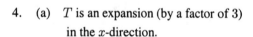

     (b) S is an expansion (by a factor of 2)
            in the y-direction.

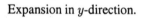

(c) $T(SD) = \begin{bmatrix} 3 & 0 \\ 0 & 1 \end{bmatrix} \left(\begin{bmatrix} 1 & 0 \\ 0 & 2 \end{bmatrix} \begin{bmatrix} 0 & 1 & 1 & 0 \\ 0 & 0 & 1 & 1 \end{bmatrix} \right)$

$= \begin{bmatrix} 3 & 0 \\ 0 & 1 \end{bmatrix} \begin{bmatrix} 0 & 1 & 1 & 0 \\ 0 & 0 & 2 & 2 \end{bmatrix}$

$= \begin{bmatrix} 0 & 3 & 3 & 0 \\ 0 & 0 & 2 & 2 \end{bmatrix}$

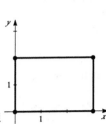

Expansion in the x-direction by a factor of 3 and expansion in the y-direction by a factor of 2.

(d) $W = TS = \begin{bmatrix} 3 & 0 \\ 0 & 1 \end{bmatrix} \begin{bmatrix} 1 & 0 \\ 0 & 2 \end{bmatrix} = \begin{bmatrix} 3 & 0 \\ 0 & 2 \end{bmatrix}$

(e) $WD = \begin{bmatrix} 3 & 0 \\ 0 & 2 \end{bmatrix} \begin{bmatrix} 0 & 1 & 1 & 0 \\ 0 & 0 & 1 & 1 \end{bmatrix} = \begin{bmatrix} 0 & 3 & 3 & 0 \\ 0 & 0 & 2 & 2 \end{bmatrix}$. It is the same as $T(SD)$.

6. (a) The data matrix $D = \begin{bmatrix} 0 & 1 & 2 & 1 & 0 & 0 \\ 0 & 0 & 2 & 4 & 4 & 0 \end{bmatrix}$

represents the figure

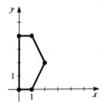

(b) $T = \begin{bmatrix} 1 & 1 \\ 0 & -1 \end{bmatrix}$,

$TD = \begin{bmatrix} 1 & 1 \\ 0 & -1 \end{bmatrix} \begin{bmatrix} 0 & 1 & 2 & 1 & 0 & 0 \\ 0 & 0 & 2 & 4 & 4 & 0 \end{bmatrix}$

$= \begin{bmatrix} 0 & 1 & 4 & 5 & 4 & 0 \\ 0 & 0 & -2 & -4 & -4 & 0 \end{bmatrix}$

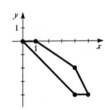

The transformation is a reflection about the x-axis and a shear in the x-direction.

(c) $T = \begin{bmatrix} 1 & 0 \\ 0 & -1 \end{bmatrix} \begin{bmatrix} 1 & 1 \\ 0 & 1 \end{bmatrix} = \begin{bmatrix} 1 & 1 \\ 0 & -1 \end{bmatrix}$,

Chapter Eight
Exercises 8.1

2. V. Only graph that opens downward.

4. I. Only graph that opens to the right.

6. IV. Graph opens downward and is wider than the graph represent in Exercise 3.

8. $x^2 = y$. Then $4p = 1 \Leftrightarrow p = \frac{1}{4}$.
Focus: $\left(0, \frac{1}{4}\right)$
Directrix: $y = -\frac{1}{4}$
Focal diameter: 1

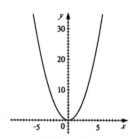

10. $y^2 = 3x$. Then $4p = 3 \Leftrightarrow p = \frac{3}{4}$.
Focus: $\left(\frac{3}{4}, 0\right)$
Directrix: $x = -\frac{3}{4}$
Focal diameter: 3

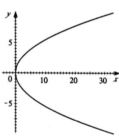

12. $y = -2x^2 \Leftrightarrow x^2 = -\frac{1}{2}y$. Then $4p = -\frac{1}{2} \Leftrightarrow p = -\frac{1}{8}$.
Focus: $\left(0, -\frac{1}{8}\right)$
Directrix: $y = \frac{1}{8}$
Focal diameter: $\frac{1}{2}$

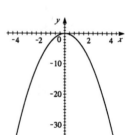

14. $x = \frac{1}{2}y^2 \Leftrightarrow y^2 = 2x$. Then $4p = 2 \Leftrightarrow p = \frac{1}{2}$.
Focus: $\left(\frac{1}{2}, 0\right)$
Directrix: $x = -\frac{1}{2}$
Focal diameter: 2

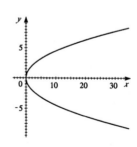

16. $x - 7y^2 = 0 \Leftrightarrow y^2 = \frac{1}{7}x$. Then $4p = \frac{1}{7} \Leftrightarrow p = \frac{1}{28}$.

Focus: $\left(\frac{1}{28}, 0\right)$

Directrix: $x = -\frac{1}{28}$

Focal diameter: $\frac{1}{7}$

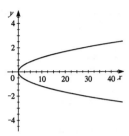

18. $8x^2 + 12y = 0 \Leftrightarrow x^2 = -\frac{3}{2}y$. Then $4p = -\frac{3}{2} \Leftrightarrow p = -\frac{3}{8}$.

Focus: $\left(0, -\frac{3}{8}\right)$

Directrix: $y = \frac{3}{8}$

Focal diameter: $\frac{3}{2}$

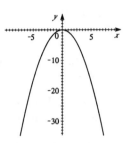

20.

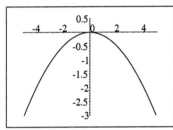

22.

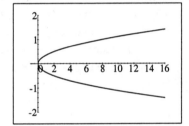

24.

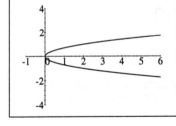

26. Since the focus is $\left(0, -\frac{1}{2}\right)$, $p = -\frac{1}{2} \Leftrightarrow 4p = -2$. So the equation of the parabola is $x^2 = -2y$.

28. Since the focus is $(5, 0)$, $p = 5 \Leftrightarrow 4p = 20$. Hence, the equation of the parabola is $y^2 = 20x$.

30. Since the directrix is $y = 6$, $p = -6 \Leftrightarrow 4p = -24$. Hence, the equation of the parabola is $x^2 = -24y$.

32. Since the directrix is $x = -\frac{1}{8}$, $p = \frac{1}{8} \Leftrightarrow 4p = \frac{1}{2}$. Hence, the equation of the parabola is $y^2 = \frac{1}{2}x$.

34. The directrix has y-intercept 6, and so $p = -6 \Leftrightarrow 4p = -24$. Therefore, the equation of the parabola is $x^2 = -24y$.

36. Since the focal diameter is 8 and the focus is on the negative y-axis, $4p = -8$. So the equation is $x^2 = -8y$.

38. The directrix is $x = -2$, and so $p = 2$ $\Leftrightarrow$ $4p = 8$. Since the parabola opens to the left, its equation is $y^2 = 8x$.

40. $p = 3$ $\Leftrightarrow$ $4p = 12$. Since the parabola opens downward, its equation is $x^2 = -12y$.

42. The focal diameter is $4p = 2(5) = 10$. Since the parabola opens upward, its equation is $x^2 = 10y$.

44. Since the directrix is $x = -p$, we have $p^2 = 16$, so $p = 4$, and the equation is $y^2 = 4px$ or $y^2 = 16x$.

46. The focus is $(0, p)$. Since the line has slope $\frac{1}{2}$, the equation of the line is $y = \frac{1}{2}x + p$. Therefore, the point where the line intersects the parabola has y-coordinate $\frac{1}{2}(2) + p = p + 1$. The parabola's equation is of the form $x^2 = 4py$, so $(2)^2 = 4p(p + 1)$ $\Leftrightarrow$ $p^2 + p - 1 = 0$ $\Leftrightarrow$ $p = \frac{-1 + \sqrt{5}}{2}$ (since $p > 0$). Hence, the equation of the parabola is $x^2 = 2(\sqrt{5} - 1)y$.

48. (a) Since the focal diameter of a parabola is $4p$ has equation $x^2 = 4py$, we have the following.
The focal diameter is $4p = 1$ so the equation is $x^2 = 1y$.
The focal diameter is $4p = 2$ so the equation is $x^2 = 2y$.
The focal diameter is $4p = 4$ so the equation is $x^2 = 4y$.
The focal diameter is $4p = 8$ so the equation is $x^2 = 8y$.

(b)

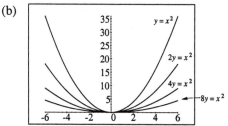

As the focal diameter increases the flatter the parabola gets.

50. The equation of the parabola has the form $x^2 = 4py$. From the diagram, the parabola passes through the point $(10, 1)$, and so $(10)^2 = 4p(1)$ $\Leftrightarrow$ $4p = 100$ and $p = 25$. Therefore, the receiver is 25 ft from the vertex.

52. The equation of the parabola has the form $x^2 = 4py$. From the diagram, the parabola passes through the point $(100, 3.79)$, and so $(100)^2 = 4p(3.79)$ $\Leftrightarrow$ $15.16p = 10000$ and $p \approx 659.63$. Therefore, the receiver is about 659.63 inches from the vertex.

54. Yes. If a cone intersects a plane that is parallel to a line on the cone, the resulting curve is a parabola (as shown on page 581 of the text).

Exercises 8.2

2. IV. Major axis is the vertical, vertices at $(0, \pm 3)$.

4. III. Major axis is the horizontal, vertices at $(\pm 5, 0)$.

6. $\dfrac{x^2}{16} + \dfrac{y^2}{25} = 1$. This ellipse has $a = 5$, $b = 4$, and so
    $c^2 = 25 - 16 = 9 \quad \Leftrightarrow \quad c = 3.$
    Vertices: $(0, \pm 5)$; foci: $(0, \pm 3)$; eccentricity: $e = \frac{c}{a} = \frac{3}{5} = 0.6$;
    length of the major axis: $2a = 10$; length of the minor axis: $2b = 8$.

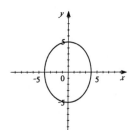

8. $4x^2 + 25y^2 = 100 \quad \Leftrightarrow \quad \dfrac{x^2}{25} + \dfrac{y^2}{4} = 1$. This ellipse has $a = 5$,
    $b = 2$, and so $c^2 = 25 - 4 = 21 \quad \Leftrightarrow \quad c = \sqrt{21}.$
    Vertices: $(\pm 5, 0)$; foci: $(\pm \sqrt{21}, 0)$; eccentricity: $e = \frac{c}{a} = \frac{\sqrt{21}}{5}$;
    length of the major axis: $2a = 10$; length of the minor axis: $2b = 4$.

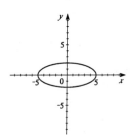

10. $4x^2 + y^2 = 16 \quad \Leftrightarrow \quad \dfrac{x^2}{4} + \dfrac{y^2}{16} = 1$. This ellipse has $a = 4$, $b = 2$,
    and so $c^2 = 16 - 4 = 12 \quad \Leftrightarrow \quad c = 2\sqrt{3}.$
    Vertices: $(0, \pm 4)$; foci: $(0, \pm 2\sqrt{3})$; eccentricity:
    $e = \frac{c}{a} = \frac{2\sqrt{3}}{4} = \frac{\sqrt{3}}{2}$; length of the major axis: $2a = 8$; length of
    the minor axis: $2b = 4$.

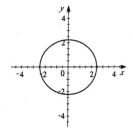

12. $5x^2 + 6y^2 = 30 \quad \Leftrightarrow \quad \dfrac{x^2}{6} + \dfrac{y^2}{5} = 1$. This ellipse has $a = \sqrt{6}$,
    $b = \sqrt{5}$, and so $c^2 = 6 - 5 = 1 \quad \Leftrightarrow \quad c = 1.$
    Vertices: $(\pm\sqrt{6}, 0)$; foci: $(\pm 1, 0)$; eccentricity:
    $e = \frac{c}{a} = \frac{1}{\sqrt{6}} = \frac{\sqrt{6}}{6}$: length of the major axis: $2a = 2\sqrt{6}$; length
    of the minor axis: $2b = 2\sqrt{5}$.

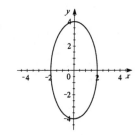

14. $9x^2 + 4y^2 = 1 \quad \Leftrightarrow \quad \dfrac{x^2}{1/9} + \dfrac{y^2}{1/4} = 1$. This ellipse has $a = \frac{1}{2}$,
    and so $c^2 = \frac{1}{4} - \frac{1}{9} = \frac{5}{36} \quad \Leftrightarrow \quad c = \frac{\sqrt{5}}{6}.$
    Vertices: $(0, \pm\frac{1}{2})$; foci: $\left(0, \pm\frac{\sqrt{5}}{6}\right)$; eccentricity:
    $e = \frac{c}{a} = \frac{\sqrt{5/6}}{1/2} = \frac{\sqrt{5}}{3}$; length of the major axis: $2a = 1$; length
    of the minor axis: $2b = \frac{2}{3}$.

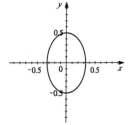

16. $x^2 = 4 - 2y^2$ $\Leftrightarrow$ $x^2 + 2y^2 = 4$ $\Leftrightarrow$ $\dfrac{x^2}{4} + \dfrac{y^2}{2} = 1$. This

ellipse has $a = 2$, $b = \sqrt{2}$, and so $c^2 = 4 - 2 = 2$ $\Leftrightarrow$ $c = \sqrt{2}$.

Vertices: $(\pm 2, 0)$; foci: $(\pm\sqrt{2}, 0)$; eccentricity: $e = \dfrac{c}{a} = \dfrac{\sqrt{2}}{2}$;

length of the major axis: $2a = 4$; length of the minor axis:

$2b = 2\sqrt{2}$.

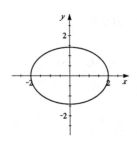

18. $20x^2 + 4y^2 = 5$ $\Leftrightarrow$ $\dfrac{x^2}{1/4} + \dfrac{y^2}{5/4} = 1$. This ellipse has $a = \dfrac{\sqrt{5}}{2}$,

$b = \frac{1}{2}$, and so $c^2 = \frac{5}{4} - \frac{1}{4} = 1$ $\Leftrightarrow$ $c = 1$.

Vertices: $\left(0, \pm\dfrac{\sqrt{5}}{2}\right)$; foci: $(0, \pm 1)$; eccentricity:

$e = \dfrac{c}{a} = \dfrac{1}{\frac{\sqrt{5}}{2}} = \dfrac{2\sqrt{5}}{5}$; length of the major axis: $2a = \sqrt{5}$;

length of the minor axis: $2b = 1$.

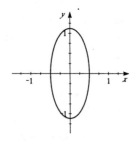

20. This ellipse has a vertical major axis with $a = 5$ and $b = 2$. Thus the equation is $\dfrac{x^2}{2^2} + \dfrac{y^2}{5^2} = 1$

   $\Leftrightarrow$ $\dfrac{x^2}{4} + \dfrac{y^2}{25} = 1$.

22. This ellipse has a vertical major axis with $a = 4$ and $c = 3$. So $c^2 = a^2 - b^2$ $\Leftrightarrow$ $9 = 16 - b^2$

   $\Leftrightarrow$ $b^2 = 7$. Thus the equation is $\dfrac{x^2}{7} + \dfrac{y^2}{4^2} = 1$ $\Leftrightarrow$ $\dfrac{x^2}{7} + \dfrac{y^2}{16} = 1$.

24. This ellipse has a vertical major axis with $b = 2$, so the equation of the ellipse is of the form

   $\dfrac{x^2}{2^2} + \dfrac{y^2}{a^2} = 1$. Substituting the point $(-1, 2)$ into the equation, we get $\dfrac{1}{4} + \dfrac{4}{a^2} = 1$ $\Leftrightarrow$

   $\dfrac{4}{a^2} = 1 - \dfrac{1}{4}$ $\Leftrightarrow$ $\dfrac{4}{a^2} = \dfrac{3}{4}$ $\Leftrightarrow$ $a^2 = \dfrac{4(4)}{3} = \dfrac{16}{3}$. Thus, the equation of the ellipse is

   $\dfrac{x^2}{4} + \dfrac{y^2}{16/3} = 1$ $\Leftrightarrow$ $\dfrac{x^2}{4} + \dfrac{3y^2}{16} = 1$.

26. $x^2 + \dfrac{y^2}{12} = 1$ $\Leftrightarrow$ $\dfrac{y^2}{12} = 1 - x^2$ $\Leftrightarrow$

   $y^2 = 12 - 12x^2$ $\Rightarrow$ $y = \pm\sqrt{12 - 12x^2}$.

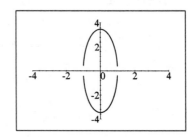

28. $x^2 + 2y^2 = 8$ ⇔ $2y^2 = 8 - x^2$ ⇔

$y^2 = 4 - \dfrac{x^2}{2}$ ⇒ $y = \pm\sqrt{4 - \dfrac{x^2}{2}}$.

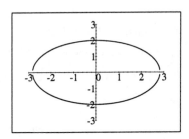

30. The foci are $(0, \pm 3)$ and the vertices are $(0, \pm 5)$. Thus, $c = 3$ and $a = 5$, and so $c^2 = a^2 - b^2$ ⇔ $9 = 25 - b^2$ ⇔ $b^2 = 25 - 9 = 16$. Therefore, the equation of the ellipse is $\dfrac{x^2}{16} + \dfrac{y^2}{25} = 1$.

32. The length of the major axis is $2a = 6$ ⇔ $a = 3$, the length of the minor axis is $2b = 4$ ⇔ $b = 2$, and the foci are on the x-axis. Therefore, the equation of the ellipse is $\dfrac{x^2}{9} + \dfrac{y^2}{4} = 1$.

34. The foci are $(\pm 5, 0)$, and the length of the major axis is $2a = 12$ ⇔ $a = 6$. Thus, $c^2 = a^2 - b^2$ ⇔ $25 = 36 - b^2$ ⇔ $b^2 = 36 - 25 = 11$. Since the foci are on the x-axis, the equation is $\dfrac{x^2}{36} + \dfrac{y^2}{11} = 1$.

36. Since the endpoints of the minor axis are $(0, \pm 3)$, we have $b = 3$. The distance between the foci is $2c = 8$, so $c = 4$. Thus, $a^2 = b^2 + c^2 = 9 + 16 = 25$, and the equation of the ellipse is $\dfrac{x^2}{25} + \dfrac{y^2}{9} = 1$.

38. Since the foci are $(0, \pm 2)$, we have $c = 2$. The eccentricity is $\dfrac{1}{9} = \dfrac{c}{a}$ ⇔ $a = 9c = 9(2) = 18$, and so $b^2 = a^2 - c^2 = 18^2 - 2^2 = 320$. Since the foci lie on the y-axis, the equation of the ellipse is $\dfrac{x^2}{320} + \dfrac{y^2}{324} = 1$.

40. Since the length of the major axis is $2a = 4$, we have $a = 2$. The eccentricity is $\dfrac{\sqrt{3}}{2} = \dfrac{c}{a} = \dfrac{c}{2}$, so $c = \sqrt{3}$. Then $b^2 = a^2 - c^2 = 4 - 3 = 1$, and since the foci are on the y-axis, the equation of the ellipse is $x^2 + \dfrac{y^2}{4} = 1$.

42. $\begin{cases} \dfrac{x^2}{16} + \dfrac{y^2}{9} = 1 \\[2mm] \dfrac{x^2}{9} + \dfrac{y^2}{16} = 1 \end{cases}$ ⇔ $\begin{cases} 9x^2 + 16y^2 = 144 \\ 16x^2 + 9y^2 = 144 \end{cases}$ ⇔

$\begin{cases} 144x^2 + 256y^2 = 2304 \\ -144x^2 - 81y^2 = -1296 \end{cases}$

Adding gives $175y^2 = 1008$ ⇒ $y = \pm\frac{12}{5}$. Substituting for y gives $9x^2 + 16\left(\frac{12}{5}\right)^2 = 144$ ⇔ $9x^2 = 144 - \frac{2304}{25} = \frac{1296}{25}$ ⇔ $x = \pm\frac{12}{5}$, and so the four points of intersection are $\left(\pm\frac{12}{5}, \pm\frac{12}{5}\right)$.

44. (a) The ellipse $x^2 + 4y^2 = 16$ $\Leftrightarrow$ $\dfrac{x^2}{16} + \dfrac{y^2}{4} = 1$ has $a = 4$ and $b = 2$. Thus, the equation of
the ancillary circle is $x^2 + y^2 = 4$.

  (b) If (s, t) is a point on the ancillary circle, then $s^2 + t^2 = 4$ $\Leftrightarrow$ $4s^2 + 4t^2 = 16$ $\Leftrightarrow$
$(2s)^2 + 4(t)^2 = 16$, which implies that $(2s, t)$ is a point on the ellipse.

46. $\dfrac{x^2}{k} + \dfrac{y^2}{4+k} = 1$ is an ellipse for $k > 0$. Then $a^2 = 4 + k$, $b^2 = k$, and so $c^2 = 4 + k - k = 4$
$\Leftrightarrow$ $c = \pm 2$. Therefore, all of the ellipses' foci are at $(0, \pm 2)$, no matter what the value of k is.

48. Using the eccentricity, $e = 0.25 = \frac{c}{a}$ $\Leftrightarrow$ $c = 0.25a$. Using the length of the minor axis,
$2b = 10{,}000{,}000{,}000$ $\Leftrightarrow$ $b = 5 \times 10^9$. Since $a^2 = c^2 + b^2$, $a^2 = (0.25a)^2 + 25 \times 10^{18}$ $\Leftrightarrow$
$\frac{15}{16}a^2 = 25 \times 10^{18}$ $\Leftrightarrow$ $a^2 = \frac{80}{3} \times 10^{18}$ $\Leftrightarrow$ $a = \sqrt{\frac{80}{3}} \times 10^9 = 4\sqrt{\frac{5}{3}} \times 10^9$. Then
$c = 0.25\left(4\sqrt{\frac{5}{3}} \times 10^9\right) = \sqrt{\frac{5}{3}} \times 10^9$. Since the Sun is at one focus of the ellipse, the distance from
Pluto to the Sun at perihelion is $a - c = 4\sqrt{\frac{5}{3}} \times 10^9 - \sqrt{\frac{5}{3}} \times 10^9 = 3\sqrt{\frac{5}{3}} \times 10^9 \approx 3.87 \times 10^9$ km;
the distance from Pluto to the Sun at aphelion is $a + c = 4\sqrt{\frac{5}{3}} \times 10^9 + \sqrt{\frac{5}{3}} \times 10^9 = 5\sqrt{\frac{5}{3}} \times 10^9$
$\approx 6.45 \times 10^9$ km.

50. Placing the origin at the center of the sheet of plywood and letting the x-axis be the long central axis,
we have $2a = 8$, so that $a = 4$, and $2b = 4$, so that $b = 2$. So $c^2 = a^2 - b^2 = 4^2 - 2^2 = 12$ $\Rightarrow$
$c = 2\sqrt{3} \approx 3.46$. So the tacks should be located $2(3.46) = 6.92$ feet apart and the string should be
$2a = 8$ feet long.

52. Have each friend hold one end of the string on the blackboard (the foci). Then use a piece of chalk
and the string to draw the ellipse.

54. The foci are $(\pm c, 0)$, where $c^2 = a^2 - b^2$. The endpoints of one latus rectum are the points $(c, \pm k)$,
and the length is $2k$. Substituting this point into the equation, we get $\dfrac{c^2}{a^2} + \dfrac{k^2}{b^2} = 1$ $\Leftrightarrow$
$\dfrac{k^2}{b^2} = 1 - \dfrac{c^2}{a^2} = \dfrac{a^2 - c^2}{a^2}$ $\Leftrightarrow$ $k^2 = \dfrac{b^2(a^2 - c^2)}{a^2}$. Since $b^2 = a^2 - c^2$, the last equation becomes
$k^2 = \dfrac{b^4}{a^2}$ $\Rightarrow$ $k = \dfrac{b^2}{a}$. Thus, the length of the latus rectum is $2k = 2\left(\dfrac{b^2}{a}\right) = \dfrac{2b^2}{a}$.

Exercises 8.3

2. IV. Opens vertically, vertices at $(0, \pm 1)$.

4. I. Opens horizontally, vertices at $(\pm 5, 0)$.

6. The hyperbola $\dfrac{y^2}{9} - \dfrac{x^2}{16} = 1$ has $a = 3$, $b = 4$, and $c^2 = 9 + 16 = 25$
   $\Rightarrow \quad c = 5$.
   Vertices: $(0, \pm 3)$; foci: $(0, \pm 5)$; asymptotes: $y = \pm \frac{3}{4} x$.

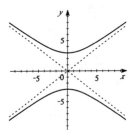

8. The hyperbola $\dfrac{x^2}{2} - \dfrac{y^2}{1} = 1$ has $a = \sqrt{2}$, $b = 1$, and $c^2 = 2 + 1 = 3$
   $\Rightarrow \quad c = \sqrt{3}$.
   Vertices: $(\pm \sqrt{2}, 0)$; foci: $(\pm \sqrt{3}, 0)$; asymptotes: $y = \pm \frac{1}{\sqrt{2}} x$.

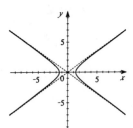

10. The hyperbola $9x^2 - 4y^2 = 36 \quad \Leftrightarrow \quad \dfrac{x^2}{4} - \dfrac{y^2}{9} = 1$ has $a = 2$,
    $b = 3$, and $c^2 = 4 + 9 = 13 \quad \Rightarrow \quad c = \sqrt{13}$.
    Vertices: $(\pm 2, 0)$; foci: $(\pm \sqrt{13}, 0)$; asymptotes: $y = \pm \frac{3}{2} x$.

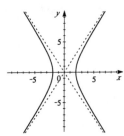

12. The hyperbola $x^2 - y^2 + 4 = 0 \quad \Leftrightarrow \quad y^2 - x^2 = 4 \quad \Leftrightarrow$
    $\dfrac{y^2}{4} - \dfrac{x^2}{4} = 1$ has $a = 2$, $b = 2$, and $c^2 = 4 + 4 = 8 = 2\sqrt{2}$.
    Vertices: $(0, \pm 2)$; foci: $(0, \pm 2\sqrt{2})$; asymptotes: $y = \pm x$.

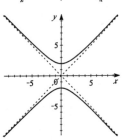

14. The hyperbola $x^2 - 2y^2 = 3 \quad \Leftrightarrow \quad \dfrac{x^2}{3} - \dfrac{y^2}{3/2} = 1$ has $a = \sqrt{3}$,
    $b = \sqrt{3/2}$, and $c^2 = 3 + \frac{3}{2} = \frac{9}{2} \quad \Rightarrow \quad c = \frac{3\sqrt{2}}{2}$.
    Vertices: $(\pm \sqrt{3}, 0)$; foci: $(\pm \sqrt{\frac{3}{2}}, 0)$; asymptotes:
    $y = \pm \dfrac{\sqrt{3/2}}{\sqrt{3}} x = \pm \dfrac{\sqrt{2}}{2} x$.

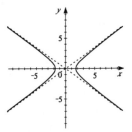

16.   The hyperbola $9x^2 - 16y^2 = 1$ $\Leftrightarrow$ $\dfrac{x^2}{1/9} - \dfrac{y^2}{1/16} = 1$ has

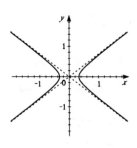

$a = \frac{1}{3}, b = \frac{1}{4}$, and $c^2 = \frac{1}{9} + \frac{1}{16} = \frac{25}{144}$ $\Rightarrow$ $c = \frac{5}{12}$.

Vertices: $(\pm\frac{1}{3}, 0)$; foci: $(\pm\frac{5}{12}, 0)$; asymptotes:

$y = \pm\dfrac{1/4}{1/3}x = \pm\frac{3}{4}x.$

18.   From the graph, the foci are $(0, \pm 13)$ and the vertices are $(0, \pm 12)$, so $c = 13$ and $a = 12$. Then $b^2 = c^2 - a^2 = 169 - 144 = 25$, and since the vertices are on the y-axis, the equation of the hyperbola is $\dfrac{y^2}{144} - \dfrac{x^2}{25} = 1.$

20.   The vertices are $(\pm 2\sqrt{3}, 0)$, so $a = 2\sqrt{3}$, so the equation of the hyperbola is of the form $\dfrac{x^2}{(2\sqrt{3})^2} - \dfrac{y^2}{b^2} = 1$. Substituting the point $(4, 4)$ into the equation, we get $\dfrac{16}{12} - \dfrac{16}{b^2} = 1$ $\Leftrightarrow$ $\dfrac{16}{b^2} = \dfrac{16}{12} - 1$ $\Leftrightarrow$ $\dfrac{16}{b^2} = \dfrac{4}{12}$ $\Leftrightarrow$ $b^2 = 48$. Thus, the equation of the hyperbola is $\dfrac{x^2}{12} - \dfrac{y^2}{48} = 1.$

22.   From the graph, the vertices are $(0, \pm 3)$, so $a = 3$. Since the asymptotes are $y = \pm 3x = \pm\dfrac{a}{b}x$, we have $\dfrac{3}{b} = 3$ $\Leftrightarrow$ $b = 1$. Since the vertices are on the x-axis, the equation is $\dfrac{y^2}{3^2} - \dfrac{x^2}{1^2} = 1$ $\Leftrightarrow$ $\dfrac{y^2}{9} - x^2 = 1.$

24. $3y^2 - 4x^2 = 24$ $\Leftrightarrow$ $3y^2 = 4x^2 + 24$
      $\Leftrightarrow$ $y^2 = \frac{4}{3}x^2 + 8$ $\Rightarrow$
      $y = \pm\sqrt{\frac{4}{3}x^2 + 8}.$

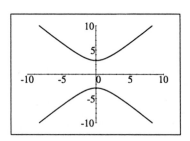

26. $\dfrac{x^2}{100} - \dfrac{y^2}{64} = 1$ $\Leftrightarrow$ $16x^2 - 25y^2 = 1600$
      $\Leftrightarrow$ $25y^2 = 16x^2 - 1600$ $\Leftrightarrow$
      $y^2 = \frac{16}{25}x^2 - 64$ $\Rightarrow$ $y = \pm\sqrt{\frac{16}{25}x^2 - 64}.$

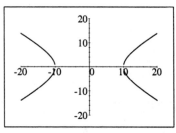

28. The foci are $(0, \pm 10)$ and the vertices are $(0, \pm 8)$, so $c = 10$ and $a = 8$. Then $b^2 = c^2 - a^2 = 100 - 64 = 36$, and since the vertices are on the y-axis, the equation of the hyperbola is $\dfrac{y^2}{64} - \dfrac{x^2}{36} = 1$.

30. The foci are $(\pm 6, 0)$, and the vertices are $(\pm 2, 0)$, so $c = 6$ and $a = 2$. Then $b^2 = c^2 - a^2 = 36 - 4 = 32$, and since the vertices are on the x-axis, the equation is $\dfrac{x^2}{4} - \dfrac{y^2}{32} = 1$.

32. The vertices are $(0, \pm 6)$, so $a = 6$. The asymptotes are $y = \pm\frac{1}{3}x = \pm\frac{a}{b}x \quad \Leftrightarrow \quad \frac{6}{b} = \frac{1}{3} \quad \Leftrightarrow$ $b = 18$. Since the vertices are on the y-axis, the equation of the hyperbola is $\dfrac{y^2}{36} - \dfrac{x^2}{324} = 1$.

34. The vertices are $(0, \pm 6)$, so $a = 6$. Since the vertices are on the y-axis, the hyperbola has an equation of the form $\dfrac{y^2}{36} - \dfrac{x^2}{b^2} = 1$. Since the hyperbola passes through the point $(-5, 9)$, we have $\dfrac{81}{36} - \dfrac{25}{b^2} = 1 \quad \Leftrightarrow \quad \dfrac{25}{b^2} = \dfrac{45}{36} \quad \Leftrightarrow \quad b^2 = 20$. Thus, the equation is $\dfrac{y^2}{36} - \dfrac{x^2}{20} = 1$.

36. The foci are $(\pm 3, 0)$, so $c = 3$. Since the vertices are on the x-axis, the equation of the hyperbola is of the form $\dfrac{x^2}{a^2} - \dfrac{y^2}{b^2} = 1$. Since the hyperbola passes through the point $(4, 1)$, we have $\dfrac{16}{a^2} - \dfrac{1}{b^2} = 1$. Using the foci, $c^2 = a^2 + b^2 = 9 \quad \Leftrightarrow \quad a^2 = 9 - b^2$, and substituting gives $\dfrac{16}{9 - b^2} - \dfrac{1}{b^2} = 1 \quad \Leftrightarrow \quad 16b^2 - (9 - b^2) = b^2(9 - b^2) \quad \Leftrightarrow \quad 17b^2 - 9 = 9b^2 - b^4 \quad \Leftrightarrow$ $b^4 + 8b^2 - 9 = 0 \quad \Leftrightarrow \quad (b^2 + 9)(b^2 - 1) = 0 \quad \Leftrightarrow \quad b^2 = 1$ or $b^2 = -9$ (never). Then we have $a^2 = 9 - b^2 = 8$. Thus, the equation of the hyperbola is $\dfrac{x^2}{8} - y^2 = 1$.

38. The foci are $(0, \pm 1)$, and the length of the transverse axis is 1, so $c = 1$ and $2a = 1 \quad \Leftrightarrow \quad a = \frac{1}{2}$. Then, $b^2 = c^2 - a^2 = 1 - \frac{1}{4} = \frac{3}{4}$, and since the foci are on the y-axis, the equation is $\dfrac{y^2}{1/4} - \dfrac{x^2}{3/4} = 1$ $\Leftrightarrow \quad 4y^2 - \dfrac{4x^2}{3} = 1$.

40. The hyperbolas $\dfrac{x^2}{a^2} - \dfrac{y^2}{b^2} = 1$ and $\dfrac{x^2}{a^2} - \dfrac{y^2}{b^2} = -1$ are conjugate to each other.

(a) $x^2 - 4y^2 + 16 = 0 \quad \Leftrightarrow \quad \dfrac{x^2}{16} - \dfrac{y^2}{4} = -1$,

and $4y^2 - x^2 + 16 = 0 \quad \Leftrightarrow \quad \dfrac{x^2}{16} - \dfrac{y^2}{4} = 1$. So the hyperbolas are conjugate to each other.

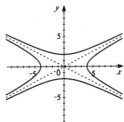

(b) They both have the same asymptotes, $y = \pm\frac{1}{2}x$.

(c)   The two general conjugate hyperbolas both have asymptotes $y = \pm\frac{b}{a}x$.

42.   (a)   The hyperbola $\dfrac{x^2}{9} - \dfrac{y^2}{16} = 1$ has $a = 3$, $b = 4$, so $c^2 = 9 + 16 = 25$, and $c = 5$. Therefore, the foci are $F_1(5, 0)$ and $F_2(-5, 0)$.

(b)   Substituting $P(5, \frac{16}{3})$ into the equation of the hyperbola, we get $\dfrac{25}{9} - \dfrac{256/9}{16} = \dfrac{25}{9} - \dfrac{16}{9} = 1$, so P lies on the hyperbola.

(c) $d(P, F_1) = \sqrt{(5-5)^2 + (16/3 - 0)^2} = \frac{16}{3}$, and $d(P, F_2) = \sqrt{(-5-5)^2 + (16/3 - 0)^2}$
       $= \frac{1}{3}\sqrt{900 + 256} = \frac{34}{3}$.

(d) $d(P, F_2) - d(P, F_1) = \frac{34}{3} - \frac{16}{3} = 6 = 2(3) = 2a$.

44. $d(AB) = 500 = 2c \quad \Leftrightarrow \quad c = 250$.

(a)   Since $\Delta t = 2640$ and $v = 980$, then $\Delta d = d(PA) - d(PB) = v\,\Delta t$
       $= (980 \text{ ft/}\mu s) \cdot (2640 \ \mu s) = 2{,}587{,}200 \text{ ft} = 490 \text{ mi}$.

(b) $c = 250$, $2a = 490 \quad \Leftrightarrow \quad a = 245$, and the foci are on the y-axis. Then,
       $b^2 = 250^2 - 245^2 = 2475$. Hence, the equation is $\dfrac{y^2}{60{,}025} - \dfrac{x^2}{2475} = 1$.

(c)   Since P is due east of A, $c = 250$ is the y-coordinate of P. Therefore, P is at $(x, 250)$, and so
       $\dfrac{250^2}{245^2} - \dfrac{x^2}{2475} = 1 \quad \Leftrightarrow \quad x^2 = 2475\left(\dfrac{250^2}{245^2} - 1\right) \approx 102.05$. Then, $x \approx 10.1$, and so P is
       approximately 10.1 miles from A.

46.   (a)   These equally spaced concentric circles can be used as a kind of measure were we count the number of rings. In the case of the red dots, the sum of the number of wave crests from each center is a constant, in this case 17. As you move out one wave crest from the left stone you move in one wave crest from the right stone. Therefore this satisfies the geometric definition of an ellipse.

(b)   Similarly, in the case of the blue dots, the difference of the number of wave crests from each center is a constant. As you move out one wave crest from the left center you also move out one wave crest from the right stone. Therefore this satisfies the geometric definition of a hyperbola.

48.   The wall is parallel to the axis of the cone of light coming from the top of the shade, so the intersection of the wall and the cone of light is a hyperbola. In the case of the flashlight, hold it parallel to the ground to form a hyperbola.

Exercises 8.4

2. The ellipse $\dfrac{(x-3)^2}{16} + (y+3)^2 = 1$ is obtained from the ellipse

   $\dfrac{x^2}{16} + y^2 = 1$ by shifting to the right 3 units and downward 3 units.

   So $a = 4$, $b = 1$, and $c = \sqrt{16-1} = \sqrt{15}$.

   Center: $(3, -3)$; foci: $(3 \pm \sqrt{15}, -3)$; vertices:
   $(3 \pm 4, -3) = (-1, -3)$ and $(7, -3)$; length of the major axis:
   $2a = 8$; length of the minor axis: $2b = 2$.

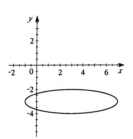

4. The ellipse $\dfrac{(x+2)^2}{4} + y^2 = 1$ is obtained from the ellipse

   $\dfrac{x^2}{4} + \dfrac{y^2}{1} = 1$ by shifting to the left 2 units. So $a = 2$, $b = 1$, and

   $c = \sqrt{4-1} = \sqrt{3}$.

   Center: $(-2, 0)$; foci: $(-2 \pm \sqrt{3}, 0)$; vertices:
   $(-2 \pm 2, 0) = (-4, 0)$ and $(0, 0)$; length of the major axis: $2a = 4$;
   length of the minor axis: $2b = 2$.

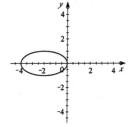

6. The parabola $(y+5)^2 = -6x + 12 = -6(x-2)$ is obtained from the
   parabola $y^2 = -6x$ by shifting to the right 2 units and down 5 units.
   So $4p = -6 \quad \Leftrightarrow \quad p = -\frac{3}{2}$.

   Vertex: $(2, -5)$; focus: $(2 - \frac{3}{2}, -5) = (\frac{1}{2}, -5)$; directrix:
   $x = 2 + \frac{3}{2} = \frac{7}{2}$.

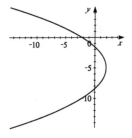

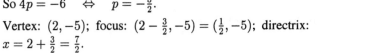

8. The parabola $y^2 = 16x - 8 = 16(x - \frac{1}{2})$ is obtained from the

   parabola $y^2 = 16x$ by shifting to the right $\frac{1}{2}$ unit. So $4p = 16$

   $\Leftrightarrow \quad p = 4$.

   Vertex: $(\frac{1}{2}, 0)$; focus: $(\frac{1}{2} + 4, 0) = (\frac{9}{2}, 0)$; directrix:

   $x = \frac{1}{2} - 4 = -\frac{7}{2}$.

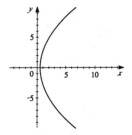

10. The hyperbola $(x-8)^2 - (y+6)^2 = 1$ is obtained from the

    hyperbola $x^2 - y^2 = 1$ by shifting to the right 8 units and

    downward 6 units. So $a = 1$, $b = 1$, and $c = \sqrt{1+1} = \sqrt{2}$.

    Center: $(8, -6)$; foci: $(8 \pm \sqrt{2}, -6)$; vertices: $(8 \pm 1, -6)$, so

    the vertices are $(7, -6)$ and $(9, -6)$; asymptotes:

    $(y+6) = \pm(x-8) \quad \Leftrightarrow \quad y = x - 14$ and $y = -x + 2$.

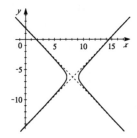

12. The hyperbola $\dfrac{(y-1)^2}{25} - (x+3)^2 = 1$ is obtained from the

hyperbola $\dfrac{y^2}{25} - x^2 = 1$ by shifting to the left 3 units and upward 1

unit. So $a = 5$, $b = 1$, and $c = \sqrt{25+1} = \sqrt{26}$.

Center: $(-3,1)$; foci: $(-3, 1 \pm \sqrt{26})$; vertices: $(-3, 1 \pm 5)$, so the
vertices are $(-3,-4)$ and $(-3,6)$; asymptotes: $(y-1) = \pm 5(x+3)$
$\Leftrightarrow \quad y = 5x + 16$ and $y = -5x - 14$.

14. This is a parabola that opens to the right with its vertex at $(-6,0)$, so its equation is of the form
$y^2 = 4p(x+6)$, with $p > 0$. Since the distance from the vertex to the directrix is $p = -6 - (-12)$
$= 6$, the equation is $y^2 = 4(6)(x+6) \quad \Leftrightarrow \quad y^2 = 24(x+6)$

16. This is an ellipse with the major axis parallel to the y-axis. From the graph the center is $(2,-3)$,
with $a = 3$ and $b = 2$. Thus, the equation is $\dfrac{(x-2)^2}{4} + \dfrac{(y+3)^2}{9} = 1$.

18. From the graph, the vertices are $(2,0)$ and $(6,0)$. The center is the midpoint between the vertices, so
the center is $\left(\frac{2+6}{2}, \frac{0+0}{2}\right) = (4,0)$. Since a is the distance from the center to a vertex, $a = 2$. Since
the vertices are on the x-axis, the equation is of the form $\dfrac{(x-4)^2}{4} - \dfrac{y^2}{b^2} = 1$. Since the point $(0,4)$

lies on the hyperbola, we have $\dfrac{(0-4)^2}{4} - \dfrac{(4)^2}{b^2} = 1 \quad \Leftrightarrow \quad \dfrac{16}{4} - \dfrac{16}{b^2} = 1 \quad \Leftrightarrow \quad 3 = \dfrac{16}{b^2}$

$\Leftrightarrow \quad b^2 = \frac{16}{3}$. Thus, the equation of the hyperbola is $\dfrac{(x-4)^2}{4} - \dfrac{y^2}{16/3} = 1 \quad \Leftrightarrow$

$\dfrac{(x-4)^2}{4} - \dfrac{3y^2}{16} = 1$.

20. $y^2 = 4(x+2y) \quad \Leftrightarrow \quad y^2 - 8y = 4x \quad \Leftrightarrow$
$y^2 - 8y + 16 = 4x + 16 \quad \Leftrightarrow \quad (y-4)^2 = 4(x+4)$. This is a
parabola that has $4p = 4 \quad \Leftrightarrow \quad p = 1$.

Vertex: $(-4,4)$; focus: $(-4+1, 4) = (-3,4)$; directrix:
$x = -4 - 1 = -5$.

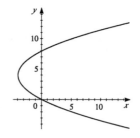

22. $x^2 + 6x + 12y + 9 = 0 \quad \Leftrightarrow \quad x^2 + 6x + 9 = -12y \quad \Leftrightarrow$
$(x+3)^2 = -12y$. This is a parabola that has $4p = -12 \quad \Leftrightarrow$
$p = -3$.

Vertex: $(-3,0)$; focus: $(-3,-3)$; directrix: $y = 3$.

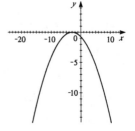

24. $2x^2 + y^2 = 2y + 1 \quad \Leftrightarrow \quad 2x^2 + (y^2 - 2y + 1) = 1 + 1 \quad \Leftrightarrow$

$2x^2 + (y-1)^2 = 2 \quad \Leftrightarrow \quad x^2 + \dfrac{(y-1)^2}{2} = 1$. This is an ellipse that

has $a = \sqrt{2}$, $b = 1$, and $c = \sqrt{2-1} = 1$.

Center: $(0,1)$; foci: $(0, 1 \pm 1) = (0,0)$ and $(0,2)$; vertices:

$(0, 1 \pm \sqrt{2})$; length of the major axis: $2a = 2\sqrt{2}$; length of the

minor axis: $2b = 2$.

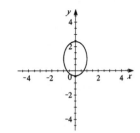

26. $4x^2 - 4x - 8y + 9 = 0 \quad \Leftrightarrow \quad 4(x^2 - x) - 8y + 9 = 0 \quad \Leftrightarrow$

$4(x^2 - x + \frac{1}{4}) = 8y - 9 + 1 \quad \Leftrightarrow \quad 4(x - \frac{1}{2})^2 = 8(y-1) \quad \Leftrightarrow$

$(x - \frac{1}{2})^2 = 2(y-1)$. This is a parabola that has $4p = 2 \quad \Leftrightarrow$

$p = \frac{1}{2}$.

Vertex: $(\frac{1}{2}, 1)$; focus: $(\frac{1}{2}, \frac{3}{2})$; directrix: $y = -\frac{1}{2} + 1 = \frac{1}{2}$.

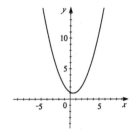

28. $x^2 - y^2 = 10(x - y) + 1 \quad \Leftrightarrow \quad x^2 - 10x - y^2 + 10y = 1 \quad \Leftrightarrow$

$(x^2 - 10x + 25) - (y^2 - 10y + 25) = 1 + 25 - 25 \quad \Leftrightarrow$

$(x-5)^2 - (y-5)^2 = 1$. This is a hyperbola that has $a = 1$, $b = 1$,

and $c = \sqrt{1+1} = \sqrt{2}$.

Center: $(5,5)$; foci: $(5 \pm \sqrt{2}, 5)$; vertices: $(5 \pm 1, 5) = (4, 5)$ and

$(6, 5)$; asymptotes: $y - 5 = \pm(x - 5) \quad \Leftrightarrow \quad y = x$ and

$y = -x + 10$.

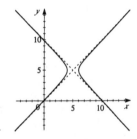

30. $x^2 + 4y^2 + 20x - 40y + 300 = 0 \quad \Leftrightarrow \quad (x^2 + 20x) + 4(y^2 - 10y) = -300 \quad \Leftrightarrow$

$(x^2 + 20x + 100) + 4(y^2 - 10y + 25) = -300 + 100 + 100 \quad \Leftrightarrow$

$(x+10)^2 + 4(y-5)^2 = -100$. Since $u^2 + v^2 \geq 0$ for all $u, v \in \mathbb{R}$, there is no (x, y) such that

$(x+10)^2 + 4(y-5)^2 = -100$. So there is no solution, and the graph is the empty set.

32. $4x^2 + 9y^2 - 36y = 0 \quad \Leftrightarrow \quad 4x^2 + 9(y^2 - 4y) = 0 \quad \Leftrightarrow$

$4x^2 + 9(y^2 - 4y + 4) = 36 \quad \Leftrightarrow$

$9(y^2 - 4y + 4) = 36 - 4x^2 \quad \Leftrightarrow \quad y^2 - 4y + 4 = 4 - \frac{4}{9}x^2$

$\Rightarrow \quad y - 2 = \pm\sqrt{4 - \frac{4}{9}x^2} \quad \Leftrightarrow \quad y = 2 \pm \sqrt{4 - \frac{4}{9}x^2}$.

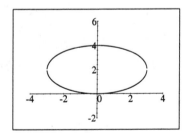

34. $x^2 - 4y^2 + 4x + 8y = 0 \quad \Leftrightarrow \quad x^2 + 4x - 4(y^2 - 2y) = 0$

$\Leftrightarrow \quad (x^2 + 4x + 4) - 4(y^2 - 2y + 1) = 0 \quad \Leftrightarrow$

$4(y-1)^2 = (x+2)^2 \quad \Rightarrow \quad 2(y-1) = \pm(x+2) \quad \Leftrightarrow$

$y - 1 = \pm\frac{1}{2}(x+2) \quad \Leftrightarrow \quad y = 1 \pm \frac{1}{2}(x+2)$.

So $y = 1 + \frac{1}{2}(x+2) = \frac{1}{2}x + 2$ and

$y = 1 - \frac{1}{2}(x+2) = -\frac{1}{2}x$. This is a degenerate conic.

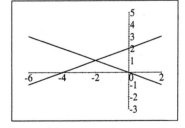

36. The parabola $x^2 + y = 100$ $\Leftrightarrow$ $x^2 = -(y - 100)$ has $4p = -1$ $\Leftrightarrow$ $p = -\frac{1}{4}$. The vertex is $(0, 100)$ and so the focus is $\left(0, 100 - \frac{1}{4}\right) = \left(0, \frac{399}{4}\right)$. Thus, one vertex of the ellipse is $(0, 100)$, and one focus is $\left(0, \frac{399}{4}\right)$. Since the second focus of the ellipse is $(0, 0)$, the second vertex is $\left(0, -\frac{1}{4}\right)$. So $2a = 100 + \frac{1}{4} = \frac{401}{4}$ $\Leftrightarrow$ $a = \frac{401}{8}$. Since $2c$ is the distance between the foci of the ellipse, $2c = \frac{399}{4}$, $c = \frac{399}{8}$, and then $b^2 = a^2 - c^2 = \frac{401^2}{64} - \frac{399^2}{64} = 25$. The center of the ellipse is $\left(0, \frac{399}{8}\right)$

and so its equation is $\dfrac{x^2}{25} + \dfrac{\left(y - \frac{399}{8}\right)^2}{\left(\frac{401}{8}\right)^2} = 1$, which simplifies to $\dfrac{x^2}{25} + \dfrac{(8y - 399)^2}{160{,}801} = 1.$

38. Since $(0, 0)$ and $(1600, 0)$ are both points on the parabola, the x-coordinate of the vertex is 800. And since the highest point it reaches is 3200, the y-coordinate of the vertex is 3200. Thus the vertex is $(800, 3200)$, and the equation is of the form $(x - 800)^2 = 4p(y - 3200)$. Substituting the point $(0, 0)$, we get $(0 - 800)^2 = 4p(0 - 3200)$ $\Leftrightarrow$ $640000 = -12800p$ $\Leftrightarrow$ $p = -50$. So the equation is $(x - 800)^2 = 4(-50)(y - 3200)$ $\Leftrightarrow$ $(x - 800)^2 = -200(y - 3200)$.

40. (a) We assume that $(0, 1)$ is the focus closer to the vertex $(0, 0)$, as shown in the figure in the text. Then the center of the ellipse will be $(0, a)$ and $1 = a - c$. So $c = a - 1$ and $(a - 1)^2 = a^2 - b^2$ $\Leftrightarrow$ $a^2 - 2a + 1 = a^2 - b^2$ $\Leftrightarrow$ $b^2 = 2a - 1$. Thus the equation will be $\dfrac{x^2}{2a - 1} + \dfrac{(y - a)^2}{a^2} = 1$. If we choose $a = 4$, then we get $\dfrac{x^2}{7} + \dfrac{(y - 4)^2}{16} = 1$. If we choose $a = 2$, then we get $\dfrac{x^2}{3} + \dfrac{(y - 2)^2}{4} = 1$. (Answers will vary, depending on your choice of $a > 1$.)

(b) Since a vertex is at $(0, 0)$ and a focus is at $(0, 1)$, we must have $c - a = 1$ ($a > 0$), and the center of the hyperbola will be at $(0, -a)$. So $c = a + 1$ and $(a + 1)^2 = a^2 + b^2$ $\Leftrightarrow$ $a^2 + 2a + 1 = a^2 + b^2$ $\Leftrightarrow$ $b^2 = 2a + 1$. Thus the equation will be $\dfrac{(y + a)^2}{a^2} - \dfrac{x^2}{2a + 1} = 1$. If we let $a = 2$, then the equation is $\dfrac{(y + 2)^2}{4} - \dfrac{x^2}{3} = 1$. If we let $a = 5$, then the equation is $\dfrac{(y + 5)^2}{25} - \dfrac{x^2}{11} = 1$. (Answers will vary, depending on your choice of a.)

(c) Since the vertex is at $(0, 0)$ and the focus is at $(0, 1)$, we must have $p = 1$. So $(x - 0)^2 = 4(1)(y - 0)$ $\Leftrightarrow$ $x^2 = 4y$, and there are no other possible parabolas.

(d) (Answers will vary, depending on your choice of a in parts (a) and (b).)

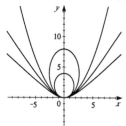

(e) The ellipses will be inside the parabola and the hyperbolas will be outside the parabola.

Review Exercises for Chapter 8

2. $x = \frac{1}{12}y^2 \iff y^2 = 12x$. This is a parabola that has
$4p = 12 \iff p = 3$.
Vertex: $(0,0)$; focus: $(3,0)$; directrix: $x = -3$.

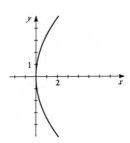

4. $2x - y^2 = 0 \iff y^2 = 2x$. This is a parabola that has
$4p = 2 \iff p = \frac{1}{2}$.
Vertex: $(0,0)$; focus: $(\frac{1}{2},0)$; directrix: $x = -\frac{1}{2}$.

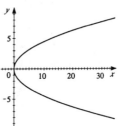

6. $2x^2 + 6x + 5y + 10 = 0 \iff 2(x^2 + 3x + \frac{9}{4}) + 5y + 10 = \frac{9}{2}$
$\iff 2(x + \frac{3}{2})^2 = -5y - \frac{11}{2} \iff 2(x + \frac{3}{2})^2 = -5(y + \frac{11}{10})$
$\iff (x + \frac{3}{2})^2 = -\frac{5}{2}(y + \frac{11}{10})$. This is a parabola that has
$4p = -\frac{5}{2} \iff p = -\frac{5}{8}$.
Vertex: $(-\frac{3}{2}, -\frac{11}{10})$; focus: $(-\frac{3}{2}, -\frac{11}{10} - \frac{5}{8}) = (-\frac{3}{2}, -\frac{44+25}{40})$
$= (-\frac{3}{2}, -\frac{69}{20})$; directrix: $y = -\frac{11}{10} + \frac{5}{8} = \frac{-44+25}{40} = -\frac{19}{40}$

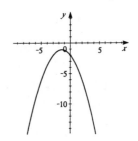

8. $x^2 = 3(x + y) \iff x^2 - 3x = 3y \iff$
$x^2 - 3x + \frac{9}{4} = 3y + \frac{9}{4} \iff (x - \frac{3}{2})^2 = 3(y + \frac{3}{4})$. This is a
parabola that has $4p = 3 \iff p = \frac{3}{4}$.
Vertex: $(\frac{3}{2}, -\frac{3}{4})$; focus: $(\frac{3}{2}, -\frac{3}{4} + \frac{3}{4}) = (\frac{3}{2}, 0)$;
directrix: $y = -\frac{3}{4} - \frac{3}{4} = -\frac{3}{2}$.

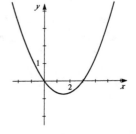

10. $\dfrac{x^2}{49} + \dfrac{y^2}{9} = 1$. This is an ellipse with $a = 7$, $b = 3$, and
$c = \sqrt{49 - 9} = 2\sqrt{10}$.
Center: $(0,0)$; vertices: $(\pm 7, 0)$; foci: $\left(\pm 2\sqrt{10}, 0\right)$;
length of the major axis: $2a = 14$;
length of the minor axis: $2b = 6$.

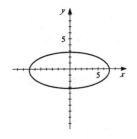

12. $9x^2 + 4y^2 = 1$ $\Leftrightarrow$ $\dfrac{x^2}{1/9} + \dfrac{y^2}{1/4} = 1$. This is an ellipse with

$a = \frac{1}{2}, b = \frac{1}{3}$, and $c = \sqrt{\frac{1}{4} - \frac{1}{9}} = \frac{1}{6}\sqrt{5}$.

Center: $(0,0)$; vertices: $\left(0, \pm\frac{1}{2}\right)$; foci: $\left(0, \pm\frac{1}{6}\sqrt{5}\right)$;

length of the major axis: $2a = 1$;

length of the minor axis: $2b = \frac{2}{3}$.

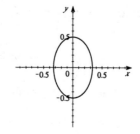

14. $\dfrac{(x-2)^2}{25} + \dfrac{(y+3)^2}{16} = 1$. This is an ellipse with $a = 5$,

$b = 4$, and $c = \sqrt{25 - 16} = 3$.

Center: $(2, -3)$; vertices: $(2 \pm 5, -3)$ which are $(-3, -3)$ and

$(7, -3)$; foci: $(2 \pm 3, -3)$ which are $(-1, -3)$ and $(5, -3)$; length

of the major axis: $2a = 10$; length of the minor axis: $2b = 8$.

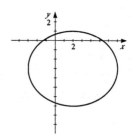

16. $2x^2 + y^2 = 2 + 4(x - y)$ $\Leftrightarrow$ $2x^2 - 4x + y^2 + 4y = 2$ $\Leftrightarrow$

$2(x^2 - 2x + 1) + (y^2 + 4y + 4) = 2 + 2 + 4$ $\Leftrightarrow$

$2(x - 1)^2 + (y + 2)^2 = 8$ $\Leftrightarrow$ $\dfrac{(x-1)^2}{4} + \dfrac{(y+2)^2}{8} = 1$.

This is an ellipse with $a = 2\sqrt{2}$, $b = 2$, and $c = \sqrt{8 - 4} = 2$.

Center: $(1, -2)$; vertices: $(1, -2 \pm 2\sqrt{2})$; foci: $(1, -2 \pm 2)$, so

the foci are $(1, 0)$ and $(1, -4)$; length of the major axis: $2a = 4\sqrt{2}$;

length of the minor axis: $2b = 4$.

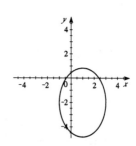

18. $\dfrac{x^2}{49} - \dfrac{y^2}{32} = 1$. This is a hyperbola that has $a = 7$, $b = 4\sqrt{2}$,

and $c = \sqrt{49 + 32} = 9$.

Center: $(0,0)$; vertices: $(\pm 7, 0)$; foci: $(\pm 9, 0)$;

asymptotes: $y = \pm\frac{4\sqrt{2}}{7}x$.

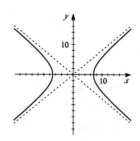

20. $x^2 - 4y^2 + 16 = 0$ $\Leftrightarrow$ $4y^2 - x^2 = 16$ $\Leftrightarrow$ $\dfrac{y^2}{4} - \dfrac{x^2}{16} = 1$.

This is a hyperbola that has $a = 2$, $b = 4$, and

$c = \sqrt{4 + 16} = 2\sqrt{5}$.

Center: $(0,0)$; vertices: $(0, \pm 2)$; foci: $(0, \pm 2\sqrt{5})$;

asymptotes: $y = \pm\frac{2}{4}x$ $\Leftrightarrow$ $y = \pm\frac{1}{2}x$.

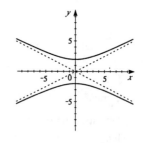

22. $\dfrac{(x-2)^2}{8} - \dfrac{(y+2)^2}{8} = 1$. This is a hyperbola that has $a = 2\sqrt{2}$,

$b = 2\sqrt{2}$ and $c = \sqrt{8+8} = 4$.

Center: $(2,-2)$; vertices: $\left(2 \pm 2\sqrt{2}, -2\right)$;

foci: $(2 \pm 4, -2)$ which are $(-2,-2)$ and $(6,-2)$;

asymptotes: $y + 2 = \pm(x-2) \quad \Leftrightarrow \quad y = -x$ and $y = x - 4$.

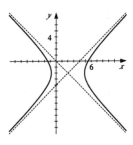

24. $y^2 = x^2 + 6y \quad \Leftrightarrow \quad (y^2 - 6y + 9) - x^2 = 9 \quad \Leftrightarrow$

$(y-3)^2 - x^2 = 9 \quad \Leftrightarrow \quad \dfrac{(y-3)^2}{9} - \dfrac{x^2}{9} = 1$. This is a

hyperbola that has $a = 3$, $b = 3$, and $c = \sqrt{9+9} = 3\sqrt{2}$.
Center: $(0,3)$; vertices: $(0, 3 \pm 3)$, so the vertices are $(0,6)$ and

$(0,0)$; foci: $(0, 3 \pm 3\sqrt{2})$; asymptotes: $y - 3 = \pm x \quad \Leftrightarrow$

$y = x + 3$ and $y = -x + 3$.

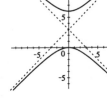

26. This is an ellipse with the center at $(0,0)$, $a = 12$, and $b = 5$. The equation is then $\dfrac{x^2}{12^2} + \dfrac{y^2}{5^2} = 1$

$\Leftrightarrow \quad \dfrac{x^2}{144} + \dfrac{y^2}{25} = 1$.

28. This is a parabola that opens to the left with its vertex at $(4,4)$, so its equation is of the form
$(y-4)^2 = 4p(x-4)$ with $p < 0$. Since $(0,0)$ is a point on this hyperbola, we must have
$(0-4)^2 = 4p(0-4) \quad \Leftrightarrow \quad 16 = -16p \quad \Leftrightarrow \quad p = -1$. Thus the equation is
$(y-4)^2 = -4(x-4)$.

30. From the graph, the center is at $(1,0)$, and the vertices are at $(0,0)$ and $(2,0)$. Since a is the
distance form the center to a vertex, $a = 1$. From the graph, the slope of one of the asymptotes is
$\dfrac{b}{a} = 1 \quad \Leftrightarrow \quad b = 1$. Thus the equation of the hyperbola is $(x-1)^2 - y^2 = 1$.

32. $\dfrac{x^2}{12} + \dfrac{y^2}{144} = \dfrac{y}{12} \quad \Leftrightarrow \quad 12x^2 + y^2 - 12y = 0 \quad \Leftrightarrow$

$12x^2 + (y^2 - 12y + 36) = 36 \quad \Leftrightarrow \quad \dfrac{x^2}{3} + \dfrac{(y-6)^2}{36} = 1$. This is an

ellipse that has $a = 6$, $b = \sqrt{3}$, and $c = \sqrt{36-3} = \sqrt{33}$.

Vertices: $(0, 6 \pm 6)$, so the vertices are $(0,0)$ and $(0,12)$; foci:
$(0, 6 \pm \sqrt{33})$.

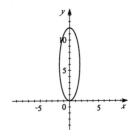

34. $x^2 + 6x = 9y^2 \quad \Leftrightarrow \quad (x^2 + 6x + 9) - 9y^2 = 9 \quad \Leftrightarrow$

$\dfrac{(x+3)^2}{9} - y^2 = 1$. This is a hyperbola that has $a = 3$, $b = 1$, and

$c = \sqrt{9+1} = \sqrt{10}$.

Vertices: $(-3 \pm 3, 0)$, so the vertices are $(-6,0)$ and $(0,0)$; foci:
$(-3 \pm \sqrt{10}, 0)$.

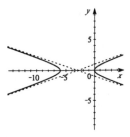

36. $3x^2 - 6(x+y) = 10 \quad \Leftrightarrow \quad 3x^2 - 6x = 6y + 10 \quad \Leftrightarrow$
$3(x^2 - 2x + 1) = 6y + 10 + 3 \quad \Leftrightarrow \quad 3(x-1)^2 = 6y + 13 \quad \Leftrightarrow$
$3(x-1)^2 = 6\left(y + \frac{13}{6}\right) \quad \Leftrightarrow \quad (x-1)^2 = 2\left(y + \frac{13}{6}\right)$. This is a
parabola that has $4p = 2 \quad \Leftrightarrow \quad p = \frac{1}{2}$.

Vertex: $\left(1, -\frac{13}{6}\right)$; focus: $\left(1, -\frac{13}{6} + \frac{1}{2}\right) = \left(1, -\frac{10}{6}\right) = \left(1, -\frac{5}{3}\right)$.

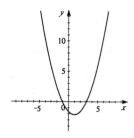

38. $2x^2 + 4 = 4x + y^2 \quad \Leftrightarrow \quad y^2 - 2x^2 + 4x = 4 \quad \Leftrightarrow$
$y^2 - 2(x^2 - 2x + 1) = 4 - 2 \quad \Leftrightarrow \quad y^2 - 2(x-1)^2 = 2 \quad \Leftrightarrow$
$\dfrac{y^2}{2} - (x-1)^2 = 1$. This is a hyperbola that has $a = \sqrt{2}$, $b = 1$, and
$c = \sqrt{2+1} = \sqrt{3}$.

Vertices: $(1, \pm\sqrt{2})$; foci: $(1, \pm\sqrt{3})$.

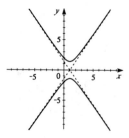

40. $36x^2 - 4y^2 - 36x - 8y = 31 \quad \Leftrightarrow \quad 36(x^2 - x) - 4(y^2 + 2y) = 31$
$\Leftrightarrow \quad 36\left(x^2 - x + \frac{1}{4}\right) - 4(y^2 + 2y + 1) = 31 + 9 - 4 \quad \Leftrightarrow$
$36\left(x - \frac{1}{2}\right)^2 - 4(y+1)^2 = 36 \quad \Leftrightarrow \quad \left(x - \frac{1}{2}\right)^2 - \dfrac{(y+1)^2}{9} = 1$.

This is a hyperbola that has $a = 1$, $b = 3$, and $c = \sqrt{1+9} = \sqrt{10}$.

Vertices: $\left(\frac{1}{2} \pm 1, -1\right)$, so the vertices are $\left(-\frac{1}{2}, -1\right)$ and $\left(\frac{3}{2}, -1\right)$;
foci: $\left(\frac{1}{2} \pm \sqrt{10}, -1\right)$.

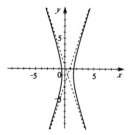

42. $x^2 + 4y^2 = 4x + 8 \quad \Leftrightarrow \quad (x^2 - 4x + 4) + 4y^2 = 8 + 4 \quad \Leftrightarrow$
$(x-2)^2 + 4y^2 = 12 \quad \Leftrightarrow \quad \dfrac{(x-2)^2}{12} + \dfrac{y^2}{3} = 1$. This is an ellipse
that has $a = 2\sqrt{3}$, $b = \sqrt{3}$, and $c = \sqrt{12 - 3} = 3$.

Vertices: $(2 \pm 2\sqrt{3}, 0)$; foci: $(2 \pm 3, 0)$, so the foci are $(-1, 0)$ and
$(5, 0)$.

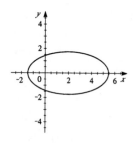

44. The ellipse has center $C(0, 4)$, foci $F_1(0, 0)$ and $F_2(0, 8)$, and major axis of length 10. Then
$2c = 8 - 0 \quad \Leftrightarrow \quad c = 4$. Also, since the length of the major axis is 10, $2a = 10 \quad \Leftrightarrow \quad a = 5$.
Therefore, $b^2 = a^2 - c^2 = 25 - 16 = 9$. Since the foci are on the y-axis, the vertices are on the y-
axis, and the equation of the ellipse is $\dfrac{x^2}{9} + \dfrac{(y-4)^2}{25} = 1$.

46. The hyperbola has center $C(2, 4)$, foci $F_1(2, 7)$ and $F_2(2, 1)$, and vertices $V_1(2, 6)$ and $V_2(2, 2)$.
Thus, $2a = 6 - 2 = 4 \quad \Leftrightarrow \quad a = 2$. Also, $2c = 7 - 1 = 6 \quad \Leftrightarrow \quad c = 3$. So $b^2 = 9 - 4 = 5$.
Since the hyperbola has center $C(2, 4)$, its equation is $\dfrac{(y-4)^2}{4} - \dfrac{(x-2)^2}{5} = 1$.

48. The parabola has vertex $V(5,5)$ and directrix the y-axis. Therefore, $-p = 0 - 5 \iff p = 5$ $\iff 4p = 20$. Since the parabola opens to the right, its equation is $(y-5)^2 = 20(x-5)$.

50. The parabola has vertex $V(-1,0)$, horizontal axis of symmetry, and crosses the y-axis where $y = 2$. Since the parabola has a horizontal axis of symmetry and $V(-1,0)$, its equation is of the form $y^2 = 4p(x+1)$. Also, since the parabola crosses the y-axis where $y = 2$, it passes through the point $(0,2)$. Substituting this point gives $(2)^2 = 4p(0+1) \iff 4p = 4$. Therefore, the equation of the parabola is $y^2 = 4(x+1)$.

52. We sketch the LORAN station on the y-axis and place the x-axis halfway between them as suggested in the exercise. This gives us the general form $\dfrac{y^2}{a^2} - \dfrac{x^2}{b^2} = 1$. Since the ship is 80 miles closer to A than to B we have $2a = 80 \iff a = 40$. Since the foci are $(0, \pm 150)$, we have $c = 150$. Thus $b^2 = c^2 - a^2 = 150^2 - 40^2 = 20900$. So this places the ship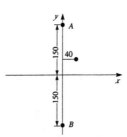

on the hyperbola given by the equation $\dfrac{y^2}{1600} - \dfrac{x^2}{20900} = 1$. When $x = 40$, we get

$\dfrac{y^2}{1600} - \dfrac{1600}{20900} = 1 \iff \dfrac{y^2}{1600} = \dfrac{225}{209} \iff y = 41.5$. (Note that $y > 0$, since A is on the positive y-axis.) Thus the ship's position is $(40, 41.5)$.

54. (a) The graphs of $y = kx^2$ for $k = \dfrac{1}{2}$, 1, 2, and 4 are shown in the figure.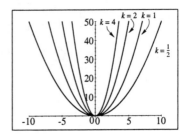

(b) $y = kx^2 \iff x^2 = \frac{1}{k}y = 4\left(\frac{1}{4k}\right)y$. Thus the foci are $\left(0, \frac{1}{4k}\right)$.

(c) As k increases, the focus gets closer to the vertex.

Focus on Modeling

2. (a) Let c_1 be the length of the bar and c_2 be the length of the string. So $c_1 < c_2$. Now
         $d(F_1, P) + d(P, A) = c_1$ and $d(F_2, P) + d(P, A) = c_2$. Subtracting we have
         $d(F_1, P) + d(P, A) - [d(F_2, P) + d(P, A)] = c_1 - c_2$. That is,
         $d(F_2, P) - d(F_2, P) = c_1 - c_2$. Since c_1 and c_2 are constants, so is their difference $c_1 - c_2$.
         Thus the curve satisfies the geometric definition of a hyperbola.

    (b) Place the pivot point at F_2 and the end of the string at F_1.

4. No.

6. The vertices of the hyperbolic cross sections get closer to each other.

8. In $\dfrac{x^2}{4} + y^2 = 1$ we have $a = 2$, $b = 1$.

    (a)

x	y
± 0	1
± 0.5	$\frac{\sqrt{15}}{4} \approx \pm 0.96825$
± 1	$\frac{\sqrt{3}}{2} \approx \pm 0.86603$
± 1.5	$\frac{\sqrt{7}}{4} \approx \pm 0.66144$
± 1.8	$\frac{\sqrt{19}}{10} \approx \pm 0.43589$

    (b) Using $y = \left(\dfrac{b^2 c}{a^2 d} \right) x + \dfrac{b^2}{d}$ with $a = 2$ and $b = 1$ we have $y = \dfrac{c}{4d} x + \dfrac{1}{d}$. Thus the equations of
         the envelope are:

(c, d)	$y = \dfrac{c}{4d} x + \dfrac{1}{d}$
$(0, 1)$	$y = 1$
$(\pm 0.5, 0.96825)$	$y = \pm 0.12910 x + 1.03280$
$(\pm 1, 0.86603)$	$y = \pm 0.28868 x + 1.15470$
$(\pm 1.5, 0.66144)$	$y = \pm 0.56695 x + 1.51186$
$(\pm 1.8, 0.43589\ )$	$y = \pm 1.03237 x + 2.29416$

    (c)

    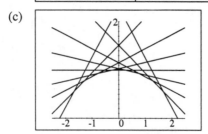

Chapter Nine

Exercises 9.1

2. $a_n = 2n + 3$. Then $a_1 = 2(1) + 3 = 5$; $a_2 = 2(2) + 3 = 7$; $a_3 = 2(3) + 3 = 9$; $a_4 = 2(4) + 3 = 11$; and $a_{100} = 2(100) + 3 = 203$.

4. $a_n = n^2 + 1$. Then $a_1 = (1)^2 + 1 = 2$; $a_2 = (2)^2 + 1 = 5$; $a_3 = (3)^2 + 1 = 10$; $a_4 = (4)^2 + 1 = 17$; and $a_{100} = (100)^2 + 1 = 10{,}001$.

6. $a_n = \dfrac{1}{n^2}$. Then $a_1 = \dfrac{1}{(1)^2} = 1$; $a_2 = \dfrac{1}{(2)^2} = \frac{1}{4}$; $a_3 = \dfrac{1}{(3)^2} = \frac{1}{9}$; $a_4 = \dfrac{1}{(4)^2} = \frac{1}{16}$; and

    $a_{100} = \dfrac{1}{(100)^2} = \frac{1}{10{,}000}$.

8. $a_n = \dfrac{(-1)^{n+1} n}{n+1}$. Then $a_1 = \dfrac{(-1)^2 \cdot 1}{1+1} = \frac{1}{2}$; $a_2 = \dfrac{(-1)^3 \cdot 2}{2+1} = -\frac{2}{3}$; $a_3 = \dfrac{(-1)^4 \cdot 3}{3+1} = \frac{3}{4}$;

    $a_4 = \dfrac{(-1)^5 \cdot 4}{4+1} = -\frac{4}{5}$; and $a_{100} = \dfrac{(-1)^{101} \cdot 100}{101} = -\frac{100}{101}$.

10. $a_n = 3$. Then $a_1 = 3$; $a_2 = 3$; $a_3 = 3$; $a_4 = 3$; and $a_{100} = 3$.

12. $a_n = \dfrac{a_{n-1}}{2}$ and $a_1 = -8$. Then $a_2 = \frac{-8}{2} = -4$; $a_3 = \frac{-4}{2} = -2$; $a_4 = \frac{-2}{2} = -1$; and

    $a_5 = \frac{-1}{2} = -\frac{1}{2}$.

14. $a_n = \dfrac{1}{1 + a_{n-1}}$ and $a_1 = 1$. Then $a_2 = \dfrac{1}{1 + (1)} = \frac{1}{2}$; $a_3 = \dfrac{1}{1 + \left(\frac{1}{2}\right)} = \frac{2}{3}$; $a_4 = \dfrac{1}{1 + \left(\frac{2}{3}\right)} = \frac{3}{5}$; and

    $a_5 = \dfrac{1}{1 + \left(\frac{3}{5}\right)} = \frac{5}{8}$.

16. $a_n = a_{n-1} + a_{n-2} + a_{n-3}$ and $a_1 = 1$; $a_2 = 1$; and $a_3 = 1$. Then $a_4 = 1 + 1 + 1 = 3$ and $a_5 = 3 + 1 + 1 = 5$.

18. (a) $a_1 = 2$ $a_2 = 6$ (b)
    $a_3 = 12$ $a_4 = 20$
    $a_5 = 30$ $a_6 = 42$
    $a_7 = 56$ $a_8 = 72$
    $a_9 = 90$ $a_{10} = 110$

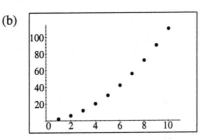

20. (a) $a_1 = 6$ $a_2 = 2$ (b)

       $a_3 = 6$ $a_4 = 2$

       $a_5 = 6$ $a_6 = 2$

       $a_7 = 6$ $a_8 = 2$

       $a_9 = 6$ $a_{10} = 2$

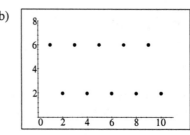

22. (a) $a_1 = 1$ $a_2 = 3$ (b)

       $a_3 = 2$ $a_4 = -1$

       $a_5 = -3$ $a_6 = -2$

       $a_7 = 1$ $a_8 = 3$

       $a_9 = 2$ $a_{10} = -1$

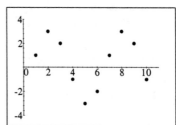

24. $-\frac{1}{3}, \frac{1}{9}, -\frac{1}{27}, \frac{1}{81}, \ldots$. The denominators are all powers of 3, and the terms alternate in sign. Thus

$$a_1 = \frac{(-1)^1}{3^1}, \; a_2 = \frac{(-1)^2}{3^2}, \; a_3 = \frac{(-1)^3}{3^3}, \; a_4 = \frac{(-1)^4}{3^4}, \ldots . \text{ So } a_n = \frac{(-1)^n}{3^n}.$$

26. $5, -25, 125, -625, \ldots$. These terms are powers of 5, and the terms alternate in sign. So

$a_1 = (-1)^2 \cdot 5^1, \; a_2 = (-1)^3 \cdot 5^2, \; a_3 = (-1)^4 \cdot 5^3, \; a_4 = (-1)^5 \cdot 5^4, \ldots .$ Thus $a_n = (-1)^{n+1} \cdot 5^n.$

28. $\frac{3}{4}, \frac{4}{5}, \frac{5}{6}, \frac{6}{7}, \ldots$. Both the numerator and the denominator increase by 1, so $a_1 = \frac{1+2}{1+3}, \; a_2 = \frac{2+2}{2+3},$

$a_3 = \frac{3+2}{3+3}, \; a_4 = \frac{4+2}{4+3}, \ldots .$ Thus $a_n = \dfrac{n+2}{n+3}.$

30. $1, \frac{1}{2}, 3, \frac{1}{4}, 5, \frac{1}{6}, \ldots$. So $a_1 = 1, \; a_2 = 2^{-1}, \; a_3 = 3^1, \; a_4 = 4^{-1}, \ldots .$ Thus $a_n = n^{(-1)^{n+1}}.$

32. $a_1 = 1^2, \; a_2 = 2^2, \; a_3 = 3^2, \; a_4 = 4^2, \ldots$. Therefore, $a_n = n^2.$ So $S_1 = 1^2 = 1;\; S_2 = 1 + 2^2 = 5;$
$S_3 = 5 + 3^2 = 5 + 9 = 14;\; S_4 = 14 + 4^2 = 14 + 16 = 30;\; S_5 = 30 + 5^2 = 30 + 25 = 55;$ and
$S_6 = 55 + 6^2 = 55 + 36 = 91.$

34. $a_1 = -1, \; a_2 = 1, \; a_3 = -1, \; a_4 = 1, \ldots$. Therefore, $a_n = (-1)^n.$ So $S_1 = -1;\; S_2 = -1 + 1 = 0;$
$S_3 = 0 - 1 = -1;\; S_4 = -1 + 1 = 0;\; S_5 = 0 - 1 = -1;$ and $S_6 = -1 + 1 = 0.$

36. $a_n = \dfrac{1}{n+1} - \dfrac{1}{n+2}.$ So $S_1 = \frac{1}{2} - \frac{1}{3};\; S_2 = \left(\frac{1}{2} - \frac{1}{3}\right) + \left(\frac{1}{3} - \frac{1}{4}\right) = \frac{1}{2} + \left(-\frac{1}{3} + \frac{1}{3}\right) - \frac{1}{4} = \frac{1}{2} - \frac{1}{4};$

$S_3 = \left(\frac{1}{2} - \frac{1}{3}\right) + \left(\frac{1}{3} - \frac{1}{4}\right) + \left(\frac{1}{4} - \frac{1}{5}\right) = \frac{1}{2} + \left(-\frac{1}{3} + \frac{1}{3}\right) + \left(-\frac{1}{4} + \frac{1}{4}\right) - \frac{1}{5} = \frac{1}{2} - \frac{1}{5};$ and

$S_4 = \left(\frac{1}{2} - \frac{1}{3}\right) + \left(\frac{1}{3} - \frac{1}{4}\right) + \left(\frac{1}{4} - \frac{1}{5}\right) + \left(\frac{1}{5} - \frac{1}{6}\right)$

$\quad = \frac{1}{2} + \left(-\frac{1}{3} + \frac{1}{3}\right) + \left(-\frac{1}{4} + \frac{1}{4}\right) + \left(-\frac{1}{5} + \frac{1}{5}\right) - \frac{1}{6} = \frac{1}{2} - \frac{1}{6}.$ Therefore,

$S_n = \left(\frac{1}{2} - \frac{1}{3}\right) + \left(\frac{1}{3} - \frac{1}{4}\right) + \cdots + \left(\dfrac{1}{n+1} - \dfrac{1}{n+2}\right)$

$\quad = \frac{1}{2} + \left(-\frac{1}{3} + \frac{1}{3}\right) + \cdots + \left(-\dfrac{1}{n+1} + \dfrac{1}{n+1}\right) - \dfrac{1}{n+2} = \dfrac{1}{2} - \dfrac{1}{n+2}.$

38. $a_n = \log\left(\dfrac{k}{k+1}\right) = [\log k - \log(k+1)]$. So $S_1 = \log 1 - \log 2 = -\log 2$,

$S_2 = (-\log 2) + (\log 2 - \log 3) = -\log 3$;

$S_3 = -\log 2 + (\log 2 - \log 3) + (\log 3 - \log 4) = (-\log 2 + \log 2) + (-\log 3 + \log 3) - \log 4$

$\quad = -\log 4$; and

$S_4 = -\log 2 + (\log 2 - \log 3) + (\log 3 - \log 4) + (\log 4 - \log 5)$

$\quad = (-\log 2 + \log 2) + (-\log 3 + \log 3) + (-\log 4 + \log 4) - \log 5 = -\log 5.$

Therefore, $S_n = -\log(n+1)$.

40. $\displaystyle\sum_{k=1}^{4} k^2 = 1 + 2^2 + 3^2 + 4^2 = 1 + 4 + 9 + 16 = 30$

42. $\displaystyle\sum_{j=1}^{100}(-1)^j = (-1)^1 + (-1)^2 + (-1)^3 + (-1)^4 + \cdots + (-1)^{99} + (-1)^{100}$

$\quad = -1 + 1 - 1 + 1 - 1 + \cdots - 1 + 1 = 0$

44. $\displaystyle\sum_{i=4}^{12} 10 = 10 + 10 + 10 + 10 + 10 + 10 + 10 + 10 + 10 = 90$

46. $\displaystyle\sum_{i=1}^{3} i\, 2^i = 1 \cdot 2^1 + 2 \cdot 2^2 + 3 \cdot 2^3 = 2 + 8 + 24 = 34$

48. 15,550   50. 0.153146   52. -0.688172

54. $\displaystyle\sum_{i=0}^{4}\dfrac{2i-1}{2i+1} = \dfrac{2\cdot 0-1}{2\cdot 0+1} + \dfrac{2\cdot 1-1}{2\cdot 1+1} + \dfrac{2\cdot 2-1}{2\cdot 2+1} + \dfrac{2\cdot 3-1}{2\cdot 3+1} + \dfrac{2\cdot 4-1}{2\cdot 4+1} = -1 + \dfrac{1}{3} + \dfrac{3}{5} + \dfrac{5}{7} + \dfrac{7}{9}$

56. $\displaystyle\sum_{k=6}^{9} k(k+3) = 6 \cdot 9 + 7 \cdot 10 + 8 \cdot 11 + 9 \cdot 12 = 54 + 70 + 88 + 108$

58. $\displaystyle\sum_{j=1}^{n}(-1)^{j+1}x^j = (-1)^2 x + (-1)^3 x^2 + (-1)^4 x^3 + \cdots + (-1)^{n+1}x^n$

$\quad = x - x^2 + x^3 - \cdots + (-1)^{n+1}x^n$

60. $2 + 4 + 6 + \cdots + 20 = \displaystyle\sum_{k=1}^{10} 2k$

62. $\dfrac{1}{2\ln 2} - \dfrac{1}{3\ln 3} + \dfrac{1}{4\ln 4} - \dfrac{1}{5\ln 5} + \cdots + \dfrac{1}{100\ln 100} = \displaystyle\sum_{k=2}^{100}\dfrac{(-1)^k}{k\ln k}$

64. $\dfrac{\sqrt{1}}{1^2} + \dfrac{\sqrt{2}}{2^2} + \dfrac{\sqrt{3}}{3^2} + \cdots + \dfrac{\sqrt{n}}{n^2} = \displaystyle\sum_{k=1}^{n}\dfrac{\sqrt{k}}{k^2}$

66. $1 - 2x + 3x^2 - 4x^3 + 5x^4 + \cdots - 100x^{99} = \sum_{k=1}^{100} (-1)^{k+1} \cdot k \cdot x^{k-1}$

68. $G_1 = 1$; $G_2 = 1$; $G_3 = 2$; $G_4 = 3$; $G_5 = 5$; $G_6 = 8$; $G_7 = 13$; $G_8 = 21$; $G_9 = 34$; $G_{10} = 55$.

70. (a) $I_1 = 0$; $I_2 = \$0.50$; $I_3 = \$1.50$; $I_4 = \$3.01$; $I_5 = \$5.03$; $I_6 = \$7.55$.

     (b) Since 5 years is 60 months, we have $I_{60} = \$977.00$.

72. (a) The amount she owes at the end of the month, A_n, is the amount she owes at the beginning of the month, A_{n-1}, plus the interest, $0.005A_{n-1}$, minus the \$200 she repay her uncle. Thus $A_n = A_{n-1} + 0.005A_{n-1} - 200$ $\Leftrightarrow$ $A_n = 1.005A_{n-1} - 200$.

     (b) $A_1 = 9850$; $A_2 = 9699.25$; $A_3 = 9547.74$; $A_4 = 9395.48$; $A_5 = 9242.46$; $A_6 = 9088.67$. Thus she owes her uncle \$9088.67 after six months.

74. (a) Let n be the years since 2002. So $P_0 = \$240,000$. Each month the median price of a house in Orange County increases to 1.06 the previous months. Thus $P_n = 1.06^n P_0$.

     (b) Since $2010 - 2002 = 8$. Thus $P_8 = 382,523.53$. Thus in 2010, the median price of a house in Orange County should be \$382,524.

76. (a) Let n be the days after she starts, so $C_0 = 4$. And each day the concentration increases by 10%. Thus after n days the brine solution is $C_n = 1.10C_{n-1}$ with $C_0 = 4$..

     (b) $C_8 = 1.10C_7 = 1.10(1.10C_6) = \cdots = 1.10^8 \cdot C_0 = 1.10^8 \cdot 4 \approx 8.6$. Thus the brine solution is 8.6 g/L of salt.

78. (a) $a_n = n^2$. Then $a_1 = 1^2 = 1$, $a_2 = 2^2 = 4$, $a_3 = 3^2 = 9$, $a_4 = 4^2 = 16$.

     (b) $a_n = n^2 + (n-1)(n-2)(n-3)(n-4)$;
          $a_1 = 1^2 + (1-1)(1-2)(1-3)(1-4) = 1 + 0(-1)(-2)(-3) = 1$;
          $a_2 = 2^2 + (2-1)(2-2)(2-3)(2-4) = 4 + 1 \cdot 0(-1)(-2) = 4$;
          $a_3 = 3^2 + (3-1)(3-2)(3-3)(3-4) = 9 + 2 \cdot 1 \cdot 0(-1) = 9$;
          $a_4 = 4^2 + (4-1)(4-2)(4-3)(4-4) = 16 + 3 \cdot 2 \cdot 1 \cdot 0 = 16$.
          Hence, the sequences agree in the first four terms. However, for the second sequence,
          $a_5 = 5^2 + (5-1)(5-2)(5-3)(5-4) = 25 + 4 \cdot 3 \cdot 2 \cdot 1 = 49$, and for the first sequence,
          $a_5 = 5^2 = 25$, and thus the sequences disagree from the fifth term on.

     (c) $a_n = n^2 + (n-1)(n-2)(n-3)(n-4)(n-5)(n-6)$ agrees with $a_n = n^2$ in the first six terms only.

     (d) $a_n = 2^n$ and $b_n = 2^n + (n-1)(n-2)(n-3)(n-4)$.

80. $a_n = a_{n-a_{n-1}} + a_{n-a_{n-2}}$; $a_1 = 1$; and $a_2 = 1$. So
     $a_3 = a_{3-1} + a_{3-1} = a_2 + a_2 = 1 + 1 = 2$; $a_4 = a_{4-2} + a_{4-1} = a_2 + a_3 = 1 + 2 = 3$;
     $a_5 = a_{5-3} + a_{5-2} = a_2 + a_3 = 1 + 2 = 3$; $a_6 = a_{6-3} + a_{6-3} = a_3 + a_3 = 2 + 2 = 4$;
     $a_7 = a_{7-4} + a_{7-3} = a_3 + a_4 = 2 + 3 = 5$; $a_8 = a_{8-5} + a_{8-4} = a_3 + a_4 = 2 + 3 = 5$;
     $a_9 = a_{9-5} + a_{9-5} = a_4 + a_4 = 3 + 3 = 6$; $a_{10} = a_{10-6} + a_{10-5} = a_4 + a_5 = 3 + 3 = 6$.

     The definition of a_n depends on the <u>values</u> of certain preceding terms. So a_n is the sum of two preceding terms whose choice depends on the <u>values</u> of a_{n-1} and a_{n-2} (not on $n-1$ and $n-2$).

Exercises 9.2

2. (a) $a_1 = 3 - 4(1 - 1) = 3$

    $a_2 = 3 - 4(2 - 1) = -1$

    $a_3 = 3 - 4(3 - 1) = -5$

    $a_4 = 3 - 4(4 - 1) = -9$

    $a_5 = 3 - 4(5 - 1) = -13$

    (c)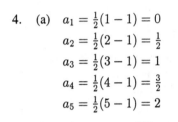

    (b) The common difference is -4.

4. (a) $a_1 = \frac{1}{2}(1 - 1) = 0$

    $a_2 = \frac{1}{2}(2 - 1) = \frac{1}{2}$

    $a_3 = \frac{1}{2}(3 - 1) = 1$

    $a_4 = \frac{1}{2}(4 - 1) = \frac{3}{2}$

    $a_5 = \frac{1}{2}(5 - 1) = 2$

    (b) The common difference is $\frac{1}{2}$.

6. $a = -6,\ d = 3,\ a_n = a + d(n - 1) = -6 + 3(n - 1)$. So $a_{10} = -6 + 3(10 - 1) = 21$.

8. $a = \sqrt{3},\ d = \sqrt{3},\ a_n = a + d(n - 1) = \sqrt{3} + \sqrt{3}(n - 1)$. So

    $a_{10} = \sqrt{3} + \sqrt{3}(10 - 1) = 10\sqrt{3}$.

10. Since $a_2 - a_1 = 6 - 3 = 3$ and $a_4 - a_3 = 13 - 9 = 4$, the terms of the sequence do not have a common difference. This sequence is not arithmetic.

12. $a_4 - a_3 = a_3 - a_2 = a_2 - a_1 = 2$. This sequence is arithmetic with the common difference 2.

14. $a_4 - a_3 = \ln 16 - \ln 8 = \ln\frac{16}{8} = \ln 2;\ a_3 - a_2 = \ln 8 - \ln 4 = \ln\frac{8}{4} = \ln 2;$

    $a_2 - a_1 = \ln 4 - \ln 2 = \ln\frac{4}{2} = \ln 2$. This sequence is arithmetic with the common difference $\ln 2$.

16. Since $a_4 - a_3 = \frac{1}{5} - \frac{1}{4} = -\frac{1}{20}$ and $a_3 - a_2 = \frac{1}{4} - \frac{1}{3} = -\frac{1}{12}$, the terms of the sequence do not have a common difference. This sequence is not arithmetic.

18. $a_1 = 4 + 2^1 = 6;\ a_2 = 4 + 2^2 = 8;\ a_3 = 4 + 2^3 = 12;\ a_4 = 4 + 2^4 = 20;\ a_5 = 4 + 2^5 = 36$. Since $a_4 - a_3 = 8$ and $a_3 - a_2 = 4$, the terms of the sequence do not have a common difference. This sequence is not arithmetic.

20. $a_1 = 1 + \frac{1}{2} = \frac{3}{2};\ a_2 = 1 + \frac{2}{2} = 2;\ a_3 = 1 + \frac{3}{2} = \frac{5}{2};\ a_4 = 1 + \frac{4}{2} = 3;\ a_5 = 1 + \frac{5}{2} = \frac{7}{2}$. This sequence is arithmetic, the common difference is $d = \frac{1}{2}$ and

    $a_n = 1 + \frac{n}{2} = 1 + \frac{1}{2}n - \frac{1}{2} + \frac{1}{2} = \frac{3}{2} + \frac{1}{2}(n - 1)$.

22. $a_1 = 3 + (-1)^1(1) = 2;\ a_2 = 3 + (-1)^2(2) = 5;\ a_3 = 3 + (-1)^3(3) = 0;\ a_4 = 3 + (-1)^4(4) = 7;$

    $a_5 = 3 + (-1)^5(5) = -2$. Since $a_4 - a_3 = 7$ and $a_3 - a_2 = -5$, the terms of the sequence do not have a common difference. This sequence is not arithmetic.

24. $1, 5, 9, 13, \ldots$. Then $d = a_2 - a_1 = 5 - 1 = 4$; $a_5 = a_4 + 4 = 13 + 4 = 17$; $a_n = 1 + 4(n-1)$; $a_{100} = 1 + 4(99) = 397$.

26. $11, 8, 5, 2, \ldots$. Then $d = a_2 - a_1 = 8 - 11 = -3$; $a_5 = a_4 - 3 = 2 - 3 = -1$; $a_n = 11 - 3(n-1)$; and $a_{100} = 11 - 3(99) = -286$.

28. $\frac{7}{6}, \frac{5}{3}, \frac{13}{6}, \frac{8}{3}, \ldots$. Then $d = a_2 - a_1 = \frac{10}{6} - \frac{7}{6} = \frac{1}{2}$; $a_5 = a_4 + \frac{1}{2} = \frac{8}{3} + \frac{1}{2} = \frac{19}{6}$; $a_n = \frac{7}{6} + \frac{1}{2}(n-1)$; $a_{100} = \frac{7}{6} + \frac{1}{2}(99) = \frac{7+297}{6} = \frac{152}{3}$.

30. $15, 12.3, 9.6, 6.9, \ldots$. Then $d = a_2 - a_1 = 12.3 - 15 = -2.7$; $a_5 = a_4 - 2.7 = 6.9 - 2.7 = 4.2$; $a_n = 15 - 2.7(n-1)$; and $a_{100} = 15 - 2.7(99) = -252.3$.

32. $-t, -t+3, -t+6, -t+9, \ldots$. Then $d = a_2 - a_1 = (-t+3) - (-t) = 3$; $a_5 = a_4 + 3 = -t + 9 + 3 = -t + 12$; $a_n = -t + 3(n-1)$; and $a_{100} = -t + 3(99) = -t + 297$.

34. $a_{12} = 32$, $a_5 = 18$, and $a_n = a + d(n-1)$. Then $a_5 = a + 4d = 18 \quad \Leftrightarrow \quad a = 18 - 4d$. Substituting for a, we have $a_{12} = a + 11d = 32 \quad \Leftrightarrow \quad (18 - 4d) + 11d = 32 \quad \Leftrightarrow \quad 7d = 14 \quad \Leftrightarrow \quad d = 2$. Then we have $a = 18 - 4 \cdot 2 = 10$, and hence, the 20th term is $a_{20} = 10 + 19(2) = 48$.

36. $a_{20} = 101$, and $d = 3$. Note that $a_{20} = a + 19d = a + 19(3) = a + 57$. Since $a_{20} = 101$ we have $a + 57 = a_{20} = 101 \quad \Leftrightarrow \quad a = 44$. Hence the nth term is $a_n = 44 + 3(n-1) = 41 + 3n$.

38.   If 11,937 is a term of an arithmetic sequence with $a_1 = 1$ and common difference 4, then $11937 = 1 + 4(n-1)$ for some integer n. Solving for n, we have $11937 = 1 + 4(n-1) = -3 + 4n \quad \Leftrightarrow \quad 11940 = 4n \quad \Leftrightarrow \quad n = 2985$. Thus 11,937 is the 2985th term of this sequence.

40. $a = 3, d = 2, n = 12$. Then $S_{12} = \frac{12}{2}[2a + (12-1)d] = \frac{12}{2}[2 \cdot 3 + 11 \cdot 2] = 168$.

42. $a = 100, d = -5, n = 8$. Then $S_8 = \frac{8}{2}[2a + (8-1)d] = \frac{8}{2}[2 \cdot 100 - 7 \cdot 5] = 660$.

44. $a_2 = 8, a_5 = 9.5, n = 15$. Thus $a_2 = a + d = 8$ and $a_5 = a + 4d = 9.5$. Subtracting the first equation from the second gives $3d = 1.5 \quad \Leftrightarrow \quad d = 0.5$. Substituting for d in the first equation gives $a + 0.5 = 8 \quad \Leftrightarrow \quad a = 7.5$. Thus $S_{15} = \frac{15}{2}[2 \cdot 7.5 + 14 \cdot 0.5] = 165$.

46. $-3 + \left(-\frac{3}{2}\right) + 0 + \frac{3}{2} + 3 + \cdots + 30$ is a partial sum of an arithmetic sequence, where $a = -3$ and $d = \left(-\frac{3}{2}\right) - (-3) = \frac{3}{2}$. The last term is $30 = a_n = -3 + \frac{3}{2}(n-1)$, so $22 = n - 1 \quad \Leftrightarrow \quad n = 23$. So, the partial sum is $S_{23} = \frac{23}{2}(-3 + 30) = \frac{621}{2} = 310.5$.

48. $-10 + (-9.9) + (-9.8) + \cdots + (-0.1)$ is a partial sum of an arithmetic sequence, where $a = -10$ and $d = 0.1$. The last term is $-0.1 = a_n = -10 + 0.1(n-1)$, so $99 = n - 1 \quad \Leftrightarrow \quad n = 100$. So, the partial sum is $S_{100} = \frac{100}{2}(-10 - 0.1) = -505$.

50. $\displaystyle\sum_{n=0}^{20}(1 - 2n)$ is a partial sum of an arithmetic sequence, where $a = 1 - 2 \cdot 0 = 1$, $d = -2$, and the last term is $a_{21} = 1 - 2 \cdot 20 = -39$. So, the partial sum is $S_{21} = \frac{21}{2}(1 - 39) = -399$.

52. $P = 10^{1/10} \cdot 10^{2/10} \cdot 10^{3/10} \cdot \ldots \cdot 10^{19/10} = 10^{(1+2+3+\ldots+19)/10}$. Now, $1 + 2 + 3 + \cdots + 19$ is an arithmetic series with $a = 1$, $d = 1$, and $n = 19$. Thus,
$1 + 2 + 3 + \cdots + 19 = S_{19} = 19\frac{(1+19)}{2} = 190$, and so $P = 10^{190/10} = 10^{19}$.

54. The two original numbers are 3 and 5. Thus, the reciprocals are $\frac{1}{3}$ and $\frac{1}{5}$, and their average is
$\frac{1}{2}\left(\frac{1}{3} + \frac{1}{5}\right) = \frac{1}{2}\left(\frac{5}{15} + \frac{3}{15}\right) = \frac{4}{15}$. Therefore, the harmonic mean is $\frac{15}{4}$.

56. Since the sequence is arithmetic, $a_4 - a_1 = 3d = 16 - 1 = 15 \quad \Leftrightarrow \quad 3d = 15 \quad \Leftrightarrow \quad d = 5$.
Now since $S_n = \frac{n}{2}[2a + (n-1)d]$ we have $2356 = \frac{n}{2}[2(1) + (n-1)5] \quad \Leftrightarrow$
$4712 = n(2 + 5n - 5) = 5n^2 - 3n \quad \Leftrightarrow \quad 5n^2 - 3n - 4712 = 0$. Using the quadratic equation
we have $n = \dfrac{3 \pm \sqrt{3^2 - 4(5)(4712)}}{2(5)} = \dfrac{3 \pm \sqrt{94249}}{10} = \dfrac{3 \pm 307}{10}$. Since n is positive, we have
$n = \frac{310}{10} = 31$. Thus 31 terms must be added.

58. The number of poles in a layer can be viewed as an arithmetic sequence, where $a_1 = 25$ and the common difference is -1. The number of poles in the first 12 layers is
$S_{12} = \frac{12}{2}[2(25) + 11(-1)] = 6 \cdot 39 = 234$.

60. The number of cars that can park in a row can be viewed as an arithmetic sequence, where $a_1 = 20$ and the common difference is 2. Thus the number of cars that can park in the 21 rows is
$S_{21} = \frac{21}{2}[2(20) + 20(2)] = 10.5 \cdot 80 = 840$.

62. The sequence is $16, 48, 80, \ldots$. This is an arithmetic sequence with $a = 16$ and $d = 48 - 16 = 32$.
 (a) The total distance after 6 seconds is $S_6 = \frac{6}{2}(32 + 5 \cdot 32) = 3 \cdot 192 = 576$ ft.
 (b) The total distance after n seconds is $S_n = \frac{n}{2}[32 + 32(n-1)] = 16n^2$ ft.

64. (a) We want an arithmetic sequence with 4 terms, so let $a_1 = 10$ and $a_4 = 18$. Since the sequence is arithmetic, $a_4 - a_1 = 3d = 18 - 10 = 8 \quad \Leftrightarrow \quad 3d = 8 \quad \Leftrightarrow \quad d = \frac{8}{3}$. Therefore,
$a_2 = 10 + \frac{8}{3} = \frac{38}{3}$ and $a_3 = 10 + 2\left(\frac{8}{3}\right) = \frac{46}{3}$ are the two arithmetic means between 10 and 18.

 (b) We want an arithmetic sequence with 5 terms, so let $a_1 = 10$ and $a_5 = 18$. Since the sequence is arithmetic, $a_5 - a_1 = 4d = 18 - 10 = 8 \quad \Leftrightarrow \quad 4d = 8 \quad \Leftrightarrow \quad d = 2$. Therefore,
$a_2 = 10 + 2 = 12$, $a_3 = 10 + 2(2) = 14$, and $a_4 = 10 + 3(2) = 16$ are the three arithmetic means between 10 and 18.

 (c) We want an arithmetic sequence with 6 terms, with the starting dosage $a_1 = 100$ and the final dosage $a_6 = 300$. Since the sequence is arithmetic, $a_6 - a_1 = 5d = 300 - 100 = 200 \quad \Leftrightarrow$
$5d = 200 \quad \Leftrightarrow \quad d = 40$. Therefore, $a_2 = 140$, $a_3 = 180$, $a_4 = 220$, $a_5 = 260$, and
$a_6 = 300$. The patient should take 140 mg, then 180 mg, then 220 mg, then 260 mg, and finally arrive at 300 mg.

Exercises 9.3

2. (a) $a_1 = 3(-4)^0 = 3$

$a_2 = 3(-4)^1 = -12$

$a_3 = 3(-4)^2 = 48$

$a_4 = 3(-4)^3 = -192$

$a_5 = 3(-4)^4 = 768$

(b) The common ratio is -4.

(c)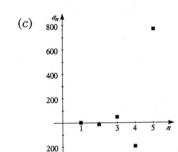

4. (a) $a_1 = 3^0 = 1$

$a_2 = 3^1 = 3$

$a_3 = 3^2 = 9$

$a_4 = 3^3 = 27$

$a_5 = 3^4 = 81$

(b) The common ratio is 3.

6. $a = -6, r = 3$. So $a_n = ar^{n-1} = -6(3)^{n-1}$ and $a_4 = -6 \cdot 3^3 = -162$.

8. $a = \sqrt{3}, r = \sqrt{3}$. So $a_n = ar^{n-1} = \sqrt{3}\left(\sqrt{3}\right)^{n-1} = \left(\sqrt{3}\right)^n$ and $a_4 = \sqrt{3} \cdot (\sqrt{3})^3 = 9$.

10. $\dfrac{a_2}{a_1} = \dfrac{6}{2} = 3$; $\dfrac{a_4}{a_3} = \dfrac{36}{18} = 2$. Since these ratios are not the same, this is not a geometric sequence.

12. $\dfrac{a_2}{a_1} = \dfrac{-9}{27} = -\dfrac{1}{3}$; $\dfrac{a_3}{a_2} = \dfrac{3}{-9} = -\dfrac{1}{3}$; $\dfrac{a_4}{a_3} = -\dfrac{1}{3}$. Since these ratios are the same, the sequence is geometric with the common ratio $-\dfrac{1}{3}$.

14. $\dfrac{a_2}{a_1} = \dfrac{e^4}{e^2} = e^2$; $\dfrac{a_3}{a_2} = \dfrac{e^6}{e^4} = e^2$; $\dfrac{a_4}{a_3} = \dfrac{e^8}{e^6} = e^2$. Since these ratios are the same, the sequence is geometric with the common ratio e^2.

16. $\dfrac{a_2}{a_1} = \dfrac{\frac{1}{4}}{\frac{1}{2}} = \dfrac{1}{2}$; $\dfrac{a_4}{a_3} = \dfrac{\frac{1}{8}}{\frac{1}{6}} = \dfrac{3}{4}$. Since these ratios are not the same, this is not a geometric sequence.

18. $a_1 = 4 + (3)^1 = 7$; $a_2 = 4 + (3)^2 = 13$; $a_3 = 4 + (3)^3 = 31$; $a_4 = 4 + (3)^4 = 85$;

$a_5 = 4 + (3)^5 = 247$. Since $\dfrac{a_2}{a_1} = \dfrac{13}{7}$ and $\dfrac{a_3}{a_2} = \dfrac{31}{13}$ are ratios different, this is not a geometric

sequence.

20. $a_1 = (-1)^1(2)^1 = -2$; $a_2 = (-1)^2(2)^2 = 4$; $a_3 = (-1)^3(2)^3 = -8$; $a_4 = (-1)^4(2)^4 = 16$;

$a_5 = (-1)^5(2)^5 = -32$. This sequence is geometric, the common ratio is $r = -2$ and

$a_n = a_1 r^{n-1} = -2(-2)^{n-1}$.

22. $a_1 = 1^1 = 1$; $a_2 = 2^2 = 4$; $a_3 = 3^3 = 54$; $a_4 = 4^4 = 256$; $a_5 = 5^5 = 3125$. Since $\dfrac{a_2}{a_1} = \dfrac{4}{1} = 4$ and $\dfrac{a_3}{a_2} = \dfrac{27}{4}$ these ratios are different, this is not a geometric sequence.

24. $7, \frac{14}{3}, \frac{28}{9}, \frac{56}{27}, \dots$. Then $r = \dfrac{a_2}{a_1} = \dfrac{\frac{14}{3}}{7} = \dfrac{2}{3}$; $a_5 = a_4 \cdot \frac{2}{3} = \frac{56}{27} \cdot \frac{2}{3} = \frac{112}{81}$; and $a_n = 7\left(\frac{2}{3}\right)^{n-1}$.

26. $1, \sqrt{2}, 2, 2\sqrt{2}, \dots$. Then $r = \dfrac{a_2}{a_1} = \dfrac{\sqrt{2}}{1} = \sqrt{2}$; $a_5 = a_4 \cdot \sqrt{2} = 2\sqrt{2} \cdot \sqrt{2} = 4$; and $a_n = \left(\sqrt{2}\right)^{n-1}$.

28. $-8, -2, -\frac{1}{2}, -\frac{1}{8}, \dots$. Then $r = \dfrac{a_2}{a_1} = \dfrac{-2}{-8} = \dfrac{1}{4}$; $a_5 = a_4 \cdot \frac{1}{4} = -\frac{1}{8} \cdot \frac{1}{4} = -\frac{1}{32}$; and $a_n = -8\left(\frac{1}{4}\right)^{n-1}$.

30. $t, \dfrac{t^2}{2}, \dfrac{t^3}{4}, \dfrac{t^4}{8}, \dots$. Then $r = \dfrac{a_2}{a_1} = \dfrac{\frac{t^2}{2}}{t} = \dfrac{t}{2}$; $a_5 = a_4 \cdot \frac{t}{2} = \frac{t^4}{8} \cdot \frac{t}{2} = \frac{t^5}{16}$; and $a_n = t\left(\frac{t}{2}\right)^{n-1}$.

32. $5, 5^{c+1}, 5^{2c+1}, 5^{3c+1}, \dots$. Then $r = \dfrac{a_2}{a_1} = \dfrac{5^{c+1}}{5} = 5^c$; $a_5 = a_4 \cdot 5^c = 5^{3c+1} \cdot 5^c = 5^{4c+1}$; and $a_n = 5(5^c)^{n-1} = 5 \cdot 5^{cn-c} = 5^{cn-c+1}$.

34. $a_1 = 3$, $a_3 = \frac{4}{3}$. Thus $r^2 = \dfrac{\frac{4}{3}}{3} = \dfrac{4}{9} \Leftrightarrow r = \pm\frac{2}{3}$. If $r = \frac{2}{3}$, then $a_5 = 3\left(\frac{2}{3}\right)^4 = \frac{16}{27}$, and if $r = -\frac{2}{3}$, then $a_5 = 3\left(-\frac{2}{3}\right)^4 = \frac{16}{27}$. Therefore, $a_5 = \frac{16}{27}$.

36. $r = \frac{3}{2}$, $a_5 = 1$. Then $1 = a_5 = a \cdot r^4 = a\left(\frac{3}{2}\right)^4 \Leftrightarrow 1 = a\left(\frac{3}{2}\right)^4 \Leftrightarrow a = \left(\frac{2}{3}\right)^4 = \frac{16}{81}$. Therefore, $a_1 = \frac{16}{81}$, $a_2 = \frac{16}{81}\left(\frac{3}{2}\right) = \frac{8}{27}$, and $a_3 = \frac{8}{27}\left(\frac{3}{2}\right) = \frac{4}{9}$.

38. $a_2 = 10$, $a_5 = 1250$. Thus, $\dfrac{1250}{10} = \dfrac{a_5}{a_2} = \dfrac{ar^4}{ar} = r^3 \Leftrightarrow r = 5$. Then $a_1 = \dfrac{a_2}{r} = \dfrac{10}{5} = 2$, so $a_n = 2 \cdot 5^{n-1}$. We seek n such that $a_n = 31{,}250 \Leftrightarrow 2 \cdot 5^{n-1} = 31{,}250 \Leftrightarrow 5^{n-1} = 15{,}625$ $\Leftrightarrow n - 1 = \log_5 15{,}625 = 6 \Leftrightarrow n = 7$. Therefore, 31,250 is the 7th term of the sequence.

40. $a = \frac{2}{3}$, $r = \frac{1}{3}$, $n = 4$. Then $S_4 = \left(\frac{2}{3}\right)\dfrac{1 - \left(\frac{1}{3}\right)^4}{1 - \frac{1}{3}} = \left(\frac{2}{3}\right)\dfrac{\frac{80}{81}}{\frac{2}{3}} = \dfrac{80}{81}$.

42. $a_2 = 0.12$, $a_5 = 0.00096$, $n = 4$. So, $r^3 = \dfrac{a_5}{a_2} = \dfrac{0.00096}{0.12} = 0.008 \Leftrightarrow r = 0.2$, and thus $a_1 = \dfrac{a_2}{r} = \dfrac{0.12}{0.2} = 0.6$. Therefore, $S_4 = (0.6)\dfrac{1 - (0.2)^4}{1 - 0.2} = 0.7488$.

44. $1 - \frac{1}{2} + \frac{1}{4} - \frac{1}{8} + \cdots - \frac{1}{512}$ is a partial sum of a geometric sequence, where $a = 1$ and $r = \dfrac{a_2}{a_1} = \dfrac{-\frac{1}{2}}{1} = -\frac{1}{2}$. The last term is a_n, where $\frac{-1}{512} = a_n = 1\left(-\frac{1}{2}\right)^{n-1}$, so $n = 10$. So, the partial

sum is $S_{10} = (1)\dfrac{1 - \left(-\frac{1}{2}\right)^{10}}{1 - \left(-\frac{1}{2}\right)} = \frac{341}{512}$.

46. $\displaystyle\sum_{j=0}^{5} 7\left(\frac{3}{2}\right)^{j}$ is a partial sum of a geometric sequence, where $a = 7$, $r = \frac{3}{2}$, and $n = 6$. So, the partial

sum is $S_6 = (7)\dfrac{1 - \left(\frac{3}{2}\right)^{6}}{1 - \frac{3}{2}} = (7)\dfrac{1 - \left(\frac{729}{64}\right)}{-\frac{1}{2}} = -14\left(1 - \frac{729}{64}\right) = \frac{4655}{32}$.

48. $1 - \frac{1}{2} + \frac{1}{4} - \frac{1}{8} + \cdots$ is an infinite geometric series with $a = 1$ and $r = -\frac{1}{2}$. Therefore, the sum of

the series is $S = \dfrac{1}{1 - \left(-\frac{1}{2}\right)} = \dfrac{1}{\frac{3}{2}} = \frac{2}{3}$.

50. $\frac{2}{5} + \frac{4}{25} + \frac{8}{125} + \cdots$ is an infinite geometric series with $a = \frac{2}{5}$ and $r = \frac{2}{5}$. Therefore, the sum of the

series is $S = \dfrac{\frac{2}{5}}{1 - \frac{2}{5}} = \dfrac{\frac{2}{5}}{\frac{3}{5}} = \frac{2}{3}$.

52. $3 - \frac{3}{2} + \frac{3}{4} - \frac{3}{8} + \cdots$ is an infinite geometric series with $a = 3$ and $r = -\frac{1}{2}$. Therefore, the sum of

the series is $S = \dfrac{3}{1 - \left(-\frac{1}{2}\right)} = 2$.

54. $\frac{1}{\sqrt{2}} + \frac{1}{2} + \frac{1}{2\sqrt{2}} + \frac{1}{4} + \cdots$ is an infinite geometric series with $a = \frac{1}{\sqrt{2}}$ and $r = \frac{1}{\sqrt{2}}$. Therefore, the

sum of the series is $S = \dfrac{\frac{1}{\sqrt{2}}}{1 - \frac{1}{\sqrt{2}}} = \dfrac{1}{\sqrt{2}-1} = \sqrt{2} + 1$.

56. $0.2535353\ldots = 0.2 + \frac{53}{1000} + \frac{53}{100,000} + \frac{53}{10,000,000} + \cdots$ is an infinite geometric series

(after the first term) with $a = \frac{53}{1000}$ and $r = \frac{1}{100}$. Thus $0.05353\ldots = \dfrac{\frac{53}{1000}}{1 - \frac{1}{100}} = \frac{53}{1000} \cdot \frac{100}{99} = \frac{53}{990}$,

and so $0.2535353\ldots = \frac{2}{10} + \frac{53}{990} = \frac{2 \cdot 99 + 53}{990} = \frac{251}{990}$.

58. $2.11252525\ldots = 2.11 + \frac{25}{10,000} + \frac{25}{1,000,000} + \frac{25}{100,000,000} + \cdots$ is an infinite geometric series

(after the first term) with $a = \frac{25}{10,000}$ and $r = \frac{1}{100}$. Thus $0.00252525\ldots = \dfrac{\frac{25}{10,000}}{1 - \frac{1}{100}} = \frac{25}{9900}$, and so

$2.11252525\ldots = \frac{211}{100} + \frac{25}{9900} = \frac{211 \cdot 99 + 25}{9900} = \frac{20,914}{9900} = \frac{10,457}{4950}$.

60. $0.123123123\ldots = \frac{123}{1000} + \frac{123}{1,000,000} + \frac{123}{1,000,000,000} + \cdots$ is an infinite geometric series with $a = \frac{123}{1000}$

and $r = \frac{1}{1000}$. Thus $0.123123123\ldots = \dfrac{\frac{123}{1000}}{1 - \frac{1}{1000}} = \frac{123}{999}$.

62. The sum is given by $(a + b) + (a^2 + 2b) + (a^3 + 3b) + \cdots + (a^{10} + 10b) =$

$(a + a^2 + a^3 + \cdots + a^{10}) + (b + 2b + 3b + \cdots + 10b) = a \cdot \dfrac{1 - a^9}{1 - a} + \dfrac{10}{2} \cdot (b + 10b)$

$= a \cdot \dfrac{1 - a^9}{1 - a} + 55b$.

64. Let a_n denote the number of ancestors a person has n generations back. Then $a_1 = 2$, $a_2 = 4$, $a_3 = 8, \ldots$. Since $\dfrac{4}{2} = \dfrac{8}{4} = 2$, etc...., this is a geometric sequence with $r = 2$. Therefore, $a_{15} = 2 \cdot 2^{14} = 2^{15} = 32{,}768$.

66. $a = 5000$, $r = 1.08$. After 1 hour, there are $5000 \cdot 1.08 = 5400$; after 2 hours, $5400 \cdot 1.08 = 5832$; after 3 hours, $5832 \cdot 1.08 = 6298.56$; after 4 hours, $6298.56 \cdot 1.08 = 6802.4448$; and after 5 hours, $6802.4448 \cdot 1.08 \approx 7347$ bacteria. After n hours, the number of bacteria is $a_n = 5000 \cdot (1.08)^n$.

68. Let a_{mc} denote the term of the geometric series that is the frequency of middle C. Then $a_{\mathrm{mc}} = 256$ and $a_{\mathrm{mc}+1} = 512$. Since this is a geometric sequence, $r = \frac{512}{256} = 2$, and so $a_{\mathrm{mc}-2} = \dfrac{a_{\mathrm{mc}}}{r^2}$
$= \dfrac{256}{2^2} = 64$.

70. We have a geometric sequence with $a = 1$ and $r = 2$. Then $S_n = (1)\dfrac{1-2^n}{1-2} = 2^n - 1$. At the end of 30 days, she will have $S_{30} = 2^{30} - 1 = 1{,}073{,}741{,}823$ cents $= \$10{,}737{,}418.23$. To become a billionaire, we want $2^n - 1 = 10^{11}$ or approximately $2^n = 10^{11}$. So $\log 2^n = \log 10^{11} \iff n = \frac{11}{\log 2} \approx 36.5$. Thus it will take 37 days.

72. (a) $\displaystyle\sum_{k=1}^{10} 50\left(\tfrac{1}{2}\right)^{k-1} = 50 \cdot \dfrac{1 - \left(\frac{1}{2}\right)^{10}}{1 - \frac{1}{2}} \approx 100(1 - 0.0009765) = 99.9023$ mg.

(b) $\displaystyle\sum_{k=1}^{\infty} 50\left(\tfrac{1}{2}\right)^{k-1} = \dfrac{50}{1 - \frac{1}{2}} = 100$ mg.

74. The time required for the ball to stop bouncing is $t = 1 + \frac{1}{\sqrt{2}} + \left(\frac{1}{\sqrt{2}}\right)^2 + \cdots$ which is an infinite geometric series with $a = 1$ and $r = \frac{1}{\sqrt{2}}$. The sum of this series is $t = \dfrac{1}{1 - \frac{1}{\sqrt{2}}} = \dfrac{\sqrt{2}}{\sqrt{2} - 1}$
$= \dfrac{\sqrt{2}}{\sqrt{2}-1} \cdot \dfrac{\sqrt{2}+1}{\sqrt{2}+1} = \dfrac{2+\sqrt{2}}{2-1} = 2 + \sqrt{2}$. Thus the time required for the ball to stop is $2 + \sqrt{2} \approx 3.41$ s.

76. Let A_n be the area of the disks of paper placed at the nth stage. Then $A_1 = \pi R^2$, $A_2 = 2 \cdot \pi\left(\frac{1}{2}R\right)^2 = \frac{\pi}{2}R^2$, $A_3 = 4 \cdot \pi\left(\frac{1}{4}R\right)^2 = \frac{\pi}{4}R^2, \ldots$. We see from this pattern that the total area is $A = \pi R^2 + \frac{1}{2}\pi R^2 + \frac{1}{4}\pi R^2 + \cdots$. Thus, the total area, A, is an infinite geometric series with $a_1 = \pi R^2$ and $r = \frac{1}{2}$. So, $A = \dfrac{\pi R^2}{1 - \frac{1}{2}} = 2\pi R^2$.

78. (a) $5, -3, 5, -3, \ldots$. Now $a_2 - a_1 = -3 - 5 = -8$, but $a_3 - a_2 = 5 - (-3) = 8$, and $\dfrac{a_2}{a_1} = \dfrac{-3}{5}$, but $\dfrac{a_3}{a_2} = \dfrac{5}{-3}$. Thus, the sequence is neither arithmetic nor geometric.

(b) $\frac{1}{3}, 1, \frac{5}{3}, \frac{7}{3}, \ldots$. Now $a_2 - a_1 = 1 - \frac{1}{3} = \frac{2}{3}$; $a_3 - a_2 = \frac{5}{3} - 1 = \frac{2}{3}$; and $a_4 - a_3 = \frac{7}{3} - \frac{5}{3} = \frac{2}{3}$. Therefore, the sequence is arithmetic with $d = \frac{2}{3}$ and $a_5 = \frac{7}{3} + \frac{2}{3} = 3$.

(c) $\sqrt{3}, 3, 3\sqrt{3}, 9, \ldots$. Now $\dfrac{a_2}{a_1} = \dfrac{3}{\sqrt{3}} = \sqrt{3}$; $\dfrac{a_3}{a_2} = \dfrac{3\sqrt{3}}{3} = \sqrt{3}$; and $\dfrac{a_4}{a_3} = \dfrac{9}{3\sqrt{3}} = \sqrt{3}$.
Therefore, the sequence is geometric with $r = \sqrt{3}$ and $a_5 = 9\sqrt{3}$.

(d) $1, -1, 1, -1, \ldots$. Now $\dfrac{a_2}{a_1} = \dfrac{-1}{1} = -1$; $\dfrac{a_3}{a_2} = \dfrac{1}{-1} = -1$; and $\dfrac{a_4}{a_3} = \dfrac{-1}{1} = -1$. Therefore, the
sequence is geometric with $r = -1$ and $a_5 = (-1)(-1) = 1$.

(e) $2, -1, \frac{1}{2}, 2, \ldots$. Now $a_2 - a_1 = -1 - 2 = -3$, but $a_3 - a_2 = \frac{1}{2} + 1 = \frac{3}{2}$, and so the
sequence is not arithmetic. Also, $\dfrac{a_2}{a_1} = \dfrac{-1}{2}$, but $\dfrac{a_4}{a_3} = \dfrac{2}{\frac{1}{2}} = 4$, and so the sequence is not
geometric. Thus, the sequence is neither arithmetic nor geometric.

(f) $x - 1, x, x + 1, x + 2, \ldots$. Now $a_2 - a_1 = x - (x - 1) = 1$; $a_3 - a_2 = (x + 1) - x = 1$;
and $a_4 - a_3 = (x + 2) - (x + 1) = 1$. Therefore, the sequence is arithmetic with $d = 1$ and
$a_5 = (x + 2) + 1 = x + 3$.

(g) $-3, -\frac{3}{2}, 0, \frac{3}{2}, \ldots$. Now $a_2 - a_1 = -\frac{3}{2} - (-3) = \frac{3}{2}$; $a_3 - a_2 = 0 - (-\frac{3}{2}) = \frac{3}{2}$; and
$a_4 - a_3 = \frac{3}{2} - 0 = \frac{3}{2}$. Therefore, the sequence is arithmetic with $d = \frac{3}{2}$, and $a_5 = \frac{3}{2} + \frac{3}{2} = 3$.

(h) $\sqrt{5}, \sqrt[3]{5}, \sqrt[6]{5}, 1, \ldots$. Now $a_2 - a_1 = \sqrt[3]{5} - \sqrt{5}$, but $a_3 - a_2 = \sqrt[6]{5} - \sqrt[3]{5}$. Thus the
sequence is not arithmetic. However, $\dfrac{a_2}{a_1} = \dfrac{\sqrt[3]{5}}{\sqrt{5}} = \dfrac{5^{1/3}}{5^{1/2}} = 5^{-1/6}$, $\dfrac{a_3}{a_2} = \dfrac{\sqrt[6]{5}}{\sqrt[3]{5}} = \dfrac{5^{1/6}}{5^{1/3}} = 5^{-1/6}$,

and $\dfrac{a_4}{a_3} = \dfrac{1}{\sqrt[6]{5}} = \dfrac{1}{5^{1/6}} = 5^{-1/6}$. Therefore, the sequence is geometric with $r = 5^{-1/6}$ and
$a_5 = 1 \cdot 5^{-1/6} = \dfrac{1}{\sqrt[6]{5}}$.

80. $a_1, a_2, a_3, \ldots$ is a geometric sequence with common ratio r. Thus $a_2 = a_1 r$, $a_3 = a_1 \cdot r^2, \ldots$,
$a_n = a_1 \cdot r^{n-1}$. Hence $\log a_2 = \log(a_1 r) = \log a_1 + \log r$, $\log a_3 = \log(a_1 \cdot r^2) = \log a_1 + \log(r^2)$
$= \log a_1 + 2\log r$, $\ldots$, $\log a_n = \log(a_1 \cdot r^{n-1}) = \log a_1 + \log(r^{n-1}) = \log a_1 + (n - 1)\log r$, and
so $\log a_1, \log a_2, \log a_3, \ldots$ is an arithmetic sequence with common difference $\log r$.

Exercises 9.4

2. $R = 500$, $n = 24$, $i = \frac{0.08}{12} \approx 0.0066667$. So $A_f = R\dfrac{(1+i)^n - 1}{i} = 500\dfrac{(1.0066667)^{24} - 1}{0.0066667}$
   $= \$12,966.59$.

4. $R = 500$, $n = 20$, $i = \frac{0.06}{2} = 0.03$. So $A_f = R\dfrac{(1+i)^n - 1}{i} = 500\dfrac{(1.03)^{20} - 1}{0.03} = \$13,435.19$.

6. $A_f = 5000$, $n = 4 \cdot 2 = 8$, $i = \frac{0.10}{4} = 0.025$. So $R = \dfrac{iA_f}{(1+i)^n - 1} = \dfrac{(0.025)(5000)}{(1.025)^8 - 1} = \$572.34$.

8. $R = 1000$, $n = 20$, $i = \frac{0.09}{2} = 0.045$. So $A_p = R\dfrac{1-(1+i)^{-n}}{i} = 1000\dfrac{1-(1.045)^{-20}}{0.045}$
   $= \$13,007.94$.

10. $A_p = 50,000$, $n = 10(2) = 20$, $i = \frac{0.08}{2} = 0.04$. So $R = \dfrac{iA_p}{1-(1+i)^{-n}} = \dfrac{(0.04)(50,000)}{1-(1.04)^{-20}}$
    $= \$3679.09$.

12. $A_p = 80,000$, $i = \frac{0.09}{12} = 0.0075$. Over a 30 year period, $n = 30(12) = 360$, and the monthly
    payment is $R = \dfrac{iA_p}{1-(1+i)^{-n}} = \dfrac{(0.0075)(80,000)}{1-(1.0075)^{-360}} = \$643.70$. Over a 15 year period,
    $n = 15 \cdot 12 = 180$, and so the monthly payment is $R = \dfrac{80,000 \cdot 0.0075}{1-(1.0075)^{-180}} = \$811.41$.

14. $R = 650$, $n = 12(30) = 360$, $i = \frac{0.09}{12} = 0.0075$. So $A_p = R\dfrac{1-(1+i)^{-n}}{i}$
    $= 650\dfrac{1-(1.0075)^{-360}}{0.0075} = \$80,783.21$. Therefore, the couple can afford a loan of $\$80,783.21$.

16. $R = 220$, $n = 12(3) = 36$, $i = \frac{0.08}{12} \approx 0.00667$. The amount borrowed is $A_p = R\dfrac{1-(1+i)^{-n}}{i}$
    $= 220\dfrac{1-(1.00667)^{-36}}{0.00667} = \$7,020.60$. So she purchased the car for
    $\$7,020.60 + \$2000 = \$9020.60$.

18. $A_p = \$12,500$, $R = \$420$, $n = 36$. We want to solve for the
    interest rate using the equation $R = \dfrac{iA_P}{1-(1+i)^{-n}}$. Let x be
    the interest rate, then $i = \dfrac{x}{12}$. So we can express R as a
    function of x as follows: $R(x) = \dfrac{\frac{x}{12} \cdot 12{,}500}{1-\left(1+\frac{x}{12}\right)^{-36}}$. We

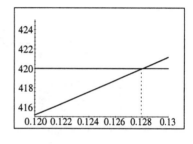

    graph $R(x)$ and $y = 420$ in the rectangle $[0.12, 0.13]$ by $[415, 425]$. The x-coordinate of the
    intersection is about 0.1280, which corresponds to an interest rate of 12.80%.

20. $A_p = \$2000 - \$200 = \$1800$, $R = \$88$, $n = 24$. We want
to solve for the interest rate using the equation

$R = \dfrac{iA_P}{1 - (1+i)^{-n}}$. Let x be the interest rate, then $i = \dfrac{x}{12}$.

So we can express R as a function of x as follows:

$$R(x) = \dfrac{\dfrac{x}{12} \cdot 1800}{1 - \left(1 + \dfrac{x}{12}\right)^{-24}}.$$

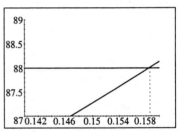

We graph $R(x)$ and $y = 88$ in the viewing rectangle $[0.14, 0.16]$ by $[87, 89]$. The x-coordinate of
the intersection is about 0.1584, which corresponds to an interest rate of 15.84%.

22. (a) The present value of the kth payment is $PV = R(1+i)^{-k} = \dfrac{R}{(1+i)^k}$. The present value of
an annuity is the sum of the present values of each of the payments of R dollars, as the time line
below shows.

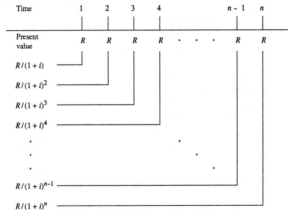

(b) $A_p = \dfrac{R}{1+i} + \dfrac{R}{(1+i)^2} + \dfrac{R}{(1+i)^3} + \cdots + \dfrac{R}{(1+i)^n}$

$= \dfrac{R}{1+i} + \left(\dfrac{R}{1+i}\right)\left(\dfrac{1}{1+i}\right) + \left(\dfrac{R}{1+i}\right)\left(\dfrac{1}{1+i}\right)^2 + \cdots + \left(\dfrac{R}{1+i}\right)\left(\dfrac{1}{1+i}\right)^{n-1}$. This is a

geometric series with $a = \dfrac{R}{1+i}$ and $r = \dfrac{1}{1+i}$. Since $S_n = a\dfrac{1 - r^n}{1 - r}$, we have

$A_p = \left(\dfrac{R}{1+i}\right)\dfrac{1 - \left[\frac{1}{(1+i)}\right]^n}{1 - \left(\frac{1}{1+i}\right)} = R\dfrac{1 - (1+i)^{-n}}{(1+i)\left[1 - \left(\frac{1}{1+i}\right)\right]} = R\dfrac{1 - (1+i)^{-n}}{(1+i) - 1} = R\dfrac{1 - (1+i)^{-n}}{i}.$

24. (a) Using the hint, we calculate the present value of the remaining 240 payment with $R = 724.17$,
$i = 0.0075$, and $n = 240$. Since $A_p = R\dfrac{1 - (1+i)^{-n}}{i} = (724.17)\dfrac{1 - (1.0075)^{-240}}{0.0075}$
$= 80{,}487.84$, they still owe $\$80{,}487.84$ on there mortgage.

(b) On their next payment $0.0075(80{,}487.84) = \$603.66$ will be interest and
$\$724.17 - 603.66 = \$120.51$ goes towards the principal.

Exercises 9.5

2. Let $P(n)$ denote the statement $1 + 4 + 7 + 10 + \cdots + (3n - 2) = \dfrac{n(3n - 1)}{2}$.

    Step 1 $P(1)$ is the statement that $1 = \dfrac{1[3(1) - 1]}{2} = \dfrac{1 \cdot 2}{2}$, which is true.

    Step 2 Assume that $P(k)$ is true; that is. $1 + 4 + 7 + \cdots + (3k - 2) = \dfrac{k(3k - 1)}{2}$. We want to use this to show that $P(k + 1)$ is true. Now,

    $$1 + 4 + 7 + 10 + \cdots + (3k - 2) + [3(k + 1) - 2] =$$
    $$\frac{k(3k - 1)}{2} + 3k + 1 = \qquad\qquad \text{induction hypothesis}$$
    $$\frac{k(3k - 1)}{2} + \frac{6k + 2}{2} = \frac{3k^2 - k + 6k + 2}{2} =$$
    $$\frac{3k^2 + 5k + 2}{2} = \frac{(k + 1)(3k + 2)}{2} = \frac{(k + 1)[3(k + 1) - 1]}{2}.$$

    Thus, $P(k + 1)$ follows from $P(k)$. So by the Principle of Mathematical Induction, $P(n)$ is true for all n.

4. Let $P(n)$ denote the statement $1^2 + 2^2 + 3^2 + \cdots + n^2 = \dfrac{n(n + 1)(2n + 1)}{6}$.

    Step 1 $P(1)$ is the statement that $1^2 = \dfrac{1 \cdot 2 \cdot 3}{6}$, which is true.

    Step 2 Assume that $P(k)$ is true; that is, $1^2 + 2^2 + 3^2 + \cdots + k^2 = \dfrac{k(k + 1)(2k + 1)}{6}$. We want to use this to show that $P(k + 1)$ is true. Now,

    $$1^2 + 2^2 + 3^2 + \cdots + k^2 + (k + 1)^2 =$$
    $$\frac{k(k + 1)(2k + 1)}{6} + (k + 1)^2 = \qquad\qquad \text{induction hypothesis}$$
    $$(k + 1)\left[\frac{k(2k + 1) + 6(k + 1)}{6}\right] =$$
    $$(k + 1)\left[\frac{2k^2 + k + 6k + 6}{6}\right] =$$
    $$(k + 1)\left[\frac{2k^2 + 7k + 6}{6}\right] =$$
    $$\frac{(k + 1)(k + 2)(2k + 3)}{6} = \frac{(k + 1)[(k + 1) + 1][2(k + 1) + 1]]}{6}.$$

    Thus $P(k + 1)$ follows from $P(k)$. So by the Principle of Mathematical Induction, $P(n)$ is true for all n.

6. Let $P(n)$ denote the statement $1 \cdot 3 + 2 \cdot 4 + 3 \cdot 5 + \cdots + n(n + 2) = \dfrac{n(n + 1)(2n + 7)}{6}$.

    Step 1 $P(1)$ is the statement that $1 \cdot 3 = \dfrac{1 \cdot 2 \cdot 9}{6}$, which is true.

<u>Step 2</u> Assume that $P(k)$ is true; that is, $1 \cdot 3 + 2 \cdot 4 + 3 \cdot 5 + \cdots + k(k+2) = \dfrac{k(k+1)(2k+7)}{6}$.

We want to use this to show that $P(k+1)$ is true. Now,

$$1 \cdot 3 + 2 \cdot 4 + 3 \cdot 5 + \cdots + k(k+2) + (k+1)[(k+1)+2] =$$

$$\dfrac{k(k+1)(2k+7)}{6} + (k+1)(k+3) = \qquad \text{induction hypothesis}$$

$$(k+1)\left[\dfrac{k(2k+7)}{6} + \dfrac{6(k+3)}{6}\right] =$$

$$(k+1)\left(\dfrac{2k^2 + 7k + 6k + 18}{6}\right) =$$

$$\dfrac{(k+1)[(k+1)+1][2(k+1)+7]}{6}.$$

Thus $P(k+1)$ follows from $P(k)$. So by the Principle of Mathematical Induction, $P(n)$ is true for all n.

8.   Let $P(n)$ denote the statement $1^3 + 3^3 + 5^3 + \cdots + (2n-1)^3 = n^2(2n^2 - 1)$.

<u>Step 1</u> $P(1)$ is the statement that $1^3 = 1^2(2 \cdot 1^2 - 1)$, which is clearly true.

<u>Step 2</u> Assume that $P(k)$ is true; that is, $1^3 + 3^3 + 5^3 + \cdots + (2k-1)^3 = k^2(2k^2 - 1)$. We want to use this to show that $P(k+1)$ is true. Now,

$$1^3 + 3^3 + 5^3 + \cdots + (2k-1)^3 + (2k+1)^3 =$$

$$k^2(2k^2 - 1) + (2k+1)^3 = \qquad \text{induction hypothesis}$$

$$2k^4 - k^2 + 8k^3 + 12k^2 + 6k + 1 =$$

$$2k^4 + 8k^3 + 11k^2 + 6k + 1 =$$

$$(k^2 + 2k + 1)(2k^2 + 4k + 1) = (k+1)^2[2(k+1)^2 - 1].$$

Thus $P(k+1)$ follows from $P(k)$. So by the Principle of Mathematical Induction, $P(n)$ is true for all n.

10.   Let $P(n)$ denote the statement $\dfrac{1}{1 \cdot 2} + \dfrac{1}{2 \cdot 3} + \dfrac{1}{3 \cdot 4} + \cdots + \dfrac{1}{n(n+1)} = \dfrac{n}{n+1}$.

<u>Step 1</u> $P(1)$ is the statement that $\dfrac{1}{1 \cdot 2} = \dfrac{1}{2}$, which is clearly true.

<u>Step 2</u> Assume that $P(k)$ is true; that is $\dfrac{1}{1 \cdot 2} + \dfrac{1}{2 \cdot 3} + \dfrac{1}{3 \cdot 4} + \cdots + \dfrac{1}{k(k+1)} = \dfrac{k}{k+1}$. We want to use this to show that $P(k+1)$ is true. Now,

$$\dfrac{1}{1 \cdot 2} + \dfrac{1}{2 \cdot 3} + \cdots + \dfrac{1}{k(k+1)} + \dfrac{1}{(k+1)(k+2)} =$$

$$\dfrac{k}{k+1} + \dfrac{1}{(k+1)(k+2)} = \qquad \text{induction hypothesis}$$

$$\dfrac{k(k+2)+1}{(k+1)(k+2)} = \dfrac{k^2 + 2k + 1}{(k+1)(k+2)} =$$

$$\dfrac{(k+1)^2}{(k+1)(k+2)} = \dfrac{k+1}{k+2} = \dfrac{k+1}{(k+1)+1}.$$

Thus $P(k+1)$ follows from $P(k)$. So by the Principle of Mathematical Induction, $P(n)$ is true for all n.

12. Let $P(n)$ denote the statement $1 + 2 + 2^2 + \cdots + 2^{n-1} = 2^n - 1$.

Step 1 $P(1)$ is the statement that $1 = 2^1 - 1$, which is clearly true.

Step 2 Assume that $P(k)$ is true; that is, $1 + 2 + 2^2 + \cdots + 2^{k-1} = 2^k - 1$. We want to use this to show that $P(k+1)$ is true. Now,

$$1 + 2 + 2^2 + \cdots + 2^{k-1} + 2^k =$$
$$2^k - 1 + 2^k = \qquad \text{induction hypothesis}$$
$$2 \cdot 2^k - 1 = 2^{k+1} - 1.$$

Thus $P(k+1)$ follows from $P(k)$. So by the Principle of Mathematical Induction, $P(n)$ is true for all n.

14. Let $P(n)$ denote the statement that $5^n - 1$ is divisible by 4.

Step 1 $P(1)$ is the statement that $5^1 - 1 = 4$ is divisible by 4, which is clearly true.

Step 2 Assume that $P(k)$ is true; that is, $5^k - 1$ is divisible by 4. We want to use this to show that $P(k+1)$ is true. Now, $5^{(k+1)} - 1 = 5 \cdot 5^k - 1 = 5 \cdot 5^k - 5 + 4 = 5(5^k - 1) + 4$ which is divisible by 4 since $5(5^k - 1)$ is divisible by 4 by the induction hypothesis. Thus $P(k+1)$ follows from $P(k)$. So by the Principle of Mathematical Induction, $P(n)$ is true for all n.

16. Let $P(n)$ denote the statement that $n^3 - n + 3$ is divisible by 3.

Step 1 $P(1)$ is the statement that $1^3 - 1 + 3 = 3$ is divisible by 3, which is true.

Step 2 Assume that $P(k)$ is true; that is, $k^3 - k + 3$ is divisible by 3. We want to use this to show that $P(k+1)$ is true. Now,
$$(k+1)^3 - (k+1) + 3 = k^3 + 3k^2 + 3k + 1 - k - 1 + 3 = k^3 - k + 3 + 3k^2 + 3k$$
$$= (k^3 - k + 3) + 3(k^2 + k), \text{ which is divisible by 3, since } k^3 - k + 3 \text{ is divisible by 3 by}$$
the induction hypothesis, and $3(k^2 + k)$ is divisible by 3. Thus $P(k+1)$ follows from $P(k)$. So by the Principle of Mathematical Induction, $P(n)$ is true for all n.

18. Let $P(n)$ denote the statement that $3^{2n} - 1$ is divisible by 8.

Step 1 $P(1)$ is the statement that $3^2 - 1 = 8$ is divisible by 8, which is clearly true.

Step 2 Assume that $P(k)$ is true; that is, $3^{2k} - 1$ is divisible by 8. We want to use this to show that $P(k+1)$ is true. Now, $3^{2(k+1)} - 1 = 9 \cdot 3^{2k} - 1 = 9 \cdot 3^{2k} - 9 + 8 = 9(3^{2k} - 1) + 8$, which is divisible by 8, since $3^{2k} - 1$ is divisible by 8 by the induction hypothesis. Thus $P(k+1)$ follows from $P(k)$. So by the Principle of Mathematical Induction, $P(n)$ is true for all n.

20. Let $P(n)$ denote the statement $(n+1)^2 < 2n^2$, for all $n \geq 3$.

Step 1 $P(3)$ is the statement that $(3+1)^2 < 2 \cdot 3^2$ or $16 < 18$, which is true.

Step 2 Assume that $P(k)$ is true; that is, $(k+1)^2 < 2k^2$, $k \geq 3$. We want to use this to show that $P(k+1)$ is true. Now,

$$(k+2)^2 =$$
$$k^2 + 4k + 4 =$$
$$(k^2 + 2k + 1) + (2k + 3) =$$
$$(k+1)^2 + (2k+1) < 2k^2 + (2k+3) \qquad \text{induction hypothesis}$$
$$< 2k^2 + (2k+3) + (2k-1) \qquad \text{because } 2k - 1 > 0 \text{ for } k \geq 3$$
$$= 2k^2 + 4k + 2 = 2(k+1)^2.$$

Thus $P(k + 1)$ follows from $P(k)$. So by the Principle of Mathematical Induction, $P(n)$ is true for all $n \geq 3$.

22. Let $P(n)$ denote the statement $100n \leq n^2$, for all $n \geq 100$.

Step 1 $P(100)$ is the statement that $100(100) \leq (100)^2$, which is true.

Step 2 Assume that $P(k)$ is true; that is, $100k \leq k^2$. We want to use this to show that $P(k+1)$ is true. Now,

$100(k + 1) =$

$100k + 100 \leq k^2 + 100$ $\qquad\qquad\qquad$ induction hypothesis

$\leq k^2 + 2k + 1 = (k + 1)^2.$ $\qquad\qquad$ because $2k + 1 \geq 100$ for $k \geq 100$

Thus $P(k + 1)$ follows from $P(k)$. So by the Principle of Mathematical Induction, $P(n)$ is true for all $n \geq 100$.

24. $a_{n+1} = 3a_n - 8$ and $a_1 = 4$. Then $a_2 = 3 \cdot 4 - 8 = 4$, $a_3 = 3 \cdot 4 - 8 = 4$, $a_4 = 3 \cdot 4 - 8 = 4, \ldots$, and the conjecture is that $a_n = 4$. Let $P(n)$ denote the statement that $a_n = 4$.

Step 1 $P(1)$ is the statement that $a_1 = 4$, which is true.

Step 2 Assume that $P(k)$ is true; that is, $a_k = 4$. We want to use this to show that $P(k+1)$ is true. Now, $a_{k+1} = 3 \cdot a_k - 8 = 3 \cdot 4 - 8 = 4$, by the induction hypothesis. This is exactly $P(k + 1)$, so by the Principle of Mathematical Induction, $P(n)$ is true for all n.

26. Let $P(n)$ be the statement that $x + y$ is a factor of $x^{2n-1} + y^{2n-1}$.

Step 1 $P(1)$ is the statement that $x + y$ is a factor of $x^1 + y^1$, which is clearly true.

Step 2 Assume that $P(k)$ is true; that is, $x + y$ is a factor of $x^{2k-1} + y^{2k-1}$. We want to use this to show that $P(k + 1)$ is true. Now, $x^{2(k+1)-1} + y^{2(k+1)-1} = x^{2k+1} + y^{2k+1}$

$= x^{2k+1} - x^{2k-1}y^2 + x^{2k-1}y^2 + y^{2k+1} = x^{2k-1}(x^2 - y^2) + (x^{2k-1} + y^{2k-1})y^2$, for which $x + y$ is a factor. This is because $x + y$ is a factor of $x^2 - y^2 = (x + y)(x - y)$ and $(x + y)$ is a factor of $x^{2k-1} + y^{2k-1}$ by our induction hypothesis. Thus $P(k + 1)$ follows from $P(k)$. So by the Principle of Mathematical Induction, $P(n)$ is true for all n.

28. Let $P(n)$ denote the statement that $F_1 + F_2 + F_3 + \cdots + F_n = F_{n+2} - 1$.

Step 1 $P(1)$ is the statement that $F_1 = F_3 - 1$. But $F_1 = 1 = 2 - 1 = F_3 - 1$, which is true.

Step 2 Assume that $P(k)$ is true; that is, $F_1 + F_2 + F_3 + \cdots + F_k = F_{k+2} - 1$. We want to use this to show that $P(k + 1)$ is true. Now, $F_{(k+1)+2} - 1 = F_{k+3} - 1 = F_{k+2} + F_{k+1} - 1$

$= (F_{k+2} - 1) + F_{k+1} = F_1 + F_2 + F_3 + \cdots + F_k + F_{k+1}$ by the induction hypothesis. Thus $P(k + 1)$ follows from $P(k)$. So by the Principle of Mathematical Induction, $P(n)$ is true for all n.

30. Let $P(n)$ denote the statement that $F_1 + F_3 + \cdots + F_{2n-1} = F_{2n}$

Step 1 $P(1)$ is the statement that $F_1 = F_2$, which is true since $F_1 = 1$ and $F_2 = 1$.

Step 2 Assume that $P(k)$ is true; that is, $F_1 + F_3 + \cdots + F_{2k-1} = F_{2k}$, for some $k \geq 1$. We want to use this to show that $P(k + 1)$ is true; that is, $F_1 + F_3 + \cdots + F_{2(k+1)-1} = F_{2(k+1)}$. Now,

$F_1 + F_3 + \cdots + F_{2k-1} + F_{2k+1} =$

$\qquad\qquad F_{2k} + F_{2k+1} = \qquad$ by induction hypothesis

$\qquad\qquad\qquad F_{2k+2} = F_{2(k+1)}. \quad$ definition of F_{2k+2}

Therefore, $P(k+1)$ is true. So by the Principle of Mathematical Induction, $P(n)$ is true for all n.

32. Let $a_1 = 1$ and $a_{n+1} = \dfrac{1}{1+a_n}$, for $n \geq 1$. Let $P(n)$ be the statement that $a_n = \dfrac{F_n}{F_{n+1}}$, for all $n \geq 1$.

Step 1 $P(1)$ is the statement that $a_1 = \dfrac{F_1}{F_2}$, which is true since $a_1 = 1$ and $\dfrac{F_1}{F_2} = \dfrac{1}{1} = 1$.

Step 2 Assume that $P(k)$ is true; that is, $a_k = \dfrac{F_k}{F_{k+1}}$. We want to use this to show that $P(k+1)$ is true. Now,

$$a_{k+1} =$$
$$\frac{1}{1+a_k} = \qquad\qquad\qquad\qquad \text{by the definition of } a_{k+1}$$
$$\frac{1}{1+\frac{F_k}{F_{k+1}}} = \qquad\qquad\qquad \text{induction hypothesis}$$
$$\frac{F_{k+1}}{F_k + F_{k+1}} = \frac{F_{k+1}}{F_{k+2}}. \qquad\qquad \text{definition of } F_{k+2}$$

Thus $P(k+1)$ follows from $P(k)$. So by the Principle of Mathematical Induction, $P(n)$ is true for all $n \geq 100$.

34. Since $100 \cdot 10 = 10^3$, $100 \cdot 11 < 11^3$, $100 \cdot 12 < 12^3$, $\ldots$ our conjecture is that $100n \leq n^3$, for all natural numbers $n \geq 10$. Let $P(n)$ denote the statement that $100n \leq n^3$, for $n \geq 10$.

Step 1 $P(10)$ is the statement that $100 \cdot 10 = 1{,}000 \leq 10^3 = 1{,}000$, which is true.

Step 2 Assume that $P(k)$ is true; that is, $100k \leq k^3$. We want to use this to show that $P(k+1)$ is true. Now,

$$100(k+1) =$$
$$100k + 100 \leq k^2 + 100 \qquad\qquad\qquad\qquad \text{induction hypothesis}$$
$$\leq k^3 + k^2 \qquad\qquad\qquad\qquad\qquad \text{because } k \geq 10$$
$$\leq k^3 + 3k^2 + 3k + 1 = (k+1)^3.$$

Thus $P(k+1)$ follows from $P(k)$. So by the Principle of Mathematical Induction, $P(n)$ is true for all $n \geq 10$.

36. The induction step fails when $k = 2$, that is, $P(2)$ does not follow from $P(1)$. If there are only two cats, Midnight and Sparky, and we remove Sparky, then only Midnight remains. So at this point, we still only know that Midnight is black. Now removing Midnight and putting Sparky back leaves Sparky alone. So the induction hypothesis does not allow us to conclude that Sparky is black.

Exercises 9.6

2. $(2x+1)^4 = (2x)^4 + 4(2x)^3 + 6(2x)^2 + 4 \cdot 2x + 1 = 16x^4 + 32x^3 + 24x^2 + 8x + 1$

4. $(x-y)^5 = x^5 - 5x^4 y + 10x^3 y^2 - 10x^2 y^3 + 5xy^4 - y^5$

6. $\left(\sqrt{a}+\sqrt{b}\right)^6 = a^3 + 6a^2\sqrt{a}\sqrt{b} + 15a^2 b + 20a\sqrt{ab}\sqrt{b} + 15ab^2 + 6\sqrt{ab^2}\sqrt{b} + b^3$
   $= a^3 + 6a^2\sqrt{ab} + 15a^2 b + 20ab\sqrt{ab} + 15ab^2 + 6b^2\sqrt{ab} + b^3$ or
   $= a^3 + 6a^{5/2}b^{1/2} + 15a^2 b + 20a^{3/2}b^{3/2} + 15ab^2 + 6a^{1/2}b^{5/2} + b^3$

8. $\left(1+\sqrt{2}\right)^6 = 1^6 + 6\cdot 1^5 \cdot \sqrt{2} + 15\cdot 1^4 \cdot 2 + 20\cdot 1^3 \cdot 2\sqrt{2} + 15\cdot 1^2 \cdot 4 + 6\cdot 1 \cdot 4\sqrt{2} + 2^3$
   $= 1 + 6\sqrt{2} + 30 + 40\sqrt{2} + 60 + 24\sqrt{2} + 8 = 99 + 70\sqrt{2}$

10. $(1+x^3)^3 = 1^3 + 3\cdot 1^2 \cdot x^3 + 3\cdot 1(x^3)^2 + (x^3)^3 = 1 + 3x^3 + 3x^6 + x^9$

12. $\left(2+\dfrac{x}{2}\right)^5 = (2)^5 + 5(2)^4\dfrac{x}{2} + 10(2)^3\left(\dfrac{x}{2}\right)^2 + 10(2)^2\left(\dfrac{x}{2}\right)^3 + 5(2)\left(\dfrac{x}{2}\right)^4 + \left(\dfrac{x}{2}\right)^5$
   $= 32 + 40x + 20x^2 + 5x^3 + \frac{5}{8}x^4 + \frac{1}{32}x^5$

14. $\dbinom{8}{3} = \dfrac{8!}{3!\,5!} = \dfrac{8\cdot 7\cdot 6\cdot 5!}{3\cdot 2\cdot 1\cdot 5!} = 8\cdot 7 = 56$

16. $\dbinom{10}{5} = \dfrac{10!}{5!\,5!} = \dfrac{10\cdot 9\cdot 8\cdot 7\cdot 6\cdot 5!}{5\cdot 4\cdot 3\cdot 2\cdot 1\cdot 5!} = 3\cdot 2\cdot 7\cdot 6 = 252$

18. $\dbinom{5}{2}\dbinom{5}{3} = \dfrac{5!}{2!\,3!} \cdot \dfrac{5!}{3!\,2!} = \dfrac{5\cdot 4\cdot 3!}{2\cdot 1\cdot 3!} \cdot \dfrac{5\cdot 4\cdot 3!}{3!\cdot 2\cdot 1} = 10\cdot 10 = 100$

20. $\dbinom{5}{0} - \dbinom{5}{1} + \dbinom{5}{2} - \dbinom{5}{3} + \dbinom{5}{4} - \dbinom{5}{5} = 1 - \dfrac{5!}{1!\,4!} + \dfrac{5!}{2!\,3!} - \dfrac{5!}{3!\,2!} + \dfrac{5!}{4!\,1!} - 1 = 0.$
    Notice that the 1st and 6th terms cancel, as do the 2nd and 5th terms and the 3rd and 4th terms.

22. $(1-x)^5 = \dbinom{5}{0}(1)^5 - \dbinom{5}{1}(1)^4 x + \dbinom{5}{2}(1)^3 x^2 - \dbinom{5}{3}(1)^2 x^3 + \dbinom{5}{4}(1)x^4 - \dbinom{5}{5}x^5$
    $= 1 - 5x + 10x^2 - 10x^3 + 5x^4 - x^5$

24. $(2A+B^2)^4 = \dbinom{4}{0}(2A)^4 + \dbinom{4}{1}(2A)^3(B^2) + \dbinom{4}{2}(2A)^2(B^2)^2 + \dbinom{4}{3}(2A)(B^2)^3$
    $+ \binom{4}{4}(B^2)^4 = 16A^4 + 32A^3 B^2 + 24A^2 B^4 + 8AB^6 + B^8$

26. The first four terms in the expansion of $(x^{1/2}+1)^{30}$ are $\dbinom{30}{0}(x^{1/2})^{30} = x^{15}$,
    $\dbinom{30}{1}(x^{1/2})^{29}(1) = 30x^{29/2}$, $\dbinom{30}{2}(x^{1/2})^{28}(1)^2 = 435x^{14}$, and $\dbinom{30}{3}(x^{1/2})^{27}(1)^3 = 4060x^{27/2}$.

28. The first three terms in the expansion of $\left(x + \dfrac{1}{x}\right)^{40}$ are $\dbinom{40}{0}x^{40} = x^{40}$, $\dbinom{40}{1}x^{39}\left(\dfrac{1}{x}\right) = 40x^{38}$, and $\dbinom{40}{2}x^{38}\left(\dfrac{1}{x}\right)^2 = 780x^{36}$.

30. The fifth term in the expansion of $(ab - 1)^{20}$ is $\dbinom{20}{4}(ab)^{16}(-1)^4 = 4845a^{16}b^{16}$.

32. The 28th term in the expansion of $(A - B)^{30}$ is $\dbinom{30}{27}A^3(-B)^{27} = -4060A^3B^{27}$.

34. The 2nd term in the expansion of $\left(x^2 - \dfrac{1}{x}\right)^{25}$ is $\dbinom{25}{1}(x^2)^{24}\left(-\dfrac{1}{x}\right) = -25x^{47}$.

36. The rth term in the expansion of $\left(\sqrt{2} + y\right)^{12}$ is $\dbinom{12}{r}\left(\sqrt{2}\right)^r y^{12-r}$. The term that contains y^3 occurs when $12 - r = 3 \iff r = 9$. Therefore, the term is $\dbinom{12}{9}\left(\sqrt{2}\right)^9 y^3 = 3520\sqrt{2}\,y^3$.

38. The rth term is $\dbinom{8}{r}(8x)^r\left(\dfrac{1}{2x}\right)^{8-r} = \dbinom{8}{r}\dfrac{8^r}{2^{8-r}} \cdot \dfrac{x^r}{x^{8-r}} = \dbinom{8}{r}\dfrac{8^r}{2^{8-r}} \cdot x^{2r-8}$. So the term that does not contain x occurs when $2r - 8 = 0 \iff r = 4$. Thus, the term is $\dbinom{8}{4}(8x)^4\left(\dfrac{1}{2x}\right)^4 = 17{,}920$.

40. $(x - 1)^5 + 5(x - 1)^4 + 10(x - 1)^3 + 10(x - 1)^2 + 5(x - 1) + 1 = [(x - 1) + 1]^5 = x^5$.

42. $x^8 + 4x^6y + 6x^4y^2 + 4x^2y^3 + y^4 = \dbinom{4}{0}(x^2)^4 + \dbinom{4}{1}(x^2)^3y + \dbinom{4}{2}(x^2)^2y^2 + \dbinom{4}{3}x^2y^3$

$+ \dbinom{4}{4}y^4 = (x^2 + y)^4$.

44. $\dfrac{(x + h)^4 - x^4}{h} = \dfrac{\dbinom{4}{0}x^4 + \dbinom{4}{1}x^3h + \dbinom{4}{2}x^2h^2 + \dbinom{4}{3}xh^3 + \dbinom{4}{4}h^4 - x^4}{h}$

$= \dfrac{x^4 + 4x^3h + 6x^2h^2 + 4xh^3 + h^4 - x^4}{h} = \dfrac{4x^3h + 6x^2h^2 + 4xh^3 + h^4}{h}$

$= \dfrac{h(4x^3 + 6x^2h + 4xh^2 + h^3)}{h} = 4x^3 + 6x^2h + 4xh^2 + h^3$.

46. $\dbinom{n}{0} = \dfrac{n!}{0!\,n!} = \dfrac{n!}{1 \cdot n!} = 1$. $\dbinom{n}{n} = \dfrac{n!}{n!\,0!} = \dfrac{n!}{n! \cdot 1} = 1$. Therefore, $\dbinom{n}{0} = \dbinom{n}{n}$.

48. $\dbinom{n}{r} = \dfrac{n!}{r!\,(n - r)!} = \dfrac{n!}{(n - r)!\,r!} = \dbinom{n}{n - r}$, for $0 \le r \le n$.

50.   Let $P(n)$ be the proposition that $\binom{n}{r}$ is an integer for the number n, $0 \leq r \leq n$.

    <u>Step 1</u> Suppose $n = 0$. If $0 \leq r \leq n$, then $r = 0$, and so $\binom{n}{r} = \binom{0}{0} = 1$, which is obviously an integer. Therefore, $P(0)$ is true.

    <u>Step 2</u>   Suppose that $P(k)$ is true. We want to use this to show that $P(k+1)$ must also be true; that is, $\binom{k+1}{r}$ is an integer for $0 \leq r \leq k+1$. But we know that

$$\binom{k+1}{r} = \binom{k}{r-1} + \binom{k}{r}$$ by the key property of binomial coefficients (see Exercise

    49). Furthermore, $\binom{k}{r-1}$ and $\binom{k}{r}$ are both integers by the induction hypothesis. Since the sum of two integers is always an integer, $\binom{k+1}{r}$ must be an integer. Thus, $P(k+1)$ is true if $P(k)$ is true. So by the Principal of Mathematical induction, $\binom{n}{r}$ is an integer for all $n \geq 0$, $0 \leq r \leq n$.

52.

$$1 + 1 = 2$$
$$1 + 2 + 1 = 4$$
$$1 + 3 + 3 + 1 = 8$$
$$1 + 4 + 6 + 4 + 1 = 16$$
$$1 + 5 + 10 + 10 + 5 + 1 = 32.$$

Conjecture: The sum is 2^n.

<u>Proof:</u>

$$2^n = (1+1)^n = \binom{n}{0} 1^0 \cdot 1^n + \binom{n}{1} 1^1 \cdot 1^{n-1} + \binom{n}{2} 1^2 \cdot 1^{n-2} + \cdots + \binom{n}{n} 1^n \cdot 1^0$$
$$= \binom{n}{0} + \binom{n}{1} + \binom{n}{2} + \cdots + \binom{n}{n}.$$

Review Exercises for Chapter 9

2. $a_n = (-1)^n \dfrac{2^n}{n}$. Then $a_1 = (-1)^1 \dfrac{2^1}{1} = -2$; $a_2 = (-1)^2 \dfrac{2^2}{2} = 2$; $a_3 = (-1)^3 \dfrac{2^3}{3} = -\dfrac{8}{3}$;

$a_4 = (-1)^4 \dfrac{2^4}{4} = 4$, and $a_{10} = (-1)^{10} \dfrac{2^{10}}{10} = \dfrac{1024}{10} = \dfrac{512}{5}$.

4. $a_n = \dfrac{n(n+1)}{2}$. Then $a_1 = \dfrac{1(1+1)}{2} = 1$; $a_2 = \dfrac{2(2+1)}{2} = 3$; $a_3 = \dfrac{3(3+1)}{2} = 6$;

$a_4 = \dfrac{4(4+1)}{2} = 10$; and $a_{10} = \dfrac{10(10+1)}{2} = 55$.

6. $a_n = \dbinom{n+1}{2}$. Then $a_1 = \dbinom{1+1}{2} = 1$; $a_2 = \dbinom{2+1}{2} = \dfrac{3!}{2!\,1!} = 3$; $a_3 = \dbinom{3+1}{2}$

$= \dfrac{4!}{2!\,2!} = 6$; $a_4 = \dbinom{4+1}{2} = \dfrac{5!}{2!\,3!} = 10$; and $a_{10} = \dbinom{10+1}{2} = \dfrac{11!}{2!\,9!} = 55$.

8. $a_n = \dfrac{a_{n-1}}{n}$ and $a_1 = 1$. Then $a_2 = \dfrac{a_1}{2} = \dfrac{1}{2}$; $a_3 = \dfrac{a_2}{3} = \dfrac{1}{6}$; $a_4 = \dfrac{a_3}{4} = \dfrac{1}{24}$; $a_5 = \dfrac{a_4}{5} = \dfrac{1}{120}$;

$a_6 = \dfrac{a_5}{6} = \dfrac{1}{720}$; and $a_7 = \dfrac{a_6}{7} = \dfrac{1}{5040}$.

10. $a_n = \sqrt{3a_{n-1}}$ and $a_1 = \sqrt{3} = 3^{1/2}$. Then $a_2 = \sqrt{3a_1} = \sqrt{3\sqrt{3}} = (3 \cdot 3^{1/2})^{1/2} = 3^{3/4}$;

$a_3 = \sqrt{3a_2} = \sqrt{3 \cdot 3^{3/4}} = \sqrt{3^{7/4}} = 3^{7/8}$; $a_4 = \sqrt{3a_3} = \sqrt{3 \cdot 3^{7/8}} = \sqrt{3^{15/8}} = 3^{15/16}$;

$a_5 = \sqrt{3a_4} = \sqrt{3 \cdot 3^{15/16}} = \sqrt{3^{31/16}} = 3^{31/32}$; $a_6 = \sqrt{3a_5} = \sqrt{3 \cdot 3^{31/32}} = \sqrt{3^{63/32}} = 3^{63/64}$;

$a_7 = \sqrt{3a_6} = \sqrt{3 \cdot 3^{63/64}} = \sqrt{3^{127/64}} = 3^{127/128}$.

12. (a) $a_1 = \dfrac{5}{2^1} = \dfrac{5}{2}$

$a_2 = \dfrac{5}{2^2} = \dfrac{5}{4}$

$a_3 = \dfrac{5}{2^3} = \dfrac{5}{8}$

$a_4 = \dfrac{5}{2^4} = \dfrac{5}{16}$

$a_5 = \dfrac{5}{2^5} = \dfrac{5}{32}$

(b)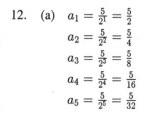

(c) This sequence is geometric, the common ratio is $\dfrac{1}{2}$.

14. (a) $a_1 = 4 - \dfrac{1}{2} = \dfrac{7}{2}$

$a_2 = 4 - \dfrac{2}{2} = 3$

$a_3 = 4 - \dfrac{3}{2} = \dfrac{5}{2}$

$a_4 = 4 - \dfrac{4}{2} = 2$

$a_5 = 4 - \dfrac{5}{2} = \dfrac{3}{2}$

(c)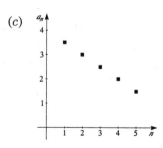

(c)   This sequence is arithmetic, the common difference is 2.

16. $1, -\dfrac{3}{2}, 2, -\dfrac{5}{2}, \ldots$. Since $-\dfrac{3}{2} - 1 \neq 2 + \dfrac{3}{2}$, and $\dfrac{-\frac{3}{2}}{1} \neq \dfrac{2}{-\frac{3}{2}}$, the series is neither arithmetic nor geometric.

18. $\sqrt{2}, 2, 2\sqrt{2}, 4, \ldots$. Since $\dfrac{2}{\sqrt{2}} = \dfrac{2\sqrt{2}}{2} = \dfrac{4}{2\sqrt{2}} = \sqrt{2}$, this is a geometric sequence with $a_1 = \sqrt{2}$ and $r = \sqrt{2}$. Then $a_5 = a_4 \cdot r = 4 \cdot \sqrt{2} = 4\sqrt{2}$.

20. $t^3, t^2, t, 1, \ldots$. Since $\dfrac{t^2}{t^3} = \dfrac{t}{t^2} = \dfrac{1}{t}$, this is a geometric sequence with $a_1 = t^3$ and $r = \dfrac{1}{t}$. Then $a_5 = a_4 \cdot r = 1 \cdot \dfrac{1}{t} = \dfrac{1}{t}$.

22. $a, 1, \dfrac{1}{a}, \dfrac{1}{a^2}, \ldots$. Since $\dfrac{1}{a} = \dfrac{\frac{1}{a}}{1}$ and $\dfrac{\frac{1}{a^2}}{\frac{1}{a}} = \dfrac{1}{a}$, this is a geometric sequence with $a_1 = a$ and $r = \dfrac{1}{a}$. Then $a_5 = a_4 \cdot r = \dfrac{1}{a^2} \cdot \dfrac{1}{a} = \dfrac{1}{a^3}$.

24.   The sequence $2, 2 + 2i, 4i, -4 + 4i, -8, \ldots$ is geometric (where $i^2 = -1$), since
$$\dfrac{a_2}{a_1} = \dfrac{2 + 2i}{2} = 1 + i; \quad \dfrac{a_3}{a_2} = \dfrac{4i}{2 + 2i} = \dfrac{4i}{2 + 2i} \cdot \dfrac{2 - 2i}{2 - 2i} = \dfrac{8 + 8i}{8} = 1 + i;$$
$$\dfrac{a_4}{a_3} = \dfrac{-4 + 4i}{4i} = -\dfrac{1}{i} + 1 = -\dfrac{1}{i} \cdot \dfrac{i}{i} + 1 = \dfrac{-i}{i^2} + 1 = 1 + i;$$
$$\dfrac{a_5}{a_4} = \dfrac{-8}{-4 + 4i} = \dfrac{-8}{-4 + 4i} \cdot \dfrac{-4 - 4i}{-4 - 4i} = \dfrac{32 + 32i}{32} = 1 + i. \text{ Thus the common ratio is } i + 1, \text{ and}$$
the first term is 2. So the nth term is $a_n = a_1 r^{n-1} = 2(1 + i)^{n-1}$.

26. $a_{20} = 96$ and $d = 5$. Then $96 = a_{20} = a + 19 \cdot 5 = a + 95 \quad \Leftrightarrow \quad a = 96 - 95 = 1$. Therefore, $a_n = 1 + 5(n - 1)$.

28. $a_2 = 10$ and $a_5 = \dfrac{1250}{27}$. Then $r^3 = \dfrac{a_5}{a_2} = \dfrac{\frac{1250}{27}}{10} = \dfrac{125}{27} \quad \Leftrightarrow \quad r = \dfrac{5}{3}$ and $a = a_1 = \dfrac{a_2}{r}$
$= \dfrac{10}{\frac{5}{3}} = 6$. Therefore, $a_n = ar^{n-1} = 6\left(\dfrac{5}{3}\right)^{n-1}$.

30.   (a) $A_n = 35{,}000 + 1{,}200(n - 1)$.

      (b) $A_8 = 35{,}000 + 1{,}200(7) = \$43{,}400$. The salary for this teacher is higher for the first 6 years is higher, but from the 7th year on, the salary of the teacher in Exercise 29 will be higher.

32.   Let d be the common difference in the arithmetic sequence $a_1, a_2, a_3, \ldots$, so that $a_n = a_1 + (n - 1)d$, $n = 1, 2, 3, \ldots$, and let e be the common difference for $b_1, b_2, b_3, \ldots$, so that $b_n = b_1 + (n - 1)e$. Then $a_n + b_n = [a_1 + (n - 1)a] + [b_1 + (n - 1)e]$
$= (a_1 + b_1) + (n - 1)(d + e)$, $n = 1, 2, 3, \ldots$. Thus $a_1 + b_1, a_2 + b_2, \ldots$ is an arithmetic sequence with first term $a_1 + b_1$ and common difference $d + e$.

34. (a) Yes. If the common difference is d, then $a_n = a_1 + (n-1)d$. So
$a_n + 2 = a_1 + 2 + (n-1)d$, and thus the sequence $a_1 + 2, a_2 + 2, a_3 + 2, \ldots$ is an arithmetic
sequence with the common difference d, but with the first term $a_1 + 2$.

   (b) Yes. If the common ratio is r, then $a_n = a_1 \cdot r^{n-1}$. So $5a_n = (5a_1) \cdot r^{n-1}$, and the sequence
$5a_1, 5a_2, 5a_3, \ldots$ is also geometric, with common ratio r, but with the first term $5a_1$.

36. (a) $2, x, y, 17, \ldots$ is arithmetic. Therefore, $15 = 17 - 2 = a_4 - a_1 = a + 3d - a = 3d$. So
$d = 5$, and hence, $x = a + d = 2 + 5 = 7$ and $y = a + 2d = 2 + 2 \cdot 5 = 12$.

   (b) $2, x, y, 17, \ldots$ is geometric. Therefore, $\frac{17}{2} = \frac{a_4}{a_1} = \frac{ar^3}{a} = r^3 \Leftrightarrow r = \left(\frac{17}{2}\right)^{1/3}$. So
$x = a_2 = ar = 2\left(\frac{17}{2}\right)^{1/3} = 2^{2/3} 17^{1/3}$ and $y = a_3 = ar^2 = 2\left[\left(\frac{17}{2}\right)^{1/3}\right]^2 = 2\left(\frac{17}{2}\right)^{2/3}$
$= 2^{1/3} 17^{2/3}$.

38. $\displaystyle\sum_{i=1}^{4} \frac{2i}{2i-1} = \frac{2 \cdot 1}{2 \cdot 1 - 1} + \frac{2 \cdot 2}{2 \cdot 2 - 1} + \frac{2 \cdot 3}{2 \cdot 3 - 1} + \frac{2 \cdot 4}{2 \cdot 4 - 1} = 2 + \frac{4}{3} + \frac{6}{5} + \frac{8}{7}$
$= \dfrac{210 + 140 + 126 + 120}{105} = \dfrac{596}{105}$

40. $\displaystyle\sum_{m=1}^{5} 3^{m-2} = 3^{-1} + 3^0 + 3^1 + 3^2 + 3^3 = \frac{1}{3} + 1 + 3 + 9 + 27 = \frac{121}{3}$

42. $\displaystyle\sum_{j=2}^{100} \frac{1}{j-1} = \frac{1}{1} + \frac{1}{2} + \frac{1}{3} + \frac{1}{4} + \frac{1}{5} + \cdots + \frac{1}{98} + \frac{1}{99}$

44. $\displaystyle\sum_{n=1}^{10} n^2 2^n = 1^2 \cdot 2^1 + 2^2 \cdot 2^2 + 3^2 \cdot 2^3 + \cdots + 9^2 \cdot 2^9 + 10^2 \cdot 2^{10}$

46. $1^2 + 2^2 + 3^3 + \cdots + 100^2 = \displaystyle\sum_{k=1}^{100} k^2$

48. $\dfrac{1}{1 \cdot 2} + \dfrac{1}{2 \cdot 3} + \dfrac{1}{3 \cdot 4} + \cdots + \dfrac{1}{999 \cdot 1000} = \displaystyle\sum_{k=1}^{999} \frac{1}{k(k+1)}$

50. $3 + 3.7 + 4.4 + \cdots + 10$ is an arithmetic series with $a = 3$ and $d = 0.7$. Then
$10 = a_n = 3 + 0.7(n-1) \Leftrightarrow 0.7(n-1) = 7 \Leftrightarrow n = 11$. So the sum of the series is
$S_{11} = \dfrac{11}{2}(3 + 10) = \dfrac{143}{2} = 71.5$

52. $\frac{1}{3} + \frac{2}{3} + 1 + \frac{4}{3} + \cdots + 33$ is an arithmetic series with $a = \frac{1}{3}$ and $d = \frac{1}{3}$. Then
$a_n = 33 = \frac{1}{3} + \frac{1}{3}(n-1) \Leftrightarrow n = 99$. So the sum is $S_{99} = \dfrac{99}{2}\left(\dfrac{2}{3} + \dfrac{99-1}{3}\right) = \dfrac{99}{2} \cdot \dfrac{100}{3} = 1650$.

54. $\displaystyle\sum_{k=0}^{8} 7 \cdot 5^{k/2}$ is a geometric series with $a = 7$, $r = 5^{1/2}$, and $n = 9$. Thus, the sum of the series is

$$S_9 = 7 \cdot \frac{1 - 5^{9/2}}{1 - 5} = \frac{7}{4}(5^{9/2} - 1) = \frac{7}{4}(625\sqrt{5} - 1).$$

56. We have a geometric series with $S_3 = 52$ and $r = 3$. Then $52 = S_3 = a + 3a + 9a = 13a \quad \Leftrightarrow$ $a = 4$, and so the first term is 4.

58. $R = 1000$, $i = 0.08$, and $n = 16$. Thus, $A = 1000\dfrac{1.08^{16} - 1}{1.08 - 1} = 12{,}500[1.08^{16} - 1] = \$30{,}324.28$.

60. $A = 60{,}000$ and $i = \dfrac{0.09}{12} = 0.0075$.

    (a) If the period is 30 years, $n = 360$ and $R = \dfrac{60{,}000 \cdot 0.0075}{1 - 1.0075^{-360}} = \$482.77$.

    (b) If the period is 15 years, $n = 180$ and $R = \dfrac{60{,}000 \cdot 0.0075}{1 - 1.0075^{-180}} = \$608.56$.

62. $0.1 + 0.01 + 0.001 + 0.0001 + \cdots$ is an infinite geometric series with $a = 0.1$ and $r = 0.1$. Therefore, the sum is $S = \frac{0.1}{1 - 0.1} = \frac{1}{9}$.

64. $a + ab^2 + ab^4 + ab^6 + \cdots$ is an infinite geometric series with first term a and common ratio b^2. Thus, the sum is $S = \dfrac{a}{1 - b^2}$.

66. Let $P(n)$ denote the statement that $\dfrac{1}{1 \cdot 3} + \dfrac{1}{3 \cdot 5} + \dfrac{1}{5 \cdot 7} + \cdots + \dfrac{1}{(2n - 1)(2n + 1)} = \dfrac{n}{2n + 1}$.

  <u>Step 1</u> $P(1)$ is the statement that $\dfrac{1}{1 \cdot 3} = \dfrac{1}{2 \cdot 1 + 1}$, which is true.

  <u>Step 2</u> Assume that $P(k)$ is true; that is, $\dfrac{1}{1 \cdot 3} + \dfrac{1}{3 \cdot 5} + \dfrac{1}{5 \cdot 7} + \cdots + \dfrac{1}{(2k - 1)(2k + 1)} = \dfrac{k}{2k + 1}$.

    We want to use this to show that $P(k + 1)$ is true. Now,

$$\dfrac{1}{1 \cdot 3} + \dfrac{1}{3 \cdot 5} + \dfrac{1}{5 \cdot 7} + \cdots + \dfrac{1}{(2k - 1)(2k + 1)} + \dfrac{1}{(2k + 1)(2k + 3)} =$$

$$\dfrac{k}{2k + 1} + \dfrac{1}{(2k + 1)(2k + 3)} = \quad \text{induction hypothesis}$$

$$\dfrac{k(2k + 3) + 1}{(2k + 1)(2k + 3)} = \dfrac{2k^2 + 3k + 1}{(2k + 1)(2k + 3)} =$$

$$\dfrac{(k + 1)(2k + 1)}{(2k + 1)(2k + 3)} = \dfrac{k + 1}{2k + 3} = \dfrac{k + 1}{2(k + 1) + 1}.$$

    Thus $P(k + 1)$ follows from $P(k)$. So by the Principle of Mathematical Induction, $P(n)$ is true for all n.

68. Let $P(n)$ denote the statement that $7^n - 1$ is divisible by 6.

  <u>Step 1</u> $P(1)$ is the statement that $7^1 - 1 = 6$ is divisible by 6, which is clearly true.

Step 2 Assume that $P(k)$ is true; that is, $7^k - 1$ is divisible by 6. We want to use this to show that $P(k+1)$ is true. Now $7^{k+1} - 1 = 7 \cdot 7^k - 1 = 7 \cdot 7^k - 7 + 6 = 7(7^k - 1) + 6$, which is divisible by 6. This is because $7^k - 1$ is divisible by 6 by the induction hypothesis, and clearly 6 is divisible by 6. Thus $P(k+1)$ follows from $P(k)$. So by the Principle of Mathematical Induction, $P(n)$ is true for all n.

70. Let $P(n)$ denote the statement that F_{4n} is divisible by 3.

Step 1 Show that $P(1)$ is true, but $P(1)$ is true since $F_4 = 3$ is divisible by 3.

Step 2 Assume that $P(k)$ is true; that is, F_{4k} is divisible by 3. We want to use this to show that $P(k+1)$ is true. Now, $F_{4(k+1)} = F_{4k+4} = F_{4k+2} + F_{4k+3}$
$$= (F_{4k} + F_{4k+1}) + (F_{4k+1} + F_{4k+2}) = F_{4k} + F_{4k+1} + F_{4k+1} + (F_{4k} + F_{4k+1})$$
$$= 2 \cdot F_{4k} + 3 \cdot F_{4k+1}, \text{ which is divisible by 3 because } F_{4k} \text{ is divisible by 3 by our induction}$$
hypothesis, and $3 \cdot F_{4k+1}$ is clearly divisible by 3. Thus, $P(k+1)$ follows from $P(k)$. So by the Principle of Mathematical Induction, $P(n)$ is true for all n.

72. $\dbinom{5}{2}\dbinom{5}{3} = \dfrac{5!}{2!3!} \cdot \dfrac{5!}{3!2!} = \dfrac{5 \cdot 4}{2} \cdot \dfrac{5 \cdot 4}{2} = 10 \cdot 10 = 100$

74. $\displaystyle\sum_{k=0}^{5}\dbinom{5}{k} = \dbinom{5}{0} + \dbinom{5}{1} + \dbinom{5}{2} + \dbinom{5}{3} + \dbinom{5}{4} + \dbinom{5}{5} = 2\left(\dfrac{5!}{0!\,5!} + \dfrac{5!}{1!\,4!} + \dfrac{5!}{2!\,3!}\right)$
$$= 2(1 + 5 + 10) = 32$$

76. $(1 - x^2)^6 = \dbinom{6}{0}1^6 - \dbinom{6}{1}1^5x^2 + \dbinom{6}{2}1^4x^4 - \dbinom{6}{3}1^3x^6 + \dbinom{6}{4}1^2x^8 - \dbinom{6}{5}x^{10} + \dbinom{6}{6}x^{12}$
$$= 1 - 6x^2 + 15x^4 - 20x^6 + 15x^8 - 6x^{10} + x^{12}$$

78. The 20th term is $\dbinom{22}{19}a^3b^{19} = 1540a^3b^{19}$.

80. The rth term in the expansion of $(A + 3B)^{10}$ is $\dbinom{10}{r}A^r(3B)^{10-r}$. The term that contains A^6 occurs when $r = 6$. Thus, the term is $\dbinom{10}{6}A^6(3B)^4 = 210A^6 81B^4 = 17{,}010A^6B^4$.

Focus on Problem Solving

2. (a) $T_n = T_{n-1} + 1.5$ with $T_1 = 5$.

   (b) $T_1 = 5$, $T_2 = T_1 + 1.5 = 5 + 1.5 = 6.5$,
       $T_3 = T_2 + 1.5 = (5 + 1.5) + 1.5 = 5 + 2 \cdot 1.5 = 8.0$,
       $T_4 = T_3 + 1.5 = (5 + 2 \cdot 1.5) + 1.5 = 5 + 3 \cdot 1.5 = 9.5$,
       $T_5 = T_4 + 1.5 = (5 + 3 \cdot 1.5) + 1.5 = 5 + 4 \cdot 1.5 = 11.0$,
       $T_6 = T_5 + 1.5 = (5 + 4 \cdot 1.5) + 1.5 = 5 + 5 \cdot 1.5 = 12.5$.

   (c) This is an arithmetic sequence with $T_n = 5 + 1.5(n - 1)$.

   (d) $T_n = 65 = 5 + 1.5n - 1.5$ $\Leftrightarrow$ $61.5 = 1.5n$ $\Leftrightarrow$ $n = 41$. So Sheila swims 65 minutes on the 41st day.

   (e) Using the partial sum of an arithmetic sequence, she will swim
       $\frac{30}{2}[2(5) + (30 - 1)1.5] = 15 \cdot 53.5 = 802.5$ minutes $= 13$ hours 22.5 minutes.

4. In each case, $P_0 = 4000$.

   (a) $P_n = 1.2P_{n-1}$. $P_5 = 1.2^5(4000) \approx 9953$.

   (b) $P_n = 1.2P_{n-1} - 600$. $P_1 = 1.2(4000) - 600 = 4200$; $P_2 = 1.2(4200) - 600 = 4440$;
       $P_3 = 1.2(4440) - 600 = 4728$; $P_4 = 1.2(4728) - 600 = 5073$;
       $P_5 = 1.2(5073) - 600 = 5488$.

   (c) $P_n = 1.2P_{n-1} + 250$. $P_1 = 1.2(4000) + 250 = 5050$; $P_2 = 1.2(5050) + 250 = 6310$;
       $P_3 = 1.2(6310) + 250 = 7822$; $P_4 = 1.2(7822) + 250 = 9636$;
       $P_5 = 1.2(9636) + 250 = 11,814$.

   (d) Since 10% is have $P_n = 1.2P_{n-1} - 0.1P_{n-1} + 300 = 1.1P_{n-1} + 300$.
       $P_1 = 1.1(4000) + 300 = 4700$; $P_2 = 1.1(4700) + 300 = 5470$;
       $P_3 = 1.1(5470) + 300 = 6317$; $P_4 = 1.1(6317) + 300 = 7249$;
       $P_5 = 1.1(7249) + 300 = 8,274$.

6. (a) $U_n = 1.05U_{n-1} + 0.10(1.05U_{n-1}) = 1.10(1.05)U_{n-1} = 1.155U_{n-1}$ with $U_0 = 5000$.

   (b) $U_0 = \$5000$, $U_1 = 1.155U_0 = \$5775$;
       $U_2 = 1.155U_1 = 1.155(1.155 \cdot 5000) = 1.155^2 \cdot 5000 = \$6670.13$;
       $U_3 = 1.155U_2 = 1.155(1.155^2 \cdot 5000) = 1.155^3 \cdot 5000 = \$7703.99$;
       $U_4 = 1.155U_3 = 1.155(1.155^3 \cdot 5000) = 1.155^4 \cdot 5000 = \$8898.11$.

   (c) Using the pattern we found in part (d) we have $U_n = 1.155^n \cdot 5000$.

   (d) $U_{10} = 1.155^{10} \cdot 5000 = \$21,124.67$.

8. (a) $T_n = T_{n-1} - 0.03(T_{n-1} - 70) = 0.97T_{n-1} + 2.1$, with $T_0 = 170$.

   (b) $T_0 = 170°F$; $T_{10} = 143.7°F$; $T_{20} = 124.4°F$; $T_{30} = 110.1°F$; $T_{40} = 99.6°F$; $T_{50} = 91.8°F$;
       $T_{60} = 86.1°F$.

(c)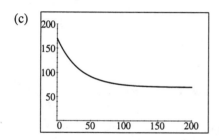

The temperature of the soup will approach 70°F.

Chapter Ten
Exercises 10.1

2. By the Fundamental Counting Principle, the possible number of 3-letter words is
$$\left(\begin{smallmatrix} \text{number of ways to} \\ \text{choose the 1st letter} \end{smallmatrix}\right) \cdot \left(\begin{smallmatrix} \text{number of ways to} \\ \text{choose the 2nd letter} \end{smallmatrix}\right) \cdot \left(\begin{smallmatrix} \text{number of ways to} \\ \text{choose the 3rd letter} \end{smallmatrix}\right).$$

  (a) Since repetitions are allowed, we have 26 choices for each letter. Thus, there are $26 \cdot 26 \cdot 26 = 17{,}576$ words.

  (b) Since repetitions are *not* allowed, we have 26 choices for the 1st letter, 25 choices for the 2nd letter, and 24 choices for the 3rd letter. Thus there are $26 \cdot 25 \cdot 24 = 15{,}600$ words.

4. (a) By the Fundamental Counting Principle, the possible number of ways 8 horses can complete a race, assuming no ties in any position, is
$$\left(\begin{smallmatrix} \text{number of ways to} \\ \text{choose the 1st finisher} \end{smallmatrix}\right) \cdot \left(\begin{smallmatrix} \text{number of ways to} \\ \text{choose the 2nd finisher} \end{smallmatrix}\right) \cdot \; \cdots \; \cdot \left(\begin{smallmatrix} \text{number of ways to} \\ \text{choose the 8th finisher} \end{smallmatrix}\right)$$
$$= 8 \cdot 7 \cdot 6 \cdot 5 \cdot 4 \cdot 3 \cdot 2 \cdot 1 = 8! = 40{,}320.$$

  (b) By the Fundamental Counting Principle, the possible number of ways the first, second, and third place can be decided, assuming no ties, is
$$\left(\begin{smallmatrix} \text{number of ways to} \\ \text{choose the 1st finisher} \end{smallmatrix}\right) \cdot \left(\begin{smallmatrix} \text{number of ways to} \\ \text{choose the 2nd finisher} \end{smallmatrix}\right) \cdot \left(\begin{smallmatrix} \text{number of ways to} \\ \text{choose the 3th finisher} \end{smallmatrix}\right) = 8 \cdot 7 \cdot 6 = 336.$$

6. The number of possible seven-digit phone numbers is
$$\left(\begin{smallmatrix} \text{number of ways to} \\ \text{choose the 1st digit} \end{smallmatrix}\right) \cdot \left(\begin{smallmatrix} \text{number of ways to} \\ \text{choose the 2nd digit} \end{smallmatrix}\right) \cdot \; \cdots \; \cdot \left(\begin{smallmatrix} \text{number of ways to} \\ \text{choose the 7th digit} \end{smallmatrix}\right).$$
Since the first digit cannot be a 0 or a 1, there are only 8 digits to choose from, while there are 10 digits to choose from for the other six digits in the phone number. Thus the number of possible seven-digit phone numbers is $8 \cdot 10 \cdot 10 \cdot 10 \cdot 10 \cdot 10 \cdot 10 = 8{,}000{,}000.$

8. By the Fundamental Counting Principle, the number of ways of seating 5 people in a row of 5 seats is $5 \cdot 4 \cdot 3 \cdot 2 \cdot 1 = 5! = 120.$

10. By the Fundamental Counting Principle, the number of ways of arranging 5 mathematics books is $5 \cdot 4 \cdot 3 \cdot 2 \cdot 1 = 5! = 120.$

12. The number of different boy-girl combinations of 4 children is
$$\left(\begin{smallmatrix} \text{number of ways to} \\ \text{choose a boy or a} \\ \text{girl for the 1st child} \end{smallmatrix}\right) \cdot \left(\begin{smallmatrix} \text{number of ways to} \\ \text{choose a boy or a} \\ \text{girl for the 2nd child} \end{smallmatrix}\right) \cdot \; \cdots \; \cdot \left(\begin{smallmatrix} \text{number of ways to} \\ \text{choose a boy or a} \\ \text{girl for the 4th child} \end{smallmatrix}\right) = 2^4 = 16.$$

14. Since each die has six different faces, the number of different outcomes when rolling a red and a white die is $6 \cdot 6 = 36.$

16. (a) Since there are 13 spades and 13 hearts, the number of ways for which the first card is a spade and the second is a heart is $(13)(13) = 169.$

  (b) Since the card is not replaced, the number of ways for which the first card is a spade and the second is a spade is $(13)(12) = 156.$

18. The number of possible ID numbers consisting of one letter followed by 3 digits is $(26)(10)(10)(10) = 26,000$.

20. The number of possible pairs of a pitcher and a catcher is $\left(\begin{smallmatrix} \text{number of ways to} \\ \text{choose a pitcher} \end{smallmatrix}\right) \cdot \left(\begin{smallmatrix} \text{number of ways to} \\ \text{choose a catcher} \end{smallmatrix}\right)$
    $= (7)(3) = 21$.

22. Since successive numbers cannot be the same, the number of possible choices for the second number in the combination is only 59. The third number in the combination cannot be the same as the second in the combination, but it can be the same as the first number, so the number of possible choices for the third number in the combination is also 59. So the number of possible combinations consisting of a number in the clockwise direction, a number in the counterclockwise direction, and then a number in the clockwise direction is $(60)(59)(59) = 208,860$.

24. The possible number of different cars is
    $\left(\begin{smallmatrix} \text{number of} \\ \text{models} \end{smallmatrix}\right) \cdot \left(\begin{smallmatrix} \text{number} \\ \text{of colors} \end{smallmatrix}\right) \cdot \left(\begin{smallmatrix} \text{number of} \\ \text{types of stereos} \end{smallmatrix}\right) \cdot \left(\begin{smallmatrix} \text{air cond.} \\ \text{or none} \end{smallmatrix}\right) \cdot \left(\begin{smallmatrix} \text{sunroof} \\ \text{or none} \end{smallmatrix}\right) = (5)(4)(3)(2)(2) = 240$.

26. The possible number of three initial monograms is
    $\left(\begin{smallmatrix} \text{number of ways to} \\ \text{chose the 1st initial} \end{smallmatrix}\right) \cdot \left(\begin{smallmatrix} \text{number of ways to} \\ \text{chose the 2nd initial} \end{smallmatrix}\right) \cdot \left(\begin{smallmatrix} \text{number of ways to} \\ \text{chose the 3}^{\text{rd}}\text{ initial} \end{smallmatrix}\right) = (26)(26)(26) = 17,576$.

28. (a) The number of different plates consisting of 1 letter followed by 5 digits is $(26)(10^5) = 2,600,000 = 2.6$ million. Since this is less than the 17 million registered cars, there are not enough different plates to go around.

    (b) The system with two letters followed by 4 digits gives $(26^2)(10^4) = 6,760,000 < 17$ million which is not enough. The system consisting of 3 letters followed by 3 digits gives $(26^3)(10^3) = 17,576,000 > 17$ million. Thus the fewest number of letters that will provide enough different plates is 3.

30. The number of ways that a president, a vice-president and a secretary can be chosen is
    $\left(\begin{smallmatrix} \text{number of ways to choose a} \\ \text{female from the 20 females} \end{smallmatrix}\right) \cdot \left(\begin{smallmatrix} \text{number of ways to choose a} \\ \text{male from the 30 males} \end{smallmatrix}\right) \cdot \left(\begin{smallmatrix} \text{number of ways to choose the} \\ \text{sect. from the remaining 48 students} \end{smallmatrix}\right)$
    $= (20)(30)(48) = 28,800$.

32. We have seven choices for the first digit and 10 choices for each of the other 8 digits. Thus, the number of Social Security numbers is $7 \cdot 10^8 = 700,000,000$.

34. Since the fourth letter must be the same as the second letter and the fifth letter must be the same as the first letter, the possible number of 5-letter palindromes is $(26)(26)(26)(1)(1) = 17,576$.

36. (a) The number of possible different 3-character code words with a letter as the first entry is $26 \cdot 36 \cdot 36 = 33,696$.

    (b) The number of possible different 3-character code words in which zero is not the first entry is $35 \cdot 36 \cdot 36 = 45,360$.

38. Since the two algebra books must be next to each other, we first consider them as one object. So we now have four objects to arrange and there are 4! ways to arrange these four objects. Now there are 2 ways to arrange the two algebra books. Thus the number of ways that 5 mathematics books may be placed on a shelf if the two algebra books are to be next to each other is $2 \cdot 4! = 48$.

40. (a) Since the numbers to be formed are less than 700, the first digit must be either 2, 4, or 5. Thus there are 3 possible ways to choose the first digit. The remaining 2 digits can be 2, 4, 5, or 7. Therefore the number of 3-digit numbers that can be formed using the digits 2, 4, 5, and 7 if the numbers are less than 700 is $(3)(4)(4) = 48$.

   (b) Since the numbers to be formed must be even, the last digit must be either 2 or 4. The first 2 digits can be 2, 4, 5, or 7. Therefore the number of even 3-digit numbers that can be formed using the digits 2, 4, 5, and 7 is $(4)(4)(2) = 32$.

   (c) Since the numbers to be formed must be divisible by 5, the last digit must be 5. The first 2 digits can be 2, 4, 5, or 7. Therefore the number of 3-digit numbers divisible by 5 that can be formed using the digits 2, 4, 5, and 7 is $(4)(4)(1) = 16$.

42. Since there are 26 letters, the possible number of combinations of the first and the last initials is $(26)(26) = 676$. Since $677 > 676$, there must be at least 2 people that have the same first and last initials in any group of 677 people.

Exercises 10.2

2. $P(9,2) = \dfrac{9!}{(9-2)!} = \dfrac{9!}{7!} = 9 \cdot 8 = 72$

4. $P(10,5) = \dfrac{10!}{(10-5)!} = \dfrac{10!}{5!} = 30{,}240$

6. $P(99,3) = = \dfrac{99!}{(99-3)!} = \dfrac{99!}{96!} = 941{,}094$

8.   Here we have 9 letters, of which 3 are A's, 3 are B's, and 3 are C's. Thus the number of distinguishable permutations is $\dfrac{9!}{3!\,3!\,3!} = \dfrac{9 \cdot 8 \cdot 7 \cdot 6 \cdot 5 \cdot 4 \cdot 3!}{3 \cdot 2 \cdot 1 \cdot 3 \cdot 2 \cdot 1 \cdot 3!} = 1680.$

10. Here we have 8 letters, of which 1 is an A, 1 is a B, 1 is a C, 3 are D's, and 2 are E's. Thus the number of distinguishable permutations is $\dfrac{8!}{1!\,1!\,1!\,3!\,2!} = 3360.$

12. Here we have 7 letters, of which 2 are X's, 2 are Y, and 3 a Z. Thus the number of distinguishable permutations is $\dfrac{7!}{2!\,2!\,3!} = \dfrac{7 \cdot 6 \cdot 5 \cdot 4 \cdot 3!}{2 \cdot 2 \cdot 3!} = 210.$

14. $C(9,2) = \dfrac{9!}{2!\,(9-2)!} = \dfrac{9!}{2!\,7!} = \dfrac{9 \cdot 8}{2 \cdot 1} = 36$

16. $C(10,5) = \dfrac{10!}{5!\,(10-5)!} = \dfrac{10!}{5!\,5!} = \dfrac{10 \cdot 9 \cdot 8 \cdot 7 \cdot 6}{5 \cdot 4 \cdot 3 \cdot 2 \cdot 1} = 252$

18. $C(99,3) = \dfrac{99!}{3!\,96!} = \dfrac{99 \cdot 98 \cdot 97}{3 \cdot 2 \cdot 1} = 156{,}849$

20. Since the order of finish is important, we want the number of permutations of 8 objects (the contestants) taken three at a time, which is $P(8,3) = \dfrac{8!}{(8-3)!} = \dfrac{8!}{5!} = 8 \cdot 7 \cdot 6 = 336.$

22. The number of ways of ordering 6 distinct objects (the 6 people) is $P(6,6) = 6! = 720.$

24. Since the word *permutations* is used in this exercise, we are looking for the number of ways to arrange these 4 letters without repeats. Thus, the number of ways is $P(4,4) = 4! = 24.$

26. The number of ways of ordering 8 pieces in order (without repeats) is $P(8,8) = 8! = 40{,}320.$

28. The number of ways of ordering three of the five distinct flags is $P(5,3) = 60.$

30. In selecting these officers, order is important and repeats are not allowed, so the number of ways of choosing 4 officers from 30 students $= P(30,4) = 657{,}720.$

32. We start by first placing Jack in the middle seat, and then we place the remaining 4 students in the remaining 4 seats. Thus the number of a these arrangements is

$$\left(\begin{smallmatrix}\text{number of ways}\\ \text{to seat Jack}\end{smallmatrix}\right) \cdot \left(\begin{smallmatrix}\text{number of ways to seat}\\ \text{the remaining 4 students}\end{smallmatrix}\right) = 1 \cdot P(4,4) = 1! \, 4! = 24.$$

34. Here we have 14 objects (the 14 balls) of which 5 are red balls, 2 are white balls, and 7 are blue balls. So the number of distinguishable permutations is $\dfrac{14!}{5! \, 2! \, 7!} = 72{,}072.$

36. The word ELEEMOSYNARY has 12 letters of which 3 are E's, 2 are Y's, and the remaining letters are distinct. So we wish to find the number of distinguishable permutations of 12 objects (the 12 letters) from like groups of size 3, 2, and seven of size 1. We get

$$\frac{12!}{3! \, 2! \, 1! \, 1! \, 1! \, 1! \, 1! \, 1! \, 1!} = 39{,}916{,}800.$$

38. This is the number of distinguishable permutations of 7 objects (the students) from like groups of size 3 (the ones who stay in the 3-person room), size 2 (the ones who stay in the 2-person room), size 1 (the one who stays in the 1-person room), and size 1 (the one who sleeps in the car). This number is $\dfrac{7!}{3! \, 2! \, 1! \, 1!} = 420.$

40. The number of distinguishable permutations of 13 objects (the total number of blocks he must travel) which can be partitioned into like groups of size 8 (the east blocks) and of size 5 (the north blocks) is $\dfrac{13!}{8! \, 5!} = 1{,}287.$

42. In this exercise, we assume that the pizza toppings cannot be repeated, so we are interested in the number of ways to select a subset of 3 toppings from a set of 12 toppings. The number of ways this can occur is $C(12,3) = \dfrac{12!}{3! \, 9!} = \dfrac{12 \cdot 11 \cdot 10}{3 \cdot 2 \cdot 1} = 220.$

44. Here we are interested in the number of ways of choosing 3 objects (the 3 members of the committee) from a set of 25 objects (the 25 members). The number of combinations of 25 objects taken 3 at a time is $C(25,3) = \dfrac{25!}{3! \, 22!} = 2300.$

46. Since order is not important in a 7-card hand, the number of combinations of 52 objects (the 52 cards) taken 7 at a time is $C(52,7) = \dfrac{52!}{7! \, 45!} = 133{,}784{,}560.$

48. In this exercise, we assume that the pizza toppings cannot be repeated, so we are interested in the number of ways to select a subset of 3 toppings from a set of 16 toppings. The number of ways this can occur is $C(16,3) = \dfrac{16!}{3! \, 13!} = 560.$

50. The order the skirts are selected is not important and no skirt is repeated. So the number of combinations of 8 skirts taken 5 at a time is $C(8,5) = \dfrac{8!}{5! \, 3!} = 56.$

52. Since Jack must go on the field trip, we first pick Jack to go on the field trip, and then select the six other students from the remaining 29 students. Since $C(29,6) = \dfrac{29!}{6!\,23!} = 475{,}020$, there are 475,020 ways to select the students to go on the field trip with Jack.

54. Since the order in which the numbers are selected is not important, the number of combinations of 49 numbers taken 6 at a time is $C(49,6) = \dfrac{49!}{6!\,43!} = 13{,}983{,}816$.

56. (a) The number of ways of choosing 5 students from the 20 students is $C(20,5) = \dfrac{20!}{5!\,15!} = 15{,}504$.

    (b) The number of ways of choosing 5 students for the committee from the 12 females is
    $$C(12,5) = \dfrac{12!}{5!\,7!} = 792.$$

    (c) We use the Fundamental Counting Principle to count the number of possible committees with 3 females and 2 males. Thus, we get
    $$\left(\begin{smallmatrix}\text{number of ways to choose the}\\ \text{3 females from the 12 females}\end{smallmatrix}\right) \cdot \left(\begin{smallmatrix}\text{number of ways to choose the}\\ \text{2 males from the 8 males}\end{smallmatrix}\right) = C(12;3) \cdot C(8,2) = (220)(28) = 6160.$$

58. We may choose any subset of the 8 available brochures. There are $2^8 = 256$ ways to do this.

60. We consider a set of 20 objects (the shoppers in the mall) and a subset that corresponds to those shoppers that enter the store. Since a set of 20 objects has $2^{20} = 1{,}048{,}576$ subsets, there are 1,048,576 outcomes to their decisions.

62. The number of ways the committee can be chosen is
    $$\left(\begin{smallmatrix}\text{number of ways to}\\ \text{choose 2 of 6 freshmen}\end{smallmatrix}\right) \cdot \left(\begin{smallmatrix}\text{number of ways to}\\ \text{choose 3 of 8 sophomores}\end{smallmatrix}\right) \cdot \left(\begin{smallmatrix}\text{number of ways to}\\ \text{choose 4 of 12 juniors}\end{smallmatrix}\right) \cdot \left(\begin{smallmatrix}\text{number of ways to}\\ \text{choose 5 of 10 seniors}\end{smallmatrix}\right)$$
    $$= C(6,2) \cdot C(8,3) \cdot C(12,4) \cdot C(10,5) = 15 \cdot 56 \cdot 495 \cdot 252 = 104{,}781{,}600.$$

64. We choose 3 forwards from the forwards, 2 defense players from the defense men, and the goalie from the two goalies. Thus the number of ways to pick the 6 starting players is
    $$\left(\begin{smallmatrix}\text{number of ways to}\\ \text{pick 3 of 12 forwards}\end{smallmatrix}\right) \cdot \left(\begin{smallmatrix}\text{number of ways to}\\ \text{pick 2 of 6 defenders}\end{smallmatrix}\right) \cdot \left(\begin{smallmatrix}\text{number of ways to}\\ \text{pick 1 of 2 goalies}\end{smallmatrix}\right) = C(12,3) \cdot C(6,2) \cdot C(2,1)$$
    $$= (220)(15)(2) = 6600.$$

66. (a) We choose 2 girls from the girls on the camping trip and the 4 boys from the boys on the camping trip. Thus the number of ways to pick the 6 to gather firewood is
    $$\left(\begin{smallmatrix}\text{number of ways to}\\ \text{choose 2 of 9 girls}\end{smallmatrix}\right) \cdot \left(\begin{smallmatrix}\text{number of ways to}\\ \text{choose 4 of 16 boys}\end{smallmatrix}\right) = C(9,2) \cdot C(16,4) = (36)(1820) = 65{,}520.$$

    (b) Method 1: We consider the number of ways of selecting the group of 6 from the 25 campers and subtract off the groups that contain no girls and those that contain one girls. Thus the number groups that contain at least two girls is
    $$C(25,6) - C(9,0) \cdot C(16,6) - C(9,1) \cdot C(16,5) = 177{,}100 - (1)(8008) - (9)(4368)$$
    $$= 129{,}780.$$
    Method 2: In the method we construct all the groups that are possible. Thus the number of groups is

$$\begin{pmatrix} \text{groups with} \\ \text{with 2 girls} \\ \text{and 4 boys} \end{pmatrix} + \begin{pmatrix} \text{groups with} \\ \text{with 3 girls} \\ \text{and 3 boys} \end{pmatrix} + \begin{pmatrix} \text{groups with} \\ \text{with 4 girls} \\ \text{and 2 boys} \end{pmatrix} + \begin{pmatrix} \text{groups with} \\ \text{with 5 girls} \\ \text{and 1 boys} \end{pmatrix} + \begin{pmatrix} \text{groups with} \\ \text{with 6 girls} \\ \text{and 0 boys} \end{pmatrix}$$

$$= C(9,2) \cdot C(16,4) + C(9,3) \cdot C(16,3) + C(9,4) \cdot C(16,2) + C(9,5) \cdot C(16,1)$$
$$+ C(9,6) \cdot C(16,0)$$

$$= (36)(1820) + (84)(560) + (126)(120) + (126)(16) + (84)(1)$$
$$= 65,520 + 47,040 + 15,120 + 2,016 + 84 = 129,780.$$

68. We count the number of arrangements where Mike and Molly are standing together and subtract the arrangements in which John and Jane are also standing together. If we consider Mike and Molly as one object, then we need the number of ways of arranging 9 objects. We then need to arrange Mike and Molly. Thus the number of arrangements where Mike and Molly are standing together is $P(9,9) \cdot P(2,2) = 9! \cdot 2! = 725,760$. To count the number of arrangements where Mike and Molly stand together and John and Jane also stand together, we treat John and Jane as one object and Mike and Molly as one object and arrange the 8 objects. So there are $P(8,8) \cdot P(2,2) \cdot P(2,2) = 8! \cdot 2! \cdot 2! = 161,280$ ways to do this. (Remember, we still needed to arrange John and Jane within their group and Mike and Molly within their group.) Hence the number of ways to arrange the students is $725,760 - 161,280 = 564,480$.

70. (a) We treat the women as one object and find the number of ways of permuting 5 objects. We then permute the 4 women. Thus the number of arrangements is $P(5,5) \cdot P(4,4) = 5! \cdot 4! = 2880$.

(b) There are two ways to solve this exercise.
Method 1: The number of ways the men and women can be seated
$$= \begin{pmatrix} \text{number of ways the} \\ \text{men can be arranged} \end{pmatrix} \cdot \begin{pmatrix} \text{number of ways the} \\ \text{women can be arranged} \end{pmatrix} \cdot \begin{pmatrix} \text{number of ways to select} \\ \text{the gender of the first seat} \end{pmatrix}$$
$$= P(4,4) \cdot P(4,4) \cdot C(2,1) = 4! \cdot 4! \cdot 2 = 1152.$$
Method 2: There are 8 choices for the first seat, 4 for the second, 3 for the third, 3 for the fourth, 2 for the fifth, 2 for the sixth, 1 for the seventh, and 1 for the eighth. Thus the number of ways of seating the men and women in the required fashion is $8 \cdot 4 \cdot 3 \cdot 3 \cdot 2 \cdot 2 \cdot 1 \cdot 1 = 1152$

72. There are many different possibilities here, so we consider the complement where NO professors are chosen for the delegates and subtract this number form the way to select 3 people from the group of 8 people which is $C(8,3)$. If the professor cannot not be selected, then we must select 3 people from a group of 5 and this can be done in $C(5,3)$ ways. Thus the number of delegations that contain a professor is $C(8,3) - C(5,3) = 56 - 10 = 46$.

74. Since two points determine a line and no three points are collinear, the exercise is how many ways can you select 2 of the 12 dots. Thus the number of possible lines is $C(12,2) = 66$.

76. When two objects are chosen from ten objects, it determines a unique set of eight objects, those not chosen. So choosing two objects from ten objects is the same a choosing eight objects from ten objects. In general, every subset of r objects chosen from a set of n objects determines a corresponding set of $(n - r)$ objects, namely, those not chosen. Therefore, the total number of combinations for each type are equal.

78. (a) $(x+y)^5 = (x+y)(x+y)(x+y)(x+y)(x+y)$

$= (x+y)(x+y)(x+y)(xx + xy + yx + yy)$

$= (x+y)(x+y)(xxx + xxy + xyx + xyy + yxx + yxy + yyx + yyy)$

$= (x+y)(xxxx + xxxy + xxyx + xxyy + xyxx + xyxy + xyyx + xyyy + yxxx$
$\quad + yxxy + yxyx + yxyy + yyxx + yyxy + yyyx + yyyy)$

$= xxxxx + xxxxy + xxxyx + xxxyy + xxyxx + xxyxy + xxyyx + xxyyy$
$\quad + xyxxx + xyxxy + xyxyx + xyxyy + xyyxx + xyyxy + xyyyx + xyyyy$
$\quad + yxxxx + yxxxy + yxxyx + yxxyy + yxyxx + yxyxy + yxyyx + yxyyy$
$\quad + yyxxx + yyxxy + yyxyx + yyxyy + yyyxx + yyyxy + yyyyx + yyyyy$

(b) There are ten terms that contain two x's and three y's. They are

$xxyyy + xyxyy + xyyxy + xyyyx + yxxyy + yxyxy + yxyyx + yyxxy + yyxyx + yyyxx$

(c) To count the number of terms with two x's, we must count the number of ways to pick two of the five positions to contain an x. This number is $C(5, 2)$.

(d) In the Binomial Theorem, the coefficient $\binom{n}{r}$ is the number of ways of picking r positions in a term with n factors to contain an x. By definition, this is $C(n, r)$.

Exercises 10.3

2. Let H stand for heads, T for tails; the numbers $1, 2, \ldots, 6$ are the faces of the die.

   (a) $S = \{H1, H2, H3, H4, H5, H6, T1, T2, T3, T4, T5, T6\}$.

   (b) Let E be the event of getting heads and rolling an even number. So $E = \{H2, H4, H6\}$, and
   $$P(E) = \frac{n(E)}{n(S)} = \tfrac{3}{12} = \tfrac{1}{4}.$$

   (c) Let F be the event of getting heads and rolling a number greater than 4. So $F = \{H5, H6\}$, and $P(F) = \dfrac{n(F)}{n(S)} = \tfrac{2}{12} = \tfrac{1}{6}.$

   (d) Let G be the event of getting tails and rolling an odd number. So $G = \{T1, T3, T5\}$, and
   $$P(G) = \frac{n(G)}{n(S)} = \tfrac{3}{12} = \tfrac{1}{4}.$$

4. (a) Let E be the event of rolling a two or a three. Then $P(E) = \dfrac{n(E)}{n(S)} = \tfrac{2}{6} = \tfrac{1}{3}.$

   (b) Let F be the event of rolling an odd number. So $F = \{1, 3, 5\}$, and $P(F) = \dfrac{n(F)}{n(S)} = \tfrac{3}{6} = \tfrac{1}{2}.$

   (c) Let G be the event of rolling a number divisible by 3. So $G = \{3, 6\}$, and $P(G) = \dfrac{n(G)}{n(S)} = \tfrac{1}{3}.$

6. (a) Let E be the event of choosing a heart. Since there are 13 hearts, $P(E) = \dfrac{n(E)}{n(S)} = \tfrac{13}{52} = \tfrac{1}{4}.$

   (b) Let F be the event of choosing a heart or a spade. Since there are 13 hearts and 13 spades,
   $$P(E) = \frac{n(E)}{n(S)} = \tfrac{26}{52} = \tfrac{1}{2}.$$

   (c) Let G be the event of choosing a heart, a diamond or a spade. Since there are 13 cards in each suit, $P(G) = \dfrac{n(G)}{n(S)} = \tfrac{39}{52} = \tfrac{3}{4}.$

8. (a) Let E be the event of selecting a white or a yellow ball. Since there are 2 white balls and 1 yellow ball, $P(E') = 1 - P(E) = 1 - \dfrac{n(E)}{n(S)} = 1 - \tfrac{3}{8} = \tfrac{5}{8}.$

   (b) Let F be the event of selecting a red, a white or a yellow ball. Since all the types of balls are in the jar, $P(E) = 1$.

   (c) Let G be the event of selecting a white ball. Since there are 2 white balls, $P(E') = 1 - P(E)$
   $$= 1 - \frac{n(E)}{n(S)} = 1 - \tfrac{2}{8} = \tfrac{6}{8} = \tfrac{3}{4}.$$

10. The spinner has 9 equal sized regions numbered 1 through 9.

(a) Let E be the event that the spinner stops at an even number. Since there are 4 even numbers, $P(E) = \frac{4}{9}$.

(b) Let F be the event that the spinner stops at an odd number or a number greater than 3. Since the only number not greater than 3 and not odd is 2, $P(E) = 1 - P(E') = 1 - \dfrac{n(E')}{n(S)}$

$= 1 - \frac{1}{9} = \frac{8}{9}$.

12. Let E be the event of dealing 5 hearts. Since there are 13 hearts, $P(E) = \dfrac{C(13,5)}{C(52,5)}$

$= \frac{1287}{2,598,960} \approx 0.000495$.

14. Let E be the event of dealing 5 face cards. Since there are 3 face cards for each suit and 4 suits, $P(E) = \dfrac{C(12,5)}{C(52,5)} = \frac{792}{2,598,960} \approx 0.000305$.

16. (a) $S = \{(1,1),(1,2),(1,3),(1,4),(1,5),(1,6),(2,1),(2,2),(2,3),(2,4),(2,5),(2,6),$
$(3,1),(3,2),(3,3),(3,4),(3,5),(3,6),(4,1),(4,2),(4,3),(4,4),(4,5),(4,6),$
$(5,1),(5,2),(5,3),(5,4),(5,5),(5,6),(6,1),(6,2),(6,3),(6,4),(6,5),(6,6)\}$.

(b) Let E be the event of getting a sum of 7. Then $E = \{(1,6),(2,5),(3,4),(4,3),(5,2),(6,1)\}$, and $P(E) = \frac{6}{36} = \frac{1}{6}$.

(c) Let F be the event of getting a sum of 9. Then $F = \{(3,6),(4,5),(5,4),(6,3)\}$, and $P(F) = \frac{4}{36} = \frac{1}{9}$.

(d) Let E be the event that the two dice show the same number. Then $P(E) = \frac{6}{36} = \frac{1}{6}$.

(e) Let E be the event that the two dice show different numbers. Then E' is the event that the two dice show the same number. Thus, $P(E) = 1 - P(E') = 1 - \frac{1}{6} = \frac{5}{6}$.

(f) Let E be the event of getting a sum of 9 or higher. Then $P(E) = \frac{10}{36} = \frac{5}{18}$.

18. Let E be the event that a 13-card bridge hand consists of all cards from the same suit. Since there are exactly 4 such hands (one for each suit), $P(E) = \dfrac{4}{C(52,13)} \approx 6.3 \times 10^{-12}$.

20. (a) Let E be the event that the toddler arranges the word "FRENCH". Since the letters are distinct, there are $P(6,6)$ ways of arranging the blocks of which only one spells the word "FRENCH". Thus $P(E) = \dfrac{1}{P(6,6)} = \frac{1}{720} \approx 0.0014$.

(b) Let E be the event that the toddler arranges the letters in alphabetical order. Since there are $P(6,6)$ ways of arranging the blocks of which only one is in alphabetical order, $P(E) = \dfrac{1}{P(6,6)} = \frac{1}{720} \approx 0.0014$.

22. Let E be the event that no women are chosen. The number of ways that no women are chosen is the same as the number of ways that only men are chosen, which is $C(11,6)$. Thus $P(E) = \dfrac{C(11,6)}{C(30,6)} \approx 0.00078$.

24. Let E be the event that the batch will be discarded. Thus, E is the event that at least one defective bulb is found. It is easier to find E', the event that no defective bulbs are found. Since there are 10 bulbs in the batch of which 8 are non-defective, $P(E') = \dfrac{C(8,3)}{C(10,3)}$. Thus $P(E) = 1 - P(E')$

$$= 1 - \frac{C(8,3)}{C(10,3)} \approx 1 - 0.4667 = 0.5333.$$

26. Let E be the event that the monkey will arrange the 6 blocks to spell "HAMLET". Then $P(E) = \frac{1}{6!} = \frac{1}{720} \approx 0.0014.$

28. Let E be the event that you predict the correct order for the horses to finish the race. Since there are eight horses, there are $P(8,8) = 8!$ ways that the horses could finish, with only one being the correct order. Thus, $P(E) = \dfrac{1}{P(8,8)} = \frac{1}{40,320} \approx 2.48 \times 10^{-5}.$

30.

		Parent 2	
		t	t
Parent 1	T	T t	T t
	t	t t	t t

   (a) Let E be the event that the offspring will be tall. Since only offspring with genotype Tt will be tall, $P(E) = \frac{2}{4} = \frac{1}{2}.$

   (b) E' is the event that the offspring will not be tall (thus, the offspring is short). So $P(E') = 1 - P(E) = 1 - \frac{1}{2} = \frac{1}{2}.$

32. (a) NO, the events are not mutually exclusive since a student can be female and wear glasses.

   (b) NO, the events are not mutually exclusive since a student can be male and have long hair.

34. (a) NO, the events are not mutually exclusive since 4 is greater than 3 and also less than 5. So $P(E \cup F) = P(E) + P(F) - P(E \cap F) = \frac{3}{6} + \frac{4}{6} - \frac{1}{6} = 1.$

   (b) YES, the events are mutually exclusive since there are only 2 numbers less than 3, namely 1 and 2, but they are not divisible by 3. So $P(E \cup F) = P(E) + P(F) = \frac{2}{6} + \frac{2}{6} = \frac{2}{3}.$

36. (a) NO, events E and F are not mutually exclusive since a king can be a club. So $P(E \cup F) = P(E) + P(F) - P(E \cap F) = \frac{13}{52} + \frac{4}{52} - \frac{1}{52} = \frac{4}{13}.$

   (b) NO, events E and F are not mutually exclusive since an ace can be a spade. So $P(E \cup F) = P(E) + P(F) - P(E \cap F) = \frac{4}{52} + \frac{13}{52} - \frac{1}{52} = \frac{4}{13}.$

38. (a) Let E be the event that the spinner stops on blue. Since only 4 of the regions are blue, $P(E) = \frac{4}{16} = \frac{1}{4}.$

   (b) Let F be the event that the spinner stops on an odd number. Since 8 of the regions are odd-numbered, $P(F) = \frac{8}{16} = \frac{1}{2}.$

   (c) Since none of the odd-numbered regions are blue, $P(E \cup F) = P(E) + P(F) = \frac{1}{4} + \frac{1}{2} = \frac{3}{4}.$

40. Let E be the event of arranging the letters to spell "TRIANGLE" and F be the event of arranging them to spell "INTEGRAL". There are $P(8,8)$ possible ways of arranging the blocks and only one way to spell each of these words. Also, these events are mutually exclusive since the blocks cannot spell both words at the same time, so
$$P(E \cup F) = P(E) + P(F) = \frac{1}{P(8,8)} + \frac{1}{P(8,8)} = \frac{2}{8!} = \frac{1}{20160} \approx 4.96 \times 10^{-5}.$$

42. Let E be the event that a player has exactly 5 winning numbers and F be the event that a player has all 6 winning numbers. These events are mutually exclusive. For a players to have exactly 5 winning numbers means that the player has 5 of the 6 winning numbers and 1 number that was not selected in the lottery. So $n(E) = C(6,5) \cdot C(43,1)$. Thus,
$$P(\text{at least 5 winning numbers}) = P(E \cup F) = P(E) + P(F) = \frac{C(6,5) \cdot C(43,1)}{C(49,6)} + \frac{1}{C(49,6)}$$
$$\approx 0.0000185.$$

44. (a) YES, because the first flip does not influence the outcome of the second flip.

   (b) The probability of showing heads on both tosses is $P(E \cap F) = P(E) \cdot P(F) = \left(\frac{1}{2}\right)\left(\frac{1}{2}\right) = \frac{1}{4}$.

46. (a) YES. What happens on spinner A does not influence what happens on spinner B.

   (b) The probability that A stops on red and B stops on yellow is
   $$P(E \cap F) = P(E) \cdot P(F) = \left(\frac{2}{4}\right)\left(\frac{2}{8}\right) = \frac{1}{8}.$$

48. Let E be the event of getting a 1 on the first toss, and let F be the event of getting a 1 on the second toss. Since the events are independent, $P(E \cap F) = P(E) \cdot P(F) = \frac{1}{6} \cdot \frac{1}{6} = \frac{1}{36}$.

50. Since the card is replaced, the selection of the first card does not influence the selection of the second card, so the events are independent.

   (a) Let E be the event of getting an ace on the first draw and F be the event of getting an ace on the second draw. Then $P(E \cap F) = P(E) \cdot P(F) = \left(\frac{4}{52}\right)\left(\frac{4}{52}\right) = \frac{1}{169}$.

   (b) Let E be the event of getting an ace on the first draw and F be the event of getting a spade on the second draw. Then $P(E \cap F) = P(E) \cdot P(F) = \left(\frac{4}{52}\right)\left(\frac{13}{52}\right) = \frac{1}{52}$.

52. The number of ways to arrange the letters E, O, K, M, N, and Y is $P(6,6) = 720$. So the probability that the monkey can arrange the letters correctly (if he is merely arranging the blocks randomly) is $\frac{1}{720}$. Also, since the monkey is arranging the blocks randomly, each arrangement is independent of every other arrangement. Hence, the probability that the monkey will be able to spell the word correctly three consecutive times is $\left(\frac{1}{720}\right)^3 = \frac{1}{373,248,000} \approx 2.68 \times 10^{-9}$.

54. The number of ways a set of six numbers can be selected from a group of 49 numbers is $C(49,6)$. Since the games are independent, the probability of winning the lottery two times in a row is
$$\left(\frac{1}{C(49,6)}\right)^2 \approx 5.11 \times 10^{-15}.$$

56. (a) The probability that the first wheel has a bar is $\frac{1}{11}$, and is the same for the second and third wheels. The events are independent, and so the probability of getting 3 bars is
   $$\frac{1}{11} \cdot \frac{1}{11} \cdot \frac{1}{11} = \frac{1}{1331}.$$

(b)   The probability of getting a number on the first wheel is $\frac{10}{11}$, the probability of getting the same number on the second wheel is $\frac{1}{11}$, and the probability of getting the same number on the third wheel is $\frac{1}{11}$. Thus, the probability of getting the same number on each wheel is $\frac{10}{11} \cdot \frac{1}{11} \cdot \frac{1}{11} = \frac{10}{1331}$.

(c)   We use the complement, NO BARS, to determine the probability of at least one bar. The probability that the first wheel does not have a bar is $\frac{10}{11}$, and is the same for the second and third wheels. Since the events are independent, the probability of getting NO BARS is $\frac{10}{11} \cdot \frac{10}{11} \cdot \frac{10}{11} = \frac{10^3}{11^3} = \frac{1000}{1331}$, and so $P(\text{at least one BAR}) = 1 - \frac{1000}{1331} = \frac{331}{1331}$.

58.   Let E be the event that at least two have a birthday in the same month, so that E' is the event that no two have a birthday in the same month. So
$$P(E') = \frac{\text{number of ways to assign 6 distinct birth months}}{\text{number of ways to assign 6 birth months}} = \frac{P(12,6)}{12^6} = \frac{12 \cdot 11 \cdot 10 \cdot 9 \cdot 8 \cdot 7}{12^6} = \frac{385}{1728}. \text{ Thus}$$
$$P(E) = 1 - P(E') = 1 - \frac{385}{1728} = \frac{1343}{1728}.$$

60.   (a)   Let E be the event that curriculum committee consists of 2 women and 4 men. So
$$P(E) = \frac{\text{number of committees with 2 women and 4 men}}{\text{number of ways to select 6 member committee}} = \frac{C(8,2) \cdot C(10,4)}{C(18,6)} = \frac{28 \cdot 210}{18564}$$
$$= \frac{490}{1547}.$$

(b)   Let F be the event that curriculum committee consists of 2 or fewer women. Then F is the event that the committee has NO women or 1 woman or 2 women.
$$P(F) = \frac{C(8,0) \cdot C(10,6)}{C(18,6)} + \frac{C(8,1) \cdot C(10,5)}{C(18,6)} + \frac{C(8,2) \cdot C(10,4)}{C(18,6)} = \frac{1 \cdot 210 + 8 \cdot 252 + 28 \cdot 210}{18564}$$
$$= \frac{8106}{18564} = \frac{193}{442}.$$

(c)   So F' is the event the at curriculum committee has more that 2 women. The
$$P(F') = 1 - P(F) = 1 - \frac{193}{442} = \frac{249}{442}.$$

62.   Let E be the event that all the boys are standing together and all the girls are standing together. If we consider all 8 boys as one object and all 12 girls as one object, then there are 2! ways to arrange these 2 objects (8 boys and 12 girls). There are 8! possible ways to arrange 8 boys and 12! possible ways to arrange 12 girls, thus there are $2! \cdot 8! \cdot 12!$ possible ways to arrange 8 boys and 12 girls in which all the boys are standing together and all the girls are standing together. Therefore
$$P(E) = \frac{2! \cdot 8! \cdot 12!}{20!} \approx 1.59 \times 10^{-5}.$$

64.   Since most families in the US have three or less children, we construct a table in which we calculate the probability that a randomly selected child is the oldest son or daughter.

Number of children in the family	Sample Space	Probability that the child is the oldest son or daughter
1	(b) (g)	$\left.\begin{array}{l} 0.5 \cdot 1 = 0.5 \\ 0.5 \cdot 1 = 0.5 \end{array}\right\} = 1$
2	(b, b) (b, g) (g, b) (g, g)	$\left.\begin{array}{l} 0.25 \cdot 0.5 = 0.125 \\ 0.25 \cdot 1 = 0.25 \\ 0.25 \cdot 1 = 0.25 \\ 0.25 \cdot 0.5 = 0.125 \end{array}\right\} = 0.75$

Number of children in the family	Sample Space	Probability that the child is the oldest son or daughter	
3	(b, b, b)	$0.125 \cdot 0.333 = 0.041625$	
	(b, b, g)	$0.125 \cdot 0.667 = 0.08325$	
	(b, g, b)	$0.125 \cdot 0.667 = 0.08325$	
	(g, b, b)	$0.125 \cdot 0.667 = 0.08325$	$= 0.58275$
	(b, g, g)	$0.125 \cdot 0.667 = 0.08325$	
	(g, b, g)	$0.125 \cdot 0.667 = 0.08325$	
	(g, g, b)	$0.125 \cdot 0.667 = 0.08325$	
	(g, g, g)	$0.125 \cdot 0.333 = 0.041625$	

As the table shows, for families of this size, the relevant probability is always greater than 0.5.

Exercises 10.4

2. $P(3 \text{ successes in } 5) = C(5,3) \cdot (0.7^3)(0.3^2) = 0.30870.$

4. $P(5 \text{ successes in } 5) = C(5,5) \cdot (0.7^5)(0.3^0) = 0.16807.$

6. $P(1 \text{ failures in } 5) = C(5,1) \cdot (0.3^1)(0.7^4) = 0.36015.$ (Note *exactly one failure* is the same as *exactly 4 successes* and $P(4 \text{ successes in } 5) = C(5,4) \cdot (0.7^4)(0.3^1) = 0.36015.$)

8. $P(\text{at least 3 successes}) = P(3 \text{ successes}) + P(4 \text{ successes}) + P(5 \text{ successes})$
    $= 0.30870 + 0.36015 + 0.16807 = 0.83692.$

10. $P(\text{at most 2 failures}) = P(0 \text{ failure}) + P(1 \text{ failure}) + P(2 \text{ failure})$
    $= P(5 \text{ successes}) + P(4 \text{ successes}) + P(3 \text{ successes})$
    $= 0.16807 + 0.36015 + 0.30870 = 0.83692.$

12. $P(\text{at most 3 failures}) = P(0 \text{ failure}) + P(1 \text{ failure}) + + P(2 \text{ failure}) + P(3 \text{ failure})$
    $= P(5 \text{ successes}) + P(4 \text{ successes}) + P(3 \text{ successes}) + P(2 \text{ successes})$
    $= 0.16807 + 0.36015 + 0.30870 + 0.13230 = 0.96922.$

14. Here $P(\text{success}) = 0.8$ and $P(\text{failure}) = 0.2.$
    (a) $P(0 \text{ successes in } 7) = C(7,0) \cdot (0.8^0)(0.2^7) = 0.0000128.$

    (b) $P(7 \text{ successes in } 7) = C(7,7) \cdot (0.8^7)(0.2^0) \approx 0.209715.$

    (c) $P(\text{he hits target more than once}) = 1 - [P(0 \text{ successes in } 7) + P(1 \text{ successes in } 7)]$
    $= 1 - [C(7,0) \cdot (0.8^0)(0.2^7) + C(7,1) \cdot (0.8^1)(0.2^6)] \approx 0.99963.$

    (d) $P(\text{at least 5 successes}) = P(5 \text{ successes}) + P(6 \text{ successes}) + P(7 \text{ successes})$
    $= C(7,5) \cdot (0.8^5)(0.2^2) + C(7,6) \cdot (0.8^6)(0.2^1) + C(7,7) \cdot (0.8^7)(0.2^0) \approx 0.85197.$

16. The complement is *none of the raccoons had rabies*
    $P(\text{at least 1 had rabies}) = 1 - P(\text{no rabies}) = 1 - C(4,0) \cdot (0.1^0)(0.9^4) = 1 - 0.6561 = 0.3439.$

18. (a) $P(12 \text{ in } 15) = C(15,12) \cdot (0.1^{12})(0.9^3) \approx 3.31695 \times 10^{-10}.$

    (b) $P(\text{at least } 12) = P(12 \text{ in } 15) + P(13 \text{ in } 15) + P(14 \text{ in } 15) + P(15 \text{ in } 15)$
    $= C(15,12) \cdot (0.1^{12})(0.9^3) + C(15,13) \cdot (0.1^{13})(0.9^2) +$
    $$C(15,14) \cdot (0.1^{14})(0.9^1) + C(15,15) \cdot (0.1^{15})(0.9^0)$$
    $\approx 3.40336 \times 10^{-10}.$

20. (a) $P(\text{at least 3 boys}) = P(3 \text{ boys}) + P(4 \text{ boys}) + P(5 \text{ boys})$
    $= C(5,3) \cdot (0.5^3)(0.5^4) + C(5,4) \cdot (0.5^4)(0.5^1) + C(5,5) \cdot (0.5^5)(0.5^0) = 0.5.$

    (b) $P(\text{at least 4 girls}) = P(4 \text{ girls}) + P(5 \text{ girls}) + P(6 \text{ girls}) + P(7 \text{ girls})$
    $= C(7,4) \cdot (0.5^4)(0.5^3) + C(7,5) \cdot (0.5^5)(0.5^2) + C(7,6) \cdot (0.5^6)(0.5^1)$
    $$+ C(7,7) \cdot (0.5^7)(0.5^0)$$
    $= 0.5.$

22. (a) $P(2 \text{ in } 12) = C(12,2) \cdot (0.2^2)(0.8^3) \approx 0.28347.$

(b) The complement of "at least 3" is "at most two", so

$P(\text{at least 3 in 12}) = 1 - P(\text{at most 2 in 12}) = 1 - [P(\text{0 in 12}) + P(\text{1 in 12}) + P(\text{2 in 12})]$

$= 1 - [C(12,0) \cdot (0.2^0)(0.8^{12}) + C(12,1) \cdot (0.2^1)(0.8^{11}) + C(12,2) \cdot (0.2^2)(0.8^{10})]$

$\approx 0.44165.$

24. $P(\text{at least 1 in 10}) = 1 - P(\text{0 in 10}) = 1 - C(10,0) \cdot (0.05^0)(0.95^{10}) \approx 0.40126.$

26. $P(\text{3 in 5 favor}) = C(5,3) \cdot (0.6^3)(0.4^2) = 0.3456.$

28. (a) $P(\text{machine breaks}) = P(\text{at least 1 component fails}) = 1 - P(\text{0 components fail})$

$= 1 - C(4,0) \cdot (0.01^0)(0.99^4) \approx 0.039404.$

(b) $P(\text{0 components fail}) = C(4,0) \cdot (0.01^0)(0.99^4) \approx 0.960596.$

(c) $P(\text{3 components fail}) = C(4,3) \cdot (0.01^3)(0.99^1) \approx 0.00000396.$

30. There are 52 cards in the deck of which 13 belong to any one suit, so the
$P(\text{heart}) = P(\text{spade}) = P(\text{diamond}) = P(\text{club}) = 0.25.$

(a) $P(\text{3 in 3 are hearts}) = C(3,3) \cdot (0.25^3)(0.75^0) = 0.015625.$

(b) $P(\text{2 in 3 are spades}) = C(3,2) \cdot (0.25^2)(0.75^1) = 0.140625.$

(c) $P(\text{0 in 3 are diamonds}) = C(3,0) \cdot (0.25^0)(0.75^3) = 0.421875.$

(d) $P(\text{at least 1 is a club}) = 1 - P(\text{0 in 3 are clubs}) = 1 - C(3,0) \cdot (0.25^0)(0.75^3) \approx 0.578125.$

32. (a) $P(\text{2 or more}) = 1 - [P(\text{0 in 100}) + P(\text{1 in 100})$

$= 1 - [C(100,0) \cdot (0.02^0)(0.98^{100}) + C(100,1) \cdot (0.02^1)(0.98^{99})] \approx 0.59673.$

(b) Since $P(\text{at least 1 interested}) = 1 - P(\text{0 interested})$ and $0.98^{35} \approx 0.507$, the telephone
consultant needs to make at least 35 calls to ensure at least a 0.5 probability of reaching one or
more interested parties.

34. (a)

Number of Heads	Probability
0	0.000020
1	0.000413
2	0.003858
3	0.021004
4	0.073514
5	0.171532
6	0.266828
7	0.266828
8	0.155650
9	0.040354

(b) Between 5 and 6 heads.

Exercises 10.5

2. The probability that Jane gets \$10 is $\frac{1}{6}$, and the probability that she loses \$1 is $\frac{5}{6}$. Thus $E = (10)\left(\frac{1}{6}\right) + (-1)\left(\frac{5}{6}\right) \approx 0.833$, and so her expectation is \$0.833.

4. The expected value of this game is $E = (3)\left(\frac{1}{2}\right) + (2)\left(\frac{1}{2}\right) = \frac{5}{2} = 2.5$. So Tim's expected winnings are \$2.50 per game.

6. The probability that Albert gets two tails is $\left(\frac{1}{2}\right)^2 = \frac{1}{4}$, the probability that Albert gets one tail and one head is $C(2,1)\left(\frac{1}{2}\right)^2 = \frac{1}{2}$, and the probability that Albert gets two heads is $\left(\frac{1}{2}\right)^2 = \frac{1}{4}$. If Albert gets two heads, he will receive \$4, if he get one head and one tail, he will get \$2 − \$1 = \$1, and if he get two tails, he will lose \$2. Thus the expected value of this game is $E = (4)\left(\frac{1}{4}\right) + (1)\left(\frac{1}{2}\right) + (-2)\left(\frac{1}{4}\right) = 1$. So Albert's expected winnings are \$1 per game.

8. Since there are 4 aces, 12 face cards and only one 8 of clubs, the expected value of this game is $E = (104)\left(\frac{4}{52}\right) + (26)\left(\frac{12}{52}\right) + (13)\left(\frac{1}{52}\right) = \$14.25$.

10. The probability of choosing 2 white balls (that is, no black balls) is $\frac{8}{10} \cdot \frac{7}{9} = \frac{56}{90}$, and the probability of not choosing 2 white balls (at least one black) is $\left(1 - \frac{56}{90}\right) = \frac{34}{90}$. Therefore, the expected value of this game is $E = (5)\left(\frac{56}{90}\right) + (0)\left(\frac{34}{90}\right) \approx 3.111$. Thus, John's expected winnings are \$3.11 per game.

12. (a) We have $P(\text{winning the first prize}) = \dfrac{1}{2 \times 10^6}$. After the first prize winner is selected, then

$P(\text{winning the second prize}) = \dfrac{1}{2 \times 10^6 - 1}$. Similarly,

$P(\text{winning the third prize}) = \dfrac{1}{2 \times 10^6 - 2}$. So, the expected value of this game is

$$E = \left(10^6\right)\left(\frac{1}{2 \times 10^6}\right) + \left(10^5\right)\left(\frac{1}{2 \times 10^6 - 1}\right) + \left(10^4\right)\left(\frac{1}{2 \times 10^6 - 2}\right) = \$0.555.$$

(b) Since we expect to win \$0.555 on the average per game, if we pay \$1.00, then our net outcome is a loss of \$0.445 per game. Hence, it is not worth playing because on the average you will lose \$0.445 per game.

14. Since the safe has a six digit combination, there are 10^6 possible combinations to the safe, of which only one is correct. The expected value of this game is $E = \left(10^6 - 1\right)\left(\frac{1}{10^6}\right) + (-1)\left(\frac{10^6 - 1}{10^6}\right) = 0$.

16. Since the wheels of the slot machine are independent, the probability that you get three watermelons is $\left(\frac{1}{11}\right)^3$. So the expected value of this game is $E = (4.75)\left(\frac{1}{11^3}\right) + (-0.25)\left(1 - \frac{1}{11^3}\right) \approx -\$0.246$.

18. Let x be the fair price to pay to play this game. Then if you win, you gain $1 - x$, and if you lose, your loss will be $-x$. So $E = 0 \quad \Leftrightarrow \quad (1-x)\left(\frac{2}{8}\right) + (-x)\left(\frac{6}{8}\right) = 0 \quad \Leftrightarrow \quad 1 - 4x = 0 \quad \Leftrightarrow \quad x = \frac{1}{4} = 0.25$. Thus, a fair price to play this game is \$0.25.

20. If you win, you win \$1 million minus the price of the stamp. If you lose, you lose only the price of the stamp (currently 34¢). So the expected value of this game is

$(999{,}999.67) \cdot \dfrac{1}{20 \times 10^6} + (-0.34) \cdot \dfrac{20 \times 10^6 - 1}{20 \times 10^6} \approx -0.29$. Thus you expect to lose 29¢ on each entry, and so it's not worth it.

Review Exercises for Chapter 10

2. (a) The number of 3-digit numbers that can be formed using the digits 1-6 if a digit can be used any number of times is $(6)(6)(6) = 216$.

    (b) The number of 3-digit numbers that can be formed using the digits 1-6 if a digit can be used only once is $(6)(5)(4) = 120$.

4. Since the order in which the people are chosen is not important and a person cannot be bumped more than once (no repetitions), the number of ways that 7 passengers can be bumped is $C(120, 7) \approx 5.9488 \times 10^{10}$.

6. There 2 ways to answer each of the 10 true-false questions and 4 ways to answer each of the 5 multiple choice questions. So the number of ways that this test can be completed is $(2^{10})(4^5) = 1{,}048{,}576$.

8. Since the order of the scoops of ice cream is not important and the scoops cannot be repeated, the number of ways to have a banana split is $C(15, 4) = 1365$.

10. Since there are $n!$ ways to arrange a group of size n and $5! = 120$, there are 5 students in this class.

12. The number ways to form a license plate consisting of 2 letters followed by 3 numbers is $(26)(26)(10)(10)(10) = 676{,}000$. Since there are fewer possible license plates than 700,000, there must be fewer than 700,000 licensed cars in the Yukon.

14. Each topping corresponds to a subset of a set with n elements. Since a set with n elements has 2^n subsets and $2^{11} = 2048$, there are 11 toppings that the pizza parlor offers.

16. Since the nucleotides can be repeated, the number of possible words of length n is 4^n. Since $4^2 = 16 < 20$ and $4^3 = 64$, the minimum length of word needed is 3.

18. (a) <u>Solution 1</u> Since the left most digit of a three digit number cannot be zero, there are 9 choices for this first digit and 10 choices for the next two digits. Thus, the number of ways to form a three digit number is $(9)(10)(10) = 900$.
    <u>Solution 2</u> Since there are 999 numbers between 1 and 999, of which the numbers between 1 and 99 do not have three digits, there are $999 - 99 = 900$ three digit numbers.

    (b) There are 1001 numbers from 0-1000. From part (a), there are 900 three digit numbers. Therefore the probability that the number chosen is a three digit number is $P(E) = \frac{900}{1001} \approx 0.899$.

20. Since MISSISSIPPI has 4 I's, 4 S's, 2 P's, and 1 M, the number of distinguishable anagrams of the word MISSISSIPPI is $\dfrac{11!}{4!\,4!\,2!\,1!} = 34{,}650$.

22. (a) The probability that the ball is red is $\frac{10}{15} = \frac{2}{3}$.

    (b) The probability that the ball is even numbered is $\frac{8}{15}$.

    (c) The probability that the ball is white and an odd number is $\frac{2}{15}$.

   (d) The probability that the ball is red or odd numbered is
$$P(\text{red}) + P(\text{odd}) - P(\text{red} \cap \text{odd}) = \tfrac{10}{15} + \tfrac{7}{15} - \tfrac{5}{15} = \tfrac{12}{15} = \tfrac{4}{5}.$$

24. (a) $S = \{HHH, HHT, HTH, HTT, THH, THT, TTH, TTT\}.$

    (b) $P(\text{HHH}) = \tfrac{1}{8}.$

    (c) $P(2 \text{ or more heads}) = P(\text{exactly 2 heads}) + P(3 \text{ heads}) = \tfrac{3}{8} + \tfrac{1}{8} = \tfrac{4}{8} = \tfrac{1}{2}.$

    (d) $P(\text{tails of the first toss}) = \tfrac{4}{8} = \tfrac{1}{2}.$

26. Since rolling a die and selecting a card is independent, the
$$P(\text{both show a six}) = P(\text{die shows a six}) \cdot P(\text{card is a six}) = \tfrac{1}{6} \cdot \tfrac{4}{52} = \tfrac{1}{78}.$$

28. (a) Since these events are independent, the probability of getting the ace of spades, a six, and a head is $\tfrac{1}{52} \cdot \tfrac{1}{6} \cdot \tfrac{1}{2} = \tfrac{1}{624}.$

    (b) The probability of getting a spade, a six, and a head is $\tfrac{13}{52} \cdot \tfrac{1}{6} \cdot \tfrac{1}{2} = \tfrac{1}{48}.$

    (c) The probability of getting a face card, a number greater than 3, and a head is $\tfrac{12}{52} \cdot \tfrac{3}{6} \cdot \tfrac{1}{2} = \tfrac{3}{52}.$

30. (a) Since there are four kings in a standard deck,
$$P(4 \text{ kings}) = \frac{C(4,4)}{C(52,4)} = \frac{1}{\frac{52\cdot51\cdot50\cdot49}{4\cdot3\cdot2\cdot1}} = \tfrac{1}{270725} \approx 3.69 \times 10^{-6}.$$

    (b) Since there are 13 spades in a standard deck,
$$P(4 \text{ spades}) = \frac{C(13,4)}{C(52,4)} = \frac{\frac{13\cdot12\cdot11\cdot10}{4\cdot3\cdot2\cdot1}}{\frac{52\cdot51\cdot50\cdot49}{4\cdot3\cdot2\cdot1}} = \tfrac{11}{4165} \approx 0.00264.$$

    (c) Since there are 26 red cards and 26 black cards,
$$P(\text{all same color}) = \frac{2 \cdot C(26,4)}{C(52,4)} = \frac{2 \cdot \frac{26\cdot25\cdot24\cdot23}{4\cdot3\cdot2\cdot1}}{\frac{52\cdot51\cdot50\cdot49}{4\cdot3\cdot2\cdot1}} = \tfrac{92}{833} \approx 0.11044.$$

32. She knows the first digit and must arrange the other four digits. Since only one of the $P(4,4) = 24$ arrangements is correct, the probability that she guesses correctly is $\tfrac{1}{24}.$

34. Using the same logic as in Exercise 29 (a), the probability that all three dice show the same number is $1 \cdot \tfrac{1}{6} \cdot \tfrac{1}{6} = \tfrac{1}{36}$, while the probability they are not all the same is $1 - \tfrac{1}{36} = \tfrac{35}{36}$. Thus, the expected value of this game is $E = (5)\left(\tfrac{1}{36}\right) + (-1)\left(\tfrac{35}{36}\right) = -\tfrac{30}{36} = -0.83$. So John's expected winnings per game are $-\$0.83$, that is, he expects to lose \$0.83 per game.

36. The number of different pizzas is the number of subsets of the set of 12 toppings, that is, $2^{12} = 4096$. The number of pizzas with anchovies is the number of ways of choosing anchovies and then choosing a subset of the 11 remaining toppings, that is, $(1) \cdot 2^{11} = 2048$. Thus, $P(\text{getting anchovies}) = \tfrac{2048}{4096} = \tfrac{1}{2}$. (Note: A probability of $\tfrac{1}{2}$ makes intuitive sense, for each pizza combination *without* anchovies there is a corresponding one *with* anchovies, so half will have anchovies and half will not.)

38. <u>Method 1</u> First choose the two forwards, then choose the 3 defense players from the remaining 7 players. Thus the number of ways of choosing a starting line up is $C(9,2) \cdot C(7,3) = 1260.$

Method 2 First choose the 3 defensive players, then choose the 2 forwards from the remaining 6 players. Thus the number of ways of choosing a starting line up is $C(9,3) \cdot C(6,2) = 1260$.

40. (a) Order is important, and repeats are possible. Thus there are 10 choices for each digit. So the number of different Zip+4 codes is $10 \cdot 10 \cdots 10 = 10^9$.

    (b) If a Zip+4 code is to be a palindrome, the first 5 digits can be chosen arbitrarily. But once chosen, the last 4 digits are determined. Since there are 10 ways to choose each of the first 5 digits, there are 10^5 palindromes.

    (c) By parts (a) and (b), the probability that a randomly chosen Zip+4 code is a palindrome is $\frac{10^5}{10^9} = 10^{-4}$.

42. Method 1 We choose the 5 states first and then one of the two senators from each state. Thus the number of committees is $C(50,5) \cdot 2^5 = 67{,}800{,}320$.

    Method 2 We choose one of 100 senators, then choose one of the remaining 98 senators (deleting the chosen senator and the other senator form that state), then choose one of the remaining 96 senators, continuing this way until the 5 senators are chosen. Finally, we need to divide by the number of ways to arrange the 5 senators. Thus the number of committees is
    $$\frac{100 \cdot 98 \cdot 96 \cdot 94 \cdot 92}{5!} = 67{,}800{,}320.$$

44. (a) $P(5 \text{ in } 5 \text{ are white flesh}) = C(5,5) \cdot (0.3^5)(0.7^0) = 0.00243$.

    (b) $P(0 \text{ in } 5 \text{ are white flesh}) = C(5,0) \cdot (0.3^0)(0.7^5) \approx 0.16807$.

    (c) $P(2 \text{ in } 5 \text{ are white flesh}) = C(5,2) \cdot (0.3^2)(0.7^3) = 0.3087$.

    (d) $P(3 \text{ or more are red flesh}) = P(3 \text{ in } 5 \text{ are red}) + P(4 \text{ in } 5 \text{ are red}) + P(5 \text{ in } 5 \text{ are red})$
    $= C(5,3) \cdot (0.3^2)(0.7^3) + C(5,4) \cdot (0.3^1)(0.7^4) + C(5,5) \cdot (0.3^0)(0.7^5) = 0.83692$.

Focus on Modeling

2. (a) You should find that you get a combination consisting of a "head" and a "tail" about 50% of the time.

    (b) The possible gender combinations are $\{BB, BG, GB, GG\}$. Thus, the probability of having one child of each sex is $\frac{2}{4} = \frac{1}{2}$.

4. (a) If you simulate 80 World Series with coin tosses, you should expect the series to end in 4 games about 10 times, in 5 games about 20 times, in 6 games about 25 times, and in 7 games about 25 times.

    (b) We first calculate the number of ways that the series can end with team A winning. (Note that a team must win the final game plus three of the preceding games to win the series.) To win in 4 games, team A must win 4 games right off the bat, and there is only 1 way this can happen. To win in 5 games, team A must win the final game plus 3 of the first 4 games, so this can happen in $C(4, 3) = 4$ ways. To win in 6 games, team A must win the final game plus 3 of the first 5 games, so this can happen in $C(5, 3) = \frac{5 \cdot 4}{2 \cdot 1} = 10$ ways. To win in 7 games, team A must win the final game plus 3 of the first 6 games, so this can happen in $C(6, 3) = \frac{6 \cdot 5 \cdot 4}{3 \cdot 2 \cdot 1} = 20$ ways. By symmetry, it is also true for team B that they can win in 4 games just 1 way, in 5 games 4 ways, in 6 games 10 ways, and in 7 games 20 ways. The probability that any particular team wins a given game is $\frac{1}{2}$; this fact, together with the assumption that the games are independent allows us to calculate the probabilities in the following table.

Series	Number of ways	Probability
4 games	2	$2 \cdot \left(\frac{1}{2} \cdot \frac{1}{2} \cdot \frac{1}{2} \cdot \frac{1}{2}\right) = \frac{1}{8}$
5 games	8	$8 \cdot \left(\frac{1}{2} \cdot \frac{1}{2} \cdot \frac{1}{2} \cdot \frac{1}{2} \cdot \frac{1}{2}\right) = \frac{1}{4}$
6 games	20	$20 \cdot \left(\frac{1}{2} \cdot \frac{1}{2} \cdot \frac{1}{2} \cdot \frac{1}{2} \cdot \frac{1}{2} \cdot \frac{1}{2}\right) = \frac{5}{16}$
7 games	40	$40 \cdot \left(\frac{1}{2} \cdot \frac{1}{2} \cdot \frac{1}{2} \cdot \frac{1}{2} \cdot \frac{1}{2} \cdot \frac{1}{2} \cdot \frac{1}{2}\right) = \frac{5}{16}$

    (c) The expected value is $\frac{1}{8} \cdot 4 + \frac{1}{4} \cdot 5 + \frac{5}{16} \cdot 6 + \frac{5}{16} \cdot 7 = 5\frac{13}{16} \approx 5.8$. Thus, on the average, we expect a World Series to end in about 5.8 games.

6. Modifying the TI-83 program in Problem 5 to:

```
PROGRAM: PROB6
:0 → P
:For(N,1,1000)
:rand → X:rand → Y
:P+(X²>Y) → P
:End
:Disp "PROBABILITY IS APPROX"
:Disp P/1000
```

You should find that the probability is very close to $\frac{1}{3}$.